POLYMER PHYSICS

Polymer Physics

MICHAEL RUBINSTEIN

*University of North Carolina,
Chapel Hill, North Carolina, USA*

and

RALPH H. COLBY

*Pennsylvania State University,
University Park, Pennsylvania, USA*

OXFORD

UNIVERSITY PRESS

OXFORD
UNIVERSITY PRESS

Great Clarendon Street, Oxford OX2 6DP

Oxford University Press is a department of the University of Oxford.
It furthers the University's objective of excellence in research, scholarship,
and education by publishing worldwide in

Oxford New York

Auckland Cape Town Dar es Salaam Hong Kong Karachi
Kuala Lumpur Madrid Melbourne Mexico City Nairobi
New Delhi Shanghai Taipei Toronto

With offices in

Argentina Austria Brazil Chile Czech Republic France Greece
Guatemala Hungary Italy Japan Poland Portugal Singapore
South Korea Switzerland Thailand Turkey Ukraine Vietnam

Oxford is a registered trade mark of Oxford University Press
in the UK and in certain other countries

Published in the United States
by Oxford University Press Inc., New York

First published 2003
Reprinted 2004 twice, 2005, 2006, 2007 (with corrections), 2008, 2009

British Library Cataloguing in Publication Data

Data available

Library of Congress Cataloging in Publication Data
Rubinstein, Michael.
Polymer physics/Michael Rubinstein and Ralph H. Colby.
Includes bibliographical references and index
1. Polymers I. Colby, Ralph H. II. Title.
QC173.4.P65 R83 2003   530.4'13–dc21   2002027401

Typeset by Newgen Imaging Systems (P) Ltd, Chennai, India
Printed and bound in Great Britain on acid-free paper by
CPI Antony Rowe, Chippenham, Wiltshire

ISBN 978-0-19-852059-7

10 9

Preface

This book introduces the reader to the fascinating field of polymer physics. It is intended to be utilized as a textbook for teaching upper level undergraduates and first year graduate students about polymers. Any student with a working knowledge of calculus, chemistry, and physics should be able to read this book. The essential tools of the polymer physical chemist or engineer are derived in this book without skipping any steps. Hence, the book is a self-contained treatise that should also serve as a useful reference for scientists and engineers working with polymers.

While the book assumes no prior knowledge of polymers, it goes far beyond introductory polymer texts in the scope of what is covered. The fundamental concepts required to fully understand polymer melts, solutions and gels in terms of both static structure and dynamics are explained in detail. Problems at the end of each chapter provide the reader with the opportunity to apply what has been learned to practice. More challenging problems are denoted by an asterisk.

The book is divided into four parts. After an introduction in Chapter 1, where the necessary concepts from a first course on polymers are summarized, the conformations of single polymer chains are treated in Part 1. Part 2 deals with the thermodynamics of polymer solutions and melts, including the conformations of chains in those states. Part 3 applies the concepts of Part 2 to the formation and properties of polymer networks. Finally, Part 4 explains the essential aspects of how polymers move in both melt and solution states. In all cases, attention is restricted to concepts that are firmly entrenched in the field, with less established uses of those concepts relegated to the problems.

The motivation for our writing this book comes from the fact that its primary antecedent, written by Paul Flory, is now 50 years old. Many of the same concepts are re-introduced in modern language. Other concepts introduced by eminent scientists over the past half-century are derived in simpler ways, with the intention of making them accessible to a broader audience. These include many of the important concepts discussed in the excellent monographs by de Gennes and by Doi and Edwards.

The book is titled Polymer Physics largely because the authors share the viewpoint of Lord Ernest Rutherford:

'Science is divided into two categories, physics and stamp-collecting.'

The foundations of this book arose from debates between the authors while they were employed for 10 glorious years at the Eastman Kodak Company. While the authors continue to debate many aspects of science, the contents of this book have emerged as the essence of what they claim to understand in polymer physics, bearing in mind the wisdom of Werner Heisenberg:

'Science progresses not only because it helps to explain newly discovered facts, but also because it teaches us over and over again what the word understanding *may mean.'*

The authors thank Jack Chang, Dennis Massa, Glen Pearson, and John Pochan for giving the authors the freedom to ponder polymer physics. The authors also thank David Boris, Andrey Dobrynin, Mark Henrichs, Christine Landry, Mike Landry, Charlie Lusignan, Don Olbris, Ravi Sharma, Yitzhak Shnidman, and Jeff Wesson for their participation in our arguments during informal weekly meetings during the Kodak years. The readers of this book are indebted to Mireille Adam, Peter Bermel, Andrey Dobrynin, Randy Duran, Brian Erwin, Liang Guo, Alexander Grosberg, Jean–Francois Joanny, Sanat Kumar, Eugenia Kumacheva, Tom Mourey, Katherine Oates, Jai Pathak, Nopparat Plucktaveesak, Jennifer Polley, Ed Samulski, Sergei Panyukov, Jay Schieber, and Sergei Sheiko for comments on the text that greatly improved the clarity of the presentation. We thank the Institute for Theoretical Physics for hospitality during the completion of the book.

Contents

Contents

III   Networks and gelation

IV   Dynamics

8   Unentangled polymer dynamics   309

Introduction

1

1.1 History of polymer science

Much of human history has been influenced by the availability of materials. In fact, history is divided into eras named after the primary materials used; the Stone Age, the Bronze Age, and the Iron Age. Similarly, we can assert that in the twentieth century we entered the Polymer Age.

Humans have used naturally occurring polymers, called biopolymers, for centuries without realizing that they were dealing with macromolecules. A prime example is caoutchouc, or natural rubber, that comes from *Hevea brasiliensis*, the rubber-tree plant. Natural rubber was used for many centuries before it was identified as polymeric.

Chemists started polymerizing synthetic macromolecules in the middle of the nineteenth century, but they did not believe that they were creating very large molecules. The standard point of view in the beginning of the twentieth century was that these materials were colloids—physically associated clusters of small molecules, with mysterious non-covalent bonds holding the clusters together. Many scientists actually measured high molar masses for these materials (of order $10^4\,\mathrm{g\,mol^{-1}}$ or even $10^5\,\mathrm{g\,mol^{-1}}$), but rejected their own measurements because the values changed systematically with concentration. We now understand such changes with concentration, and the true molar mass, obtained by extrapolation to zero concentration, would have been even larger.

In 1920, Staudinger proposed the **macromolecular hypothesis**: polymers are molecules made of covalently bonded elementary units, called monomers. In this view, the colloidal properties of polymers were attributed entirely to the sizes of these large molecules, called **macromolecules** or **polymers**. In contrast to colloids, macromolecules exhibit colloidal properties in all solvents in which they dissolve, strongly suggesting that covalent bonds hold polymers together. Although this hypothesis initially met with strong resistance, its gradual acceptance during the 1920s allowed for substantial progress in the field in subsequent years. By 1929, Carothers had synthesized a variety of polymers with well-defined structures, and the Polymer Age was born.

During the following 30 years (1930–60), the main concepts of polymer science were established. Polymer synthesis tools were developed and refined during this period. Also, most of the foundations of polymer physics that are discussed in this textbook were introduced during this time period. These include the work of Kuhn on macromolecular sizes (Chapter 2), the work of Flory on swelling a single chain in a good solvent (Chapter 3), the work of Huggins and Flory on thermodynamics (Chapter 4), the work of Flory and Stockmayer on gelation (Chapter 6) and the work of Kuhn, James, and Guth on rubber elasticity (Chapter 7). The single-molecule models of polymer dynamics, were also developed during this period by Rouse and Zimm (Chapter 8).

In the subsequent 20 years (1960–80), the main principles of modern polymer physics were developed. These include the Edwards model of the polymer chain and its confining tube (Chapters 7 and 9), the modern view of semidilute solutions established by des Cloizeaux and de Gennes (Chapter 5), and the reptation theory of chain diffusion developed by de Gennes (Chapter 9) that led to the Doi–Edwards theory for the flow properties of polymer melts.

There are of course, many facets of polymers for which our understanding is far from complete. Polymers with associating groups bonded to their chains, polymer crystallization, liquid crystalline polymers and charged polymers are examples of areas of active research in polymer physics. These four particular examples are also very pertinent to understanding the functions of important biopolymers, such as DNA, RNA, proteins, and polysaccharides. By learning the fundamentals of chain conformations, thermodynamics, elasticity, and mobility, the readers of this book should be ready to consider these more challenging facets.

1.2   Polymer microstructure

The word (poly)-(mer) means (many)-(parts) and refers to molecules consisting of many elementary units, called **monomers**.[1] Monomers are structural repeating units of a polymer that are connected to each other by covalent bonds. Since 'monomer' can mean anything that repeats along the chain, it is by definition ambiguous. In this book, two types of monomers are important. Chemical monomers are the repeating unit that corresponds to the small molecules that were linked together to make the polymer chain. The repeating unit that will be most important for our discussions is a longer section of chain called the Kuhn monomer, that will be defined in Chapter 2. Here we focus on the chemical monomer.

The entire structure of a polymer is generated during **polymerization**, the process by which elementary units (chemical monomers) are covalently bonded together. The number of monomers in a polymer molecule is called its **degree of polymerization** N. The **molar mass** M of a polymer is equal to

---

[1] Chemists use the term 'monomer' to indicate an unreacted small molecule that is capable of polymerizing. Since this book is concerned with polymers, we often use 'monomer' to describe the repeating unit in a polymer chain, essentially a short form of 'reacted monomer'.

its degree of polymerization N times the molar mass M_{mon} of its chemical monomer:[2]

$$M = NM_{mon}. \qquad (1.1)$$

Consider, for example, the general structure of vinyl monomers and polymers shown in Fig. 1.1, where R represents different possible chemical moieties.

If the R group in Fig. 1.1 is hydrogen, the polymer is polyethylene. The repeating unit is $-CH_2-CH_2-$ and the polymer is called polyethylene because polymers are traditionally named after the monomers used in their synthesis (in this case ethylene, $CH_2=CH_2$), even though polymethylene with repeating unit $-CH_2-$ has an identical structure. To avoid potential complications arising from different monomers creating the same polymer, we often discuss the **number n of backbone bonds** instead of the degree of polymerization N, which is the number of monomers in the chain. If the R group in Fig. 1.1 is chlorine the polymer is poly(vinyl chloride), with repeating unit $-CH_2-CHCl-$, prepared from polymerization of vinyl chloride ($CH_2=CHCl$). If the R group in Fig. 1.1 is a benzene ring, the polymer is polystyrene.

The conventional way to describe the mass of a polymer chain is the **molar mass**: the mass of one mole (equal to **Avogadro's number** $\mathcal{N}_{Av} \cong$ 6.02×10^{23} molecules mol^{-1}) of these molecules. For example, a polyethylene molecule consisting of $N = 1000$ chemical monomers, each with molar mass $M_{mon} = 28$ g mol^{-1}, has a molar mass $M = 28\,000$ g mol^{-1}. This means that the mass of $\mathcal{N}_{Av}$ such molecules is 28 000 g or the mass of one molecule is

$$M/\mathcal{N}_{Av} = 28\,000 \text{ g mol}^{-1}/(6.02 \times 10^{23} \text{ molecules mol}^{-1})$$
$$\cong 4.65 \times 10^{-20} \text{ g molecule}^{-1}.$$

It is more convenient and therefore, customary to report the mass of such a polymer as $M = 28\,000$ g mol^{-1} rather than 4.65×10^{-20} g molecule^{-1}. A related measure of the mass is the polymer **molecular weight**, with units of *Daltons* (Da) defined as 12 times the ratio of the polymer molar mass and the molar mass of ^{12}C. Hence, in the above example, the molecular weight is 28 000 Da.

The chemical identity of monomers is one of the main factors determining the properties of polymeric systems. Another major factor is the polymer's **microstructure**, which is the organization of atoms along the chain that is fixed during the polymerization process. In Fig. 1.1, once the double bond polymerizes, a variety of different isomers are possible for the repeating units along the chain. Polymer microstructure cannot be changed without breaking covalent chain bonds. Below we describe three

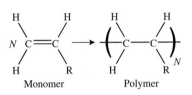

Fig. 1.1
Polymerization of vinyl monomers.

---

[2] End monomers typically have slightly different molar masses M'_{mon} and M''_{mon}. Therefore $M = M_{mon}(N-2) + M'_{mon} + M''_{mon}$. However, the approximation $M \cong M_{mon}N$ is usually very good for large degrees of polymerization. For example, in polyethylene $M_{mon} = 28$ g mol^{-1}, $M'_{mon} = M''_{mon} = 29$ g mol^{-1} and the error for $N = 100$ is 0.07%.

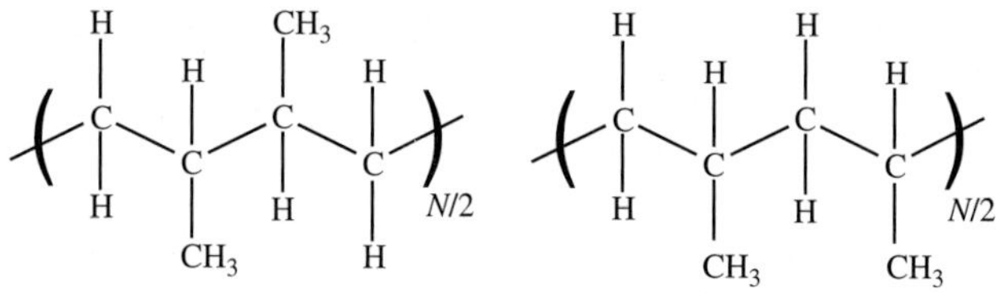

Fig. 1.2
The two sequence isomers of
polypropylene.

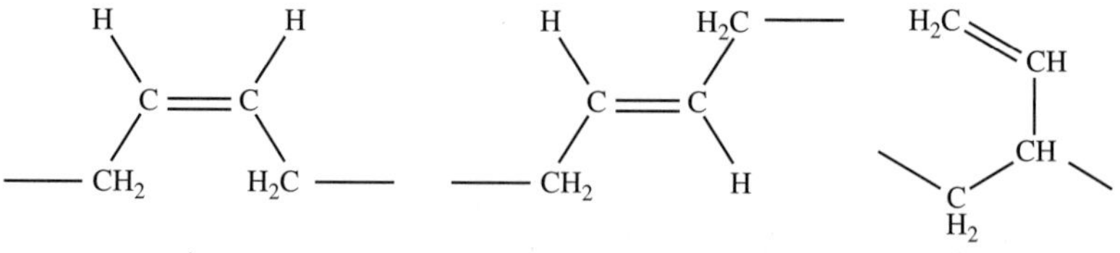

Head-to-head polypropylene   Head-to-tail polypropylene

Fig. 1.3
The three structural isomers of
polybutadiene.

cis   *trans*   vinyl

different categories of isomers: sequence, structural, and stereo isomerism. Examples of **sequence isomerism** are shown in Fig. 1.2 for polypropylene. In the **head-to-head isomer**, two adjacent monomers have their CH_3 groups attached to adjacent carbons along the chain's backbone, whereas the **head-to-tail isomer** has a CH_2 in the backbone between the CH_3 groups of adjacent monomers. Head-to-tail is the more common microstructure, but the properties change significantly with the fraction of head-to-head isomers present.

Polymers that contain a double bond in their backbone (that cannot rotate) can exhibit **structural isomerism**. Such polymers have distinct structural isomers, such as *cis*-, *trans*-, and vinyl-polybutadiene shown in Fig. 1.3. These isomers result from the different ways that dienes, such as butadiene, can polymerize and many synthetic polymers have mixtures of *cis* and *trans* structural isomers along their chains. A particular mixture reflects the probabilities of various ways that monomers add to the growing chain.

Another isomeric variation that is locked-in during polymerization of vinyl monomers is **stereoisomerism**. The four single bonds, emanating from a carbon atom, have a tetragonal structure. If all backbone carbon atoms of a polymer are arranged in a zig-zag conformation along the same plane, adjacent monomers can either have their R group on the same or different sides of this plane, as shown in Fig. 1.4. This type of stereoisomeric variation is described by the polymer's **tacticity**. If all of the R groups of a vinyl polymer are on the same side of the chain, the polymer is **isotactic**. On the other hand, if the R groups alternate regularly, the polymer is **syndiotactic**. Another possibility is that the placement of the R groups is completely random and such polymers are **atactic**. Vinyl polymers always have single C–C bonds along their backbone that allow rotations, but these rotations never change the locked-in nature of the polymer's tacticity. Many synthetic vinyl polymers do not correspond to one of the simple tacticities

Isotactic

Syntactic

Atactic

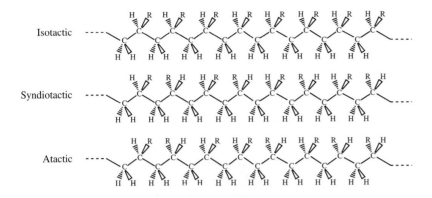

Fig. 1.4
Tacticities of vinyl polymers, illustrated with all backbone carbons in the plane of the page and with H and R groups pointing either into or out from the page.

shown in Fig. 1.4. A more general case is described by a probability p_m for a monomer to add to the growing chain with its R group on the same side as the previous monomer (and probability $1 - p_m$ that the monomer adds with its R group on the opposite side). Often polymers have p_m near $1/2$, for example if R is Cl, commercial poly(vinyl chloride) has $p_m \cong 0.45$ and if R is benzene, commercial polystyrene has $p_m \cong 0.50$. Isotactic polymers correspond to $p_m = 1$, syndiotactic macromolecules have $p_m = 0$, and atactic ones have $p_m = 1/2$, but any probability $0 \leq p_m \leq 1$ is possible. Different polymerization schemes for a given monomer often result in a different p_m. Isomeric variations, such as sequence isomerism, structural isomerism and stereoisomerism are typically measured using NMR spectroscopy.

Physical properties of polymeric systems are strongly affected by chain microstructure. For example, it is much easier to crystallize isotactic and syndiotactic polymers than atactic ones.

1.3 Homopolymers and heteropolymers

Macromolecules that contain monomers of only one type are called **homopolymers**.

$$\cdots-A-A-A-A-A-A-A-A-A-A-A-A-\cdots$$

Homopolymers are made from the same monomer, but may differ by their microstructure, degree of polymerization or architecture. Examples of different microstructure of homopolymers such as tacticity, structural or sequence isomerisms were described in Section 1.2.

Throughout this book we demonstrate that the degree of polymerization N (or the number of backbone bonds n) of macromolecules is a major factor determining many properties of polymeric systems. If a molecule consists of only a small number of monomers (generally, less than 20) it is called an **oligomer**. Linear polymers contain between 20 and 10 billion (for the longest known chromosome) monomers. As monomers are linked together, the physical properties of molecules change. Both the boiling point and the melting point increase rapidly with the number of backbone

Introduction

Table 1.1 Properties and applications of alkane hydrocarbons (following Sperling)

Number of C atoms	State at 25 °C	Example	Uses
1–4	Simple gas	Propane	Gaseous fuels
5–15	Low-viscosity liquid	Gasoline	Liquid fuels and solvents
16–25	High-viscosity liquid	Motor oil	Oils and greases
20–50	Simple soft solid	Paraffin wax	Candles and coatings
>1000	Tough plastic solid	Polyethylene	Bottles and toys

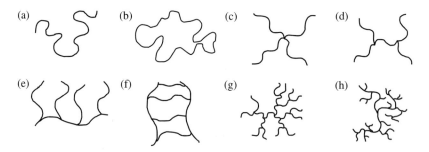

Fig. 1.5
Examples of polymer architectures:
(a) linear; (b) ring; (c) star; (d) H;
(e) comb; (f) ladder; (g) dendrimer;
(h) randomly branched.

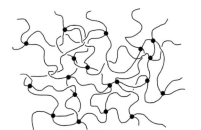

Fig. 1.6
Schematic architecture of a polymer
network, with the dots indicating
crosslinks.

bonds, resulting in different uses of these molecules, as shown in Table 1.1 for alkane hydrocarbons.

Another important feature controlling the properties of polymeric systems is polymer architecture. Types of polymer architectures include **linear, ring, star-branched, *H*-branched, comb, ladder, dendrimer,** or **randomly branched** as sketched in Fig. 1.5. Random branching that leads to structures like Fig. 1.5(h) has particular industrial importance, for example in bottles and film for packaging. A high degree of crosslinking can lead to a macroscopic molecule, called a **polymer network**, sketched in Fig. 1.6. Randomly branched polymers and the formation of network polymers will be discussed in Chapter 6. The properties of networks that make them useful as soft solids (erasers, tires) will be discussed in Chapter 7.

Combining several different types of monomers into a single chain leads to new macromolecules, called **heteropolymers**, with unique properties. The properties of heteropolymers depend both on composition (the fraction of each type of monomers present) and on the sequence in which these different monomers are combined into the chain. Macromolecules containing two different monomers are called **copolymers**. Copolymers can be **alternating, random, block,** or **graft** depending on the sequence in which their monomers are bonded together, as shown in Fig. 1.7. Polymers containing two blocks are called **diblock copolymers**. Chains with three blocks are called **triblock copolymers**. Polymers with many alternating blocks are called **multiblock copolymers** (Fig. 1.7).

Polymers containing three types of monomers are called **terpolymers** (Fig. 1.8). Examples of random terpolymers are **polyampholytes** containing positive, negative and neutral monomers. An example of block terpolymers are **ABC triblocks** shown in Fig. 1.8.

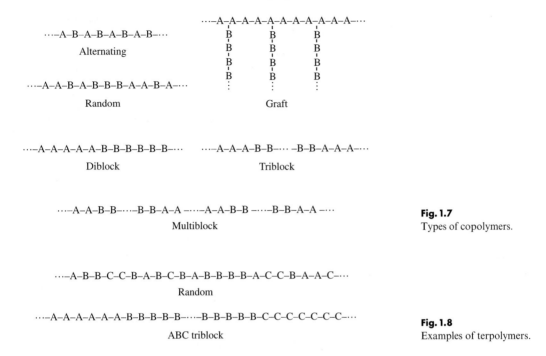

Fig. 1.7
Types of copolymers.

Fig. 1.8
Examples of terpolymers.

Many biopolymers are heteropolymers. **DNA** is a heteropolymer consisting of four different types of monomers (nucleotides), while natural **proteins** are heteropolymers commonly consisting of 20 different types of monomers (amino acids).

1.4 Fractal nature of polymer conformations

The polymer characteristics, described above—microstructure, architecture, degree of polymerization, chemical composition of heteropolymers—are all fixed during polymerization and cannot be changed without breaking covalent chemical bonds. However, after polymerization, a single flexible macromolecule can adopt many different **conformations**. A conformation is the spatial structure of a polymer determined by the relative locations of its monomers. Thus, a conformation can be specified by a set of n **bond vectors** between neighbouring backbone atoms. The conformation that a polymer adopts depends on three characteristics: flexibility of the chain, interactions between monomers on the chain, and interactions with surroundings. The inherent flexibility of the chain plays a vital role. Some chains are stiff like a piano wire, while others are quite flexible like a silk thread. There can be either attractive or repulsive interactions between monomers on the chain. The monomers also interact with their surroundings (either other chains or solvent) and the relative strengths of these various interactions can change with temperature. By tuning these effects, chain conformations change drastically, as will be explained in detail in Chapters 2–5.

To illustrate the magnitude of variations in chain dimensions, consider a chain of $n = 10^{10}$ bonds (one of the largest DNA molecules). The size of a bond is of the order of Angstroms and therefore, the contour length along the entire DNA molecule is of the order of meters. In order to get a better feeling for the range of scales involved, we magnify all lengths by the factor 10^8 to bring individual bond lengths onto a familiar size scale $l \approx 1$ cm. With strong attraction between monomers, the conformation of the polymer is a dense object, called a collapsed globule, occupying volume $V \approx nl^3 \approx 10^{10}\,\text{cm}^3 \approx 10^4\,\text{m}^3$ and densely filling a large classroom of typical linear dimension $R \approx V^{1/3} \approx n^{1/3}l \approx 20$ m (see Fig. 1.9). If there are no interactions between monomers, in Chapter 2 it will be shown that the chain conformation is a random walk with size $R \approx n^{1/2}l \approx 1$ km, a typical campus dimension. Conformations of a polymer with excluded volume repulsions (to be described in detail in Chapter 3) are those of a self-avoiding walk with $R \approx n^{3/5}l \approx 10$ km, a typical city dimension. A polymer with long-range (such as electrostatic) repulsions adopts an extended conformation with size $R \approx nl \approx 10^5$ km, of the order of the distance to the

(a)

(b)

(c)

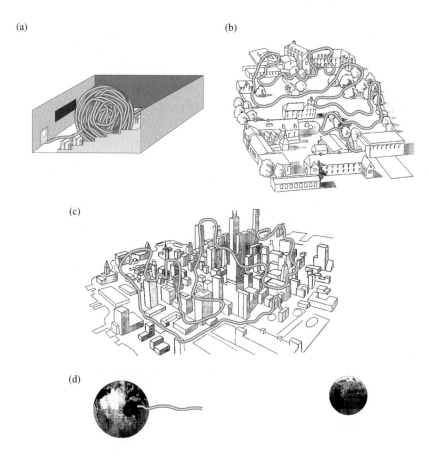

(d)

Fig. 1.9

A polymer's conformation is dictated by its interactions, here illustrated using a chain with 10^{10} monomers of size 1 cm and four types of interaction between monomers. (a) Attractive interaction—the chain fits in a classroom. (b) Zero effective interaction—the chain is the size of a campus. (c) Short-range repulsion—the chain is the size of a city. (d) Long-range repulsion—the size of a chain is a quarter of the distance to the Moon.

Moon! These astronomical variations of chain sizes make polymers a unique class of materials.

Another special feature of polymer conformations is that most of them are **self-similar (fractal)** over a wide range of length scales. In order to explain this important concept, we start with familiar objects, such as a solid ball of radius R. The volume V of this ball is approximately equal to the cube of its radius and proportional to its mass:

$$V = \frac{4\pi}{3} R^3 \cong 4.2R^3 \approx R^3 \sim \text{mass}. \tag{1.2}$$

Throughout this book we will use the sign '$\cong$' to indicate a numerical approximation (i.e. $4\pi/3 \cong 4.2$) and the sign '$\approx$' to indicate that two quantities are proportional to each other up to a dimensionless prefactor of order unity ($4\pi/3$ in the example above). We will also use the notation '$\sim$' to indicate that the two quantities are proportional to each other up to a dimensional constant. The units of mass and volume are different. The meaning of proportionality '$\sim$' is that if the radius of the ball increases by a factor of 2, its mass increases by the factor $2^3 = 8$. The exponent in Eq. (1.2) is the dimension of the ball, $d = 3$. Most of the objects we are familiar with are 3-dimensional. Relations similar to Eq. (1.2) are valid not only for the whole object, but for smaller parts of it as well. Indeed, consider cutting a small sphere of radius r out of this ball (Fig. 1.10). The mass m of this smaller ball is also proportional to the cube of its radius:

$$m \sim r^3. \tag{1.3}$$

Other dimensions we are familiar with are $d = 2$ and $d = 1$. An example of an almost two-dimensional object is a sheet of paper with uniform thickness and density (see Fig. 1.10). The mass m of the circle cut out of the piece of paper is proportional to the square of the radius r of this circle:

$$m \sim r^2. \tag{1.4}$$

Note that we have dropped all prefactors (both dimensionless and dimensional) in Eqs (1.3) and (1.4). A long wire is an example of an almost one-dimensional object (Fig. 1.11). The mass m of a piece of a wire is

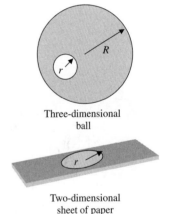

Three-dimensional
ball

Two-dimensional
sheet of paper

Fig. 1.10
Examples of regular objects.

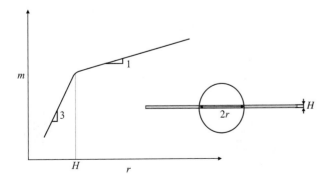

Fig. 1.11
Mass m of the part of the wire inside a sphere of radius r. Both axes have logarithmic scales.

Introduction

proportional to the length $2r$ of this piece (if the diameter and density are uniform):

$$m \sim r. \tag{1.5}$$

In order to explain what we mean by the word 'almost', imagine a sphere of radius r around the wire. As long as the radius of the sphere r is much larger than the diameter H of the wire, the mass of material inside such an imaginary sphere would follow Eq. (1.5). But as soon as the radius of the sphere becomes smaller than the diameter H of the wire, its three-dimensional nature becomes important and the mass of the piece of wire inside the sphere changes proportionally to the cube of its radius [Eq. (1.3)]. The dependence of mass m of the part of the wire inside a sphere of radius r on the size of this sphere for a wire of diameter H is sketched in Fig. 1.11. Thus, we can say that the wire is one-dimensional on length scales much larger than its diameter $r \gg H$ and three-dimensional on smaller length scales $r \ll H$.

As the first example of a self-similar object, consider a regular fractal, called a triadic **Koch curve** (Fig. 1.12). We start from a section of straight line and divide it into three equal subsections (hence the name triadic) [Fig. 1.12(a)]. On the top of the middle subsection we draw an equilateral triangle and erase its bottom side (the original middle subsection of the line). Thus, we end up with four segments of equal length instead of the three original ones [Fig. 1.12(b)]. We repeat the above procedure for each of these four segments—divide each of them into three equal subsections and replace the middle subsections with the two opposite sides of equilateral triangles. At the end of the second step, we obtain a line with each of the four sections consisting of four smaller subsections [Fig. 1.12(c)]. This process can continue as long as your patience allows [Fig. 1.12(d) and (e)]. It is usually limited by the resolution of the computer screen or of the printer.

In order to calculate the dependence of the mass of the triadic Koch curve on the length scale, let us draw circles of diameter $2r$ equal to the lengths of the segments of two consecutive generations [Fig. 1.12(f)]. As we compare circles drawn around the segments of the consecutive generations of the curve, the radius of the circles changes by the factor of 3 ($r_1 = 3r_2$), while the mass m of the section of the curve inside these circles changes by the factor of 4 ($m_1 = 4m_2$). We are looking for an exponent $\mathcal{D}$ defined by the relation

$$m \sim r^{\mathcal{D}}, \tag{1.6}$$

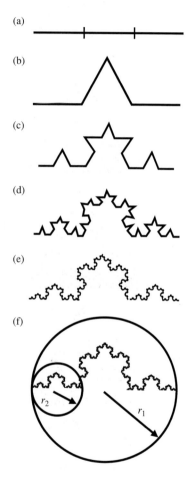

(a)

(b)

(c)

(d)

(e)

(f)

Fig. 1.12
Construction of a triadic Koch curve.

similar to Eqs (1.3)–(1.5) above. This exponent $\mathcal{D}$ in Eq. (1.6) is called the **fractal dimension**. The fractal dimension for a triadic Koch curve can be determined from the fact that we have two ways to calculate m_1 in terms of r_2,

$$m_1 = Ar_1^{\mathcal{D}} = A(3r_2)^{\mathcal{D}}, \tag{1.7}$$

$$m_1 = 4m_2 = 4Ar_2^{\mathcal{D}}, \tag{1.8}$$

where A is the proportionality constant in Eq. (1.6). Equations (1.7) and (1.8) require

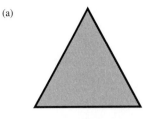

$$(3r_2)^D = 4r_2^D, \qquad (1.9)$$

which can be solved for the fractal dimension of the triadic Koch curve.

$$D = \frac{\log 4}{\log 3} \cong 1.26. \qquad (1.10)$$

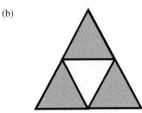

Identical reasoning can be applied to any two consecutive generations of the Koch curve, making the exponent D [Eq. (1.10)] valid on all length scales. The self-similar nature of the Koch curve is clear from the fact that if a small piece of the curve is magnified, it looks exactly like the larger piece.

Another example of a regular fractal is a **Sierpinski gasket** shown in Fig. 1.13. Start with a filled equilateral triangle [Fig. 1.13(a)], draw the three medians that divide it into four smaller equilateral triangles and cut out the middle one [Fig. 1.13(b)]. In the second step, repeat the same procedure with each of the three remaining equilateral triangles, obtaining nine still smaller ones [Fig. 1.13(c)], and so on [Fig. 1.13(d) and (e)]. The fractal dimension of this Sierpinski gasket is calculated by the same method as for the Koch curve above [Fig. 1.13(f)]. As the radius of the circle around a section of the Sierpinski gasket doubles, the number of triangles (the mass of the gasket inside the circle) triples.

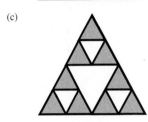

$$D = \frac{\log 3}{\log 2} \cong 1.58. \qquad (1.11)$$

This approach can be applied to any fractal. If the size (the radius of a sphere) changes by the factor C_r

$$r_1 = C_r r_2, \qquad (1.12)$$

while the mass inside this sphere changes by the factor C_m,

$$m_1 = C_m m_2, \qquad (1.13)$$

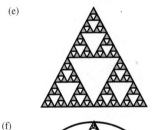

the fractal dimension is the ratio of the logarithms of these factors.

$$D = \frac{\log C_m}{\log C_r}. \qquad (1.14)$$

There are many beautiful examples of regular fractals described and drawn in numerous books on this subject.

Polymers are random fractals, quite different from Koch curves and Sierpinski gaskets, which are examples of regular fractals. Consider, for example, a single conformation of an ideal chain, shown in Fig. 1.14. As will be discussed in detail in Chapter 2, the mean-square end-to-end distance of an ideal chain is proportional to its degree of polymerization.

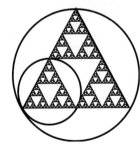

$$N \sim \langle R^2 \rangle. \qquad (1.15)$$

Fig. 1.13
Construction of a Sierpinski gasket.

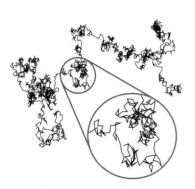

Fig. 1.14
Fractal structure of an ideal chain with fractal dimension $\mathcal{D} = 2$ obtained by computer simulation (courtesy of Q. Liao).

The brackets refer to averages over different possible conformations of the ideal chain. This relation should not be taken literally to be valid for each possible conformation, but rather on average for a distribution of conformations of a given type. A similar relation holds for any subsection of the ideal chain with g monomers and size r:

$$g \sim \langle r^2 \rangle. \tag{1.16}$$

Therefore, the fractal dimension of an ideal chain is $\mathcal{D} = 2$. The ideal chain is a self-similar object because if its smaller sections are magnified, they look like the whole chain (see Fig. 1.14). Unlike the regular fractals, such as the Koch curve, the magnified sections do not look exactly like the whole chain, but only on average (they have the same statistical properties, as will be explained in Chapter 2). Another distinction between regular fractals, such as the Sierpinski gasket, and polymeric fractals is that regular fractals are self-similar on *all* length scales. Polymeric fractals are self-similar on a finite, though possibly quite large, range of length scales. There is a natural cutoff of self-similarity on small length scales—the length l of the bond— and on large scales—the size R of the polymer. Thus, Eq. (1.16) is valid for $l < r < R$. The fractal dimension $\mathcal{D}$ of any polymer is defined through the relation between the number of monomers g in any section of this polymer and the root-mean-square size $\sqrt{\langle r^2 \rangle}$ of this section:

$$g \sim \left(\sqrt{\langle r^2 \rangle} \right)^{\mathcal{D}}. \tag{1.17}$$

In Chapters 2, 3, 5, and 6, the fractal dimension $\mathcal{D}$ of polymers will be derived in different conditions. Examples of the fractal dimensions of polymers are shown in Table 1.2.

Table 1.2   Fractal dimensions of polymers

Architecture	Interactions	Space dimension d	$\mathcal{D}$
Linear	None	Any	2
Linear	Short-range repulsion	$d = 2$	4/3
Linear	Short-range repulsion	$d = 3$	1.7
Randomly branched	None	Any	4
Randomly branched	Short-range repulsion	$d = 2$	8/5
Randomly branched	Short-range repulsion	$d = 3$	2.0
Incipient gel	Partially screened repulsion	$d = 2$	91/48
Incipient gel	Partially screened repulsion	$d = 3$	2.5

1.5   Types of polymeric substances

1.5.1   Polymer liquids

There are two types of **polymer liquids**: polymer melts and polymer solutions. **Polymer solutions** can be obtained by dissolving a polymer in a solvent. Examples of polymer solutions are wood protectants (varnish and polyurethane coatings) and floor shines. Polymer solutions are classified as dilute or semidilute (Fig. 1.15) depending on the polymer **mass concentration** c, the ratio of the total mass of polymer dissolved in a solution and the volume of the solution. An alternative measure of concentration is the **volume fraction** ϕ, the ratio of occupied volume of the polymer in the solution and the volume of the solution. These two concentrations are related through the polymer density ρ:

$$\phi = \frac{c}{\rho} = c\frac{v_{mon}\mathcal{N}_{Av}}{M_{mon}}. \tag{1.18}$$

We used the fact that the polymer density is the ratio of monomer molar mass M_{mon} and monomer molar volume $v_{mon}\mathcal{N}_{Av}$ (v_{mon} is the occupied volume of a single chemical monomer).

$$\rho = \frac{M_{mon}}{v_{mon}\mathcal{N}_{Av}}. \tag{1.19}$$

A typical monomer volume is $v_{mon} \approx 100\,\text{Å}^3$ and the corresponding monomer molar volume is $v_{mon}\mathcal{N}_{Av} \approx 60\,\text{cm}^3$.

The **pervaded volume** V is the volume of solution spanned by the polymer chain

$$V \approx R^3, \tag{1.20}$$

where R is the size of the chain. This volume is typically orders of magnitude larger than the occupied volume of the chain $v_{mon}N$ (see Fig. 1.14). The fractal nature of polymers ($N \sim R^D$) with typical fractal dimension $D < 3$, means most of the pervaded volume is filled with solvent or other chains. The volume fraction of a single molecule inside its pervaded volume is called the **overlap volume fraction** ϕ^* or the corresponding **overlap concentration** c^*:

$$\phi^* = \frac{Nv_{mon}}{V} \qquad c^* = \frac{\rho Nv_{mon}}{V} = \frac{M}{V\mathcal{N}_{Av}}. \tag{1.21}$$

If the volume fraction ϕ of the polymer solution is equal to the overlap volume fraction ϕ^*, the pervaded volumes of macromolecules densely fill space and chains are just at overlap ($\phi = \phi^*$) (see Fig. 1.15).[3]

If the polymer volume fraction ϕ in solution is below the overlap volume fraction ϕ^*, the solution is called **dilute** ($\phi < \phi^*$). The average distance between chains in dilute solutions is larger than their size. Therefore, polymer coils in dilute solutions are far from each other swimming happily in surrounding solvent. Most properties of dilute solutions are very similar to pure solvent with slight modifications due to the presence of the polymer.

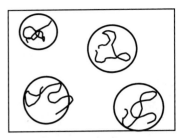

Dilute ($\phi < \phi^*$)

Overlap ($\phi = \phi^*$)

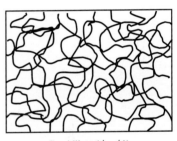

Semidilute ($\phi > \phi^*$)

Fig. 1.15
Solution regimes of flexible polymers.

---

[3] In the definition of the overlap volume fraction ϕ^*, the pervaded volume V is taken at ϕ^*, since polymer size and hence its pervaded volume may change with concentration.

Solutions are called **semidilute** at polymer volume fractions above overlap (for $\phi > \phi^*$) (Fig. 1.15). This name comes from the fact that the actual values of volume fractions in these solutions are very low ($\phi \ll 1$). Most of the volume of a semidilute solution is occupied by the solvent. However, polymer coils overlap and dominate most of the physical properties of semidilute solutions (such as viscosity). Thus, adding a very small amount of polymer to a solvent can create a liquid with drastically different properties than the solvent. This unique feature of polymer overlap is due to their open conformations. Linear polymers in solution are fractals with fractal dimension $\mathcal{D} < 3$. In semidilute solutions, both solvent and other chains are found in the pervaded volume of a given coil. The **overlap parameter** P is the average number of chains in a pervaded volume that is randomly placed in the solution:

$$P = \frac{\phi V}{N v_{\mathrm{mon}}}. \tag{1.22}$$

At the overlap volume fraction (for $\phi = \phi^*$) $P \equiv 1$, and as the concentration of linear chains is increased P steadily grows, reflecting the presence of additional chains inside the pervaded volume of each molecule. Notice that the overlap parameter counts the number of whole chains that share the pervaded volume. In reality, small parts of numerous chains are within each chain's pervaded volume in semidilute solution, and the overlap parameter counts these parts as though they were connected together into chains of N monomers. Use of Eq. (1.22) in semidilute solutions requires care because the chain size and hence, the pervaded volume V may change with concentration, as will be discussed in Chapter 5. The pervaded volume of a chain is not defined precisely, which makes ideas about overlap rather vague, meaning that both P and ϕ^* are somewhat imprecise. In practice this ambiguity arises because polymer overlap occurs over a range of concentrations.

In the absence of solvent, macromolecules can form a bulk liquid state, called a **polymer melt**. Polymer melts are neat polymeric liquids above their glass transition and melting temperatures. A macroscopic piece of a polymer melt remembers its shape and has elasticity on short time scales, but exhibits liquid flow (with a high viscosity) at long times. Such time-dependent mechanical properties are termed **viscoelastic** because of the combination of viscous flow at long times and elastic response at short times (viscoelasticity will be discussed in Chapters 8 and 9). A familiar example of a polymer melt is Silly Putty®. On short time scales (of order seconds), a sphere of Silly Putty resembles a soft elastic solid that bounces when dropped on the floor. However, if left on a table top for an hour, Silly Putty flows into a puddle like a liquid. In a polymer melt the overlap parameter is large ($P \gg 1$) and the strong overlap with neighbouring chains leads to entanglement that greatly slows the motion of polymers. However, individual chains in a polymer melt do move over large distances on long time scales, a property characteristic of fluids.

1.5.2   Polymer solids

There are several different types of polymeric solids. If a polymer melt is cooled, it can either transform into a **semicrystalline** solid below its **melting temperature** T_m or into a polymeric **glass** below its **glass transition temperature** T_g. Semicrystalline solids consist of crystalline regions, called **lamellae**, in which sections of chains are packed parallel to each other, and of amorphous regions between these lamellae (see Fig. 1.16). This multiphase nature makes semicrystalline polymers opaque, but also deformable and tough, when used at temperatures above the T_g of the amorphous phase (such as for polyethylene and polypropylene at room temperature). Macromolecules with regular configurations, such as isotactic and syndiotactic homopolymers often crystallize easily. Macromolecules with more random configurations, such as atactic homopolymers and random copolymers, tend to transform upon cooling into a transparent yet brittle glassy state (such as polymethylmethacrylate and polystyrene). However, there are technologically important exceptions to this rule. Polycarbonate, for example, is a tough glassy polymer at room temperature, making it the polymer of choice for transparent structural applications such as greenhouses and skylights.

If the chains of a polymer melt are reacted with each other to form covalent crosslinks between chains, a polymer network can be formed (Fig. 1.6). Polymer networks are solids and have a preferred shape determined during their preparation by crosslinking. Above their T_g, the chains between crosslinks in a polymer network can move locally, but not globally. Therefore, polymer networks above T_g are called soft solids. Rubbers or elastomers are crosslinked polymer networks with T_g' below room temperature. Examples are vulcanized natural rubber (crosslinked polyisoprene) and silicone caulks (crosslinked polydimethylsiloxane). A polymer **gel** is a polymer network that is swollen in a solvent. The gel becomes progressively softer as more solvent is added, but always remains a solid owing to the permanent bonds that connect the chains. Examples of common polymer gels are Jello$^{®}$, which is a mixture of water and gelatin (a denatured form of the protein collagen), and superabsorbers derived from poly(acrylic acid) used in disposable diapers.

1.5.3   Liquid crystal polymers

A variety of states with order intermediate between crystalline solids and amorphous liquids are also possible for polymers that contain sufficiently rigid rodlike monomers, known as **mesogens**. These mesogens can be attached to chemical monomers as a side group (the R group in Fig. 1.1) or they may be incorporated within the backbone of the polymer. Polymers with exclusively rigid rod-like mesogens as their monomers are usually intractable because they start to decompose below their crystalline melting points. However, alternating copolymers (Fig. 1.7) of rigid rodlike mesogens and flexible segments often are able to be melt processed and have interesting properties. In particular, in a temperature range between their

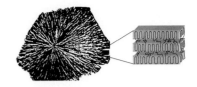

Fig. 1.16
Crystallization of polymer melts creates semicrystalline material consisting of folded chains in lamellae packed into a larger spherulitic structure, coexisting with amorphous regions.

melting point and the temperature at which they become isotropic liquids, these polymers can exhibit any of a number of phases with intermediate order.

A nematic phase, where the mesogens preferentially align in the same direction locally, is the least ordered and most common liquid crystalline phase. Often the alignment of the mesogens allows the molecules to slide past one another more easily, making the viscosity of the nematic phase lower than the isotropic liquid viscosity. A variety of smectic phases are also possible, where the mesogens form layered structures. These anisotropic liquid crystalline phases can occur in melts, solutions and networks. The physical properties of liquid crystal polymers are anisotropic as a result of the order. Additionally, electric fields, magnetic fields, and flow fields can be used to align this class of materials. It is important to realize that the polymer liquids and solids discussed in the remainder of this book are always assumed to be isotropic.

1.6   Molar mass distributions

One distinguishing feature of most synthetic polymers is that they are **polydisperse**. The entire polymer sample is made up of individual molecules that have a distribution of degrees of polymerization, determined by the particular synthesis method used. If all polymers in a given sample have the same number of monomers, the sample is **monodisperse**. There are many examples of natural polymers (such as proteins) that are perfectly monodisperse, but such perfection is very rare in synthetic polymers. The molar mass of a monodisperse polymer with degree of polymerization N is given by Eq. (1.1). $M = NM_{mon}$

The polydispersity of a sample is described by its **molar mass distribution**. Polydisperse and monodisperse distributions are sketched in Fig. 1.17. A distribution is shown as n_N, the **number fraction** (or mole fraction) of molecules containing N monomers each, plotted as a function of molar mass $M_N = M_{mon}N$ of the molecules.

In practice, it is often more convenient to deal with the **weight fraction** w_N of molecules with molar mass M_N:

$$w_N = \frac{n_N M_N}{\sum_N n_N M_N} = \frac{n_N N}{\sum_N n_N N}.$$ (1.23)

The summation $\sum_N$ is a shorthand notation for a sum over all possible values of N (i.e. $\sum_{N=1}^{\infty}$). The weight fraction w_N is related to the mass concentrations of various species (c_N is the mass of molecules with degree of polymerization N per unit volume)

$$w_N = \frac{c_N}{c},$$ (1.24)

where c is the total mass concentration.

It is convenient to define the kth **moment** of the number fraction distribution as the sum of the products of the number fraction n_N of molecules

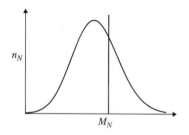

Fig. 1.17

Comparison of monodisperse and polydisperse molar mass distributions.

with degree of polymerization N and their molar mass raised to the kth power:

$$m_k = \sum_N n_N M_N{}^k. \tag{1.25}$$

The zeroth ($k=0$) moment is equal to unity because the number fraction distribution is normalized:

$$m_0 = \sum_N n_N = 1. \tag{1.26}$$

In order to characterize the molar mass distribution, several average molar masses are defined that emphasize different parts of this distribution. The **number-average molar mass** M_n is defined as the ratio of the first ($k=1$) to zeroth ($k=0$) moments of the number fraction distribution:

$$M_n \equiv \frac{m_1}{m_0} = \frac{\sum_N n_N M_N}{\sum_N n_N} = \sum_N n_N M_N. \tag{1.27}$$

The number-average is the common average used, for example, to determine the average denomination of the coins in your pocket. The number fraction of each type of coin (n_N) is multiplied by its denomination (M_N).

Substituting Eq. (1.27) into Eq. (1.23) shows that the ratio of the molar mass of a polymer with N monomers and the number-average molar mass M_n relates the number fraction and weight fraction of molecules:

$$w_N = \frac{M_N}{M_n} n_N = \frac{N}{N_n} n_N. \tag{1.28}$$

The final relation was obtained by introducing the number-average degree of polymerization $N_n \equiv M_n/M_{mon}$. The number-average is the quantity that is directly controlled by polymerization chemistry. The total number density of chains is the sum of the number density of all species:

$$\frac{c\mathcal{N}_{Av}}{M_n} = \sum_N \frac{c_N \mathcal{N}_{Av}}{M_N}. \tag{1.29}$$

Solving for M_n gives an alternative expression for calculating the number-average molar mass:

$$M_n = \frac{c}{\sum_N c_N/M_N} = \frac{1}{\sum_N w_N/M_N}. \tag{1.30}$$

The final relation was obtained using Eq. (1.24). For strictly linear polymers, each chain has exactly two ends, so the number-average molar mass can be measured by counting end groups using spectroscopy. However, many polymer properties are controlled by the longer chains in the molar mass distribution, making higher-order averages useful.

The **weight–average molar mass** M_w is the ratio of the second and the first moments of the number fraction distribution:

$$M_w \equiv \frac{m_2}{m_1} = \frac{\sum\limits_N n_N M_N^2}{\sum\limits_N n_N M_N} = \frac{\sum\limits_N n_N M_N^2}{M_n} = \sum_N w_N M_N$$

$$= \sum_N \frac{c_N}{c} M_N. \tag{1.31}$$

The last set of relations was obtained using the connection between weight and number fractions [Eqs (1.28) and (1.24)]. The weight-average is the molar mass obtained by randomly choosing the monomer. For example, consider a mixture of different length strings in a box. The weight-average length of the strings can be measured by reaching into the box (with eyes closed), pulling out one of the strings, measuring its length (with eyes open), putting it back into the box, mixing the strings and repeating the procedure many times. The probability of pulling a particular string out of the box is proportional to that string's length, since each section of string has the same probability of being selected.

In Section 1.7.2, we will see that the weight-average molar mass can be measured by light scattering from a dilute polymer solution. The viscosity of polymer liquids correlates well with the weight-average molar mass.

The **polydispersity index** is defined as the ratio of the weight-average and number-average molar masses M_w/M_n. Monodisperse samples with $M_w = M_n$ have polydispersity index $M_w/M_n = 1$. Larger polydispersity indices correspond to samples with broader molar mass distributions.

The **z-average molar mass** M_z is defined as the ratio of the third to the second moments of the number fraction distribution:

$$M_z \equiv \frac{m_3}{m_2} = \frac{\sum\limits_N n_N M_N^3}{\sum\limits_N n_N M_N^2} = \frac{\sum\limits_N w_N M_N^2}{\sum\limits_N w_N M_N} = \frac{\sum\limits_N c_N M_N^2}{\sum\limits_N c_N M_N}. \tag{1.32}$$

Similarly, the $(z+1)$-average molar mass is the ratio of the fourth to the third moments of the number fraction distribution:

$$M_{z+1} \equiv \frac{m_4}{m_3} = \frac{\sum\limits_N n_N M_N^4}{\sum\limits_N n_N M_N^3} = \frac{\sum\limits_N w_N M_N^3}{\sum\limits_N w_N M_N^2} = \frac{\sum\limits_N c_N M_N^3}{\sum\limits_N c_N M_N^2}. \tag{1.33}$$

In general, the $(z+k)$-average molar mass is defined as

$$M_{z+k} \equiv \frac{m_{k+3}}{m_{k+2}} = \frac{\sum\limits_N n_N M_N^{k+3}}{\sum\limits_N n_N M_N^{k+2}} = \frac{\sum\limits_N w_N M_N^{k+2}}{\sum\limits_N w_N M_N^{k+1}} = \frac{\sum\limits_N c_N M_N^{k+2}}{\sum\limits_N c_N M_N^{k+1}}. \tag{1.34}$$

Higher-order molar mass averages, such as M_z and M_{z+1} emphasize the high molar mass tail of the molar mass distribution. Molecular theories of polymer dynamics predict these higher-order averages are important, but currently available characterization methods for measuring them have insufficient precision to be useful.

Political example. In order to better understand the difference between the number- and weight-averages, let us calculate the average population of a state in the USA. One possible way of obtaining an average is to ask each US senator for a population of their respective state (there are two senators from each of the 50 states) and average these 100 answers. The result would be the number-average population of a state. If, instead of getting answers from senators, who are too busy, we ask congressmen and average their answers, we would obtain the weight-average state population. The reason is that the number of congressmen from each state is proportional to the population of the corresponding state. In 2001, the number-average population of each state is 6×10^6 people per state and the weight-average state population is 12×10^6 people per state, making the polydispersity index 2.

1.6.1 Binary distributions

In this section, two examples of **binary mixtures** of two different mono-disperse chain lengths are used to better understand the various molar mass averages.

Example 1

Consider a mixture containing number fraction $n_A = 1/2$ of the protein gelatin with molar mass $M_A = 10^5 \, \text{g mol}^{-1}$ and number fraction $n_B = 1 - n_A = 1/2$ of gelatin dimers with molar mass $M_B = 2 \times 10^5 \, \text{g mol}^{-1}$. What are the number- and weight-average molar masses of this sample and its polydispersity index?

The number-average molar mass of the sample is

$$M_n = m_1 = \sum_N n_N M_N = n_A M_A + n_B M_B = 1.5 \times 10^5 \, \text{g mol}^{-1}. \quad (1.35)$$

The second moment of the distribution is calculated in a similar fashion:

$$m_2 = \sum_N n_N M_N^2 = n_A M_A^2 + n_B M_B^2 \quad (1.36)$$

The weight-average molar mass is the ratio of the second and first moments:

$$M_w = \frac{m_2}{m_1} = \frac{n_A M_A^2 + n_B M_B^2}{n_A M_A + n_B M_B} \simeq 1.67 \times 10^5 \, \text{g mol}^{-1}. \quad (1.37)$$

The polydispersity index of this binary mixture is the ratio of M_w and M_n:

$$\frac{M_w}{M_n} = \frac{n_A M_A^2 + n_B M_B^2}{[n_A M_A + n_B M_B]^2} = \frac{10}{9}. \quad (1.38)$$

Notice that even though the chain lengths in the binary mixture differ by a factor of 2, the polydispersity index $M_w/M_n = 10/9 \cong 1.11$ is only slightly larger than its monodisperse value of $M_w/M_n = 1$.

Example 2

Consider an example similar to the one above, but with the binary mixture containing *weight* fraction $w_A = 1/2$ (rather than number fraction) of gelatin molecules with molar mass $M_A = 10^5 \, \text{g mol}^{-1}$ and weight fraction $w_B = 1 - w_A = 1/2$ of gelatin dimers with molar mass $M_B = 2 \times 10^5 \, \text{g mol}^{-1}$. What are the number- and weight-average molar masses of this sample and its polydispersity index?

Eqs (1.30) and (1.31) can be used to calculate number-average molar mass

$$M_n = \frac{1}{\sum_N w_N/M_N} = \frac{1}{(w_A/M_A) + (w_B/M_B)} \cong 1.33 \times 10^5 \, \text{g mol}^{-1} \quad (1.39)$$

and weight-average molar mass

$$M_w = \sum_N w_N M_N = w_A M_A + w_B M_B = 1.5 \times 10^5 \, \text{g mol}^{-1}. \quad (1.40)$$

The polydispersity index of this binary mixture is the ratio of M_w and M_n:

$$\frac{M_w}{M_n} = [w_A M_A + w_B M_B]\left(\frac{w_A}{M_A} + \frac{w_B}{M_B}\right) = \frac{9}{8}. \quad (1.41)$$

This polydispersity index is larger than for the binary mixture in Example 1. Note that the arithmetic average of M_A and M_B ($1.5 \times 10^5 \, \text{g mol}^{-1}$) is the number-average molar mass M_n for Example 1 with equal number fractions and is the weight-average molar mass M_w for Example 2 with equal weight fractions.

1.6.2   Linear condensation polymers

Consider linear condensation polymerization of monomers AB. An unreacted A group of any monomer is capable of forming a bond with any unreacted B group of any other monomer. Thus, an unreacted B group at the end monomer of an N-mer (molecule containing N monomers) can react with an unreacted A group at the end monomer of a K-mer, forming an $(N + K)$-mer:

$$(AB)_N + (AB)_K \rightarrow (AB)_{N+K}. \quad (1.42)$$

The ratio of the number of formed bonds to the maximum possible number of bonds in a reaction is called the **extent of reaction** p. If we select any group (A or B) randomly, p is the probability that the group has reacted. In linear condensation polymerization, each chain has one unreacted A group at one end of the chain and one unreacted B group at the other end. Flory showed long ago that the reactivity of these groups is independent of chain length. Therefore, every A group of the original AB monomers has the same probability p to be reacted with some B group.

Next, we estimate the probability of a given A group to be at the end of an N-mer. To be at the end of the N-mer, this A group must be unreacted, with probability $1-p$. The probability that the B group of this end monomer has reacted and formed a bond with the A group of the second monomer is p. There are $N-1$ such polymerization bonds between monomers in any N-mer. The probability of forming these $N-1$ independent bonds is p^{N-1}. The probability that the B group at the other end of the N-mer is unreacted is $1-p$. Therefore, the probability that a given A group is at the end of an N-mer with $N-1$ polymerization bonds and two unreacted groups at its ends is given by the product:

$$n(p, N) = p^{N-1}(1-p)^2. \tag{1.43}$$

This probability is the number of N-mers *per monomer*: the number of N-mers in the sample at extent of reaction p divided by the total number of monomers in the system. The number of N-mers per monomer $n(p, N)$ is the ratio of the number density of N-mers $c_N \mathcal{N}_{Av}/(M_{mon}N)$ and the number density of all monomers $c\mathcal{N}_{Av}/M_{mon}$ in the sample

$$n(p, N) = \frac{c_N(p)}{cN} = \frac{w_N(p)}{N}, \tag{1.44}$$

where $c_N(p)$ is the mass concentration of N-mers at extent of reaction p and $w_N(p)$ is their weight fraction [Eq. (1.24)].

It is important to clarify the difference between the number of N-mers per monomer $n(p, N)$ and the number fraction $n_N(p)$ of N-mers (the number of N-mers per polymer chain) at extent of reaction p. The number fraction (or mole fraction) of N-mers was discussed in detail in the previous section. If we reach into the polymerization reactor at extent of reaction p and randomly select a *chain*, the probability that it has degree of polymerization N is the number fraction of N-mers:

$$n_N(p) = \frac{n(p, N)}{\sum\limits_{N=1}^{\infty} n(p, N)} = N_n n(p, N) \tag{1.45}$$

The sum in the denominator is equal to the total number of molecules per monomer and hence, is the reciprocal of the number-average degree of polymerization $1/N_n$, thereby providing the final result in Eq. (1.45)

$$N_n = \frac{1}{\sum\limits_{N=1}^{\infty} n(p, N)} = \frac{1}{(1-p)^2 \sum\limits_{N=1}^{\infty} p^{N-1}} = \frac{1}{(1-p)^2 \sum\limits_{k=0}^{\infty} p^k}. \tag{1.46}$$

The last equality was obtained by defining $k = N - 1$. The sum of the geometric series

$$\sum_{k=0}^{\infty} p^k = 1 + p + p^2 + p^3 + \cdots \tag{1.47}$$

can be determined by multiplying it by p

$$p \sum_{k=0}^{\infty} p^k = p + p^2 + p^3 + \cdots \tag{1.48}$$

leading to the equation

$$\sum_{k=0}^{\infty} p^k - p \sum_{k=0}^{\infty} p^k = 1 \tag{1.49}$$

with solution

$$\sum_{k=0}^{\infty} p^k = \frac{1}{1-p}. \tag{1.50}$$

Therefore, the number-average degree of polymerization has a simple form:

$$N_n = \frac{1}{\displaystyle\sum_{N=1}^{\infty} n(p, N)} = \frac{1}{1-p}. \tag{1.51}$$

Substituting this result into Eq. (1.45) we obtain the number fraction distribution for linear condensation polymers:

$$n_N(p) = \frac{n(p, N)}{\displaystyle\sum_{N=1}^{\infty} n(p, N)} = p^{N-1}(1-p). \tag{1.52}$$

This number fraction distribution [Eq. (1.52)] is shown as dashed line in Fig. 1.18 for the extent of reaction $p = 0.991$.

The weight fraction of N-mers is determined from Eqs (1.23) and (1.28):

$$w_N(p) = \frac{N}{N_n} n_N(p) = N p^{N-1}(1-p)^2. \tag{1.53}$$

If we reach into the polymerization reactor at extent of reaction p and randomly select a *monomer*, the weight fraction distribution $w_N(p)$ is the probability that the randomly chosen monomer is part of a chain with degree of polymerization N. Note that in general, the k-moment of the number fraction distribution is related to the $(k-1)$-moment of the weight fraction distribution (see Problem 1.31):

$$\sum_N n_N M_N^k = M_n \sum_N w_N M_N^{k-1}. \tag{1.54}$$

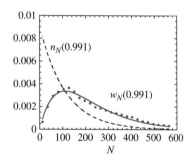

Fig. 1.18

Comparison of the most-probable weight fraction distribution $w_N (0.991)$ with experimental nylon 66 data from G. B. Taylor, *J. Am. Chem. Soc.* **69**, 638 (1947). The dashed line is the number fraction distribution $n_N (0.991)$.

Equations (1.52) and (1.53) are expressions of the number and weight fractions for the **most-probable distribution** of molecules expected for linear condensation polymerization. The most-probable weight fraction distribution $w_N(p)$ is compared with experimental data in Fig. 1.18. While the number fraction for the most-probable distribution is a monotonic function, the weight fraction has a maximum. The position of the maximum in the most-probable weight fraction distribution [Eq. (1.53) and the solid curve in Fig. 1.18)] is close to the number-average degree of polymerization

$$N_n = M_n/M_{mon} = 1/(1-p), \qquad (1.55)$$

$N_n = 111$ for $p = 0.991$.

The number- and the weight-average molar masses are calculated from the moments of the molar mass distribution. The first moment of the number fraction distribution is the number-average molar mass M_n:

$$M_n = m_1 = M_{mon} \sum_{N=1}^{\infty} N n_N(p) = M_{mon}(1-p) \sum_{N=1}^{\infty} N p^{N-1}. \qquad (1.56)$$

The summation can be carried out as follows:

$$\sum_{N=1}^{\infty} N p^{N-1} = \frac{d}{dp} \sum_{N=1}^{\infty} p^N = \frac{d}{dp} \left[\sum_{N=0}^{\infty} p^N - 1 \right]$$
$$= \frac{d}{dp} \left(\frac{1}{1-p} \right) = \frac{1}{(1-p)^2}. \qquad (1.57)$$

This leads to the number-average molar mass

$$M_n = m_1 = \frac{M_{mon}}{1-p}, \qquad (1.58)$$

in agreement with the number-average degree of polymerization N_n [Eqs (1.51) and (1.55)]. The number-average molar mass M_n is larger than the monomer molar mass M_{mon}, and as p approaches unity (where all monomers are in a single chain) M_n gets extremely large (see Fig. 1.19).

The second moment of the number fraction distribution is obtained from Eq. (1.25) with $k = 2$ and Eq. (1.52):

$$m_2 = M_{mon}^2 \sum_{N=1}^{\infty} N^2 n_N(p) = M_{mon}^2(1-p) \sum_{N=1}^{\infty} N^2 p^{N-1}. \qquad (1.59)$$

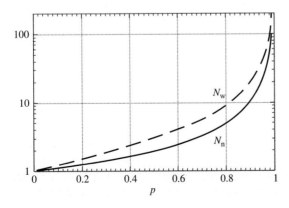

Fig. 1.19
Number-average (solid curve) and weight-average (dashed curve) degrees of polymerization as functions of extent of reaction for linear condensation polymerization.

The summation can be done by a procedure similar to Eq. (1.57):

$$\sum_{N=1}^{\infty} N^2 p^{N-1} = \frac{d}{dp}\left[p\frac{d}{dp}\sum_{N=0}^{\infty}p^N\right] = \frac{d}{dp}\left[\frac{p}{(1-p)^2}\right] = \frac{1+p}{(1-p)^3}. \tag{1.60}$$

Therefore, the second moment of the distribution is

$$m_2 = M_{mon}^2 \frac{1+p}{(1-p)^2}, \tag{1.61}$$

and the weight-average molar mass for linear condensation polymers is the ratio of Eqs (1.61) and (1.58):

$$M_w = \frac{m_2}{m_1} = M_{mon}\frac{1+p}{1-p}. \tag{1.62}$$

The weight-average molar mass for linear condensation polymers also diverges as the extent of reaction $p \to 1$ (see Fig. 1.19). The polydispersity index of the linear condensation polymers,

$$\frac{M_w}{M_n} = 1 + p, \tag{1.63}$$

approaches $M_w/M_n = 2$ for samples with high conversion ($p \to 1$) and high molar mass. Linear polymers prepared by condensation chemistry typically have $M_w/M_n \cong 2$ with a most-probable distribution of chain lengths.

The most-probable weight fraction distribution [Eq. (1.53)] can be approximated for large number-average degree of polymerization N_n by an exponential representation, utilizing the expansion of the logarithm for p near unity $\ln p \cong p - 1$ and Eq. (1.55) for N_n:

$$p^N = \exp\left(N \ln p\right) \cong \exp\left(-N\left(1 - p\right)\right) = \exp\left(-\frac{N}{N_n}\right). \tag{1.64}$$

Thus, the most-probable weight fraction distribution can be approximated by a linear function with an exponential cutoff

$$w_N(p) = Np^{N-1}(1-p)^2 \cong \frac{N}{N_n^2}\exp\left(-\frac{N}{N_n}\right), \tag{1.65}$$

for large extents of reaction $p \to 1$ ($N_n \gg 1$). The most-probable number fraction distribution [Eq. (1.52)] for linear condensation polymers can also be approximated by an exponential form for large N_n:

$$n_N(p) = \frac{N_n}{N}w_N(p) \cong \frac{1}{N_n}\exp\left(-\frac{N}{N_n}\right). \tag{1.66}$$

This is the first of many examples where the molar mass distributions are products of simple functions and exponential cutoffs.

1.6.3 Linear addition polymers

In addition polymerization, monomers are added one at a time to a growing chain by propagation of a free radical through a liquid of monomers:

$$A \xrightarrow{+A} A_2 \xrightarrow{+A} A_3 \xrightarrow{+A} \cdots \xrightarrow{+A} A_{N-1} \xrightarrow{+A} A_N.$$

In the case of addition polymerization without termination, the number fraction distribution function (the probability that a given chain has degree of polymerization N) is given by the **Poisson distribution** function:

$$n_N = \frac{(N_n - 1)^{N-1}}{(N-1)!} \exp(1 - N_n). \tag{1.67}$$

The weight fraction distribution function for addition polymerization without termination is determined using Eq. (1.28):

$$w_N = \frac{N}{N_n} n_N = \frac{N(N_n - 1)^{N-1}}{N_n(N-1)!} \exp(1 - N_n). \tag{1.68}$$

The polydispersity index for the Poisson distribution is quite narrow, since there is no termination (see Problem 1.37):

$$\frac{N_w}{N_n} = 1 + \frac{1}{N_n} - \frac{1}{N_n^2} \tag{1.69}$$

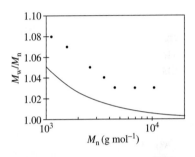

Fig. 1.20
Polydispersity index for polybutadienes polymerized using s-butyl lithium initiator in cyclohexane. The curve is the Poisson distribution prediction [Eq. (1.69)]. The ordinate data appear discretized because M_w/M_n can only be determined to three significant figures.

Many addition polymerization reactions with very low concentrations of impurities have propagation rates much faster than initiation rates and have essentially no termination. Such reactions produce narrow molar mass distributions that can be approximated by the Poisson distribution. Comparison of the polydispersity index of anionically polymerized butadiene with Eq. (1.69) is shown in Fig. 1.20.

The experimental data lie above the prediction of Eq. (1.69) because the experimental determination of polydispersity index by size exclusion chromatography (Section 1.7.4) systematically overestimates M_w/M_n. The low molar mass polydispersity index does follow the trend expected by Eq. (1.69). The prediction of Eq. (1.69) that $M_w/M_n \cong 1$ for chains with $N_n \gtrsim 1000$ (or $M_n \gtrsim 10^5$) is never realized in practice because real polymerization reactions always have some impurities present.

Many addition polymerizations that involve free radicals at chain ends have termination reactions when two growing chain ends meet, and more generally termination can occur when a growing chain end meets an impurity. For addition polymerization with termination the **Shultz distribution** is used,

$$w_N = \frac{1}{N_n \Gamma(s)} \left(\frac{sN}{N_n}\right)^s \exp\left(-\frac{sN}{N_n}\right), \tag{1.70}$$

where $\Gamma(s)$ is the Gamma function.[4] Note that for $s = 1$ the most-probable distribution is recovered [compare with Eq. (1.65)]. The polydispersity index of the Shultz distribution [Eq. (1.70)] is

$$\frac{N_{\mathrm{w}}}{N_{\mathrm{n}}} = \frac{s+1}{s},\qquad(1.71)$$

which leads to $N_{\mathrm{w}}/N_{\mathrm{n}} = 2$ for the most-probable distribution, but allows for broader ($N_{\mathrm{w}}/N_{\mathrm{n}} > 2$ for $s < 1$) and narrower ($1 < N_{\mathrm{w}}/N_{\mathrm{n}} < 2$ for $s > 1$) distributions.

1.7 Molar mass measurements

There are a variety of experimental methods available for determining different average molar masses and molar mass distributions. These methods utilize dilute solutions, and often require measurements at several concentrations so that an extrapolation to the limit of zero concentration can be made. Different methods are applicable to different ranges of polymer molar masses. Table 1.3 summarizes common characterization methods. Here we discuss only the four most common methods in detail.

Table 1.3 Molar mass measurement methods

Method	Absolute	Relative	M_{n}	M_{w}	A_2	Range (g mol^{-1})
End group analysis	×		×			$M_{\mathrm{n}} < 10\,000$
Vapor pressure osmometry	×		×		×	$M_{\mathrm{n}} < 30\,000$
Cryoscopy	×		×		×	$M_{\mathrm{n}} < 30\,000$
Ebulliometry	×		×		×	$M_{\mathrm{n}} < 30\,000$
Membrane osmometry	×		×		×	$20\,000 < M_{\mathrm{n}}$
Light scattering (LS)	×			×	×	$10^4 < M_{\mathrm{w}} < 10^7$
Intrinsic viscosity (IV)		×				$M < 10^6$
SEC^a with c detector		×	×	×		$10^3 < M < 10^7$
SEC^a with c and LS detectors	×			×		$10^4 < M < 10^7$
SEC^a with c and IV detectors		×	×	×		$10^3 < M < 10^6$
MALDI-TOF-MS^b	×		×	×		$M < 10\,000$

^{a}SEC, size exclusion chromatography. ^{b}MALDI-TOF-MS, matrix-assisted laser desorption/ionization time-of-flight mass spectroscopy.

1.7.1 Measuring M_{n} by osmotic pressure

Number-average molar mass can be determined directly by end group analysis, typically using infrared or nuclear magnetic resonance spectroscopies. Colligative solution properties (sensitive to the number of polymers

---

[4] The Gamma function is defined as $\Gamma(a) = \int_0^\infty \mathrm{e}^{-x} x^{a-1}\,\mathrm{d}x$. Gamma functions of different arguments are related by $\Gamma(a+1) = a\Gamma(a)$. For integer values of a, the Gamma function is simply a factorial $\Gamma(a+1) = a!$.

present) also determine M_n from dilute solutions of the polymer. These include osmotic pressure, lowering of solvent vapor pressure, ebulliometry (elevation of boiling point), or cryoscopy (depression of freezing point). The most common method to measure M_n is osmotic pressure.

Osmotic pressure is a thermodynamic colligative property that measures the free energy difference between a polymer solution and a pure solvent [see Eq. (4.62) for the proper definition of osmotic pressure]. In practice, the two are separated by a membrane that allows solvent to pass through easily, but restricts polymer to stay on one side, as shown schematically in Fig. 1.21.

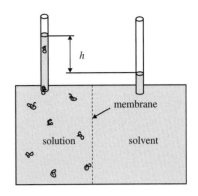

Fig. 1.21
Schematic representation of a membrane osmometer. The osmotic pressure is determined from the height difference h as $\Pi = \rho g h$, where ρ is the solvent density and g is the gravitational acceleration.

There is a free energy gain in mixing polymer with solvent (discussed in detail in Chapter 4) that makes more solvent flow into the solution with the polymer. This solvent flow continues until a pressure difference across the membrane makes the chemical potential of solvent the same on both sides of the membrane. This pressure difference is the osmotic pressure, Π. In the limit of very dilute solutions, individual polymer coils do not interact with each other, and the osmotic pressure in sufficiently dilute solutions is equivalent to the pressure of an ideal gas. Polymer molecules cannot pass through the membrane and impose pressure on it similar to the pressure on the walls of the container by the molecules of an ideal gas. In both cases the pressure is equal to the **thermal energy** kT ($k = 1.38 \times 10^{-23}$ J K^{-1} is the Boltzmann constant and T is absolute temperature) times the number density of molecules. For a monodisperse polymeric sample, the number density of chains is $c\mathcal{N}_{Av}/M$, leading to the **van't Hoff Law**:

$$\lim_{c \to 0} \frac{\Pi}{c} = kT\frac{\mathcal{N}_{Av}}{M} = \frac{\mathcal{R}T}{M}, \tag{1.72}$$

$\mathcal{R} = k\mathcal{N}_{Av} \cong 8.314$ J mol^{-1} K^{-1} is the gas constant. Since osmotic pressure is a colligative property, it is simply proportional to the number density of solute molecules that cannot cross the membrane, regardless of the length of the chains. For a polydisperse sample, the contribution to the osmotic pressure from polymers with different molar masses M_i and concentrations c_i are simply added:

$$\lim_{c \to 0} \frac{\Pi}{c} = \frac{\mathcal{R}T}{c}\sum_i \frac{c_i}{M_i} = \frac{\mathcal{R}T}{M_n}. \tag{1.73}$$

The final relation was obtained using Eq. (1.30). Hence, *dilute osmotic pressure measurements give the number-average molar mass of a polydisperse polymer sample.*

In order to obtain the number-average molar mass of a particular sample, **osmotic coefficient** (Π/c) data, measured at various low concentrations, must be extrapolated to the zero concentration limit. In addition to the ideal gas contribution [Eq. (1.73)] that arises from individual polymers, the osmotic pressure also has a contribution from polymer–polymer interactions. The contribution to osmotic pressure from interaction between species i and j is represented by $A_{ij}c_ic_j\mathcal{R}T$, where A_{ij} is called the **second virial coefficient** between species i and j. The total contribution from pairwise interactions to the osmotic pressure is the sum over all pairs of species i and j:

$$\Pi = \mathcal{R}T\left(\frac{c}{M_n} + \sum_i\sum_j A_{ij}c_ic_j + \cdots\right)$$
$$= \mathcal{R}T\left(\frac{c}{M_n} + A_{2,w}c^2 + \cdots\right). \tag{1.74}$$

The coefficient $A_{2,w}$ in front of the c^2 term is the weight-average second virial coefficient (to be discussed in Chapter 3):

$$A_{2,w} \equiv \frac{1}{c^2}\sum_i\sum_j A_{ij}c_ic_j = \sum_i\sum_j A_{ij}w_iw_j, \tag{1.75}$$

where w_i is a weight fraction of species i ($w_i = c_i/c$). The subscript w is often dropped and the second virial coefficient is usually denoted by A_2. The sign of the second virial coefficient A_2 indicates repulsion or attraction between chains. Positive values of A_2 result in increased osmotic pressure and correspond to repulsion between polymers. Conversely, negative values ($A_2 < 0$) correspond to attraction between chains. Strong attraction between chains may lead to phase separation. Molar masses of polymers *cannot* be obtained from osmotic pressure measurements in a phase separated sample.

By plotting $\Pi/c\mathcal{R}T$ against concentration (Fig. 1.22) the number-average molar mass is determined as the reciprocal of the intercept and the weight-average second virial coefficient is the slope:

$$\frac{\Pi}{c\mathcal{R}T} = \frac{1}{M_n} + A_{2,w}c + \cdots \tag{1.76}$$

At higher concentrations, the higher-order virial terms have to be taken into account and extrapolation to zero concentration becomes more difficult.

Fig. 1.22
Concentration dependence of osmotic coefficient for three poly(α-methylstyrene) samples in toluene at 25 °C. The data corresponding to dilute solutions for these three samples are shown, with lines fit to the lowest concentration data. Data over wider ranges of concentration and chain length are shown in Fig. 5.7 [data from I. Noda, N. Kato, T. Kitano and M. Nagasawa, *Macromolecules* **16**, 668 (1981)].

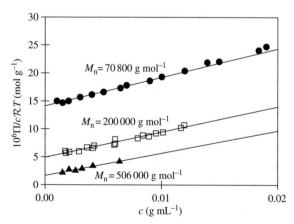

1.7.2 Measuring M_w by scattering

Weight-average molar mass can be measured by scattering and by ultra-centrifugation. Recent developments in photodetector technology have allowed light scattering to completely displace the ultracentrifuge as the method of choice for determining M_w. This section will focus on the scattering of visible light, although the same basic principles apply to scattering of other forms of radiation. Other forms of radiation used for studying polymers include X-rays (where the **scattering contrast** arises from electron density differences) and neutrons (where the scattering contrast comes from differences in atomic nuclei). Light scattering, on the other hand, relies on differences in refractive index n for scattering contrast (polarization of electron clouds). To measure the molar mass of a polymer chain, a dilute solution is prepared in a solvent with sufficiently different refractive index than the polymer. Usually a solvent can be found with refractive index differing from the polymer by $\Delta n \gtrsim 0.1$, making the **refractive index increment** $dn/dc \gtrsim 0.1 \, \mathrm{ml \, g^{-1}}$.

1.7.2.1 Scattering from gases

In order to understand the principle of **light scattering**, consider a gas of N_{tot} relatively small (compared to the wavelength of light λ) non-interacting molecules in a **scattering volume** V. The scattering volume is the portion of the sample that is illuminated by the incident beam and seen by the detector. In the simplest scattering geometry, the light source, the sample and the detector are all aligned in the horizontal yz plane (Fig. 1.23). An incident wave (a vertically polarized laser beam) travels from left to right along the z-axis. Light scattered by an angle θ in the horizontal yz plane is analysed by a detector.

The incident light produces oscillating electric and magnetic fields in the transverse direction (in the xy plane) at every point along the beam. In a typical case of a vertically polarized laser light, the electric field oscillates

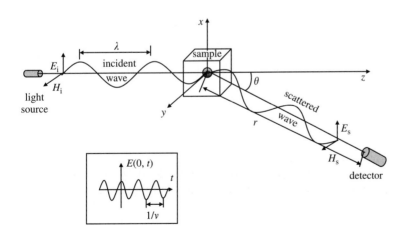

Fig. 1.23
Schematic geometry of Rayleigh scattering. Insert: oscillations in time of the electric field, $E(0, t)$, felt by a molecule at the origin.

along the vertical x axis, while the magnetic field oscillates along the y axis. The electric field at location z at time t along the wave is sinusoidal:

$$E(z, t) = E_i \cos\left[2\pi\left(\nu t - \frac{z}{\lambda}\right)\right]. \tag{1.77}$$

The oscillations of this electric field have amplitude E_i and frequency ν. The wavelength λ is the period of the wave in space. It is related to the period of oscillations in time $1/\nu$ through the speed of light $c \cong 2.998 \times 10^8 \, \text{ms}^{-1}$:

$$\nu = \frac{c}{\lambda}. \tag{1.78}$$

The same relation is valid in a dielectric medium, such as a polymer solution, where both the speed c/n and the wavelength λ/n of light are reduced by the refractive index n of the medium.

Consider a molecule, located at the origin. It experiences time oscillations of the electric field from the beam (see insert in Fig. 1.23)

$$E(0, t) = E_i \cos\left(2\pi \frac{c}{\lambda} t\right). \tag{1.79}$$

This field causes the electrons in the molecule to oscillate, resulting in an induced dipole moment p proportional to the applied field:[5]

$$p = \alpha E = \alpha E_i \cos\left(2\pi \frac{c}{\lambda} t\right). \tag{1.80}$$

Here α is the **polarizability** of the molecule. An oscillating dipole emits electromagnetic waves in *all directions* with electric field proportional to the acceleration of charges $(d^2 p / dt^2)$ and decaying reciprocally with the distance r from the molecule. For a detector located in the horizontal yz plane at distance $r = \sqrt{y^2 + z^2}$ from the origin (see Fig. 1.23) the scattered wave has electric field

$$E_s = \frac{1}{c^2}\frac{1}{r}\frac{d^2 p}{dt^2}\bigg|_{t'=t-r/c} = -\frac{4\pi^2}{\lambda^2 r}\alpha E_i \cos\left[2\pi\frac{c}{\lambda}\left(t - \frac{r}{c}\right)\right], \tag{1.81}$$

where the delay r/c is the time it takes the wave to travel from the oscillating dipole at the origin to the detector.

The **intensity** I_s of the wave scattered by a single molecule is proportional to the mean-square average of the field and is therefore related to the intensity I_i of the incident wave:

$$I_s = \frac{16\pi^4}{\lambda^4 r^2}\alpha^2 I_i. \tag{1.82}$$

The inverse fourth power dependence of the scattered intensity I_s on the wavelength λ was first understood by Lord Rayleigh. It explains the blue color of the sky since molecules in the Earth's atmosphere scatter the solar

---

[5] Our discussion of electromagnetic radiation uses the cgs system of units.

light of shorter wavelengths (blue) with greater intensity than the longer wavelengths.

The intensity of light scattered by all $N_{tot}/V = c\mathcal{N}_{Av}/M$ molecules in a unit volume is the sum of intensities from each individual molecule,

$$\bar{I} = \frac{c\mathcal{N}_{Av}}{M} \frac{16\pi^4}{\lambda^4 r^2} \alpha^2 I_i, \tag{1.83}$$

where c is the mass concentration and M is the molar mass of the molecules. This simple addition of intensities is valid because there is no coherence between light scattered by non-interacting molecules. If we also assume that the distance r from all scattering particles to the detector is approximately the same, then the measured intensity is the product of the scattering volume V and the intensity scattered per unit scattering volume $\bar{I}$. In practice, this assumption is improved by moving the detector further from the sample or by making the scattering volume smaller, but either choice decreases the scattered intensity.

1.7.2.2 Scattering from dilute polymer solutions

The above results can be generalized to scattering from solutions, as long as there is contrast (difference in refractive indices) between solute and solvent. However, polymers can only be treated as point source scatterers in the limit of small scattering angles.[6] Adding molecules with small number density $c\mathcal{N}_{Av}/M \ll 1/\alpha$ to a solvent with refractive index n_0 leads to a linear change in dielectric constant, and hence a linear change in the square of the **refractive index** n of the solution:[7]

$$n^2 = n_0^2 + 4\pi\alpha \frac{c\mathcal{N}_{Av}}{M}. \tag{1.84}$$

Differentiating both sides of Eq. (1.84) with respect to concentration c, provides the expression for the molecular polarizability (with units of volume):

$$\alpha = \frac{n}{2\pi} \frac{dn}{dc} \frac{M}{\mathcal{N}_{Av}}. \tag{1.85}$$

Substituting this expression for the molecular polarizability α into Eq. (1.83) gives the intensity per unit scattering volume of vertically polarized light scattered in the horizontal plane:

$$\bar{I} = \frac{4\pi^2 n^2}{\lambda^4 r^2} \left(\frac{dn}{dc}\right)^2 \frac{cM}{\mathcal{N}_{Av}} I_i. \tag{1.86}$$

The scattered intensity per unit scattering volume $\bar{I}$ normalized by the incident intensity I_i and corrected for the $1/r^2$ distance dependence is called the **Rayleigh ratio**.[8]

$$R_\theta \equiv \frac{\bar{I}r^2}{I_i} \tag{1.87}$$

---

[6] Precisely how small the scattering angle must be is discussed in Section 2.8.
[7] The square of the refractive index is the dielectric constant of the medium.
[8] The pure solvent always has some small scattering intensity due to density fluctuations. In practice, this solvent background scattering is subtracted from the Rayleigh ratio.

The unit of the Rayleigh ratio is inverse length (m^{-1}) since I_i is the incident intensity and $\bar{I}$ is the scattered intensity per unit volume. The Rayleigh ratio describes the attenuation of the incident beam upon passing through a medium. The Rayleigh ratio for vertically polarized light scattered in the horizontal plane is

$$R_\theta = \frac{\bar{I}r^2}{I_i} = \frac{4\pi^2 n^2}{\lambda^4}\left(\frac{dn}{dc}\right)^2 \frac{cM}{\mathcal{N}_{Av}} = KcM, \tag{1.88}$$

where we have defined an **optical constant**

$$K = \frac{4\pi^2 n^2}{\lambda^4 \mathcal{N}_{Av}}\left(\frac{dn}{dc}\right)^2. \tag{1.89}$$

Expression (1.88) is valid for very low concentrations ($c < c^*$) or for non-interacting molecules. In a polydisperse solution, each species j with molar mass M_j contributes $Kc_j M_j$ to the Rayleigh ratio. Therefore, the total Rayleigh ratio for a polydisperse system is given by the sum

$$R_\theta = K\sum_j c_j M_j = K\, cM_w. \tag{1.90}$$

The final relation was obtained using Eq. (1.31). Hence, *dilute scattering measurements give the weight-average molar mass of a polydisperse polymer sample.*

At low concentrations, polymer interactions make a contribution to the scattering proportional to the second virial coefficient, just as in the case of osmotic pressure [see Eq. (1.74)]. The contrast required for scattering from polymer solutions primarily comes from concentration fluctuations, which are controlled by the rate of change of osmotic pressure with concentration. For wavelengths much larger than the molecules, the scattering intensity is related to the **osmotic compressibility** ($c\partial\Pi/\partial c$).

$$\frac{Kc}{R_\theta} = \frac{1}{RT}\left(\frac{\partial\Pi}{\partial c}\right)_T = \frac{1}{M} + 2A_2 c + \cdots \tag{1.91}$$

The Rayleigh ratio for monodisperse samples can then be rewritten as:

$$R_\theta = \frac{KcM}{1 + 2A_2 cM + \cdots}. \tag{1.92}$$

In a polydisperse sample there are contributions to scattering from all species, which can be rewritten using the fact that $1/(1+x) \cong 1 - x$ for small x:

$$R_\theta = K\sum_i \frac{c_i M_i}{1 + 2\sum_j A_{ij} c_j M_j + \cdots} \cong K\sum_i c_i M_i\left(1 - 2\sum_j A_{ij} c_j M_j\right)$$

$$\cong K\left(\sum_i c_i M_i - 2\sum_i\sum_j A_{ij} c_i M_i c_j M_j + \cdots\right)$$

$$= K\left[cM_w - 2A_{2,z}(cM_w)^2 + \cdots\right] = KcM_w(1 - 2A_{2,z}cM_w + \cdots) \tag{1.93}$$

$$\cong K\frac{cM_{\mathrm{w}}}{1+2A_{2,z}cM_{\mathrm{w}}+\cdots}, \qquad (1.94)$$

where A_{ij} is the second virial coefficient between species i and j, discussed in Section 1.7.1. We have already demonstrated [see Eq. (1.90)] that in scattering the weight-average molar mass, M_{w}, is measured in polydisperse samples at low concentrations. It is a higher moment of the distribution than the number-average molar mass measured by osmotic pressure. Similarly, the z-average second virial coefficient $A_{2,z}$, measured by light scattering,

$$A_{2,z} \equiv \frac{\sum_i \sum_j A_{ij}c_i M_i c_j M_j}{(cM_{\mathrm{w}})^2} = \frac{\sum_i \sum_j A_{ij}c_i M_i c_j M_j}{\left(\sum_i c_i M_i\right)^2}$$

$$= \frac{1}{M_{\mathrm{w}}^2}\sum_i \sum_j A_{ij}w_i M_i w_j M_j, \qquad (1.95)$$

is a higher moment than the weight-average second virial coefficient $A_{2,\mathrm{w}}$, measured by osmotic pressure [Eq. (1.75)]. Measurement of scattered intensity at different dilute concentrations allows determination of the weight-average molar mass M_{w} and the z-average second virial coefficient $A_{2,z}$.

$$\frac{Kc}{R_\theta} = \frac{1}{M_{\mathrm{w}}} + 2A_{2,z}c + \cdots \qquad (1.96)$$

This procedure is analogous to determination of the number-average molar mass M_{n} and weight-average second virial coefficient $A_{2,\mathrm{w}}$ from the measurements of osmotic pressure Π at different concentrations [Eq. (1.76) and Fig. 1.22].

It is important to stress that the above treatment is only valid if the wavelength of light λ is much larger than the size of the molecules. Otherwise the angular dependence of the scattered light $I(\theta)$ contains a contribution from the coherent scattering between different parts of the same molecule (called the form factor). In this case, a more sophisticated method of analysis is used, called a Zimm plot, that allows not only determination of the molar mass, but also the size of the molecules. The angular dependence of light scattering is discussed in Section 2.8. Equation (1.96) is recovered only in the small-angle limit.

1.7.3 Intrinsic viscosity

Each polymer coil in a solution contributes to viscosity. In very dilute solutions, the contribution from different coils is additive and solution viscosity η increases above the solvent viscosity η_{s} linearly with polymer concentration c. The effective 'virial expansion' for viscosity at low concentration is of the same form as Eq. (1.76) for osmotic pressure and Eq. (1.96) for light scattering:

$$\eta = \eta_{\mathrm{s}}(1 + [\eta]c + k_{\mathrm{H}}[\eta]^2 c^2 + \cdots). \qquad (1.97)$$

The term that is linear in concentration contains the **intrinsic viscosity** $[\eta]$ and the quadratic term includes the **Huggins coefficient** k_H, which plays the role of the second virial coefficient for viscosity. Intrinsic viscosity $[\eta]$ is the initial slope of a plot of relative viscosity η/η_s against concentration. Since relative viscosity is dimensionless, the intrinsic viscosity has units of reciprocal concentration. It is proportional to the reciprocal concentration of monomers in a chain's pervaded volume V [Eq. (1.21)]:

$$[\eta] \approx \frac{1}{c^*} \sim \frac{V}{M} \approx \frac{R^3}{M}. \tag{1.98}$$

Intrinsic viscosity is related to the linear size of the coil R and the molar mass M by the **Fox–Flory equation**:

$$[\eta]M \sim V \sim R^3. \tag{1.99}$$

This equation will be derived in Chapter 8. Dilute solution viscosity measurements are important characterization tools for polymers because the Fox–Flory relation shows that the product $[\eta]M$ is proportional to the pervaded volume of the polymer coil.[9] Since polymers are fractals, their molar mass and their size are related by a power law ($M \sim R^{\mathcal{D}}$), leading to the **Mark–Houwink equation**:

$$[\eta] = KM^a. \tag{1.100}$$

K and a are tabulated for nearly all linear polymers in various solvents, which means that intrinsic viscosity provides a simple effective measure of molar mass. From Eq. (1.98) it is clear that the Mark–Houwink exponent $a = (3/\mathcal{D}) - 1$. Representative values of Mark–Houwink coefficients and Huggins coefficients are shown in Table 1.4.

Table 1.4 Representative Mark-Houwink and Huggins coefficients of linear polymers

Polymer	Solvent	$K[(\text{dL g}^{-1})(\text{mol g}^{-1})^a]$	a	k_H
Polybutadiene	Tetrahydrofuran* at 25 °C	2.88×10^{-4}	0.726	
Polybutadiene	Dioxane† at 26.5 °C	1.78×10^{-3}	0.50	
Polystyrene	Tetrahydrofuran* at 25 °C	1.10×10^{-4}	0.725	0.35
Polystyrene	Cyclohexane† at 34.5 °C	8.46×10^{-3}	0.50	0.5–0.8
Polyethylene	Xylene at 81 °C	1.05×10^{-3}	0.63	0.38
Polypropylene	Xylene at 85 °C	9.6×10^{-4}	0.63	

*Good solvents typically have $0.7 < a < 0.8$ and $0.3 < k_H < 0.4$. †θ–solvents have $a = 0.5$ and have k_H increasing with molar mass with $0.5 < k_H < 1.5$.

In practice, dilute solution viscosity is measured at multiple concentrations and two different forms of Eq. (1.97) are used to extrapolate to zero concentration. One form is the **Huggins equation**

$$\frac{\eta - \eta_s}{\eta_s c} = [\eta] + k_H[\eta]^2 c + \cdots, \tag{1.101}$$

[9] The appropriate volume appears to be a combination of static and dynamic coil sizes, as will be discussed in Chapter 8.

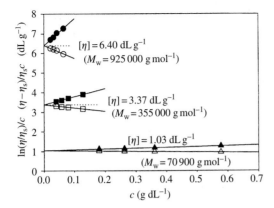

Fig. 1.24
Determination of the intrinsic viscosity
for three different polybutadiene
samples in tetrahydrofuran at 25°C.
Each sample exhibits linear Huggins
(filled symbols) and Kraemer (open
symbols) plots that extrapolate to the
intrinsic viscosity at zero concentration.
All three polymers have Huggins
coefficients of 0.37.

where $(\eta - \eta_s)/\eta_s c$ is plotted against mass concentration c. The intercept of
the Huggins plot (at $c = 0$) is the intrinsic viscosity and the slope is $k_H[\eta]^2$.
Solving for η/η_s and taking the natural logarithm allows the Huggins
equation to be rewritten as :

$$\ln(\eta/\eta_s) = \ln(1 + [\eta]c + k_H[\eta]^2 c^2 + \cdots)$$
$$= [\eta]c + \left(k_H - \tfrac{1}{2}\right)[\eta]^2 c^2 + \cdots$$

The second result was obtained from expansion of the logarithm
$[\ln(1 + x + k_H x^2) = x + (k_H - 1/2)x^2$ for small $x]$. Dividing this last result by
c gives the second extrapolation form, known as the **Kraemer equation**,
having the same intercept but a different slope:

$$\frac{\ln(\eta/\eta_s)}{c} = [\eta] + \left(k_H - \frac{1}{2}\right)[\eta]^2 c + \cdots \qquad (1.102)$$

Figure 1.24 shows experimental determinations of the intrinsic viscosity.
When both the Huggins and Kraemer equations provide the same intrinsic
viscosity and Huggins coefficient, the higher-order terms in these equations
can be safely ignored. On the other hand, if the Huggins and Kraemer plots
are curved and do not give the same intercept, viscosity measurements need
to be made at lower concentrations.

Owing to the superb precision of viscosity measurements, the intrinsic
viscosity can easily be measured to three significant figures, which makes it
by far the most precise polymer characterization method. However, care
must be taken to control the temperature precisely, and polymers with
large molar mass ($M \gtrsim 10^6$ g mol^{-1}) can shear thin in conventional capil-
lary viscometers.

1.7.4 Size-exclusion chromatography

The entire molar mass distribution, including the higher-order average
molar masses, such as M_z, M_{z+1}, etc. can be measured by either **size-
exclusion chromatography** (SEC) or ultracentrifuge sedimentation. Owing

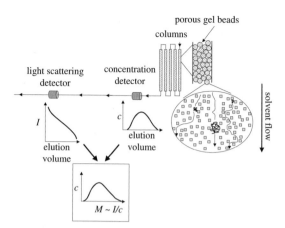

Fig. 1.25
Schematic view of size exclusion
chromatography.

to the development of excellent chromatographic columns for SEC, it has
completely displaced the ultracentrifuge for characterization of molar
mass distribution. Furthermore, modern SEC is highly automated and
reasonably precise, making it the most popular method of polymer
characterization.

Size exclusion chromatography, also called **gel permeation chromatography** (GPC) continually pumps solvent through a series of columns filled
with porous beads (see Fig. 1.25). Often, the beads are polymer gels swollen
with solvent. The beads are intentionally made with a variety of pore sizes
that span the range of the sizes of macromolecules to be separated. A dilute
polymer solution in the same solvent is injected into a small volume of the
flowing solvent stream entering the columns. As the polymer solution
passes through the columns, the largest polymers are excluded from all but
the largest pores, and elute from the columns first. Progressively smaller
polymers can explore progressively smaller pores and therefore, larger
volumes of the column, and consequently elute later. Thus the separation
of molecules in SEC occurs by polymer size rather than by polymer mass.

After separation, the solution passes through a variety of detectors,
depending on the information needed for a particular sample. Common
detectors include a differential refractometer (for measuring concentration), absorption spectrophotometric detection (such as ultraviolet and
infrared), light scattering photometer (for measuring M_w of each eluent),
and viscometer (for measuring $[\eta]$ of each eluent). With proper calibration
using narrow molar mass distribution standards, SEC can in principle
determine the full molar mass distribution, including higher-order averages. However, practical limitations make determination of averages that
are higher-order than M_z unreliable. In the best of circumstances with
modern SEC equipment and a full compliment of detectors, M_z is determined to $\pm 10\%$, whereas M_w can be determined to $\pm 5\%$ from light
scattering.

As an analogy describing the SEC process, consider an art museum with
an entry on one side and an exit on the other side of the building. On a

sunny Sunday morning the museum opens at 10:00 AM and different groups of tourists enter the museum. Some of them come in big buses on 5 h city tours. These large groups stop at a couple of major paintings and exit the museum less than an hour later. Smaller groups come in vans on special tours of city museums. They visit all the floors of the museum and spend a couple of hours in it. But there are many individual art lovers who visit all of the rooms in the museum and spend a long time in front of many paintings and leave the museum late in the afternoon.

Size exclusion chromatography separates polymers by their size in solution. A given polymer coil can only enter the pores of the column that are larger than the coil. This idea led Benoit to propose a scheme for **universal calibration** of SEC involving the pervaded volume of the coil. The volume within the columns that is accessible to a polymer determines how long the polymer stays in the columns. The amount of solvent that exits the columns between the time the polymer is injected and when it exits the columns is termed the **elution volume**. Benoit's idea is simply that there is a unique relation between the pervaded volume of the coil and it's elution volume, for a given set of SEC columns. Hence, both polystyrene and poly(vinyl chloride) with the same pervaded volume will elute from the SEC at the same elution volume. Most modern SEC columns are designed to give a linear relation between the logarithm of pervaded volume [experimentally measured as $[\eta]M$, see Eq. (1.99)] plotted against elution volume over a wide range, as shown in Fig. 1.26. For both very short chains and very long chains, Fig. 1.26 shows departures from the roughly linear calibration curve. When chains are too short, their pervaded volume is smaller than all of the pores in the columns and the SEC no longer separates such short chains because they experience all of the available volume of the columns. This is the downturn in Fig. 1.26 at large elution volume. Similarly, chains that are too long have a pervaded volume that is larger than any of the pores in the column. These long chains pass through the column in the interstitial spaces between beads and hence are also not separated (the upturn at small elution volume in Fig. 1.26).

For strictly linear chains, universal calibration is extremely useful because the Mark–Houwink coefficients have been tabulated for all common linear polymers. The calibration curve allows $[\eta]M$ to be determined from the elution volume. The Mark–Houwink equation [Eq. (1.100)] then allows the SEC measure of $[\eta]M$ to determine the molar mass of the polymer:

$$[\eta]M = KM^{a+1} \quad \Rightarrow \quad M = \left(\frac{[\eta]M}{K}\right)^{1/(a+1)}. \qquad (1.103)$$

In practice, SEC columns are often calibrated using linear monodisperse polystyrene standards, generating a calibration curve like Fig. 1.26. Then any linear polymer that is soluble in the same solvent, for which a Mark–Houwink equation is known, can have its molar mass determined by this calibrated SEC experiment. The elution volume of the polymer determines $[\eta]M$ from the calibration curve and the Mark–Houwink

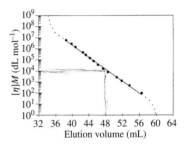

Fig. 1.26
Universal calibration curve for size exclusion chromatography. Data points are polybutadienes in tetrahydrofuran at 25 °C in crosslinked polystyrene columns. There is a reasonably linear region of the calibration curve spanning the data (solid line) but the linear region has its limits (dotted curves).

equation subsequently determines the molar mass from $[\eta]M$. For branched polymers, universal calibration still works. A branched polymer will elute at the same elution volume as a linear polymer with the same pervaded volume. However, unless the Mark–Houwink relation for the branched polymer is known (which is rare) there is no means to convert $[\eta]M$ into a molar mass.

For branched polymers and for linear polymers that do not have an established Mark–Houwink equation, SEC is typically used in conjunction with light scattering and viscosity detectors. These detectors measure the weight-average molar mass and viscosity of each elution volume. In principle, this experiment directly determines both M_w and $[\eta]$ for *any* polymer that is soluble in the SEC solvent, but in practice the M_w determination also requires that the polymer have an appreciable difference in refractive index from the solvent so that in Eq. (1.88), $dn/dc \gtrsim 0.05\,\mathrm{mL\,g^{-1}}$. Often a new polymerization is not understood sufficiently to know for certain whether the polymer produced will have branched chains present or not, and SEC with multiple detectors is the method of choice for characterizing such polymers. Multiple-detector SEC provides an enormous amount of information in an automated fashion. The weight-average molar mass and intrinsic viscosity of each individual elution volume are determined quite precisely because the concentrations are low enough that second virial and Huggins coefficient terms make negligible contributions.

It is very important to realize that the size separation in SEC is far from perfect. If a perfectly monodisperse sample were instantaneously injected into the columns, the sample would *not* all elute in the same elution volume. This is because the polymer diffuses randomly in the solvent while the solution proceeds through the column. This broadening always makes the polydispersity index measured by SEC larger than the real polydispersity index of the sample, thereby accounting for the discrepancy between theory and experiment at low molar masses in Fig. 1.20.

1.8　Summary

Polymers are formed by repetitive covalent bonding of chemical monomers. The number of monomers in a macromolecule N is its degree of polymerization. Some polymer characteristics, such as degree of polymerization, microstructure, architecture and chemical composition are fixed during polymerization. These characteristics control many important properties of polymeric materials.

A macromolecule can adopt many conformations, defined by relative locations of its monomers in space. Polymer conformations are often self-similar (fractal) with pervaded volume V much larger than their occupied volume $N v_{\mathrm{mon}}$ where v_{mon} is the monomer volume. The overlap volume fraction

$$\phi^* = \frac{N v_{\mathrm{mon}}}{V}, \tag{1.104}$$

is much less than unity. The overlap parameter P is the average number of chains in a pervaded volume that is randomly placed in the solution. At volume fractions below overlap ($\phi < \phi^*$) the solution is called dilute and the overlap parameter is less than unity ($P < 1$). At volume fractions above overlap ($\phi > \phi^*$) the solution is called semidilute and the overlap parameter $P > 1$.

Synthetic polymers are often polydisperse, containing a mixture of molecules with different molar masses. This mixture is described by a distribution—number fraction (or mole fraction) n_N of molecules with molar mass M_N. The distribution is characterized by its moments, with the kth moment of the number fraction distribution function defined as a sum over the distribution:

$$m_k = \sum_N n_N M_N^k. \tag{1.105}$$

Different molar mass averages are defined as ratios of consecutive moments of the molar mass distribution. The number-average molar mass is the ratio of the first to zeroth moment of the number fraction distribution function (which is equal to the first moment because the normalization of the number fraction distribution makes the zeroth moment unity):

$$M_n \equiv \frac{m_1}{m_0} = \sum_N n_N M_N. \tag{1.106}$$

The weight-average molar mass is the ratio of the second to the first moment of the number fraction distribution function and is equal to the first moment of the weight fraction distribution function:

$$M_w \equiv \frac{m_2}{m_1} = \frac{\sum\limits_N n_N M_N^2}{\sum\limits_N n_N M_N} = \sum_N w_N M_N. \tag{1.107}$$

The polydispersity index is defined as the ratio of the weight- to number-average molar masses M_w/M_n. The polydispersity index is equal to unity for monodisperse polymers (samples with only one molar mass). A larger value of the polydispersity index corresponds to polymeric systems with broader molar mass distributions. Linear condensation polymers typically have $M_w/M_n \cong 2$, while linear addition polymers can have virtually any polydispersity index, depending on the relative rates of polymerization and termination.

Molar masses and molar mass distributions are usually measured in dilute solutions. Osmotic pressure measurements determine M_n and light scattering measurements determine M_w. The entire molar mass distribution can be measured using properly calibrated SEC.

Problems

Section 1.2

1.1 Consider a 'true macromolecule'—a chunk of polybutadiene network of mass 100 g. How many monomers all covalently bonded together does it contain if the molar mass of a monomer is $M_{mon} = 54\,g\,mol^{-1}$? What is the molar mass of this macromolecule?

Section 1.3

1.2 If two different monomers A and B prefer to react with each other than react with their own kind, explain how you would synthesize the following copolymers:

  (i) alternating copolymer
  (ii) diblock copolymer
  (iii) triblock copolymer

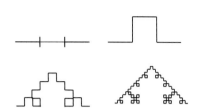

Fig. 1.27
A Koch curve based on squares.

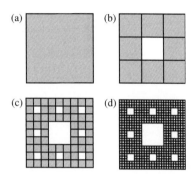

Fig. 1.28
Sierpinski carpet construction:
Start with a square (a), divide it into nine equal squares and remove the center one (b). With each of the remaining eight squares, repeat the process (c) and so on (d).

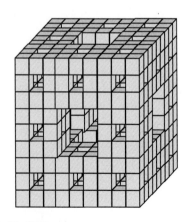

Fig. 1.29
Menger sponge.

Section 1.4

1.3 Consider a dense globule of polyethylene with molar mass $M = 10^6\,g\,mol^{-1}$ in a non-solvent. What is the radius of the globule if the density inside the globule is $\rho = 0.784\,g\,cm^{-3}$?

1.4 Consider an ideal polyethylene chain with molar mass $M = 10^6\,g\,mol^{-1}$. Its mean-square end-to-end distance is given by

$$\langle R^2 \rangle = Cb^2 N,$$

where the monomer length is $b = 2.5\,\text{Å}$ and the coefficient $C = 5.5$. Estimate its root-mean-square end-to-end distance $\sqrt{\langle R^2 \rangle}$ if the molar mass of the monomer is $M_{mon} = 28\,g\,mol^{-1}$.

1.5 What is the maximum length

$$R_{max} = bN$$

of a polyethylene chain with molar mass $M = 10^6\,g\,mol^{-1}$ and monomer length $b = 2.5\,\text{Å}$?

1.6 Calculate the fractal dimension of the Koch curve in Fig. 1.27 with the center third of each segment replaced by three sides of a square (instead of two sides of a triangle as discussed in Section 1.4).

1.7 Determine the fractal dimension of a Sierpinski carpet (see Fig. 1.28), constructed by dividing solid squares into 3×3 arrays and removing their centers.

1.8 Calculate the fractal dimension of a Menger sponge (see Fig. 1.29), a three-dimensional version of the Sierpinski carpet. A solid cube is divided into $3 \times 3 \times 3$ cubes and the body-center cube along with the six face-center cubes are removed. The same procedure is repeated for each of the remaining 20 cubes, etc.

1.9 A polymer in a melt is in its ideal state, which is a fractal with fractal dimension $\mathcal{D} = 2$. Consider two such chains, a longer one with degree of polymerization $N_1 = 1000$ and a shorter one with degree of polymerization $N_2 = 250$. What is the ratio of their sizes R_1/R_2?

1.10 A linear polymer in a good solvent is a fractal with fractal dimension $\mathcal{D} \cong 1.7$. What fraction of a chain has size (average distance between its two end monomers) equal to half of the average distance between two ends of the whole chain?

1.11 An ideal randomly branched polymer is a fractal object with fractal dimension $\mathcal{D} = 4$. In Chapter 6, we will learn how this polymer can fit into three-dimensional space. What is the ratio of molar masses M_1/M_2 of two ideal randomly branched polymers if the ratio of their sizes is $R_1/R_2 = 3$?

1.12 In Chapter 3, we will learn that a linear polymer confined to an air–water interface is a fractal object with fractal dimension $\mathcal{D} = 4/3$. What is the ratio of sizes R_1/R_2 of two linear polymers at the air–water interface if the ratio of their molar masses is $M_1/M_2 = 16$?

1.13 Give additional examples of: (i) regular fractals; (ii) fractals in nature.

Section 1.5

1.14 The density of a 1,4-polybutadiene melt at 298 K is $\rho = 0.895\,\mathrm{g\,cm^{-3}}$. What is the monomer volume v_{mon}, if the mass of the monomer is $M_{mon} = 54\,\mathrm{g\,mol^{-1}}$?

1.15 Consider a polystyrene solution with concentration $c = 1\,\mathrm{g\,L^{-1}}$ in a solvent with density $\rho = 0.9\,\mathrm{g\,cm^{-3}}$. Estimate the volume of a polystyrene monomer in this solution if the density of bulk polystyrene is $\rho = 1\,\mathrm{g\,cm^{-3}}$ and the mass of the monomer is $M_{mon} = 104\,\mathrm{g\,mol^{-1}}$. What is the volume fraction of polystyrene in this solution?

1.16 What is the volume fraction of $1\,\mathrm{mg\,mL^{-1}}$ poly(vinyl chloride) in solution if the volume of each monomer is $v_{mon} = 75\,\mathrm{Å^3}$ and the molar mass of each monomer is $M_{mon} = 62\,\mathrm{g\,mol^{-1}}$?

1.17 Calculate the overlap volume fraction ϕ^* of a polymer with degree of polymerization $N = 10^4$ and monomer volume $v_{mon} = 100\,\mathrm{Å^3}$, if its pervaded volume is a sphere with radius $200\,\mathrm{Å}$.

1.18 Consider a solution of rod polymers with degree of polymerization $N = 100$ and end-to-end distance $L = Nb$ with monomer length $b = 5.5\,\mathrm{Å}$ and monomer mass $M_{mon} = 75\,\mathrm{g\,mol^{-1}}$.

   (i) What is the pervaded volume of this polymer?
   (ii) Is a solution with concentration $10^{-3}\,\mathrm{g\,cm^{-3}}$ dilute or semidilute?

1.19 Consider a polymer solution with degree of polymerization $N = 500$, volume fraction $\phi = 10^{-2}$, monomer volume $v_{mon} = 90\,\mathrm{Å^3}$ and pervaded volume $10^4\,\mathrm{nm^3}$. What is the overlap parameter P of this solution?

1.20 Consider a 10 mL solution obtained by mixing 20 mg of dry polymer with solvent. The bulk density of the dry polymer is $\rho = 0.8\,\mathrm{g\,cm^{-3}}$. What is the volume fraction of polymer in this solution? Assume no change of volume upon mixing.

1.21 A polymer with molar mass $M = 10^5\,\mathrm{g\,mol^{-1}}$ is at overlap in a solution with concentration $c^* = 1.67 \times 10^{-2}\,\mathrm{g\,cm^{-3}}$. What is the pervaded volume V of each polymer chain?

1.22 Estimate the overlap parameter P for polymers with fractal dimension D in the melt ($\phi = 1$) if the degree of polymerization is N. Estimate the overlap parameter for an ideal chain (with $D = 2$) in a melt with $N = 10^4$ monomer segments.

Section 1.6

1.23 Consider five textbooks from the polymer bookshelf; by P. J. Flory consisting of 672 pages, by P. G. de Gennes consisting of 324 pages, by M. Doi and S. F. Edwards consisting of 391 pages, by A. Yu. Grosberg and A. R. Khokhlov consisting of 350 pages and by J. des Cloizeaux and G. Jannink consisting of 896 pages.

   (i) What is the number-average number of pages per textbook?
   (ii) What is the weight-average number of pages per textbook?
   (iii) What is the polydispersity index?

1.24 Science Fiction: Three Planets.

   (i) On the planet Demos, all major decisions are made by votes of all inhabitants. All votes are counted with equal weight. What kind of average decision is achieved on the planet Demos?

(ii) On the planet Fatos all major decisions are also made by votes of all inhabitants. Votes on Fatos are counted proportional to the weight of the corresponding inhabitant. What kind of average decision is achieved by this weighted voting on the planet Fatos?

(iii) On the planet Thinos all major decisions are also made by votes of all inhabitants. Votes on Thinos are counted inversely proportional to the weight of the corresponding inhabitant. What kind of average decision is achieved by this weighted voting on the planet Thinos?

1.25 Consider a system consisting of one elephant with mass $M_1 = 10^4$ kg and nine mosquitoes riding on its back with mass $M_2 = 0.1$ g each.

(i) Calculate the number-average mass of this system.
(ii) Calculate the weight-average mass of this system.
(iii) Calculate the polydispersity index of this system.
(iv) Which average is appropriate for calculating the damage to your Land Rover in a collision with this system?

1.26 Consider the following distribution of polymer chains:

10 chains with degree of polymerization 100
100 chains with degree of polymerization 1000
10 chains with degree of polymerization 10 000

(i) Calculate the number-average degree of polymerization N_n of this distribution.
(ii) What is the weight-average degree of polymerization N_w of this distribution?
(iii) What is the polydispersity index of this distribution?

1.27 Consider a blend obtained by mixing 1 g of a polymer with molar mass $M_A = 1 \times 10^5$ g mol^{-1} and 2 g of the same type of polymer with molar mass $M_B = 2 \times 10^5$ g mol^{-1}.

(i) Calculate the number-average molar mass M_n of this blend.
(ii) What is the weight-average molar mass M_w of this blend?
(iii) What is the polydispersity index of this polymer blend?

1.28 A protein sample consists of 80% by weight material with $M = 5 \times 10^4$ g mol^{-1} and 20% by weight of dimer with molar mass 10^5 g mol^{-1}. Calculate M_n, M_w, and polydispersity index.

1.29 (i) Is it possible for a number fraction n_N of a chain of N monomers to be larger than the weight fraction w_N of this species? What can you state about the molar mass M_N of this species?

(ii) Is it possible for a number fraction n_N of a chain of N monomers in a polydisperse sample to be equal to its weight fraction w_N? What can you state about the molar mass M_N of this species?

1.30 The number fraction (or mole fraction) of a protein with molar mass $M_A = 10^5$ g mol^{-1} in an unknown mixture of different protein species is $n_A = 0.1$. The weight fraction of this protein in the same mixture is $w_A = 0.2$.

(i) What is the number-average molar mass M_n of the mixture?
(ii) What is the weight-average molar mass M_w of this mixture?

1.31 Show that the k-moment of the number fraction distribution n_N is related to the $(k-1)$-moment of the weight fraction distribution w_N [derive Eq. (1.54)].

1.32 Consider the condensation polymerization of aminocaproic acid to make nylon 6:

$$n\text{H}_2\text{N}(\text{CH}_2)_5\text{COOH} \rightarrow [(\text{CH}_2)_5\text{CONH}]_n + n\text{H}_2\text{O}$$

(i) What is the number-average molar mass M_n at the extent of reaction $p = 0.99$?

(ii) What is the weight-average molar mass M_w and polydispersity index at the same extent of reaction $p = 0.99$?

1.33 At what extent of reaction p is the polydispersity index of a linear condensation polymerization sample equal to $M_w/M_n = 1.5$?

1.34* Prove that the weight-average molar mass is never smaller than the number-average molar mass and therefore the polydispersity index is never less than unity:

$$\frac{M_w}{M_n} \geq 1.$$

1.35 Calculate the degree of polymerization $N_{max}(p)$ corresponding to the maximum of the most-probable weight fraction $w_N(p)$ [Eq. (1.53)] at the extent of reaction p. Is this value $N_{max}(p)$ better approximated for small $(1-p)$ by the number-average N_n or by the weight-average N_w degree of polymerization? *Hint*: Expand the logarithm for small $(1-p)$.

1.36* Consider condensation polymerization of f-arm stars. Each star molecule contains one multifunctional monomer B_f and $N-1$ bifunctional monomers AB with condensation reaction possible only between unreacted A and B groups. Arms of the star are polydisperse with each of f arms containing between 0 and $N-1$ monomers AB.

(i) Demonstrate that the number fraction distribution function of N-mers is

$$n_N(p) = \frac{(N+f-2)!}{(f-1)!(N-1)!}p^{N-1}(1-p)^f,$$

where N is the total degree of polymerization of all arms of the star.

(ii) Show that the number-average degree of polymerization is

$$N_n = \frac{(f-1)p+1}{1-p}.$$

(iii) Calculate the weight-average degree of polymerization and show that as extent of reaction $p \rightarrow 1$, the polydispersity index decreases with the number of arms as

$$\frac{N_w}{N_n} \cong 1 + \frac{1}{f}.$$

1.37* Addition polymerization

(i) Calculate the weight-fraction distribution function w_N for addition polymerization without termination from the Poisson number fraction distribution function n_N [Eq. (1.67)] using

$$w_N = \frac{Nn_N}{\sum\limits_{N=1}^{\infty} Nn_N}.$$

(ii) Prove that the number-average degree of polymerization of the Poisson distribution is N_n.

(iii) Calculate the weight-average degree of polymerization N_w and polydispersity index for addition polymerization without termination. Does N_w/N_n increase or decrease as the reaction proceeds?

(iv) The number-average degree of polymerization of an addition polymerization sample is $N_n = 100$. What is the polydispersity index of this sample?

---

* Throughout this book, problems marked with an asterisk are more challenging than those that are not.

1.38 Prove that the Shultz distribution w_N [Eq. (1.70)] has a maximum at $N = N_n$ for any value of $s > 0$.

1.39 List the necessary conditions for formation of a narrow molar mass distribution in addition polymerization.

Section 1.7

1.40 The osmotic pressure of a polymer solution at temperature $T = 23\,°C$ was measured at several concentrations and is reported in the table below.

c (g cm^{-3})	2×10^{-3}	4×10^{-3}	6×10^{-3}	8×10^{-3}	10^{-2}
Π (dyn cm^{-2})	508	1.04×10^3	1.58×10^3	2.15×10^3	2.74×10^3

Determine the number-average molar mass M_n of the polymer and the second virial coefficient of the solution A_2.

1.41 Light scattering measurements at scattering angle $\theta = 3°$ were performed on a dilute polymer solution using a laser with a wavelength $\lambda = 5500\,\text{Å}$. The refractive index of the solvent is $n_0 = 1.4$ and the refractive index increment of the solution is $dn/dc = 0.1\,\text{cm}^3\,\text{g}^{-1}$. The following data for the Rayleigh ratio were obtained:

c (g cm^{-3})	5×10^{-4}	10^{-3}	1.5×10^{-3}	2×10^{-3}	2.5×10^{-3}
R_θ (cm^{-1})	4.9×10^{-4}	8.4×10^{-4}	1.1×10^{-3}	1.3×10^{-3}	1.5×10^{-3}

(i) What is the value of the optical constant K if one can assume that in dilute solutions $n \cong n_0$?

(ii) Plot Kc/R_θ as a function of the concentration c and determine the weight-average molar mass M_w and the second virial coefficient A_2.

1.42* Modern light scattering uses a polarized laser, but since much of the older literature used unpolarized light sources, it is useful to understand them. Unpolarized light can be represented as a combination of a vertically and horizontally polarized waves—the first one with electric field oscillating along the vertical x axis and the second one—along the horizontal y axis. The intensities of these two parts of the incident light are $I_x = I_y = I_i/2$. The intensity per unit scattering volume of the vertically polarized scattered wave is $\bar{I}_{\text{polar}}/2$.

(i) Show that the intensity at scattering angle θ, per unit scattering volume of the horizontally polarized scattered wave, is $(\bar{I}_{\text{polar}}/2)\cos^2 \theta$.

(ii) Show that the intensity of the scattered light per unit scattering volume using an unpolarized light source valid for any radial position of the detector with scattering angle θ is

$$\bar{I}_{\text{unpolar}} = \frac{2\pi^2 n^2}{\lambda^4 r^2}(1 + \cos^2 \theta)\left(\frac{dn}{dc}\right)^2 \frac{cM}{\mathcal{N}_{\text{Av}}} I_i. \tag{1.108}$$

(iii) Demonstrate that the Rayleigh ratio from an unpolarized light source [Eq. (1.87)] is equal to

$$R_\theta^{\text{unpolar}} \equiv \frac{\bar{I}_{\text{unpolar}} r^2}{I_i} = \frac{2\pi^2 n^2}{\lambda^4}\left(\frac{dn}{dc}\right)^2 \frac{cM}{\mathcal{N}_{\text{Av}}}(1 + \cos^2 \theta)$$

$$= KcM\frac{1 + \cos^2 \theta}{2}, \tag{1.109}$$

where K is the optical ratio defined in Eq. (1.89).

1.43 The intensity of the incident beam decreases due to scattering following **Beer's Law**:

$$I_i(z) = I_0 \exp[-\tau(z - z_0)], \tag{1.110}$$

where τ is the **turbidity** of the sample and $z - z_0$ is the length of the path of the incident beam between the entry into the sample z_0 (with intensity I_0 at z_0) and the point with coordinate z. Turbidity can be obtained from the Rayleigh ratio by integrating the radiated energy over all scattering directions. Using the Rayleigh ratio for an unpolarized light source [Eq. (1.109)], show that the turbidity is

$$\tau = \frac{8\pi}{3} KcM. \tag{1.111}$$

1.44 Size exclusion chromatography

A fraction of a polystyrene sample elutes in tetrahydrofuran at $25\,°\text{C}$ at $48\,\text{mL}$. Estimate the molar mass of this fraction using the universal calibration for this set of columns presented in Fig. 1.26 and the Mark–Houwink coefficients listed in Table 1.4.

1.45 Determine what moment of the molar mass distribution is measured by intrinsic viscosity of a polydisperse sample.

Bibliography

Billmeyer, F. W. Jr. *Textbook of Polymer Science*, 2nd edition (Wiley, New York, 1971).

Chu, B. *Laser Light Scattering: Basic Principles and Practice* (Academic Press, New York, 1991).

Flory, P. J. *Principles of Polymer Chemistry* (Cornell University Press, Ithaca, New York, 1953).

Grosberg, A. Yu. and Khokhlov, A. R. *Giant Molecules: Here, There, and Everywhere* (Academic Press, New York, 1997).

Guinier, A. *X-ray Diffraction in Crystals, Imperfect Crystals, and Amorphous Bodies* (W. H. Freeman, New York, 1963).

Higgins, J. S. and Benoit, H. C. *Polymers and Neutron Scattering* (Clarendon Press, Oxford, 1994).

Mandelbrot, B. B. *The Fractal Geometry of Nature* (W. H. Freeman, New York, 1982).

Morawetz, H. *Macromolecules in Solution*, 2nd edition (Wiley, New York, 1975).

Morawetz, H. *Polymers: The Origins and Growth of a Science* (Wiley, New York, 1985).

Painter, P. C. and Coleman, M. M. *Fundamentals of Polymer Science*, 2nd edition (Technomic, Lancaster, 1997).

Peebles, L. H. *Molecular Weight Distributions in Polymers* (Wiley, New York, 1971).

Sperling, L. H. *Introduction of Physical Polymer Science*, 3rd edition (Wiley, New York, 2001).

Single chain conformations

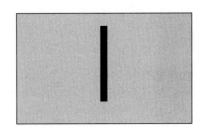

Ideal chains

2

In this chapter, we consider the conformations of chains with no interactions between monomers that are far apart along the chain, even if they approach each other in space. Such chains are called **ideal chains**. This situation is never completely realized for real chains, but there are several types of polymeric systems with nearly ideal chains. Real chains interact with both their solvent and themselves. The relative strength of these interactions determines whether the monomers effectively attract or repel one another. In Chapter 3, we will learn that real chains in a solvent at low temperatures can be found in a collapsed conformation due to a dominance of attractive over repulsive interactions between monomers. At high temperatures, chains swell due to dominance of repulsive interactions. At a special intermediate temperature, called the θ-**temperature**, chains are in nearly ideal conformations because the attractive and repulsive parts of monomer–monomer interactions cancel each other. This θ-temperature is analogous to the Boyle temperature of a gas, where the ideal gas law happens to work at low pressures. Even more importantly, linear polymer melts and concentrated solutions have practically ideal chain conformations because the interactions between monomers are almost completely screened by surrounding chains.

The conformation of an ideal chain, with no interactions between monomers, is the essential starting point of most models in polymer physics. In this sense, the role of the ideal chain is similar to the role of the harmonic oscillator or the hydrogen atom in other branches of physics.

2.1 Flexibility mechanisms

In order to understand the multitude of conformations available for a polymer chain, consider an example of a polyethylene molecule. The distance between carbon atoms in the molecule is almost constant $l = 1.54\,\text{Å}$. The fluctuations in the bond length (typically $\pm 0.05\,\text{Å}$) do not affect chain conformations. The angle between neighbouring bonds, called the **tetrahedral angle** $\theta = 68°$ is also almost constant.

The main source of polymer flexibility is the variation of **torsion angles** [see Fig. 2.1(a)]. In order to describe these variations, consider a plane defined by three neighbouring carbon atoms C_{i-2}, C_{i-1}, and C_i. The bond

Ideal chains

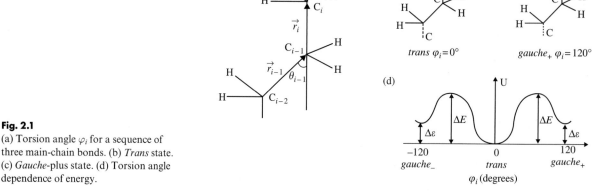

Fig. 2.1
(a) Torsion angle φ_i for a sequence of three main-chain bonds. (b) *Trans* state. (c) *Gauche*-plus state. (d) Torsion angle dependence of energy.

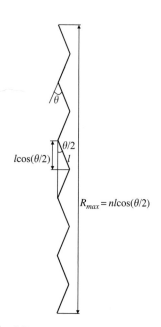

Fig. 2.2
All-*trans* (zig-zag) conformation of a short polymer with $n = 10$ main-chain bonds.

vector $\vec{r}_i$ between atoms C_{i-1} and C_i defines the axis of rotation for the bond vector $\vec{r}_{i+1}$ between atoms C_i and C_{i+1} at constant bond angle θ_i. The zero value of the torsion angle φ_i corresponds to the bond vector $\vec{r}_{i+1}$ being colinear to the bond vector $\vec{r}_{i-1}$ and is called the **trans** state (*t*) of the torsion angle φ_i [Fig. 2.1(b)].

The *trans* state of the torsion angle φ_i is the lowest energy conformation of the four consecutive CH_2 groups. The changes of the torsion angle φ_i lead to the energy variations shown in Fig. 2.1(d). These energy variations are due to changes in distances and therefore interactions between carbon atoms and hydrogen atoms of this sequence of four CH_2 groups. The two secondary minima corresponding to torsion angles $\varphi_i = \pm 120°$ are called **gauche-plus** (*g+*) [Fig. 2.1(c)] and **gauche-minus** (*g−*). The energy difference between *gauche* and *trans* minima $\Delta\varepsilon$ determines the relative probability of a torsion angle being in a *gauche* state in thermal equilibrium. In general, this probability is also influenced by the values of torsion angles of neighbouring monomers. These correlations are included in the rotational isomeric state model (Section 2.3.4). The value of $\Delta\varepsilon$ for polyethylene at room temperature is $\Delta\varepsilon \cong 0.8kT$. The energy barrier ΔE between *trans* and *gauche* states determines the dynamics of conformational rearrangements.

Any section of the chain with consecutive *trans* states of torsion angles is in a rod-like zig-zag conformation (see Fig. 2.2). If all torsion angles of the whole chain are in the *trans* state (Fig. 2.2), the chain has the largest possible value of its end-to-end distance R_{max}. This largest end-to-end distance is determined by the product of the number of skeleton bonds n and their projected length $l\cos(\theta/2)$ along the contour, and is referred to as the **contour length** of the chain:

$$R_{\text{max}} = nl \cos \frac{\theta}{2}. \tag{2.1}$$

Gauche states of torsion angles lead to flexibility in the chain conformation since each *gauche* state alters the conformation from the all-*trans* zig-zag of

Fig. 2.2. In general, there will be a variable number of consecutive torsion angles in the *trans* state. Each of these all-*trans* rod-like sections will be broken up by a *gauche*. The chain is rod-like on scales smaller than these all-*trans* sections, but is flexible on larger length scales. Typically, all-*trans* sections comprise fewer than ten main-chain bonds and most synthetic polymers are quite flexible.

A qualitatively different mechanism of flexibility of many polymers, such as double-helix DNA is uniform flexibility over the whole polymer length. These chains are well described by the worm-like chain model (see Section 2.3.2).

2.2 Conformations of an ideal chain

Consider a flexible polymer of $n+1$ backbone atoms A_i (with $0 \le i \le n$) as sketched in Fig. 2.3. The bond vector $\vec{r}_i$ goes from atom A_{i-1} to atom A_i The backbone atoms A_i may all be identical (such as polyethylene) or may be of two or more atoms [Si and O for poly(dimethyl siloxane)]. The polymer is in its ideal state if there are no net interactions between atoms A_i and A_j that are separated by a sufficient number of bonds along the chain so that $|i - j| \gg 1$.

The **end-to-end vector** is the sum of all n bond vectors in the chain:

$$\vec{R}_n = \sum_{i=1}^{n} \vec{r}_i. \tag{2.2}$$

Different individual chains will have different bond vectors and hence different end-to-end vectors. The distribution of end-to-end vectors shall be discussed in Section 2.5. It is useful to talk about average properties of this distribution. The average end-to-end vector of an isotropic collection of chains of n backbone atoms is zero:

$$\langle \vec{R}_n \rangle = 0. \tag{2.3}$$

The **ensemble average** $\langle \rangle$ denotes an average over all possible states of the system (accessed either by considering many chains or many

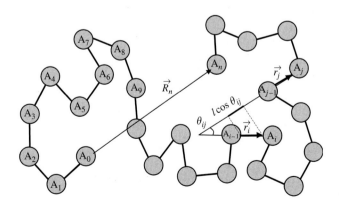

Fig. 2.3
One conformation of a flexible polymer.

Ideal chains

different conformations of the same chain). In this particular case the ensemble average corresponds to averaging over an ensemble of chains of n bonds with all possible bond orientations. Since there is no preferred direction in this ensemble, the average end-to-end vector is zero [Eq. (2.3)]. The simplest non-zero average is the mean-square end-to-end distance:

$$\langle R^2 \rangle \equiv \langle \vec{R}_n^2 \rangle = \langle \vec{R}_n \cdot \vec{R}_n \rangle = \left\langle \left(\sum_{i=1}^{n} \vec{r}_i \right) \cdot \left(\sum_{j=1}^{n} \vec{r}_j \right) \right\rangle$$

$$= \sum_{i=1}^{n} \sum_{j=1}^{n} \langle \vec{r}_i \cdot \vec{r}_j \rangle. \tag{2.4}$$

If all bond vectors have the same length $l = |\vec{r}_i|$, the scalar product can be represented in terms of the angle θ_{ij} between bond vectors $\vec{r}_i$ and $\vec{r}_j$ as shown in Fig. 2.3:

$$\vec{r}_i \cdot \vec{r}_j = l^2 \cos \theta_{ij}. \tag{2.5}$$

The mean-square end-to-end distance becomes a double sum of average cosines:

$$\langle R^2 \rangle = \sum_{i=1}^{n} \sum_{j=1}^{n} \langle \vec{r}_i \cdot \vec{r}_j \rangle = l^2 \sum_{i=1}^{n} \sum_{j=1}^{n} \langle \cos \theta_{ij} \rangle. \tag{2.6}$$

One of the simplest models of an ideal polymer is the **freely jointed chain model** with a constant bond length $l = |\vec{r}_i|$ and no correlations between the directions of different bond vectors, $\langle \cos \theta_{ij} \rangle = 0$ for $i \neq j$. There are only n non-zero terms in the double sum ($\cos \theta_{ij} = 1$ for $i = j$). The mean-square end-to-end distance of a freely jointed chain is then quite simple:

$$\langle R^2 \rangle = n l^2. \tag{2.7}$$

In a typical polymer chain, there are correlations between bond vectors (especially between neighbouring ones) and $\langle \cos \theta_{ij} \rangle \neq 0$. But in an ideal chain there is no interaction between monomers separated by a great distance along the chain contour. This implies that there are no correlations between the directions of distant bond vectors.

$$\lim_{|i-j| \to \infty} \langle \cos \theta_{ij} \rangle = 0. \tag{2.8}$$

It can be shown (see Section 2.3.1) that for any bond vector i, the sum over all other bond vectors j converges to a finite number, denoted by C_i':

$$C_i' \equiv \sum_{j=1}^{n} \langle \cos \theta_{ij} \rangle. \tag{2.9}$$

Therefore, Eq. (2.6) reduces to

$$\langle R^2 \rangle = l^2 \sum_{i=1}^{n} \sum_{j=1}^{n} \langle \cos \theta_{ij} \rangle = l^2 \sum_{i=1}^{n} C'_i = C_n n l^2, \qquad (2.10)$$

where the coefficient C_n, called Flory's **characteristic ratio**, is the average value of the constant C'_i over all main-chain bonds of the polymer:

$$C_n = \frac{1}{n} \sum_{i=1}^{n} C'_i. \qquad (2.11)$$

The main property of ideal chains is that $\langle R^2 \rangle$ is proportional to the product of the number of bonds n and the square of the bond length l^2 [Eq. (2.10)].

An infinite chain has a C'_i value for all i given by C_∞. A real chain has a cutoff in the sum [Eq. (2.9)] at finite j that results in a smaller C'_i. This effect is more pronounced near chain ends.

The characteristic ratio is larger than unity ($C_n > 1$) for all polymers. The physical origins of these local correlations between bond vectors are restricted bond angles and steric hindrance. All models of ideal polymers ignore steric hindrance between monomers separated by many bonds and result in characteristic ratios saturating at a finite value C_∞ for large numbers of main-chain bonds ($n \to \infty$) (see Fig. 2.4). Thus, the mean-square end-to-end distance [Eq. (2.10)] can be approximated for long chains:

$$\langle R^2 \rangle \cong C_\infty n l^2. \qquad (2.12)$$

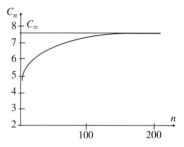

Fig. 2.4
Flory's characteristic ratio C_n saturates at C_∞ for long chains.

The numerical value of Flory's characteristic ratio depends on the local stiffness of the polymer chain with typical numbers of 7–9 for many flexible polymers. The values of the characteristic ratios of some common polymers are listed in Table 2.1. There is a tendency for polymers with bulkier side groups to have higher C_∞, owing to the side groups sterically hindering bond rotation (as in polystyrene), but there are many exceptions to this general tendency (such as polyethylene).

Flexible polymers have many universal properties that are independent of local chemical structure. A simple unified description of all ideal

Table 2.1 Characteristic ratios, Kuhn lengths, and molar masses of Kuhn monomers for common polymers at 413 K

Polymer	Structure	C_∞	b (Å)	ρ (g cm^{-3})	M_0 (g mol^{-1})
1,4-Polyisoprene (PI)	$-(CH_2CH=CHCH(CH_3))-$	4.7	8.4	0.830	120
1,4-Polybutadiene (PB)	$-(CH_2CH=CHCH_2)-$	5.5	9.9	0.826	113
Polypropylene (PP)	$-(CH_2CH_2(CH_3))-$	6.0	11	0.791	183
Poly(ethylene oxide) (PEO)	$-(CH_2CH_2O)-$	6.7	11	1.064	137
Poly(dimethyl siloxane) (PDMS)	$-(OSi(CH_3)_2)-$	6.8	13	0.895	381
Polyethylene (PE)	$-(CH_2CH_2)-$	7.4	14	0.784	150
Poly(methyl methacrylate) (PMMA)	$-(CH_2C(CH_3)(COOCH_3))-$	8.2	15	1.13	598
Atactic polystyrene (PS)	$-(CH_2CHC_6H_5)-$	9.5	18	0.969	720

polymers is provided by an **equivalent freely jointed chain**. The equivalent chain has the same mean-square end-to-end distance $\langle R^2 \rangle$ and the same maximum end-to-end distance R_{max} as the actual polymer, but has N freely-jointed effective bonds of length b. This effective bond length b is called the **Kuhn length**. The contour length of this equivalent freely jointed chain is

$$Nb = R_{max}, \tag{2.13}$$

and its mean-square end-to-end distance is

$$\langle R^2 \rangle = Nb^2 = bR_{max} = C_\infty nl^2. \tag{2.14}$$

Therefore, the equivalent freely jointed chain has

$$N = \frac{R_{max}^2}{C_\infty nl^2} \tag{2.15}$$

equivalent bonds (**Kuhn monomers**) of length

$$b = \frac{\langle R^2 \rangle}{R_{max}} = \frac{C_\infty nl^2}{R_{max}}. \tag{2.16}$$

Example: Calculate the Kuhn length b of a polyethylene chain with $C_\infty = 7.4$, main-chain bond length $l = 1.54\,\text{Å}$, and bond angle $\theta = 68°$.

Substituting the maximum end-to-end distance from Eq. (2.1) into Eq. (2.16) determines the Kuhn length:

$$b = \frac{C_\infty l^2 n}{nl\cos(\theta/2)} = \frac{C_\infty l}{\cos(\theta/2)}. \tag{2.17}$$

For polyethylene $b \cong 1.54\,\text{Å} \times 7.4/0.83 \cong 14\,\text{Å}$.

The values of the Kuhn length b and corresponding molar mass of a Kuhn monomer M_0 for various polymers are listed in Table 2.1. Throughout this book, we will use the equivalent freely jointed chain to describe all flexible polymers and will call N the 'degree of polymerization' or number of 'monomers' (short for Kuhn monomers) and call b the monomer length (instead of the Kuhn monomer length) and

$$R_0 = \sqrt{\langle R^2 \rangle} = bN^{1/2}, \tag{2.18}$$

the root-mean-square end-to-end distance (the subscript 0 refers to the ideal state). This is not to be confused with the chemical definitions of the degree of polymerization and of monomer size. By renormalizing the monomer, Eq. (2.18) holds for *all* flexible linear polymers in the ideal state with $N \gg 1$, with all chemical-specific characteristics contained within that monomer size (Kuhn length).

2.3 Ideal chain models

Below we describe several models of ideal chains. Each model makes different assumptions about the allowed values of torsion and bond angles. However, every model ignores interactions between monomers separated by large distance along the chain and is therefore a model of an ideal polymer. The chemical structure of polymers determines the populations of torsion and bond angles. Some polymers (like 1,4-polyisoprene) are very flexible chains while others (like double-stranded DNA) are locally very rigid, becoming random walks only on quite large length scales.

2.3.1 Freely rotating chain model

As the name suggests, this model ignores differences between the probabilities of different torsion angles and assumes all torsion angles $-\pi < \varphi_i \leq \pi$ to be equally probable. Thus, the **freely rotating chain model** ignores the variations of the potential $U(\varphi_i)$. This model assumes all bond lengths and bond angles are fixed (constant) and all torsion angles are equally likely and independent of each other.

To calculate the mean-square end-to-end distance [Eq. (2.4)]

$$\langle R^2 \rangle = \langle \vec{R}_n \cdot \vec{R}_n \rangle = \sum_{i=1}^{n} \sum_{j=1}^{n} \langle \vec{r}_i \cdot \vec{r}_j \rangle, \tag{2.19}$$

the correlation between bond vectors $\vec{r}_i$ and $\vec{r}_j$ must be determined. This correlation is passed along through the chain of bonds connecting bonds $\vec{r}_i$ and $\vec{r}_j$. For the freely rotating chain, the component of $\vec{r}_j$ normal to vector $\vec{r}_{j-1}$ averages out to zero due to free rotations of the torsion angle φ_j (see Fig. 2.5). The only correlation between the bond vectors that is transmitted down the chain is the component of vector $\vec{r}_j$ along the bond vector $\vec{r}_{j-1}$. The value of this component is $l \cos \theta$. Bond vector $\vec{r}_{j-1}$ passes this correlation down to vector $\vec{r}_{j-2}$, but again only the component along $\vec{r}_{j-2}$ survives due to free rotations of torsion angle φ_{j-1}. The leftover memory of the vector $\vec{r}_j$ at this stage is $l(\cos \theta)^2$. The correlations from bond vector $\vec{r}_j$ at bond vector $\vec{r}_i$ are reduced by the factor $(\cos \theta)^{|j-i|}$ due to independent free rotations of $|j - i|$ torsion angles between these two vectors. Therefore, the correlation between bond vectors $\vec{r}_i$ and $\vec{r}_j$ is

$$\langle \vec{r}_i \cdot \vec{r}_j \rangle = l^2 (\cos \theta)^{|j-i|}. \tag{2.20}$$

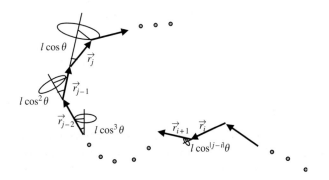

Fig. 2.5
All torison angles are equally likely in a freely rotating chain.

The mean-square end-to-end distance of the freely rotating chain can now be written in terms of cosines:

$$\langle R^2 \rangle = \sum_{i=1}^{n} \sum_{j=1}^{n} \langle \vec{r}_i \cdot \vec{r}_j \rangle = \sum_{i=1}^{n} \left(\sum_{j=1}^{i-1} \langle \vec{r}_i \cdot \vec{r}_j \rangle + \langle \vec{r}_i^2 \rangle + \sum_{j=i+1}^{n} \langle \vec{r}_i \cdot \vec{r}_j \rangle \right)$$

$$= \sum_{i=1}^{n} \langle \vec{r}_i^2 \rangle + l^2 \sum_{i=1}^{n} \left(\sum_{j=1}^{i-1} (\cos \theta)^{i-j} + \sum_{j=i+1}^{n} (\cos \theta)^{j-i} \right)$$

$$= nl^2 + l^2 \sum_{i=1}^{n} \left(\sum_{k=1}^{i-1} \cos^k \theta + \sum_{k=1}^{n-i} \cos^k \theta \right). \tag{2.21}$$

Note that $(\cos \theta)^{|j-i|}$ decays rapidly as the number of bonds between bond vectors $\vec{r}_i$ and $\vec{r}_j$ is increased.

$$(\cos \theta)^{|j-i|} = \exp[|j - i| \ln(\cos \theta)] = \exp\left[-\frac{|j - i|}{s_p}\right]. \tag{2.22}$$

The final relation defines s_p as the number of main-chain bonds in a persistence segment, which is the scale at which local correlations between bond vectors decay:

$$s_p = -\frac{1}{\ln(\cos \theta)}. \tag{2.23}$$

Since the decay is so rapid, the summation in Eq. (2.21) can be replaced by an infinite series over k:

$$\sum_{i=1}^{n} \left(\sum_{k=1}^{i-1} \cos^k \theta + \sum_{k=1}^{n-i} \cos^k \theta \right) \cong 2 \sum_{i=1}^{n} \sum_{k=1}^{\infty} \cos^k \theta = 2n \sum_{k=1}^{\infty} \cos^k \theta$$

$$= 2n \frac{\cos \theta}{1 - \cos \theta}. \tag{2.24}$$

The mean-square end-to-end distance of the freely rotating chain is a simple function of the number of bonds in the chain backbone n, the length of each backbone bond l and the bond angle θ:

$$\langle R^2 \rangle = nl^2 + 2nl^2 \frac{\cos \theta}{1 - \cos \theta} = nl^2 \frac{1 + \cos \theta}{1 - \cos \theta}. \tag{2.25}$$

Polymers with carbon single bonds making up their backbone have a bond angle of $\theta = 68°$.

$$C_\infty = \frac{1 + \cos \theta}{1 - \cos \theta} \cong 2 \quad \text{and} \quad s_p \cong 1. \tag{2.26}$$

Polymer chains are never as flexible as the freely rotating chain model predicts, since the most flexible polymers with $\theta = 68°$ have $C_\infty > 4$

(see Table 2.1). This is because there is steric hindrance to bond rotation in all polymers.

2.3.2 Worm-like chain model

The **worm-like chain model** (sometimes called the Kratky–Porod model) is a special case of the freely rotating chain model for very small values of the bond angle. This is a good model for very stiff polymers, such as double-stranded DNA for which the flexibility is due to fluctuations of the contour of the chain from a straight line rather than to *trans–gauche* bond rotations. For small values of the bond angle ($\theta \ll 1$), the $\cos\theta$ in Eq. (2.23) can be expanded about its value of unity at $\theta = 0$:

$$\cos\theta \cong 1 - \frac{\theta^2}{2}. \tag{2.27}$$

For small x, $\ln(1-x) \cong -x$.

$$\ln(\cos\theta) \cong -\frac{\theta^2}{2}. \tag{2.28}$$

Since θ is small, the persistence segment of the chain [Eq. (2.23)] contains a large number of main-chain bonds.

$$s_p = -\frac{1}{\ln(\cos\theta)} \cong \frac{2}{\theta^2}. \tag{2.29}$$

The **persistence length** is the length of this persistence segment:

$$l_p \equiv s_p l = l\frac{2}{\theta^2}. \tag{2.30}$$

The Flory characteristic ratio of the worm-like chain is very large:

$$C_\infty = \frac{1+\cos\theta}{1-\cos\theta} \cong \frac{2-(\theta^2/2)}{(\theta^2/2)} \cong \frac{4}{\theta^2}. \tag{2.31}$$

The corresponding Kuhn length [see Eq. (2.16)] is twice the persistence length:

$$b = l\frac{C_\infty}{\cos(\theta/2)} \cong l\frac{4}{\theta^2} = 2l_p. \tag{2.32}$$

For example, the persistence length of a double-helical DNA $l_p \approx 50\,\text{nm}$ and the Kuhn length is $b \approx 100\,\text{nm}$.

The combination of parameters l/θ^2 enters in the expressions of the persistence length l_p and the Kuhn length b. The worm-like chain is defined as the limit $l \to 0$ and $\theta \to 0$ at constant persistence length l_p (constant l/θ^2) and constant chain contour length $R_{max} = nl\cos(\theta/2) \cong nl$.

The mean-square end-to-end distance of the worm-like chain can be evaluated using the exponential decay of correlations between tangent vectors along the chain [Eq. (2.22)]:

$$\langle R^2 \rangle = l^2 \sum_{i=1}^{n} \sum_{j=1}^{n} \langle \cos \theta_{ij} \rangle = l^2 \sum_{i=1}^{n} \sum_{j=1}^{n} (\cos \theta)^{|j-i|}$$

$$= l^2 \sum_{i=1}^{n} \sum_{j=1}^{n} \exp\left(-\frac{|j-i|}{l_p} l\right). \tag{2.33}$$

The summation over bonds can be changed into integration over the contour of the worm-like chain:

$$l\sum_{i=1}^{n} \rightarrow \int_0^{R_{max}} du \quad \text{and} \quad l\sum_{j=1}^{n} \rightarrow \int_0^{R_{max}} dv. \tag{2.34}$$

$$\langle R^2 \rangle = \int_0^{R_{max}} \left[\int_0^{R_{max}} \exp\left(-\frac{|u-v|}{l_p}\right) dv \right] du$$

$$= \int_0^{R_{max}} \left[\left(\exp\left(-\frac{u}{l_p}\right) \int_0^{u} \exp\left(\frac{v}{l_p}\right) dv \right. \right.$$

$$\left. \left. + \exp\left(\frac{u}{l_p}\right) \int_u^{R_{max}} \exp\left(-\frac{v}{l_p}\right) dv \right) \right] du$$

$$= l_p \int_0^{R_{max}} \left[\exp\left(-\frac{u}{l_p}\right) \left(\exp\left(\frac{u}{l_p}\right) - 1 \right) \right.$$

$$\left. + \exp\left(\frac{u}{l_p}\right) \left(-\exp\left(-\frac{R_{max}}{l_p}\right) + \exp\left(-\frac{u}{l_p}\right) \right) \right] du$$

$$= l_p \int_0^{R_{max}} \left[2 - \exp\left(-\frac{u}{l_p}\right) - \exp\left(-\frac{R_{max}}{l_p}\right) \exp\left(\frac{u}{l_p}\right) \right] du$$

$$= l_p \left[2R_{max} + l_p \left(\exp\left(-\frac{R_{max}}{l_p}\right) - 1 \right) \right.$$

$$\left. - l_p \exp\left(-\frac{R_{max}}{l_p}\right) \left(\exp\left(\frac{R_{max}}{l_p}\right) - 1 \right) \right]$$

$$= 2l_p R_{max} - 2l_p^2 \left(1 - \exp\left(-\frac{R_{max}}{l_p}\right) \right). \tag{2.35}$$

There are two simple limits of this expression. The ideal chain limit is for worm-like chains much longer than their persistence length.

$$\langle R^2 \rangle \cong 2l_p R_{max} = b R_{max} \quad \text{for } R_{max} \gg l_p. \tag{2.36}$$

The rod-like limit is for worm-like chains much shorter than their persistence length. The exponential in Eq. (2.35) can be expanded in this limit:

$$\exp\left(-\frac{R_{max}}{l_p}\right) \cong 1 - \frac{R_{max}}{l_p} + \frac{1}{2}\left(\frac{R_{max}}{l_p}\right)^2 - \frac{1}{6}\left(\frac{R_{max}}{l_p}\right)^3 + \cdots \text{ for } R_{max} \ll l_p, \tag{2.37}$$

$$\langle R^2 \rangle \cong R_{max}^2 - \frac{R_{max}^3}{3l_p} + \cdots \quad \text{for } R_{max} \ll l_p. \tag{2.38}$$

The mean-square end-to-end distance of the worm-like chain [Eq. (2.35)] is a smooth crossover between these two simple limits.

The important difference between freely jointed chains and worm-like chains is that each bond of Kuhn length b of the freely jointed chain is assumed to be completely rigid. Worm-like chains are also stiff on length scales shorter than the Kuhn length, but are not completely rigid and can fluctuate and bend. These bending modes lead to a qualitatively different dependence of extensional force on elongation near maximum extension, as will be discussed in Section 2.6.2.

2.3.3 Hindered rotation model

The **hindered rotation model** also assumes bond lengths and bond angles are constant and torsion angles are independent of each other. As its name suggests, the torsion angle rotation is taken to be hindered by a potential $U(\varphi_i)$ [see Fig. 2.1(d)]. The probability of any value of the torsion angle φ_i is taken to be proportional to the **Boltzmann factor** $\exp[-U(\varphi_i)/kT]$. Most of the torsion angles are in low energy states [near the minima in Fig. 2.1(d)] but for ordinary temperatures there are some torsion angles corresponding to high energy states as well. The Boltzmann factor ensures that states with higher energy are progressively less likely to be populated.

The hindered rotation model assumes independent but hindered rotations of torsion angles at constant bond lengths and bond angles with different potential profiles $U(\varphi_i)$ corresponding to different polymers. The hindered rotation model predicts the mean-square end-to-end distance

$$\langle R^2 \rangle = C_\infty l^2 n, \tag{2.39}$$

with the characteristic ratio (see problem 2.9)

$$C_\infty = \left(\frac{1 + \cos\theta}{1 - \cos\theta} \right) \left(\frac{1 + \langle\cos\varphi\rangle}{1 - \langle\cos\varphi\rangle} \right), \tag{2.40}$$

where $\langle\cos\varphi\rangle$ is the average value of the cosine of the torsion angle with probabilities determined by Boltzmann factors, $\exp[-U(\varphi_i)/kT]$:

$$\langle\cos\varphi\rangle = \frac{\int_0^{2\pi} \cos\varphi \exp(-U(\varphi)/kT)\, d\varphi}{\int_0^{2\pi} \exp(-U(\varphi)/kT)\, d\varphi}. \tag{2.41}$$

2.3.4 Rotational isomeric state model

This is the most successful ideal chain model used to calculate the details of conformations of different polymers. In this model, bond lengths l and bond angles θ are fixed (constant).

For a relatively high barrier between *trans* and *gauche* states $\Delta E \gg kT$ the values of the torsion angles φ_i are close to the minima (t, g_+, g_-) [see Fig. 2.1(d)]. In the **rotational isomeric state model** each molecule is assumed to exist only in discrete torsional states corresponding to the potential energy minima. The fluctuations about these minima are ignored. A conformation of a chain with n main-chain bonds is thus represented by a sequence of $n-2$ torsion angles:

$$\cdots t\,g_-\,t\,t\,g_+\,t\,g_+\,t\,t\,g_-\,t\,\cdots. \qquad (2.42)$$

Each of these $n-2$ torsion angles can be in one of three states (t, g_+, g_-) and therefore the whole chain has 3^{n-2} rotational isomeric states. For example, n-pentane, with $n=4$ main-chain bonds and $n-2=2$ torsion angles, has $3^2=9$ rotational isomeric states:[1]

$$tt,\ tg_+,\ tg_-,\ g_+t,\ g_-t,\ g_+g_+,\ g_+g_-,\ g_-g_+,\ g_-g_-. \qquad (2.43)$$

In the rotational isomeric state model, these states are *not* equally probable. Correlations between neighbouring torsional states are included in the model. For example, a consecutive sequence of g_+ and g_- has high energy due to overlap between atoms and therefore is taken to have very low probability in the rotational isomeric state model. The relative probabilities of the states of neighboring torsional angles are used to calculate the mean-square end-to-end distance and C_∞ [Eq. (2.12)].

Table 2.2 summarizes the assumptions of the ideal chain models. The worm-like chain model is a special case of the freely rotating chain with a small value of the bond angle θ. Moving from left to right in Table 2.2, the models become progressively more specific (and more realistic). As more constraints are adopted, the chain becomes stiffer, reflected in larger C_∞.

Table 2.2 Assumptions and predictions of ideal chain models: FJC, freely jointed chain; FRC, freely rotating chain; HR, hindered rotation; RIS, rotational isomeric state

Models	FJC	FRC	HR	RIS
Bond length l	Fixed	Fixed	Fixed	Fixed
Bond angle θ	Free	Fixed	Fixed	Fixed
Torsion angle φ	Free	Free	Controlled by $U(\varphi)$	$t, g+, g-$
Next φ independent?	Yes	Yes	Yes	No
C_∞	1	$\dfrac{1+\cos\theta}{1-\cos\theta}$	$\left(\dfrac{1+\cos\theta}{1-\cos\theta}\right)\left(\dfrac{1+\langle\cos\varphi\rangle}{1-\langle\cos\varphi\rangle}\right)$	Specific

2.4 Radius of gyration

The size of linear chains can be characterized by their mean-square end-to-end distance. However, for branched or ring polymers this quantity is not well defined, because they either have too many ends or no ends at all. Since all objects possess a radius of gyration, it can characterize the size of polymers of any architecture. Consider, for example, the branched polymer sketched in Fig. 2.6. The square **radius of gyration** is defined as the average square distance between monomers in a given conformation (position vector $\vec{R}_i$) and the polymer's centre of mass (position vector $\vec{R}_{cm}$):

$$R_g^2 \equiv \frac{1}{N}\sum_{i=1}^{N}(\vec{R}_i - \vec{R}_{cm})^2. \qquad (2.44)$$

---

[1] Six of the nine rotational isomers are distinguishable.

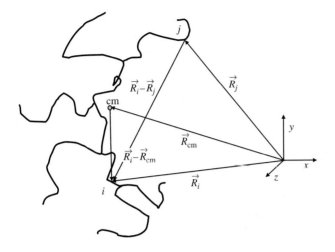

Fig. 2.6
One conformation of a randomly
branched polymer and its centre of mass,
denoted by cm.

The position vector of the centre of mass of the polymer is the number-average of all monomer position vectors:[2]

$$\vec{R}_{\text{cm}} \equiv \frac{1}{N}\sum_{j=1}^{N}\vec{R}_j. \qquad (2.45)$$

Substituting the definition of the position vector of the centre of mass [Eq. (2.45)] into Eq. (2.44) gives an expression for the square radius of gyration as a double sum of squares over all inter-monomer distances:

$$R_{\text{g}}^2 = \frac{1}{N}\sum_{i=1}^{N}(\vec{R}_i^2 - 2\vec{R}_i\vec{R}_{\text{cm}} + \vec{R}_{\text{cm}}^2)$$

$$= \frac{1}{N}\sum_{i=1}^{N}\left[\vec{R}_i^2\frac{1}{N}\sum_{j=1}^{N}1 - 2\vec{R}_i\frac{1}{N}\sum_{j=1}^{N}\vec{R}_j + \left(\frac{1}{N}\sum_{j=1}^{N}\vec{R}_j\right)^2\right]. \qquad (2.46)$$

The last term in the sum can be rewritten as

$$\frac{1}{N}\sum_{i=1}^{N}\left(\frac{1}{N}\sum_{j=1}^{N}\vec{R}_j\right)^2 = \left(\frac{1}{N}\sum_{j=1}^{N}\vec{R}_j\right)^2 = \left(\frac{1}{N}\sum_{i=1}^{N}\vec{R}_i\right)\left(\frac{1}{N}\sum_{j=1}^{N}\vec{R}_j\right)$$

$$= \frac{1}{N^2}\sum_{i=1}^{N}\sum_{j=1}^{N}\vec{R}_i\vec{R}_j.$$

Therefore, the expression for the square radius of gyration takes the form

$$R_{\text{g}}^2 = \frac{1}{N^2}\sum_{i=1}^{N}\sum_{j=1}^{N}(\vec{R}_i^2 - 2\vec{R}_i\vec{R}_j + \vec{R}_i\vec{R}_j) = \frac{1}{N^2}\sum_{i=1}^{N}\sum_{j=1}^{N}(\vec{R}_i^2 - \vec{R}_i\vec{R}_j).$$

---

[2] In general, the mass of the monomers M_j should be included in the definitions of the radius of gyration and of the centre of mass. For example, the proper centre of mass definition is

$$\vec{R}_{\text{cm}} \equiv \frac{\sum_{j=1}^{N}M_j\vec{R}_j}{\sum_{j=1}^{N}M_j}.$$

We assume that all the monomers have the same mass $M_j = M_0$ for all j.

Ideal chains

This expression does not depend on the choice of summation indices and can be rewritten in a symmetric form:

$$R_g^2 = \frac{1}{N^2} \sum_{i=1}^{N} \sum_{j=1}^{N} (\vec{R}_i^2 - \vec{R}_i \vec{R}_j)$$

$$= \frac{1}{2} \left[\frac{1}{N^2} \sum_{i=1}^{N} \sum_{j=1}^{N} (\vec{R}_i^2 - \vec{R}_i \vec{R}_j) + \frac{1}{N^2} \sum_{j=1}^{N} \sum_{i=1}^{N} (\vec{R}_j^2 - \vec{R}_j \vec{R}_i) \right]$$

$$= \frac{1}{2N^2} \sum_{i=1}^{N} \sum_{j=1}^{N} (\vec{R}_i^2 - 2\vec{R}_i \vec{R}_j + \vec{R}_j^2)$$

$$= \frac{1}{2N^2} \sum_{i=1}^{N} \sum_{j=1}^{N} (\vec{R}_i - \vec{R}_j)^2. \tag{2.47}$$

Each pair of monomers enters twice in the double sum of Eq. (2.47). Alternatively, this expression for the square radius of gyration can be written with each pair of monomers entering only once in the double sum:

$$R_g^2 = \frac{1}{N^2} \sum_{i=1}^{N} \sum_{j=i}^{N} (\vec{R}_i - \vec{R}_j)^2. \tag{2.48}$$

For polymers and other fluctuating objects, the square radius of gyration is usually averaged over the ensemble of allowed conformations giving the mean-square radius of gyration:

$$\langle R_g^2 \rangle = \frac{1}{N} \sum_{i=1}^{N} \langle (\vec{R}_i - \vec{R}_{cm})^2 \rangle = \frac{1}{N^2} \sum_{i=1}^{N} \sum_{j=i}^{N} \langle (\vec{R}_i - \vec{R}_j)^2 \rangle. \tag{2.49}$$

For non-fluctuating (solid) objects such averaging is unnecessary. The expression with the centre of mass is useful only if the position of the centre of mass $\vec{R}_{cm}$ of the object is known or is easy to evaluate. Otherwise the expression for the radius of gyration in terms of the average square distances between all pairs of monomers is used.

2.4.1 Radius of gyration of an ideal linear chain

To illustrate the use of Eq. (2.48), we now calculate the mean-square radius of gyration for an ideal linear chain. For the linear chain, the summations over the monomers can be changed into integrations over the contour of the chain, by replacing monomer indices i and j with continuous coordinates u and v along the contour of the chain:

$$\sum_{i=1}^{N} \rightarrow \int_0^N du \quad \text{and} \quad \sum_{j=i}^{N} \rightarrow \int_u^N dv. \tag{2.50}$$

This transformation results in the integral form for the mean-square radius of gyration

$$\langle R_g^2 \rangle = \frac{1}{N^2} \int_0^N \int_u^N \langle (\vec{R}(u) - \vec{R}(v))^2 \rangle \, dv \, du, \qquad (2.51)$$

where $\vec{R}(u)$ is the position vector corresponding to the contour coordinate u. The mean-square distance between points u and v along the contour of the chain can be obtained by treating each section of $v - u$ monomers as a shorter ideal chain. The outer sections of u and of $N - v$ monomers do not affect the conformations of this inner section. The mean-square end-to-end distance for an ideal chain of $v - u$ monomers is given by Eq. (2.18):

$$\langle (\vec{R}(u) - \vec{R}(v))^2 \rangle = (v - u)b^2. \qquad (2.52)$$

The mean-square radius of gyration is then calculated by a simple integration using the change of variables $v' \equiv v - u$ and $u' \equiv N - u$:

$$\langle R_g^2 \rangle = \frac{b^2}{N^2} \int_0^N \int_u^N (v - u) \, dv \, du = \frac{b^2}{N^2} \int_0^N \int_0^{N-u} v' \, dv' \, du$$

$$= \frac{b^2}{N^2} \int_0^N \frac{(N-u)^2}{2} \, du = \frac{b^2}{2N^2} \int_0^N (u')^2 \, du' = \frac{b^2}{2N^2} \frac{N^3}{3} = \frac{Nb^2}{6} \qquad (2.53)$$

Comparing this result with Eq. (2.18), we obtain the classic Debye result relating the mean-square radius of gyration and the mean-square end-to-end distance of an ideal linear chain:

$$\langle R_g^2 \rangle = \frac{b^2 N}{6} = \frac{\langle R^2 \rangle}{6}. \qquad (2.54)$$

The radius of gyration of other shapes of flexible ideal chains can be calculated in a similar way and examples of the results are given in Table 2.3.

Table 2.3 Mean-square radii of gyration of ideal polymers with N Kuhn monomers of length b: linear chain, ring, f-arm star with each arm containing N/f Kuhn monomers, and H-polymer with all linear sections containing $N/5$ Kuhn monomers

Ideal chains	Linear	Ring	f-arm star	H-polymer
$\langle R_g^2 \rangle$	$Nb^2/6$	$Nb^2/12$	$[(N/f)b^2/6](3 - 2/f)$	$(Nb^2/6)\, 89/125$

2.4.2 Radius of gyration of a rod polymer

Consider a rod polymer of N monomers of length b, with end-to-end distance $L = Nb$. It is convenient to calculate the radius of gyration of a rod polymer using the original definition, Eq. (2.44), written in integral form:

$$R_g^2 \equiv \frac{1}{N} \int_0^N \left[(\vec{R}(u) - \vec{R}_{cm})^2 \right] du. \qquad (2.55)$$

A rigid rod polymer has only one conformation with the distance between coordinate u along the chain and its centre of mass (coordinate $N/2$):

$$\left| \vec{R}(u) - \vec{R}_{cm} \right| = \left| u - \frac{N}{2} \right| b. \tag{2.56}$$

Therefore, no averaging is needed for calculation of the radius of gyration of a rod. The square radius of gyration of the rod polymer is calculated by a simple integration

$$R_g^2 \equiv \frac{b^2}{N} \int_0^N \left(u - \frac{N}{2} \right)^2 du = \frac{b^2}{N} \int_{-N/2}^{N/2} x^2\, dx = \frac{N^2 b^2}{12}, \tag{2.57}$$

where the change of variables $x = u - N/2$ has been used. Note that the relation between the end-to-end distance and the radius of gyration for a rod polymer is different from that for an ideal linear chain [Eq. (2.54)]:

$$R_g^2 = \frac{N^2 b^2}{12} = \frac{L^2}{12}. \tag{2.58}$$

Examples of the radii of gyration of other rigid objects are listed in Table 2.4.

Table 2.4 Square radii of gyration of rigid objects: uniform thin disc of radius R, uniform sphere of radius R, thin rod of length L, and uniform right cylinder of radius R and length L

Rigid objects	Disk	Sphere	Rod	Cylinder
R_g^2	$R^2/2$	$3R^2/5$	$L^2/12$	$(R^2/2) + (L^2/12)$

2.4.3 Radius of gyration of an ideal branched polymer (Kramers theorem)

Consider an ideal molecule that contains an arbitrary number of branches, but no loops. This molecule consists of N freely jointed segments (Kuhn monomers) of length b. The mean-square radius of gyration of this molecule is calculated using Eq. (2.48):

$$\langle R_g^2 \rangle = \frac{1}{N^2} \sum_{i=1}^N \sum_{j=i}^N \langle (\vec{R}_j - \vec{R}_i)^2 \rangle. \tag{2.59}$$

The vector $\vec{R}_j - \vec{R}_i$ between monomers i and j can be represented by the sum over the bond vectors $\vec{r}_k$ of a linear strand connecting these two monomers:

$$\vec{R}_j - \vec{R}_i = \sum_{k=i+1}^j \vec{r}_k. \tag{2.60}$$

Since we have assumed freely jointed chain statistics with no correlations between different segments,

$$\langle \vec{r}_k \vec{r}_{k'} \rangle = 0 \quad \text{if } k \neq k', \tag{2.61}$$

the mean-square distance between monomers i and j can be rewritten:

$$\langle (\vec{R}_j - \vec{R}_i)^2 \rangle = \sum_{k=i+1}^{j} \sum_{k'=i+1}^{j} \langle \vec{r}_k \vec{r}_{k'} \rangle = \sum_{k=i+1}^{j} (\vec{r}_k)^2. \tag{2.62}$$

Each segment of a linear strand connecting monomers i and j contributes $(\vec{r}_k)^2 = b^2$ to the double sum in Eq. (2.59). There is only one such strand connecting each pair of monomers because the molecule is assumed to have no loops. Therefore, the contribution of each segment of the molecule to the double sum in Eq. (2.59) is equal to b^2 times the number of strands between different monomers i and j that pass through this segment. Consider, for example, segment k in Fig. 2.7. It divides the molecule into two tree-like parts. The lower part contains N_1 monomers and the upper part contains $N - N_1$ monomers. Monomer i could be any one of $N - N_1$ monomers of the upper part, while monomer j could be any one of N_1 monomers of the lower part of the molecule. Therefore, there are $N_1(N - N_1)$ different strands between all pairs of monomers i and j passing through segment k. Thus, the segment k contributes $N_1(k)[N - N_1(k)]b^2$ to the double sum in Eq. (2.59).

The radius of gyration can be expressed as the sum over all N molecular bonds, of the product of the number of monomers of the two branches $N_1(k)$ and $N - N_1(k)$ that each bond k divides the molecule into:

$$\langle R_g^2 \rangle = \frac{b^2}{N^2} \sum_{k=1}^{N} N_1(k)[N - N_1(k)]. \tag{2.63}$$

The average value of this product is:

$$\langle N_1(N - N_1) \rangle = \frac{1}{N} \sum_{k=1}^{N} N_1(k)[N - N_1(k)]. \tag{2.64}$$

The **Kramers theorem** is expressed in terms of this average over all possible ways of dividing the molecule into two parts:

$$\langle R_g^2 \rangle = \frac{b^2}{N} \langle N_1(N - N_1) \rangle. \tag{2.65}$$

This expression is valid for a linear polymer with the average evaluated by integration.

$$\langle N_1(N - N_1) \rangle = \frac{1}{N} \int_0^N N_1(N - N_1)\, dN_1$$
$$= \int_0^N N_1\, dN_1 - \frac{1}{N} \int_0^N N_1^2\, dN_1$$
$$= \frac{N^2}{2} - \frac{N^2}{3} = \frac{N^2}{6}. \tag{2.66}$$

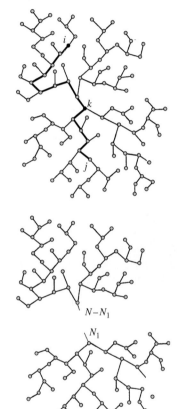

Fig. 2.7
The Kramers theorem effectively cuts a randomly branched polymer with N monomers into two parts, with N_1 and $N - N_1$ monomers.

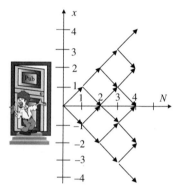

Fig. 2.8
A two-dimensional random walk on a square lattice. The direction of each step is randomly chosen from four possible diagonals.

Fig. 2.9
A one-dimensional random walk of a drunk in an alley, showing all possible trajectories up to $N = 4$ steps.

Table 2.5 The number of trajectories $W(N, x)$ for one-dimensional random walks of N steps that start at the origin and end at position x

	$N=1$	$N=2$	$N=3$	$N=4$
$x=-4$	0	0	0	1
$x=-3$	0	0	1	0
$x=-2$	0	1	0	4
$x=-1$	1	0	3	0
$x=0$	0	2	0	6
$x=1$	1	0	3	0
$x=2$	0	1	0	4
$x=3$	0	0	1	0
$x=4$	0	0	0	1

Substituting this average [Eq. (2.66)] into the Kramers theorem [Eq. (2.65)] recovers the classical result for the radius of gyration of an ideal linear chain [Eq. (2.54)]. In Section 6.4.6, we apply the Kramers theorem [Eq. (2.65)] to ideal randomly branched polymers. In this case the average is not only over different ways of dividing a molecule into two parts, but also over different branched molecules with the same degree of polymerization N.

2.5 Distribution of end-to-end vectors

A polydisperse collection of polymers can be described by an average molar mass (such as the number-average or weight-average discussed in Chapter 1). Much more information is contained in the whole molar mass distribution than in any of its moments or averages. Similarly, the average polymer conformation can be described by the mean-square end-to-end distance (or mean-square radius of gyration). Much more information is contained in the distribution of end-to-end vectors than in the mean-square end-to-end vector. In this section, we derive the distribution of end-to-end vectors for an ideal chain.

Every possible conformation of an ideal chain can be mapped onto a **random walk**. A particle making random steps defines a random walk. If the length of each step is constant and the direction of each step is independent of all previous steps, the trajectory of this random walk is one conformation of a freely jointed chain. Hence, random walk statistics and ideal chain statistics are similar.

Consider a particular random walk on a lattice with each step having independent Cartesian coordinates of either $+1$ or -1. The projection of this three-dimensional random walk onto each of the Cartesian coordinate axes is an independent one-dimensional random walk of unit step length (see Fig. 2.8 for an example of a two-dimensional projection). The fact that the one-dimensional components are independent of each other is an important property of any random walk (as well as any ideal polymer chain).

An example of a one-dimensional random walk is a drunk in a dark narrow alley. Let the drunk start at the doors of the pub at the origin of the one-dimensional coordinate system and make unit steps randomly up and down the alley. Figure 2.9 represents random wandering of the drunk up and down the alley as a function of the number of steps taken. Let $W(N, x)$ be the number of different possible trajectories for a drunk to get from the pub to the position x in N steps. For example, after the first step he could have reached either position $x = +1$ or $x = -1$, making $W(1, 1) = W(1, -1) = 1$. The numbers of different trajectories $W(N, x)$ for the first four steps of the drunk are shown in Table 2.5.

A general expression for $W(N, x)$ can be obtained in the following way. Any trajectory of our drunk consists of N_+ steps up the alley and N_- steps down the alley. The total number of steps made by the drunk is $N = N_+ + N_-$ and his final position is $x = N_+ - N_-$. The numbers of steps up N_+ and down N_- the alley uniquely specify both the total number of steps N and the final position x. Therefore, the total number of trajectories $W(N, x)$ is equal to the number of combinations of N_+ steps up and N_- steps down, that reach x in a total of N steps, which is a binomial coefficient:

$$W(N, x) = \frac{(N_+ + N_-)!}{N_+! N_-!} = \frac{N!}{[(N+x)/2]![(N-x)/2]!}. \qquad (2.67)$$

The factorial is defined as $N! = 1 \cdot 2 \cdot 3 \cdot 4 \cdots N$.

The total number of N-step walks is 2^N because on each step the drunk has two possibilities, which are independent from step to step. All of these 2^N walks are equally likely (if there is no wind or stairway in the alley) and therefore, the probability to find the drunk at position x after N steps is $W(N, x)$ divided by 2^N:

$$\frac{W(N, x)}{2^N} = \frac{1}{2^N} \frac{N!}{[(N+x)/2]![(N-x)/2]!}. \qquad (2.68)$$

This is an exact probability distribution for a one-dimensional random walk. However, it is not convenient to use for large N because of the difficulty of calculating factorials for large N (try your calculator for $N = 100$). For any N, the probability of finding the drunk is highest at the pub (at $x = 0$ for even N and at $x = \pm 1$ for odd N). This probability falls off very fast for large $|x|$ and it is therefore convenient to use the Gaussian approximation of the distribution function, valid for $x \ll N$, derived next.

First, take the natural logarithm of the distribution function:

$$\ln\left(\frac{W(N, x)}{2^N}\right) = -N \ln 2 + \ln(N!) - \ln\left(\frac{N+x}{2}\right)! - \ln\left(\frac{N-x}{2}\right)!. \qquad (2.69)$$

Each of the last two terms can be rewritten using the definition of the factorial function:

$$\ln\left(\frac{N+x}{2}\right)! = \ln\left[\left(\frac{N}{2}\right)!\left(\frac{N}{2}+1\right)\left(\frac{N}{2}+2\right)\cdots\left(\frac{N}{2}+\frac{x}{2}\right)\right]$$

$$= \ln\left(\frac{N}{2}\right)! + \sum_{s=1}^{x/2} \ln\left(\frac{N}{2}+s\right), \qquad (2.70)$$

$$\ln\left(\frac{N-x}{2}\right)! = \ln\left(\frac{N}{2}\right)! - \sum_{s=1}^{x/2} \ln\left(\frac{N}{2}+1-s\right). \qquad (2.71)$$

The logarithm of the probability distribution can now be rewritten as

$$\ln\left(\frac{W(N, x)}{2^N}\right) = -N \ln 2 + \ln(N!) - \ln\left(\frac{N}{2}\right)! - \sum_{s=1}^{x/2} \ln\left(\frac{N}{2}+s\right)$$

$$- \ln\left(\frac{N}{2}\right)! + \sum_{s=1}^{x/2} \ln\left(\frac{N}{2}+1-s\right)$$

$$= -N \ln 2 + \ln(N!) - 2 \ln\left(\frac{N}{2}\right)! - \sum_{s=1}^{x/2} \ln\left(\frac{(N/2)+s}{(N/2)+1-s}\right).$$

The logarithm in the last term can be expanded for $s \ll N/2$ up to a linear term ($\ln (1+y) \cong y$ for $|y| \ll 1$). This expansion is the essence of the Gaussian approximation.

$$
\begin{aligned}
\ln\left(\frac{(N/2)+s}{(N/2)+1-s}\right) &= \ln\left(\frac{1+(2s/N)}{1-(2s/N)+(2/N)}\right) \\
&= \ln\left(1+\frac{2s}{N}\right) - \ln\left(1-\frac{2s}{N}+\frac{2}{N}\right) \\
&\cong \frac{4s}{N} - \frac{2}{N}.
\end{aligned}
\tag{2.72}
$$

The logarithm of the probability distribution can be simplified using this approximation.

$$
\begin{aligned}
\ln\left(\frac{W(N,x)}{2^N}\right) &\cong -N\ln 2 + \ln(N!) - 2\ln\left(\frac{N}{2}\right)! - \sum_{s=1}^{x/2}\left(\frac{4s}{N}-\frac{2}{N}\right) \\
&\cong -N\ln 2 + \ln(N!) - 2\ln\left(\frac{N}{2}\right)! - \frac{4}{N}\sum_{s=1}^{x/2}s + \frac{2}{N}\sum_{s=1}^{x/2}1 \\
&\cong -N\ln 2 + \ln(N!) - 2\ln\left(\frac{N}{2}\right)! - \frac{4}{N}\frac{(x/2)(x/2+1)}{2} + \frac{x}{N} \\
&\cong -N\ln 2 + \ln(N!) - 2\ln\left(\frac{N}{2}\right)! - \frac{x^2}{2N}.
\end{aligned}
\tag{2.73}
$$

This gives the Gaussian approximation of the probability distribution:

$$
\frac{W(N,x)}{2^N} \cong \frac{1}{2^N}\frac{N!}{(N/2)!(N/2)!}\exp\left(-\frac{x^2}{2N}\right).
\tag{2.74}
$$

Using Stirling's approximation of $N!$ for large N

$$
N! \cong \sqrt{2\pi N}\left(\frac{N}{e}\right)^N,
\tag{2.75}
$$

the coefficient in front of the exponential can be rewritten:

$$
\frac{1}{2^N}\frac{N!}{(N/2)!(N/2)!} \cong \frac{1}{2^N}\frac{\sqrt{2\pi N}N^N\exp(-N)}{\left(\sqrt{\pi N}(N/2)^{N/2}\exp(-N/2)\right)^2} = \sqrt{\frac{2}{\pi N}}.
\tag{2.76}
$$

The final expression for the Gaussian approximation of the probability distribution is quite simple:

$$
\frac{W(N,x)}{2^N} \cong \sqrt{\frac{2}{\pi N}}\exp\left(-\frac{x^2}{2N}\right).
\tag{2.77}
$$

Recall from Table 2.5 that $W(N,x)$ is non-zero only either for even or odd x (depending on whether N is even or odd). Therefore, the spacing between non-zero values of $W(N,x)$ is equal to 2 along the x axis. The **probability** distribution function $P_{1d}(N, x)$ is defined as the probability $P(N, x)\,dx$ that the drunk will be found in the interval dx along the x axis. Thus, the probability distribution function differs from Eq. (2.77) by a factor of 2:

$$P_{1d}(N, x) = \frac{1}{\sqrt{2\pi N}}\exp\left(-\frac{x^2}{2N}\right). \qquad (2.78)$$

The square of the typical distance of the drunk from the pub after N steps is determined from the mean-square displacement averaged over all the walks the drunk makes day after day:

$$\langle x^2 \rangle = \int_{-\infty}^{\infty} x^2 P_{1d}(N, x)\,dx = \frac{1}{\sqrt{2\pi N}}\int_{-\infty}^{\infty} x^2 \exp\left(-\frac{x^2}{2N}\right)dx = N. \quad (2.79)$$

Therefore, the probability distribution function can be rewritten in terms of this mean-square displacement:

$$P_{1d}(N, x) = \frac{1}{\sqrt{2\pi\langle x^2 \rangle}}\exp\left(-\frac{x^2}{2\langle x^2 \rangle}\right). \qquad (2.80)$$

This function has a maximum at $x = 0$ and decays fast for distances larger than the root-mean-square displacement $x > \sqrt{\langle x^2 \rangle}$ as can be seen from Fig. 2.10.

This probability distribution function for the displacement of a one-dimensional random walk can be easily generalized to three-dimensional random walks. The probability of a walk, starting at the origin of the coordinate system, to end after N steps, each of size b, within a volume $dR_x\,dR_y\,dR_z$ of the point with displacement vector $\vec{R}$ is $P_{3d}(N, \vec{R})$ $dR_x\,dR_y\,dR_z$ (see Fig. 2.11). Since the three components of a three-dimensional random walk along the three Cartesian coordinates are independent of each other, the three-dimensional probability distribution function is a product of the three one-dimensional distribution functions:

$$P_{3d}(N, \vec{R})\,dR_x\,dR_y\,dR_z = P_{1d}(N, R_x)\,dR_x P_{1d}(N, R_y)\,dR_y P_{1d}(N, R_z)\,dR_z. \qquad (2.81)$$

The mean-square displacement of a random walk from the origin is equal to the mean-square end-to-end vector of a freely jointed chain with the number of monomers N equal to the number of steps of the walk and the monomer length b equal to the step size $\langle \vec{R}^2 \rangle = Nb^2$. This mean-square displacement is composed of three mean-square displacements of the three independent one-dimensional walks:

$$\langle \vec{R}^2 \rangle = \langle R_x^2 \rangle + \langle R_y^2 \rangle + \langle R_z^2 \rangle = Nb^2. \qquad (2.82)$$

Since each of the three Cartesian axes are equivalent, the mean-square displacement along each of them must be one-third of the total:

$$\langle R_x^2 \rangle = \langle R_y^2 \rangle = \langle R_z^2 \rangle = \frac{Nb^2}{3}. \qquad (2.83)$$

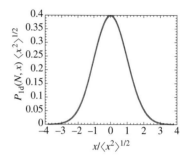

Fig. 2.10
Normalized one-dimensional Gaussian probability distribution function for occupying position x after random N steps from the origin ($x = 0$).

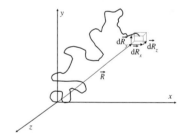

Fig. 2.11
One conformation of an ideal chain with one end at the origin and the other end within volume $dR_x\,dR_y\,dR_z$ of position $\vec{R}$.

The one-dimensional probability distribution function for the components of a random walk along each of these three axes can be obtained by substituting these mean-square displacements into Eq. (2.80)

$$P_{1d}(N, R_x) = \frac{1}{\sqrt{2\pi\langle R_x^2\rangle}} \exp\left(-\frac{R_x^2}{2\langle R_x^2\rangle}\right)$$

$$= \sqrt{\frac{3}{2\pi Nb^2}} \exp\left(-\frac{3R_x^2}{2Nb^2}\right). \tag{2.84}$$

The probability distribution function for the end-to-end vector $\vec{R}$ of an ideal linear chain of N monomers is the product of the three independent distribution functions [Eq. (2.81)]:

$$P_{3d}(N, \vec{R}) = \left(\frac{3}{2\pi Nb^2}\right)^{3/2} \exp\left(-\frac{3(R_x^2 + R_y^2 + R_z^2)}{2Nb^2}\right)$$

$$= \left(\frac{3}{2\pi Nb^2}\right)^{3/2} \exp\left(-\frac{3\vec{R}^2}{2Nb^2}\right). \tag{2.85}$$

As a function of each Cartesian component R_i of the end-to-end vector $\vec{R}$, this probability distribution function looks the same as sketched in Fig. 2.10. The average of each component is $\langle R_i\rangle = 0$. As a function of the end-to-end distance $R = |\vec{R}|$ this probability distribution can be rewritten in the spherical coordinate system:

$$P_{3d}(N, R)4\pi R^2 \, dR = 4\pi \left(\frac{3}{2\pi Nb^2}\right)^{3/2} \exp\left(-\frac{3R^2}{2Nb^2}\right) R^2 \, dR. \tag{2.86}$$

The probability distribution for the end-to-end distance R is the probability for the end-to-end vector $\vec{R}$ to be in the spherical shell with radius between R and $R+dR$. This probability of the end-to-end distance [Eq. (2.86)] is shown in Fig. 2.12. The Gaussian approximation is valid only for end-to-end vectors much shorter than the maximum extension of the chain (for $|\vec{R}| \ll R_{max} = Nb$). For $|\vec{R}| > Nb$, Eq. (2.85) predicts finite (though exponentially small) probability, which is physically unreasonable. For real chains $P_{3d}(N, R) \equiv 0$ for $R > Nb$ and this strong stretching is treated properly in Section 2.6.2.

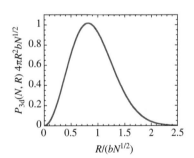

Fig. 2.12
Normalized distribution function of end-to-end distances for an ideal linear chain.

2.6 Free energy of an ideal chain

The **entropy** S is the product of the Boltzmann constant k and the logarithm of the number of states Ω:

$$S = k \ln \Omega. \tag{2.87}$$

Denote $\Omega(N, \vec{R})$ as the number of conformations of a freely jointed chain of N monomers with end-to-end vector $\vec{R}$. The entropy is then a function of N and $\vec{R}$:

$$S\left(N, \vec{R}\right) = k \ln \Omega \left(N, \vec{R}\right). \tag{2.88}$$

The probability distribution function is the fraction of all conformations that actually have an end-to-end vector $\vec{R}$ between $\vec{R}$ and $\vec{R} + d\vec{R}$:

$$P_{3d}\left(N, \vec{R}\right) = \frac{\Omega\left(N, \vec{R}\right)}{\int \Omega\left(N, \vec{R}\right) d\vec{R}}. \tag{2.89}$$

The entropy of an ideal chain with N monomers and end-to-end vector $\vec{R}$ is thus related to the probability distribution function:

$$S\left(N, \vec{R}\right) = k \ln P_{3d}\left(N, \vec{R}\right) + k \ln\left[\int \Omega\left(N, \vec{R}\right) d\vec{R}\right]. \tag{2.90}$$

Equation (2.85) for the probability distribution function determines the entropy:

$$S\left(N, \vec{R}\right) = -\frac{3}{2}k\frac{\vec{R}^2}{Nb^2} + \frac{3}{2}k \ln\left(\frac{3}{2\pi Nb^2}\right) + k \ln\left[\int \Omega\left(N, \vec{R}\right) d\vec{R}\right]. \tag{2.91}$$

The last two terms of Eq. (2.91) depend only on the number of monomers N, but not on the end-to-end vector $\vec{R}$ and can be denoted by $S(N, 0)$:

$$S\left(N, \vec{R}\right) = -\frac{3}{2}k\frac{\vec{R}^2}{Nb^2} + S(N, 0). \tag{2.92}$$

The **Helmholtz free energy** of the chain F is the energy U minus the product of absolute temperature T and entropy S:

$$F\left(N, \vec{R}\right) = U\left(N, \vec{R}\right) - TS\left(N, \vec{R}\right). \tag{2.93}$$

The energy of an ideal chain $U(N, \vec{R})$ is independent of the end-to-end vector $\vec{R}$, since the monomers of the ideal chain have no interaction energy.[3] The free energy can be written as

$$F\left(N, \vec{R}\right) = \frac{3}{2}kT\frac{\vec{R}^2}{Nb^2} + F(N, 0), \tag{2.94}$$

where $F(N, 0) = U(N, 0) - TS(N, 0)$ is the free energy of the chain with both ends at the same point. As was demonstrated above, the largest number of chain conformations correspond to zero end-to-end vector. The number of conformations decreases with increasing end-to-end vector, leading to the decrease of polymer entropy and increase of its free energy. The free energy of an ideal chain $F(N, \vec{R})$ increases quadratically with the magnitude of the end-to-end vector $\vec{R}$. This implies that the entropic elasticity of an ideal

---

[3] The ideal chain never has long-range interactions, but short-range interactions are possible, and their consequences are discussed in problem 7.19.

chain satisfies Hooke's law. To hold the chain at a fixed end-to-end vector $\vec{R}$, would require equal and opposite forces acting on the chain ends that are proportional to $\vec{R}$. For example, to separate the chain ends by distance R_x in x direction, requires force f_x:

$$f_x = \frac{\partial F\left(N, \vec{R}\right)}{\partial R_x} = \frac{3kT}{Nb^2} R_x. \tag{2.95}$$

The force to hold chain ends separated by a general vector $\vec{R}$ is linear in $\vec{R}$, like a simple elastic spring:

$$\vec{f} = \frac{3kT}{Nb^2} \vec{R}. \tag{2.96}$$

The coefficient of proportionality $3kT/(Nb^2)$ is the **entropic spring constant** of an ideal chain. It is easier to stretch polymers with larger numbers of monomers N, larger monomer size b, and at *lower* temperature T. The fact that the spring constant is proportional to temperature is a signature of entropic elasticity. The entropic nature of elasticity in polymers distinguishes them from other materials. Metals and ceramics become softer as temperature is raised because their deformation requires displacing atoms from their preferred positions (energetic instead of entropic elasticity).

The force increases as the chain is stretched because there are fewer possible conformations for larger end-to-end distances. The linear entropic spring result for the stretching of an ideal chain [Eq. (2.96)] is extremely important for our subsequent discussions of rubber elasticity and polymer dynamics. This linear dependence [Hooke's law for an ideal chain, Eq. (2.96)] is due to the Gaussian approximation, valid only for $|\vec{R}| \ll R_{\max} = Nb$. If the chain is stretched to the point where its end-to-end vector approaches the maximum chain extension $|\vec{R}| \leq R_{\max}$, the dependence becomes strongly non-linear, with the force diverging at $|\vec{R}| = R_{\max}$, as will be discussed in Section 2.6.2.

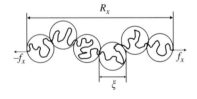

Fig. 2.13
An elongated chain is only stretched on its largest length scales. Inside the tension blob, the conformation of the chain is essentially unperturbed by the stretch.

2.6.1 Scaling argument for chain stretching

The linear relation between force and end-to-end distance can also be obtained by a very simple scaling argument. The key to understanding the scaling description is to recognize that most of the conformational entropy of the chain arises from local conformational freedom on the smallest length scales. For this reason, the random walks that happen to have end-to-end distance $R > bN^{1/2}$ can be visualized as a sequential array of smaller sections of size ξ that are essentially unperturbed by the stretch, as shown in Fig. 2.13.

The stretched polymer is subdivided into sections of g monomers each. We assume that these sections are almost undeformed so that the mean-square projection of the end-to-end vector of these sections of g monomers onto any of the coordinate axes obeys ideal chain statistics [Eq. (2.83)]:

$$\xi^2 \approx b^2 g. \tag{2.97}$$

There are N/g such sections and in the direction of elongation they are assumed to be arranged sequentially:

$$R_x \approx \xi \frac{N}{g} \approx \frac{Nb^2}{\xi}. \qquad (2.98)$$

This can be solved for the size ξ of the unperturbed sections and the number of monomers g in each section:

$$\xi \approx \frac{Nb^2}{R_x}, \qquad (2.99)$$

$$g \approx \frac{N^2 b^2}{R_x^2}. \qquad (2.100)$$

The number of monomers g and the size ξ of these sections were specially chosen so that the polymer conformation changes from that of a random walk on smaller size scales to that of an elongated chain on larger length scales. Such sections of stretched polymers are called **tension blobs**. Being extended on only its largest length scales allows the chain to maximize its conformational entropy.

The physical meaning of a tension blob is the length scale ξ at which external tension changes the chain conformation from almost undeformed on length scales smaller than ξ to extended on length scales larger than ξ. The trajectory of the stretched chain (Fig. 2.13) shows that each tension blob is forced to go in a particular direction along the x axis (rather than in a random direction as in an unperturbed chain). Therefore one degree of freedom is restricted per tension blob and the free energy of the chain increases by kT per blob:[4]

$$F \approx kT \frac{N}{g} \approx kT \frac{R_x^2}{Nb^2}. \qquad (2.101)$$

In comparing Eqs (2.94) and (2.101), we see that the scaling method gets the correct result within a prefactor of order unity. This is the character of all scaling calculations: they provide a simple means to extract the essential physics but do not properly determine numerical coefficients.

Equation (2.101) is the first of many instances where the free energy stored in the chain is of the order of kT per blob, because the blobs generally describe a length scale at which the conformation of the chain changes and is the elementary unit of deformation. In the case of stretching, the free energy is F/N per monomer. On length scales smaller than the tension blob, the thermal energy kT that randomizes the conformation is larger than the cumulative stretching energy, and the conformation is essentially unperturbed. On length scales larger than the tension blob, the cumulative stretching energy is larger than kT, and the ideal chain gets strongly stretched (see Fig. 2.13). Similar arguments apply to other problems involving conformational changes beyond a particular length scale, making the free energy of order kT per blob quite general.

---

[4] This is the consequence of the equipartition theorem.

Ideal chains

The force needed to stretch the chain is given by the derivative of the free energy:

$$f_x = \frac{\partial F}{\partial R_x} \approx kT \frac{R_x}{Nb^2} \approx \frac{kT}{\xi}. \qquad (2.102)$$

The tension blobs provide a simple framework for visualizing the chain stretching (Fig. 2.13) and provide simple relations for calculating the stretching force and free energy. They define the length scale at which elastic energy is of order kT. Since the force has dimensions of energy divided by length, Eq. (2.102) immediately follows from a dimensional analysis with length scale of tension blob ξ corresponding to kT of stored elastic energy.

The stretching along the x axis, shown in Fig. 2.13, makes the stretched conformation of an ideal chain a **directed random walk** of tension blobs. This conformation is sequential in the x direction, but the y and z directions have the usual random walk statistics that are unaffected by the stretching. The mean-square components of the end-to-end vector orthogonal to the stretching direction are obtained from one-dimensional random walks of N/g sections of step length ξ:

$$\langle R_y^2 \rangle = \langle R_z^2 \rangle \approx \xi^2 \frac{N}{g} \approx Nb^2. \qquad (2.103)$$

The linear relation between force f_x and end-to-end distance R_x (Hooke's law) is valid as long as there are many Kuhn monomers in each tension blob. As the end-to-end distance R_x approaches a significant fraction of its maximal value R_{max}, a deviation from Hooke's law is expected. Note that the Gaussian approximation assumes $R_x \ll R_{max}$ and always leads to Hooke's Law. Below we derive the non-linear relation between force and elongation for strongly stretched chains. The limit of the linear regime corresponds to a force of the order of

$$\frac{kT}{b} \cong \frac{1.38 \times 10^{-23}\,\mathrm{J\,K^{-1}} \times 295\,\mathrm{K}}{1 \times 10^{-9}\,\mathrm{m}} \cong 4 \times 10^{-12}\,\mathrm{N} \qquad (2.104)$$

for a chain with Kuhn length $b = 1$ nm at room temperature. Stiffer chains with larger Kuhn length get nearly fully stretched at weaker extension forces. For double-helical DNA with Kuhn length $b \cong 100$ nm (persistence length $l_p \cong 50$ nm) the force corresponding to the linear response limit is 100 times smaller (4×10^{-14} N).

2.6.2 Langevin dependence of elongation on force

Consider a freely jointed chain of N bonds subject to a constant elongational force f applied to its ends along the z axis. An example could be a chain with two opposite charges $+q$ and $-q$ at its ends in a constant electric field $\vec{E}$ applied along the z axis as sketched in Fig. 2.14. If the direct

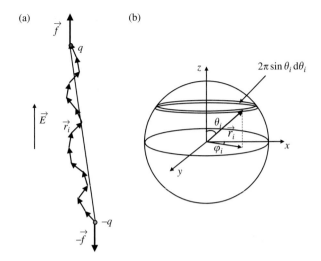

Fig. 2.14
(a) Freely jointed chain elongated by a pair of forces applied to its ends. (b) Spherical coordinate system for orientation of a bond.

Coulomb interaction between the charges is ignored, there is a constant force $\vec{f} = q\vec{E}$ acting along the z axis on the positive charge and an opposite force $-\vec{f}$ acting on the negative charge. Different chain conformations are no longer equally likely, because they correspond to different energy of the chain in the external electric field. The energy of the chain is proportional to the projection of the end-to-end vector on the direction of the field:

$$U = -q\vec{E} \cdot \vec{R} = -\vec{f} \cdot \vec{R} = -fR_z. \tag{2.105}$$

This energy is equal to the work done by the chain upon separation of the charges by vector $\vec{R}$ in an external electric field $\vec{E}$. The direction of the end-to-end vector $\vec{R}$ of the chain is chosen from the negative to the positive charge at its ends. Displacement of the positive charge down the field with respect to the negative charge, lowers the electrostatic energy of the chain and corresponds to a more favorable conformation. Thus, different chain conformations have different statistical Boltzmann factors $\exp(-U/kT)$ that depend on their corresponding energy U [Eq. (2.105)].

The sum of the Boltzmann factors over all possible conformations of the chain is called the **partition function**:

$$Z = \sum_{\text{states}} \exp\left(-\frac{U}{kT}\right) = \sum_{\text{states}} \exp\left(\frac{fR_z}{kT}\right). \tag{2.106}$$

The partition function is useful because we will calculate the free energy from it in Eq. (2.111). States with higher energy make a smaller contribution to the partition function because their Boltzmann factor determines that those states are less likely.

Different conformations in the freely jointed chain model correspond to different sets of orientations of bond vectors $\vec{r}_i$ in space [see Fig. 2.14(a)]. The orientation of each bond vector $\vec{r}_i$ can be defined by the two angles of the spherical coordinate system θ_i and φ_i [Fig. 2.14(b)]. Therefore, the sum over all possible conformations of a freely jointed chain corresponds to the integral over all possible orientations of all bond vectors of the chain:

$$Z = \sum_{\text{states}} \exp\left(\frac{fR_z}{kT}\right) = \int \exp\left(\frac{fR_z}{kT}\right) \prod_{i=1}^{N} \sin\theta_i \, d\theta_i \, d\varphi_i. \quad (2.107)$$

The notation $\prod_{i=1}^{N}$ denotes the product of N terms. The z component of the end-to-end vector can be represented as the sum of the projections of all bond vectors onto the z axis:

$$R_z = \sum_{i=1}^{N} b \cos\theta_i. \quad (2.108)$$

Therefore, the partition function [Eq. (2.107)] becomes a product of N identical integrals:

$$Z(T,f,N) = \int \exp\left(\frac{fb}{kT}\sum_{i=1}^{N}\cos\theta_i\right) \prod_{i=1}^{N} \sin\theta_i \, d\theta_i \, d\varphi_i$$

$$= \left[\int_0^\pi 2\pi \sin\theta_i \exp\left(\frac{fb}{kT}\cos\theta_i\right) d\theta_i\right]^N$$

$$= \left[\frac{2\pi}{fb/(kT)}\left[\exp\left(\frac{fb}{kT}\right) - \exp\left(-\frac{fb}{kT}\right)\right]\right]^N \quad (2.109)$$

$$= \left[\frac{4\pi \sinh(fb/(kT))}{fb/(kT)}\right]^N. \quad (2.110)$$

The Gibbs free energy G can be directly calculated from the partition function:[5]

$$G(T,f,N) = -kT \ln Z(T,f,N)$$

$$= -kTN\left[\ln\left(4\pi \sinh\left(\frac{fb}{kT}\right)\right) - \ln\left(\frac{fb}{kT}\right)\right]. \quad (2.111)$$

The average end-to-end distance corresponding to a given force can be obtained as the derivative of the free energy:

$$\langle R\rangle = -\frac{\partial G}{\partial f} = bN\left[\coth\left(\frac{fb}{kT}\right) - \frac{1}{fb/(kT)}\right]. \quad (2.112)$$

The expression in the square brackets of Eq. (2.112) is called the **Langevin function**:

$$\mathcal{L}(\beta) = \coth(\beta) - \frac{1}{\beta}. \quad (2.113)$$

The Langevin function relates average chain elongation $\langle R\rangle/R_{\max}$ and normalized extensional force $\beta = fb/(kT)$ for a freely jointed chain, as sketched in Fig. 2.15.

<hr>

[5] The Gibbs free energy is used here because the ensemble of chains corresponds to constant force f, not constant end-to-end distance R (analogous to the isothermal–isobaric ensemble, which has constant pressure instead of constant volume).

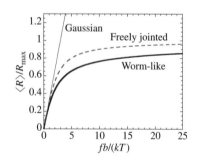

Fig. 2.15

Average end-to-end distance as a function of stretching force for a Gaussian chain [Eq. (2.95), thin line], a freely jointed chain [Langevin function, Eq. (2.112), dashed line], and a worm-like chain [Eq. (2.119), thick line].

For small relative elongations ($\langle R \rangle \ll R_{max} = bN$) the dependence is approximately linear,

$$\mathcal{L}(\beta) \cong \frac{\beta}{3} \quad \text{for } \beta \ll 1, \tag{2.114}$$

and follows Hooke's law derived above [Eq. (2.96)] $\langle R \rangle / (bN) = fb/(3kT)$. For larger relative elongations, the Langevin function significantly deviates from linear dependence and saturates at unity (see Fig. 2.15). For large extensional force $f \gg kT/b$, the Langevin function has another simple limit:

$$\mathcal{L}(\beta) \cong 1 - \frac{1}{\beta} \quad \text{for } \beta \gg 1. \tag{2.115}$$

This means that the extension for strong stretching has a simple form

$$\frac{\langle R \rangle}{R_{max}} \cong 1 - \frac{kT}{fb},$$

where $R_{max} = Nb$. The extensional force of the equivalent freely jointed chain diverges reciprocally proportional to $R_{max} - \langle R \rangle$:

$$\frac{fb}{kT} \cong \frac{R_{max}}{R_{max} - \langle R \rangle} \quad \text{for } 1 - \frac{\langle R \rangle}{R_{max}} \ll 1. \tag{2.116}$$

In the case of the worm-like chain model (Section 2.3.2), the extensional force diverges reciprocally proportional to the square of $R_{max} - \langle R \rangle$:

$$\frac{fb}{kT} \cong \frac{1}{2} \left(\frac{R_{max}}{R_{max} - \langle R \rangle} \right)^2 \quad \text{for } 1 - \frac{\langle R \rangle}{R_{max}} \ll 1. \tag{2.117}$$

The differences between divergences of force near maximum extension [Eqs (2.116) and (2.117)] are due to bending modes on length scales shorter than Kuhn length b. These modes do not exist in freely jointed chains because sections of length b are assumed to be absolutely rigid. In the worm-like chain model these bending modes with wavelength $\xi \ll b$ lead to much stronger divergence of the force [Eq. (2.117)].

At small relative extensions ($\langle R \rangle \ll R_{max}$) worm-like chains behave as Hookean springs:

$$\frac{fb}{kT} \cong \frac{3\langle R \rangle}{R_{max}} \quad \text{for } \langle R \rangle \ll R_{max}. \tag{2.118}$$

There is no simple analytical solution for the worm-like chain model at all extensions, but there is an approximate expression valid both for small and for large relative extensions:[6]

$$\frac{fb}{kT} \cong \frac{2\langle R \rangle}{R_{max}} + \frac{1}{2} \left(\frac{R_{max}}{R_{max} - \langle R \rangle} \right)^2 - \frac{1}{2}. \tag{2.119}$$

[6] J. F. Marko and E. D. Siggia, *Macromolecules* **28**, 8759 (1995).

Strong deviations from linear elasticity have been measured in polymer networks at large elongation (see Chapter 7). Optical tweezers and atomic force microscopy have been used to measure the dependence of the force applied to the ends of isolated chains on their elongation. In the optical tweezer experiments, beads were attached to the ends of long DNA segments. DNA is a biopolymer that exists as a double-stranded helix. Such stiff chains are best described by the worm-like chain model. The chain length of DNA is typically described in terms of the number of base pairs along the helix. The positions of the beads at the ends of DNA chains were manipulated by a focused laser beam (hence the name 'optical tweezers'). The force exerted on the chain ends was measured by the calibrated relative displacement of the beads with respect to the optical traps. In another type of nano-manipulation experiment a 97 kilobase λ-DNA dimer was chemically attached by one of its ends to a glass slide and by the other end to a small (3 μm) magnetic bead. The DNA was stretched by applying a known magnetic and hydrodynamic force to the bead. The stretching was measured by observing the position of the bead in an optical microscope. The extension of DNA as a function of applied force is compared with predictions of freely jointed and worm-like chain models in Fig. 2.16. The worm-like chain model is in excellent agreement with the experimental data.

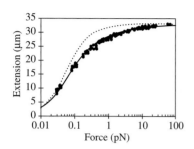

Fig. 2.16

Comparison of experimental force for 97 kilobase λ-DNA dimers with the worm-like chain model [solid curve is Eq. (2.119) with $R_{max} = 33$ μm and $b = 100$ nm]. The dotted curve corresponds to the Langevin function of the freely jointed chain model [Eq. (2.112)]. Data are from R. H. Austin *et al.*, *Phys. Today*, Feb. 1997, p. 32.

2.7 Pair correlations of an ideal chain

Consider a monomer of an ideal polymer trying to reach fellow monomers of the same chain via a CB radio (see Fig. 2.17). The number of monomers it can call depends on the range r of its transmitter. It can contact any monomer within the sphere of radius r of itself. The number of monomers m that can be reached via a CB radio with range r is given by random walk statistics:

$$m \approx \left(\frac{r}{b}\right)^2.$$ (2.120)

Fig. 2.17

A monomer can only reach other monomers with its CB radio if they are within the range of the radio.

The number density of these monomers within the volume r^3 is m/r^3. The probability of finding a monomer in a unit volume at a distance r from a given monomer is called the **pair correlation function** $g(r)$. It can be approximated by the average number density within the volume r^3:

$$g(r) \approx \frac{m}{r^3} \approx \frac{1}{rb^2}.$$ (2.121)

The exact calculation of the pair correlation function of an ideal chain leads to an additional factor $3/\pi$:

$$g(r) = \frac{3}{\pi r b^2}.$$ (2.122)

Note that the pair correlation function decreases with increasing distance r. It is less likely to find a monomer belonging to the chain further away from a given monomer because the average number density of monomers within the sphere of radius r decreases. Large polymer coils are almost empty.

To illustrate this concept, divide the cube of size R, containing an ideal polymer with N monomers of size b and end-to-end distance $R = bN^{1/2}$, into smaller cubes of size r (Fig. 2.18). There will be $(R/r)^3$ such smaller cubes. But only $(R/r)^2$ of these smaller cubes contain monomers of the chain. The average number of monomers in each of these occupied smaller cubes is $m \approx (r/b)^2$. The remaining $(R/r)^3 - (R/r)^2$ smaller cubes are empty. The local density of monomers strongly fluctuates from cell to cell. There are holes of all sizes inside a polymer chain.

This description is a manifestation of the self-similarity (fractal nature) of polymers, discussed in Section 1.4. The fractal nature of ideal chains leads to the power law dependence of the pair correlation function $g(r)$ on distance r. This treatment for the ideal chain can be easily generalized to a linear chain with any fractal dimension $\mathcal{D}$. The number of monomers within range r is $m \sim r^{\mathcal{D}}$. We use the proportionality sign '$\sim$' if the dimensional coefficient (in the particular case above $\approx 1/b^{\mathcal{D}}$) is dropped from the relation. The pair correlation function is still proportional to the ratio of m and r^3:

$$g(r) \approx \frac{m}{r^3} \sim r^{\mathcal{D}-3}. \qquad (2.123)$$

Hence, Eq. (2.121) is a special case of this result, with $\mathcal{D} = 2$ for an ideal chain. The fractal dimension of a rod polymer is $\mathcal{D} = 1$ and the pair correlation function is $g(r) \sim r^{-2}$.

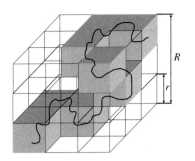

Fig. 2.18
Fractal nature of an ideal chain.

2.8   Measurement of size by scattering

Polymer conformations are studied by various scattering experiments (light, small-angle X-ray and neutron scattering). These techniques are based on the contrast between the polymer and the surrounding media (solvent in the case of polymer solutions and other polymers in the case of polymer melts or blends). The contrast in light scattering arises from differences in refractive index between polymer and solvent, and the scattered intensity is proportional to the square of the refractive index increment dn/dc [see Eq. (1.86)].

While neutron sources are not available in most laboratories, small-angle neutron scattering (SANS) has become a routine characterization method for polymer research using large-scale national and multinational facilities. To obtain the contrast needed for neutron scattering, some of the chains in a polymer melt have their hydrogen atoms replaced by deuterium. In a polymer solution, the solvent is often deuterated. This deuterium labelling appears to not alter the conformations of polymers.

2.8.1   Scattering wavevector

Consider an incident laser beam with wavelength λ illuminating a polymer sample, represented by the large circle in Fig. 2.19, along the direction with

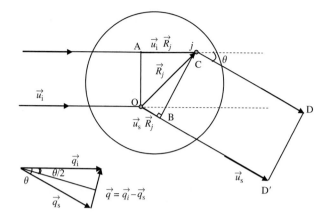

Fig. 2.19
Radiation scattered through angle θ from two distinct parts of the sample.

unit vector $\vec{u}_i$. This incident beam can be characterized by the **incident wavevector**:

$$\vec{q}_i \equiv \frac{2\pi n}{\lambda} \vec{u}_i, \tag{2.124}$$

where n is the refractive index of the solution. The incident light is scattered through angle θ and leaves the sample along the direction with unit vector $\vec{u}_s$. The scattered beam is characterized by the **scattered wavevector**:

$$\vec{q}_s \equiv \frac{2\pi n}{\lambda} \vec{u}_s. \tag{2.125}$$

The incident beam is coherent, meaning that all photons are in-phase. When the incident beam enters the sample, monomers absorb the radiation and re-emit it in all directions. The difference in optical paths between the light scattered by different monomers makes the scattered beam incoherent, meaning that the scattered photons are no longer in-phase.[7] In the example sketched in Fig. 2.19, the difference in optical paths of the radiation scattered by the monomer j at position $\vec{R}_j$ (at point C) and by the monomer at the origin O is easily calculated:

$$AC + CD - OD' = AC - OB. \tag{2.126}$$

The section AC is the projection of the vector $\vec{R}_j$ onto the incident direction and has length $\vec{u}_i \cdot \vec{R}_j$. The section OB is the projection of the vector $\vec{R}_j$ onto the scattered direction and has length $\vec{u}_s \cdot \vec{R}_j$. Thus, the difference in the optical paths can be written in terms of these vectors:

$$AC - OB = \vec{u}_i \cdot \vec{R}_j - \vec{u}_s \cdot \vec{R}_j = (\vec{u}_i - \vec{u}_s) \cdot \vec{R}_j. \tag{2.127}$$

This difference in optical paths results in the phase difference φ_j, which is $2\pi n/\lambda$ times the optical path difference [see Eq. (1.77) with λ replaced by the wavelength in the dielectric medium λ/n].

$$\varphi_j = \frac{2\pi n}{\lambda} (\vec{u}_i - \vec{u}_s) \cdot \vec{R}_j = (\vec{q}_i - \vec{q}_s) \cdot \vec{R}_j = \vec{q} \cdot \vec{R}_j. \tag{2.128}$$

---

[7] It is assumed that there is no multiple scattering, although this is not always a valid assumption.

The **scattering wavevector** $\vec{q}$ is defined as the difference of the incident and scattered wavevectors:

$$\vec{q} \equiv \vec{q}_i - \vec{q}_s. \tag{2.129}$$

From their definitions in Eqs (2.124) and (2.125), the magnitudes of the incident and scattered wavevectors are the same:

$$|\vec{q}_i| = |\vec{q}_s| = \frac{2\pi n}{\lambda}. \tag{2.130}$$

The isosceles triangle of wavevectors in Fig. 2.19 shows that half of the magnitude of the scattering wavevector is equal to the magnitude of wavevectors $\vec{q}_i$ or $\vec{q}_s$ times the sinc of half the angle θ between them:

$$q \equiv |\vec{q}| = 2|\vec{q}_i| \sin\left(\frac{\theta}{2}\right) = \frac{4\pi n}{\lambda} \sin\left(\frac{\theta}{2}\right). \tag{2.131}$$

2.8.2 Form factor

We concentrate here on light scattering, but similar results are valid for small-angle X-ray and neutron scattering. We describe scattering from a single molecule, assuming that the solution is dilute, which is the relevant regime for determining the size and shape of individual coils.

The electric field of light scattered by the jth segment is

$$E_j = E_i A \cos(2\pi\nu t - \varphi_j), \tag{2.132}$$

where φ_j is the phase difference [(Eq. (2.128)], ν is the frequency, E_i is the amplitude of the incident electric field [Eq. (1.77)] and the coefficient A contains the factors such as polarizability α, the distance r to the detector, the wavelength of light λ, etc. [see Eq. (1.81)]. Summing over the N monomers gives the electric field scattered by an isolated polymer coil:

$$E_s = E_i \sum_{j=1}^{N} A \cos(2\pi\nu t - \varphi_j). \tag{2.133}$$

The intensity of scattered light is the energy of radiation that falls onto a unit area per unit time. It is proportional to the square of the electric field averaged over one oscillation period $1/\nu$:

$$
\begin{aligned}
I_s &= 2I_i A^2 \nu \int_0^{1/\nu} \left[\sum_{j=1}^{N} \cos(2\pi\nu t - \varphi_j) \right]^2 dt \\
&= 2I_i A^2 \nu \int_0^{1/\nu} \left[\sum_{j=1}^{N} \sum_{k=1}^{N} \cos(2\pi\nu t - \varphi_j) \cos(2\pi\nu t - \varphi_k) \right] dt \\
&= I_i A^2 \nu \int_0^{1/\nu} \left[\sum_{j=1}^{N} \sum_{k=1}^{N} (\cos(4\pi\nu t - \varphi_j - \varphi_k) + \cos(\varphi_k - \varphi_j)) \right] dt.
\end{aligned}
$$

$$\tag{2.134}$$

The final result used the equation for the product of cosines:

$$\cos\alpha\cos\beta = \frac{\cos(\alpha+\beta)+\cos(\alpha-\beta)}{2}. \qquad (2.135)$$

The first term in Eq. (2.134) oscillates exactly two full periods (4π) and its integral over the time interval $0 \leq t \leq 1/\nu$ is thus equal to zero. The second term (cosine of the phase difference) is time independent and determines the intensity of light scattered by the molecule

$$I_s = I_i A^2 \sum_{k=1}^{N}\sum_{j=1}^{N}\cos(\varphi_k-\varphi_j), \qquad (2.136)$$

where the phases φ_j are determined by the positions $\vec{R}_j$ of the corresponding monomers and the scattering wavevector $\vec{q}$ [Eq. (2.128)].

The dependence of the scattered intensity on the size and the shape of the polymer is usually described by the **form factor** defined as the ratio of intensity scattered at angle θ (scattering wavevector $\vec{q}$) to that extrapolated to zero angle ($\theta \to 0$) and therefore, zero scattering wavevector ($|\vec{q}| \to 0$):

$$P(\vec{q}) \equiv \frac{I_s(\vec{q})}{I_s(0)}. \qquad (2.137)$$

All optical paths are the same at zero scattering angle ($q=0$) and there is no phase shift ($\varphi_j=0$ for all j) because the scattering wavevector $q=0$ [Eq. (2.131)]. The intensity of light scattered by the molecule at zero angle,

$$I_s(0) = I_i A^2 \sum_{k=1}^{N}\sum_{j=1}^{N}1 = I_i A^2 N^2, \qquad (2.138)$$

leads directly to the form factor, defined by Eq. (2.137):

$$P(\vec{q}) = \frac{1}{N^2}\sum_{i=1}^{N}\sum_{j=1}^{N}\cos(\varphi_i-\varphi_j)$$

$$= \frac{1}{N^2}\sum_{i=1}^{N}\sum_{j=1}^{N}\cos[\vec{q}\cdot(\vec{R}_i-\vec{R}_j)]. \qquad (2.139)$$

It is important to stress that only the relative position of monomers

$$\vec{R}_{ij} \equiv \vec{R}_i - \vec{R}_j \qquad (2.140)$$

enters into the form factor.

The form factor in Eq. (2.139) is defined for a specific orientation of the molecule with respect to the scattering wavevector $\vec{q}$. Often (but not always!), the system is isotropic with equal probabilities of all molecular orientations in space. In this case, Eq. (2.139) can be averaged over all these orientations. This averaging can be carried out in the spherical coordinate system with the z axis along the scattering wavevector $\vec{q}$ and the angle between $\vec{q}$ and $\vec{R}_{ij}$ denoted by α. The polar angle in this spherical coordinate system is denoted by β. The scalar product of wavevector $\vec{q}$ and the relative vector between monomers i and j is

$$\vec{q} \cdot \left(\vec{R}_i - \vec{R}_j\right) = qR_{ij}\cos\alpha, \qquad (2.141)$$

where R_{ij} is the distance between monomers i and j. Averaging the cosine of this scalar product over all orientations of the molecule leads to

$$\langle\cos[\vec{q} \cdot (\vec{R}_i - \vec{R}_j)]\rangle = \frac{1}{4\pi}\int_0^{2\pi}\left[\int_0^\pi \cos(qR_{ij}\cos\alpha)\sin\alpha\,d\alpha\right]d\beta$$

$$= \frac{1}{2}\int_{-1}^{1}\cos(qR_{ij}x)\,dx = \frac{\sin(qR_{ij})}{qR_{ij}}, \qquad (2.142)$$

where the integral was taken by the change of variables $x = \cos\alpha$. Thus, the form factor for any isotropic system is quite simple:

$$P(q) = \frac{1}{N^2}\sum_{i=1}^{N}\sum_{j=1}^{N}\frac{\sin(qR_{ij})}{qR_{ij}}. \qquad (2.143)$$

2.8.3 Measuring R_g^2 by scattering at small angles

One important property of the form factor in dilute solutions is that at low scattering angle ($qR_g < 1$) it becomes independent of any assumption about the shape of the molecule. Using the Taylor series expansion,

$$\frac{\sin x}{x} = 1 - \frac{x^2}{3!} + \frac{x^4}{5!} - \cdots \qquad (2.144)$$

the form factor at low angles can be rewritten, as

$$P(q) = \frac{1}{N^2}\sum_{i=1}^{N}\sum_{j=1}^{N}\left[1 - \frac{(qR_{ij})^2}{3!} + \cdots\right]$$

$$= 1 - \frac{q^2}{6N^2}\sum_{i=1}^{N}\sum_{j=1}^{N}R_{ij}^2 + \cdots \quad \text{for } qR_g < 1, \qquad (2.145)$$

$$P(q) = 1 - \frac{1}{3}q^2\langle R_g^2\rangle + \cdots \quad \text{for } qR_g < 1. \qquad (2.146)$$

In the final relation, Eq. (2.47) was used for R_g^2 and the average $\langle\cdots\rangle$ is over different polymer conformations contributing to scattering [see Eq. (2.49)]. Substituting the relation (2.131) between scattering wavevector q and scattering angle θ provides the low-angle expression of the form factor in light scattering:

$$P(q) = 1 - \frac{16\pi^2 n^2}{3\lambda^2}\langle R_g^2\rangle\sin^2\left(\frac{\theta}{2}\right) + \cdots \qquad (2.147)$$

Recall from Chapter 1 that it is convenient to plot the reciprocal Rayleigh ratio times concentration and optical constant K [Eq. (1.96)] to determine the weight-average molar mass from the zero concentration limit. In Chapter 1, we considered the Rayleigh ratio in the zero wavevector limit, and since the Rayleigh ratio is a normalized intensity [Eq. (1.87)] it has the same q dependence as the form factor [see Eq. (2.137)].

$$\left(\frac{Kc}{R_\theta}\right)_{c\to 0} = \frac{1}{M_w P(q)} = \frac{1}{M_w}\left[1 + \frac{16\pi^2 n^2}{3\lambda^2}\left\langle R_g^2\right\rangle \sin^2\left(\frac{\theta}{2}\right) + \cdots\right]. \quad (2.148)$$

Note that the plus sign in front of the $\langle R_g^2\rangle$ term arises because $(1-x)^{-1} \cong 1+x$ for small values of x. Thus, extrapolation of the ratio Kc/R_θ to zero concentration plotted as a function of $\sin^2(\theta/2)$ allows determination of the radius of gyration of the polymer (or any other scattering object) from the slope for low scattering angles ($qR_g < 1$) and the mass of the object from the y-intercept. For polydisperse samples, this method leads to the weight-average molar mass M_w and the z-average square radius of gyration $\langle R_g^2\rangle_z$. In order to understand this, we write the Rayleigh ratio for a mixture of different species with molar mass M_N, mean-square radius of gyration $\langle R_N^2\rangle$, and concentration c_N:

$$
\begin{aligned}
R_\theta &= K\left[\sum_N c_N M_N - \frac{q^2}{3}\sum_N c_N M_N \langle R_N^2\rangle + \cdots\right]\\
&= K\sum_N c_N \frac{\sum_N c_N M_N}{\sum_N c_N}\left[1 - \frac{q^2}{3}\frac{\sum_N c_N M_N \langle R_N^2\rangle}{\sum_N c_N M_N} + \cdots\right]\\
&= KcM_w\left[1 - \frac{q^2}{3}\langle R_g^2\rangle_z + \cdots\right].
\end{aligned}
\quad (2.149)
$$

The z-average mean-square radius of gyration is defined as

$$\langle R_g^2\rangle_z \equiv \frac{\sum_N c_N M_N \langle R_N^2\rangle}{\sum_N c_N M_N}, \quad (2.150)$$

and the low-concentration limit of the scattering expansion for a polydisperse solution takes a form similar to Eq. (2.148):

$$\left(\frac{Kc}{R_\theta}\right)_{c\to 0} = \frac{1}{M_w}\left[1 + \frac{16\pi^2 n^2}{3\lambda^2}\langle R_g^2\rangle_z \sin^2\left(\frac{\theta}{2}\right) + \cdots\right]. \quad (2.151)$$

An expansion similar to Eq. (2.151) for non-zero concentrations is the basis for the Zimm plot (see Problem 2.47).

The coil size of chains in dilute solution is typically measured by light scattering, using a laser. Visible light has a wavelength of order $\lambda \approx 500$ nm, which is much longer than the wavelength of neutrons ($\lambda \approx 0.3$ nm). This means that much smaller scattering wavevectors q are realized in light scattering than in neutron scattering. The limit $qR_g < 1$, required for the expansion of the form factor in Eq. (2.146), is satisfied for all but the highest molar mass chains in small-angle light scattering. At low scattering angles (for $qR_g < 1$), the form factor can also be approximated by an exponential:

$$P(q) \cong \exp\left(-\frac{q^2 R_g^2}{3}\right) \quad \text{for } qR_g < 1. \tag{2.152}$$

This last result is known as the **Guinier function** and is the basis for determining the radius of gyration from small-angle scattering experiments for objects with unknown form factor.

2.8.4 Debye function

Debye first calculated the form factor for scattering from an ideal chain in 1947. This form factor will be useful in interpretation of a wide variety of scattering experiments on polymers. The form factor for an ideal linear chain is obtained by averaging the form factor of isotropic scatterers [Eq. (2.143)] over the probability distribution for distances R_{ij} between monomers i and j on the ideal chain:

$$P(q) = \frac{1}{N^2} \sum_{i=1}^{N} \sum_{j=1}^{N} \int_0^\infty \frac{\sin(qR_{ij})}{qR_{ij}} P_{3d}(|i-j|, R_{ij}) \, 4\pi R_{ij}^2 \, dR_{ij}. \tag{2.153}$$

The probability distribution function $P_{3d}(|i-j|, R_{ij})$ is given by Eq. (2.86):

$$P_{3d}(|i-j|, R_{ij})4\pi R_{ij}^2 \, dR_{ij} = 4\pi \left(\frac{3}{2\pi|i-j|b^2}\right)^{3/2} \exp\left(-\frac{3R_{ij}^2}{2|i-j|b^2}\right) R_{ij}^2 \, dR_{ij}. \tag{2.154}$$

The integral over R_{ij} can be evaluated by converting it into a Gaussian integral (by writing the complex representation of sine and completing the square in the exponent):

$$\int_0^\infty R_{ij} \sin(qR_{ij}) \exp\left(-\frac{R_{ij}^2}{x}\right) dR_{ij} = \frac{\pi^{1/2}qx^{3/2}}{4} \exp\left(-\frac{q^2 x}{4}\right). \tag{2.155}$$

The variable

$$x = \frac{2|i-j|b^2}{3} \tag{2.156}$$

was defined for convenience and the form factor for an ideal linear chain becomes a double sum of exponentials:

$$P(q) = \frac{1}{N^2} \sum_{i=1}^{N} \sum_{j=1}^{N} \exp\left(-\frac{q^2 b^2 |i-j|}{6}\right). \tag{2.157}$$

Ideal chains

Replacing the summations over monomer indices by integrations provides the integral form of the form factor of an ideal chain:

$$P(q) = \frac{1}{N^2} \int_0^N \left[\int_0^N \exp\left(-\frac{q^2 b^2}{6} |u - v| \right) du \right] dv. \tag{2.158}$$

Next, we change the variables of integration $s \equiv u/N$ and $t \equiv v/N$ and denote the coefficient in the exponent by $Q \equiv q^2 b^2 N/6 = q^2 \langle R_g^2 \rangle$ [we have used Eq. (2.54) for the radius of gyration of an ideal linear chain].

$$
\begin{aligned}
P(q) &= \int_0^1 \left[\int_0^1 \exp(-Q|s - t|) \, ds \right] dt \\
&= \int_0^1 \left[\int_0^t \exp[-Q\,(t - s)] \, ds + \int_t^1 \exp[-Q\,(s - t)] \, ds \right] dt \\
&= \int_0^1 \left[\exp(-Qt) \int_0^t \exp(Qs) \, ds + \exp(Qt) \int_t^1 \exp(-Qs) \, ds \right] dt \\
&= \frac{1}{Q} \int_0^1 \left[\exp(-Qt)\,(\exp(Qt) - 1) - \exp(Qt)\,(\exp(-Q) - \exp(-Qt)) \right] dt \\
&= \frac{1}{Q} \int_0^1 \left[2 - \exp(-Qt) - \exp(-Q)\exp(Qt) \right] dt \\
&= \frac{1}{Q} \left[2 + \frac{\exp(-Q) - 1}{Q} - \exp(-Q) \frac{\exp(Q) - 1}{Q} \right] \\
&= \frac{2}{Q^2} \left[\exp(-Q) - 1 + Q \right]. \tag{2.159}
\end{aligned}
$$

This form factor of an ideal linear polymer is called the **Debye function** and can be rewritten in terms of the product of the square of scattering wavevector q^2 and the mean-square radius of gyration of the chain $\langle R_g^2 \rangle$:

$$P(q) = \frac{2}{(q^2 \langle R_g^2 \rangle)^2} \left[\exp(-q^2 \langle R_g^2 \rangle) - 1 + q^2 \langle R_g^2 \rangle \right]. \tag{2.160}$$

The Debye function is plotted in Fig. 2.20.

In the limit of small scattering angles, where $qR_g < 1$, the exponential can be expanded to simplify the Debye function:

$$
\begin{aligned}
P(q) &\cong \frac{2}{(q^2 \langle R_g^2 \rangle)^2} \\
&\quad \times \left[1 - q^2 \langle R_g^2 \rangle + \frac{(q^2 \langle R_g^2 \rangle)^2}{2} - \frac{(q^2 \langle R_g^2 \rangle)^3}{6} + \cdots - 1 + q^2 \langle R_g^2 \rangle \right] \\
&\cong \left(1 - \frac{q^2 \langle R_g^2 \rangle}{3} + \cdots \right) \quad \text{for } q\sqrt{\langle R_g^2 \rangle} < 1. \tag{2.161}
\end{aligned}
$$

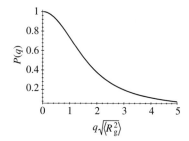

Fig. 2.20
The Debye function is the form factor of an ideal linear chain.

Note that we recover Eq. (2.146) for a general form factor for small values of $q\sqrt{\langle R_{\mathrm{g}}^2\rangle}$.

At large scattering angles, the form factor describes the conformations of smaller sections of the chain on length scales $1/q < \sqrt{\langle R_{\mathrm{g}}^2\rangle}$:

$$P(q) = \frac{2}{\left(q^2\langle R_{\mathrm{g}}^2\rangle\right)^2}\left[\exp\left(-q^2\langle R_{\mathrm{g}}^2\rangle\right) - 1 + q^2\langle R_{\mathrm{g}}^2\rangle\right]$$

$$\cong \frac{2}{q^2\langle R_{\mathrm{g}}^2\rangle} \quad \text{for } q\sqrt{\langle R_{\mathrm{g}}^2\rangle} > 1. \tag{2.162}$$

This power law character of the form factor is related to the power law decay of the pair correlation function of an ideal chain [Eq. (2.121)]. Quite generally, the form factor is related to the Fourier transform of the intramolecular pair correlation function $g(r)$:

$$P(\vec{q}) - \frac{1}{N}\left[1 + \int g(\vec{r})\exp(i\vec{q}\cdot\vec{r})\,\mathrm{d}^3r\right] \tag{2.163}$$

Equation (2.162) is a special case for a form factor of a fractal (with fractal dimension $\mathcal{D} = 2$). For any fractal, the wavevector dependence of the form factor gives a direct measure of the fractal dimension $\mathcal{D}$:

$$P(q) \sim \left(q\sqrt{\langle R_{\mathrm{g}}^2\rangle}\right)^{-\mathcal{D}} \quad \text{for } q\sqrt{\langle R_{\mathrm{g}}^2\rangle} > 1. \tag{2.164}$$

The reciprocal form factor $1/P(q)$ for a ideal linear chain [Eq. (2.160)] is shown in Fig. 2.21 as a function of $q^2\langle R_{\mathrm{g}}^2\rangle$ (medium line) and is compared with the reciprocal form factors for a rigid rod (thin line) and for a solid sphere (thick line). The form factors of a rod,

$$P(q) = \frac{1}{\sqrt{3}qR_{\mathrm{g}}}\int_0^{\sqrt{12}qR_{\mathrm{g}}}\frac{\sin t}{t}\,\mathrm{d}t - \left(\frac{\sin(\sqrt{3}qR_{\mathrm{g}})}{\sqrt{3}qR_{\mathrm{g}}}\right)^2, \tag{2.165}$$

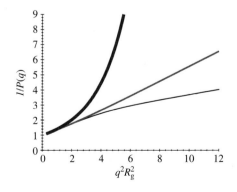

Fig. 2.21
Reciprocal form factors for simple objects: a solid sphere [Eq. (2.166), thick curve], an ideal linear chain [reciprocal Debye function, Eq. (2.160), medium line], and a rigid rod [Eq. (2.165), thin curve].

and a sphere,

$$P(q) = \left(\frac{3}{(\sqrt{5/3}qR_g)^3} [\sin(\sqrt{5/3}qR_g) - (\sqrt{5/3}qR_g)\cos(\sqrt{5/3}qR_g)] \right)^2,$$

(2.166)

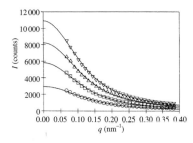

Fig. 2.22
Small-angle neutron scattering data fit to the Debye function multiplied by a zero wavevector scattering. Data are for 0.31% (circles), 0.63% (squares), 0.93% (triangles), and 1.19% (upside down triangles) PMMA with $M_w = 250\,000\,\mathrm{g\,mol^{-1}}$ in a melt of perdeuterated PMMA, from R. Kirste *et al.*, *Polymer* **16**, 120 (1975).

are derived in Problems 2.44 and 2.45. Since all curves are plotted in Fig. 2.21 as functions of $q^2\langle R_g^2\rangle$, the initial slope of all of them is the same and is equal to 1/3 [see Eq. (2.146)].

The Debye function describes the q dependence of scattering data from dilute solutions of ideal chains. Such dilute solutions can either be obtained in a θ-solvent or by having a dilute solution of ordinary chains in a melt of perdeuterated chains. Small-angle neutron scattering data for four dilute concentrations of poly(methyl methacrylate) (PMMA with $M_w = 250\,000\,\mathrm{g/mol^{-1}}$) in perdeuterated PMMA are shown to be fit by the Debye function in Fig. 2.22. Two parameters are used in the fits, $R_g = 13.3$ nm and a multiplicative intensity scale factor.

2.9 Summary of ideal chains

Polymers with no interactions between monomers separated by many bonds along the chain are called ideal chains. Chains are nearly ideal in polymer solutions at a special compensation temperature (the θ-temperature) as well as in polymer melts.

The mean-square end-to-end distance for an ideal chain with n main-chain bonds of length l is $\langle R^2\rangle = C_n nl^2$, where C_n is called Flory's characteristic ratio. For long chains, this characteristic ratio converges to C_∞, leading to a simple expression for the mean-square end-to-end distance of any long ideal linear chain:

$$\langle R^2\rangle \cong C_\infty nl^2.$$

(2.167)

It is convenient to define the Kuhn monomer of length b and the number of Kuhn monomers N such that the mean-square end-to-end distance of an ideal linear chain is a freely jointed chain of Kuhn monomers:

$$\langle R^2\rangle = Nb^2.$$

(2.168)

The mean-square radius of gyration is defined as the averaged square distance from all monomers to the center of mass of the polymer [Eq. (2.44)] and is related to the averaged square distance between all pairs of monomers [Eq. (2.48)]. The mean-square radius of gyration of an ideal linear polymer is one-sixth of its mean-square end-to-end distance:

$$\langle R_g^2\rangle = Nb^2/6.$$

(2.169)

The radius of gyration of ideal branched polymers can be calculated using the Kramers theorem [Eq. (2.65)].

The probability distribution of the end-to-end vector of an ideal chain is well described by the Gaussian function:

$$P_{3d}(N, \vec{R}) = \left(\frac{3}{2\pi Nb^2}\right)^{3/2} \exp\left(-\frac{3\vec{R}^2}{2Nb^2}\right) \quad \text{for } |\vec{R}| \ll R_{max} = Nb.$$

(2.170)

The free energy of an ideal chain is purely entropic and changes quadratically with the end-to-end vector:

$$F = \frac{3}{2}kT\frac{\vec{R}^2}{Nb^2} \quad \text{for } |\vec{R}| \ll Nb.$$

(2.171)

The quadratic form of the free energy implies a linear relationship between force and the end-to-end vector, that is valid for small extensions:

$$\vec{f} = \frac{3kT}{Nb^2}\vec{R} \quad \text{for } |\vec{R}| \ll Nb.$$

(2.172)

Thus, the ideal chain can be thought of as an entropic spring and obeys Hooke's law for elongations much smaller than the maximum elongation ($|\vec{R}| \ll R_{max} = bN$). For stronger deformations, the Langevin function [Eq. (2.112)] for freely jointed chains or Eq. (2.119) for worm-like chains can be used to describe the non-linear relation between force and elongation.

The probability to find a monomer within a distance r of a given monomer is called the pair correlation function $g(r)$. For ideal linear chains, $g(r)$ is reciprocally proportional to the distance r:

$$g(r) = \frac{3}{\pi r b^2}.$$

(2.173)

The radius of gyration of any polymer can be determined from the wavevector q dependence of the scattering intensity at low angles ($qR_g < 1$) in the limit of zero concentration:

$$P(q) = 1 - \frac{1}{3}q^2\langle R_g^2\rangle + \cdots \quad \text{for } q\sqrt{\langle R_g^2\rangle} < 1.$$

(2.174)

Distributions of monomers and correlations between them inside the chain can be determined from the angular dependence of scattering intensity in the range of higher wavevectors $q\sqrt{\langle R_g^2\rangle} > 1$:

$$P(q) \sim \left(q\sqrt{\langle R_g^2\rangle}\right)^{-\mathcal{D}} \quad \text{for } q\sqrt{\langle R_g^2\rangle} > 1,$$

(2.175)

where $\mathcal{D}$ is the fractal dimension of the polymer ($\mathcal{D} = 2$ for ideal chains). The Debye function is the form factor for scattering from an ideal linear chain:

$$P(q) = \frac{2}{(q^2\langle R_g^2\rangle)^2}\left[\exp(-q^2\langle R_g^2\rangle) - 1 + q^2\langle R_g^2\rangle\right].$$

(2.176)

Problems

Section 2.2

2.1   Prove that $\langle \cos \theta_{ij} \rangle = 0$ for the angle θ_{ij} between two bonds i and j if there are no correlations between bond vectors (see Fig. 2.3).

2.2   Calculate the mean-square end-to-end distance of atactic polystyrene with degree of polymerization 100 assuming that it is an ideal chain with characteristic ratio $C_\infty = 9.5$. (Note that the characteristic ratio is defined in terms of the main-chain bonds of length $l = 1.54\,\text{Å}$ rather than monomers.)

2.3   Calculate the root-mean-square end-to-end distance for polyethylene with $M = 10^7\,\text{g mol}^{-1}$ in an ideal conformation with $C_\infty = 7.4$. Compare the end-to-end distance with the contour length of this polymer.

Section 2.3

2.4   Calculate Flory's characteristic ratio C_n for a freely rotating chain consisting of n bonds of length l with bond angle θ. Plot C_n/C_∞ as a function of n for bond angles $\theta = 68°$ and $10°$.

2.5   Calculate the Kuhn monomer length b and number of Kuhn monomers N of a freely rotating chain consisting of n bonds of length l with angle θ.

2.6   Consider a restricted random walk on a square lattice. Let us assume that a walker is not allowed to step back (but can go forward, turn right, or turn left with equal probability). Calculate the mean-square end-to-end distance for such a restricted n-step random walk. What is the characteristic ratio C_∞ for this walk? The lattice constant is equal to l.

2.7   Consider a restricted random walk on a 3D cubic lattice. Let us assume that a walker is not allowed step back (but can go forward, turn up, down, right, or left with equal probability). The lattice constant is equal to l.

(i) Calculate the mean-square end-to-end distance for such a restricted n-step random walk.

(ii) What is the C_∞ for this walk?

Hint: Recall for a freely rotating chain $C_\infty = (1 + \cos \theta)/(1 - \cos \theta)$.

2.8*   Demonstrate that the mean-fourth end-to-end distance

$$\langle R^4 \rangle \equiv \left\langle \left(\vec{R}_n \cdot \vec{R}_n \right) \left(\vec{R}_n \cdot \vec{R}_n \right) \right\rangle$$

$$= \left\langle \left[\left(\sum_{i=1}^{n} \vec{r}_i \right) \cdot \left(\sum_{j=1}^{n} \vec{r}_j \right) \right] \left[\left(\sum_{i'=1}^{n} \vec{r}_{i'} \right) \cdot \left(\sum_{j'=1}^{n} \vec{r}_{j'} \right) \right] \right\rangle$$

$$= \sum_{i'=1}^{n} \sum_{j'=1}^{n} \sum_{i=1}^{n} \sum_{j=1}^{n} \left\langle \left(\vec{r}_i \cdot \vec{r}_j \right) \left(\vec{r}_{i'} \cdot \vec{r}_{j'} \right) \right\rangle \qquad (2.177)$$

of a worm-like chain with contour length $R_{\max}$ and persistence length l_p is

$$\langle R^4 \rangle = \frac{20}{3} R_{\max}^2 l_p^2 - \frac{208}{9} R_{\max} l_p^3 - \frac{8}{27} l_p^4 \left(1 - \exp\left(-\frac{3R_{\max}}{l_p} \right) \right)$$

$$+ 32 l_p^4 \left(1 - \exp\left(-\frac{R_{\max}}{l_p} \right) \right) - 8 R_{\max} l_p^3 \exp\left(-\frac{R_{\max}}{l_p} \right). \qquad (2.178)$$

2.9*   Derive the characteristic ratio of the hindered rotation model [Eq. (2.40)].

2.10   What are the common features of all models for ideal linear chains?

Section 2.4

2.11   The radius of gyration of a polystyrene molecule ($M_w = 3 \times 10^7\,\text{g mol}^{-1}$) was found to be $R_g = 1010\,\text{Å}$. Estimate the overlap concentration c^* in

$g\,cm^{-3}$, assuming that the pervaded volume of the chain is a sphere of radius R_g.

2.12 Consider a polymer containing N Kuhn monomers (of length b) in a dilute solution at the θ-temperature, where ideal chain statistics apply.

Answer questions (i)–(vi) symbolically before substituting numerical values.

(i) What is the mean-square end-to-end distance R_0^2 of the polymer?

(ii) What is its fully extended length R_{max}?

(iii) What is the mean-square radius of gyration R_g^2 of this polymer?

The molar mass of the polymer is M

(iv) Estimate the overlap concentration c^* for this polymer, assuming that the pervaded volume of the chain is a sphere of radius R_g. (*Hint*: It is of the order of the concentration inside the coil.)

(v) How does this overlap concentration depend on the degree of polymerization?

(vi) What is the ratio of its fully extended length to the average (root-mean-square) end-to-end distance R_{max}/R_0?

(vii) Consider an example of a polymer with molar mass $M = 10^4\,g\,mol^{-1}$ consisting of $N = 100$ Kuhn monomers (of length $b = 10\,\text{Å}$) and determine R_0, R_g, R_{max}, c^* and R_{max}/R_0.

2.13 One property of an ideal chain is that its subsections are also ideal. Derive the general relation between the end-to-end distance of the chain R, the end-to-end distance of the section ξ, the number of monomers in the chain N and the number of monomers in the section g.

2.14 The previous problem showed that the equivalent freely jointed chain follows random walk statistics even if the effective monomer is renormalized to be larger than b. What is the smallest effective monomer size for which this renormalization works?

2.15 Calculate the radius of gyration of a rod polymer with N monomers of length b using Eq. (2.51).

2.16 Calculate the radius of gyration of a uniform disc of radius R and negligible thickness.

2.17 Calculate the radius of gyration of a uniform sphere of radius R.

2.18 Calculate the radius of gyration of a uniform right cylinder of radius R and length L.

2.19 Consider a fractal line with fractal dimension $\mathcal{D}$. The mean-square distance between monomers u and v along this line is

$$\langle (\vec{R}(u) - \vec{R}(v))^2 \rangle = b^2(v - u)^{2/\mathcal{D}}. \qquad (2.179)$$

Calculate the mean-square end-to-end distance R^2 and radius of gyration R_g^2 for this fractal line. Determine the ratio R^2/R_g^2 symbolically and then calculate this ratio for fractal dimensions $\mathcal{D} = 1$, 1.7 and 2 for $v - u = 1000$.

2.20 Show that the mean-square radius of gyration of a worm-like chain is

$$\langle R_g^2 \rangle = \frac{1}{3}R_{max}l_p - l_p^2 + \frac{2l_p^3}{R_{max}} - \frac{2l_p^4}{R_{max}^2}\left(1 - \exp\left(-\frac{R_{max}}{l_p}\right)\right). \qquad (2.180)$$

Verify that in the two simple limits (ideal chains $R_{max} \gg l_p$ and rigid rods $R_{max} \ll l_p$) the correct limiting expressions for the radius of gyration are recovered.

2.21 Calculate the radius of gyration of an ideal symmetric f-arm star polymer with N monomers of length b. *Hint*: Each arm of a symmetric star polymer can be treated as an ideal chain of N/f monomers.

2.22 Calculate the radius of gyration of an ideal H-polymer with all five sections containing equal number $(N/5)$ of Kuhn monomers with length b.

2.23
(i) Calculate the radius of gyration of an asymmetric three arm star polymer with a short arm consisting of $n = N/4$ Kuhn monomers of length b and two equal long arms containing $3N/8$ Kuhn monomers each.
(ii) Evaluate R_g of this asymmetric star for $N = 1000$, $n = 250$, and $b = 3$ Å.
(iii) What length of the asymmetric arm n corresponds to the largest and smallest radii of gyration of a star polymer for constant N and b?

2.24   Consider an ideal f-arm star with n_j Kuhn monomers in the jth arm $(j = 1, 2, \ldots, f)$ and with Kuhn length b. The total number of Kuhn monomers in a molecule is $N = \sum_{j=1}^{f} n_j$. Show that the mean-square radius of gyration of this star is

$$\langle R_g^2 \rangle = N b^2 \left(\frac{1}{2N^2} \sum_{j=1}^{f} n_j^2 - \frac{1}{3N^3} \sum_{j=1}^{f} n_j^3 \right). \tag{2.181}$$

2.25*   Consider a tree polymer consisting of f branches (but no loops). Each of these branches contains N/f Kuhn monomers with Kuhn length b. Let v_{ij} be the number of branches along the linear chain connecting branch i and branch j. Demonstrate that the mean-square radius of gyration of this polymer is

$$\langle R_g^2 \rangle = N b^2 \left(\frac{1}{2f} - \frac{1}{3f^2} + \frac{1}{f^3} \sum_{i=1}^{f} \sum_{j=i+1}^{f} v_{ij} \right). \tag{2.182}$$

2.26*   Calculate the radius of gyration of an ideal ring polymer with N Kuhn monomers of length b, and compare it to the radius of gyration of a linear chain with the same number of monomers.

2.27   The radius of gyration of a spherical globule containing a single polymer and some solvent is 450 Å. Calculate the polymer density inside this globule if the molar mass of the polymer is $M = 2.6 \times 10^7 \, \text{g mol}^{-1}$.

Section 2.5

2.28   Derive Stirling's approximation for large N:

$$N! \cong \sqrt{2\pi N} N^N \exp(-N). \tag{2.183}$$

2.29   Demonstrate that the Gaussian probability distribution function of a one-dimensional random walk is normalized to unity:

$$\int_{-\infty}^{\infty} P_{1d}(N, x) \, dx = \frac{1}{\sqrt{2\pi N}} \int_{-\infty}^{\infty} \exp\left(\frac{-x^2}{2N}\right) dx = 1. \tag{2.184}$$

2.30   Show that the mean-square displacement of a one-dimensional random walker is

$$\int_{-\infty}^{\infty} x^2 P_{1d}(N, x) \, dx = \frac{1}{\sqrt{2\pi N}} \int_{-\infty}^{\infty} x^2 \exp\left(-\frac{x^2}{2N}\right) dx = N. \tag{2.185}$$

2.31   Suppose a person walks from the origin in one dimension forward or backward. The probability for a step in each direction is $1/2$. What is the probability of finding the person five steps $(x = 5)$ forward from the origin after $N = 25$ steps.

2.32   Calculate the location of the maximum of the distribution of end-to-end distances (Fig. 2.12) of an ideal chain with N Kuhn monomers of length b.

2.33 Calculate the average end-to-end distance of an ideal linear chain with N Kuhn monomers of length b.

2.34* Demonstrate that the higher moments of the end-to-end vector of an ideal chain with N Kuhn monomers of length b is

$$\langle \vec{R}^{2p} \rangle = \frac{(2p+1)!}{6^p p!} (Nb^2)^p \tag{2.186}$$

within the Gaussian approximation [with probability distribution Eq. (2.85)].

2.35* Show that the mean-square distance of the jth monomer from the centre of mass of an ideal chain with N Kuhn monomers of length b is

$$\langle (\vec{R}_j - \vec{R}_{cm})^2 \rangle = \frac{Nb^2}{3} \left[1 - \frac{3j(N-j)}{N^2} \right] \tag{2.187}$$

within the Gaussian approximation. What are the maximum and minimum values of this mean-square distance?

2.36* The mean-square radius of gyration is the second moment of the distribution of monomers around the centre of mass of the chain [Eq. (2.44)]. The mean-square radius of gyration of an ideal linear chain with N Kuhn monomers of length b is related to its mean-square end-to-end vector [Eq. (2.54)]:

$$\langle R_g^2 \rangle = \frac{1}{N} \sum_{i=1}^{N} \langle (\vec{R}_i - \vec{R}_{cm})^2 \rangle = \frac{1}{6} \langle \vec{R}^2 \rangle. \tag{2.188}$$

(i) Show that higher moments of the distribution of monomers around the centre of mass are related to the corresponding higher moments of the end-to-end vector:

$$\frac{1}{N} \sum_{i=1}^{N} \langle (\vec{R}_i - \vec{R}_{cm})^4 \rangle = \frac{1}{18} (Nb^2)^2 = \frac{1}{30} \langle \vec{R}^4 \rangle, \tag{2.189}$$

$$\frac{1}{N} \sum_{i=1}^{N} \langle (\vec{R}_i - \vec{R}_{cm})^6 \rangle = \frac{29}{972} (Nb^2)^3 = \frac{29}{3780} \langle \vec{R}^6 \rangle. \tag{2.190}$$

(ii) Demonstrate that higher moments of the radius of gyration are

$$\left\langle \left[\frac{1}{N} \sum_{i=1}^{N} (\vec{R}_i - \vec{R}_{cm})^2 \right]^2 \right\rangle = \frac{19}{540} (Nb^2)^2 = \frac{19}{15} \langle R_g^2 \rangle^2, \tag{2.191}$$

$$\left\langle \left[\frac{1}{N} \sum_{i=1}^{N} (\vec{R}_i - \vec{R}_{cm})^2 \right]^3 \right\rangle = \frac{631}{68040} (Nb^2)^3 = \frac{631}{315} \langle R_g^2 \rangle^3. \tag{2.192}$$

2.37* Consider an ideal linear chain with N Kuhn monomers of length b and fixed end-to-end vector $\vec{R}$ directed along the x axis. Demonstrate that the mean-square projection of the radius of gyration onto the direction of its end-to-end vector is

$$\frac{1}{N} \sum_{i=1}^{N} \langle (\vec{R}_i - \vec{R}_{cm})_x^2 \rangle = \frac{1}{36} Nb^2 \left(1 + \frac{3\vec{R}^2}{Nb^2} \right), \tag{2.193}$$

while the mean-square projection of the radius of gyration onto the perpendicular direction is independent of the magnitude of the end-to-end vector

$$\frac{1}{N}\sum_{i=1}^{N}\langle(\vec{R}_i - \vec{R}_{\text{cm}})_y^2\rangle = \frac{1}{N}\sum_{i=1}^{N}\langle(\vec{R}_i - \vec{R}_{\text{cm}})_z^2\rangle = \frac{1}{36}Nb^2. \qquad (2.194)$$

Note that for $|\vec{R}| = 0$ the mean-square radius of gyration of a ring polymer $\langle R_g^2\rangle = Nb^2/12$ is recovered, for $|\vec{R}| = bN^{1/2}$ the mean-square radius of gyration of an ideal linear chain $\langle R_g^2\rangle = Nb^2/6$ is recovered, and for $|\vec{R}| = bN$ the mean-square radius of gyration of a rod $\langle R_g^2\rangle \cong (Nb)^2/12$ is recovered. It is interesting to point out that the asymmetry of the ideal linear chain,

$$\left(1 + \frac{3\vec{R}^2}{Nb^2}\right),$$

is quite large and a typical shape is better represented by an elongated ellipsoid than by a sphere.

2.38* Show that the one-dimensional probability of finding a monomer of an ideal chain whose ends are fixed at positions X_1 and X_2 is the following Gaussian function.

$$P_{1d}(s, x) = \sqrt{\frac{3}{2\pi K_s b^2}} \exp\left(-\frac{3(x - x_s)^2}{2K_s b^2}\right). \qquad (2.195)$$

Written in this way, the position of the monomer that is s monomers from the end of the chain at X_1 is described as though that monomer was the end monomer of a single 'effective chain' of

$$K_s = \frac{1}{1/s + 1/(N - s)} = \frac{s(N - s)}{N} \qquad (2.196)$$

monomers, whose other end is at position

$$x_s = X_1 \frac{N - s}{N} + X_2 \frac{s}{N}. \qquad (2.197)$$

2.39 Consider a linear chain consisting of $m + N$ monomers. The ends of this chain are fixed in space. The x coordinates of the ends are X_1' and X_2', while the junction point fluctuates with x coordinate R'. What is the mean-square x coordinate of the end-to-end vector of the section containing N monomers $\langle(R' - X_2')^2\rangle$?

Section 2.6

2.40 Consider an ideal chain with N Kuhn monomers of length b. The chain is carrying a positive charge $+e$ at one end and a negative charge $-e$ at the other end. What will be its average end-to-end distance R_x in an electric field $E = 10^4$ V cm^{-1} acting along the x axis at room temperature, if $N = 10^4$ and $b = 6$ Å? What is the ratio of this average distance R_x and root-mean-square projection R_{x0} of the end-to-end vector along the x axis in the absence of the field? Ignore the direct Coulomb interaction between the charges.

2.41 Consider an ideal chain with N Kuhn monomers of length b. The chain has two multivalent positive charges $+Ze$ at both of its ends (it is called a telechelic polymer). These two charges repel each other and stretch the polymer.

(i) What is the expression for the average distance R between the chain ends in a polar solvent with dielectric constant ϵ at temperature T (in terms of Z, e, N, b, T, ϵ, etc.), within the Gaussian approximation?

(ii) What is the ratio of this average distance R and the root-mean-square projection of the end-to-end vector along the x axis R_{x0} in the absence of Coulomb interaction for the chain with $N = 100$ Kuhn monomers of length $b = 3\,\text{Å}$ in water (dielectric constant $\epsilon = 80$) at room temperature ($20\,°C$) for charges of valency $Z = 10$?

Hint: The Coulomb force between two charges $+Ze$ separated by distance R in a solution with dielectric constant ϵ is $(Ze)^2/(\epsilon R^2)$.

In order to avoid complicated conversions of units note that a combination of variables, called the Bjerrum length, is $l_B = e^2/(\epsilon kT) \cong 7\,\text{Å}$ in water at room temperature.

2.42* Demonstrate that the probability distribution function of the end-to-end distance R of a freely jointed chain can be expressed in terms of the inverse Langevin function $\mathcal{L}^{-1}(x)$ of the ratio $x = R/R_{max}$ of the end-to-end distance R to its maximum value $R_{max} = Nb$:

$$P(R) = \frac{[\mathcal{L}^{-1}(x)]^2}{(2\pi Nb^2)^{3/2} x\{1 - [\mathcal{L}^{-1}(x)\,\text{csch}\,\mathcal{L}^{-1}(x)]^2\}^{1/2}}$$

$$\times \left\{ \frac{\sinh \mathcal{L}^{-1}(x)}{\mathcal{L}^{-1}(x)\exp[x\mathcal{L}^{-1}(x)]} \right\}^N. \qquad (2.198)$$

Compare this distribution function with the Gaussian approximation [Eq. (2.85)].

2.43 What is the difference between the probability distribution function and the pair correlation function?

Section 2.8

2.44 Calculate the form factor of a uniform sphere of radius R.
2.45 Calculate the form factor of a long thin rod of length L.
2.46 (i) Use the tabulated small-angle neutron scattering data[5] for a 1% solution of $M = 254\,000\,\text{g mol}^{-1}$ deuterium-labelled polystyrene in $M = 110\,000\,\text{g mol}^{-1}$ ordinary polystyrene to determine the radius of gyration by fitting the data to the Debye function [Eq. (2.160)].
(ii) Why is the Guinier limit [Eq. (2.146) or (2.152)] not useful for determining R_g for these data?
(iii) If Eq. (2.146) were used to estimate R_g from the five lowest q data points, is R_g overestimated or underestimated? Why?

$q\ (1/\text{Å})$	0.00980	0.0128	0.0158	0.0188	0.0218	0.0248	0.0278
$I(q)$	4.43	3.37	2.55	2.05	1.58	1.23	0.966
$q\ (1/\text{Å})$	0.0308	0.0338	0.0368	0.0399	0.0429	0.0459	0.0504
$I(q)$	0.804	0.758	0.592	0.509	0.445	0.370	0.275
$q\ (1/\text{Å})$	0.0564	0.0624	0.0684	0.0744	0.0804	0.0864	0.0940
$I(q)$	0.246	0.197	0.175	0.139	0.128	0.107	0.081
$q\ (1/\text{Å})$	0.1060						
$I(q)$	0.077						

[5] Data from M. R. Landry.

2.47   In Chapter 1, we learned how to determine weight-average molar mass M_w and second viral coefficient A_2 from the concentration dependence of light scattered at a very small angle from a polymer solution. In this chapter, we learned that the angular dependence of light scattering gives information about the radius of gyration R_g of the polymer coil. In practice, these analyses are often combined to obtain M_w, A_2, and R_g using a **Zimm plot**. The following equation is the basis of the Zimm plot and was obtained by combining the concentration expansion of Eq. (1.96) with the angular expansion of Eq. (2.147):

$$\frac{Kc}{R_\theta} = \left[\frac{1}{M_w} + 2A_2c + \cdots \right]\left[1 + \frac{16\pi^2 n^2}{3\lambda^2} R_g^2 \sin^2\left(\frac{\theta}{2}\right) + \cdots \right]. \qquad (2.199)$$

Use the following table of data for Kc/R_θ of a polystyrene in benzene at four concentrations and five angles, to construct a Zimm plot by plotting Kc/R_θ against $100c + \sin^2(\theta/2)$ with c in g mL^{-1} and extrapolating to $c \to 0$ and $\theta \to 0$ to determine M_w, A_2, and R_g.

Table of $10^6 Kc/R_\theta$ (in mol g^{-1}) for a polystyrene in benzene:

c (mg mL^{-1})	$\theta = 30°$	$\theta = 45°$	$\theta = 60°$	$\theta = 75°$	$\theta = 90°$
0.5	1.92	1.98	2.16	2.33	2.51
1.0	2.29	2.37	2.53	2.66	2.85
1.5	2.73	2.81	2.94	3.08	3.27
2.0	3.18	3.25	3.45	3.56	3.72

For the laser used, $\lambda = 546\,\text{nm}$ and the refractive index for light of this wavelength travelling though benzene is $n = 1.5014$. After plotting the 20 data points, the data at each angle must be extrapolated to zero concentration, making a $c = 0$ line of five points (corresponding to the five angles) whose slope determines R_g. The data at each concentration must be extrapolated to zero angle, making a $\theta = 0$ line of four points (corresponding to the four concentrations) whose slope determines A_2. Both the $c = 0$ line and the $\theta = 0$ line should have the same intercept, which is the reciprocal of the weight-average molar mass M_w.

2.48   Calculate the radius of gyration and the mean-square end-to-end distance of an ideal linear diblock copolymer consisting of N_1 Kuhn monomers of length b_1 connected at one end to N_2 Kuhn monomers of length b_2.

Bibliography

Doi, M. *Introduction to Polymer Physics* (Clarendon Press, Oxford, 1996).

Flory, P. J. *Statistical Mechanics of Chain Molecules* (Wiley, New York, 1969).

Higgins, J. S. and Benoit, H. C. *Polymers and Neutron Scattering* (Clarendon Press, Oxford, 1994).

Yamakawa, H. *Modern Theory of Polymer Solutions* (Harper and Row, New York, 1971).

Real chains

3

In Chapter 2, we studied the conformations of an ideal chain that ignore interactions between monomers separated by many bonds along the chain. In this chapter we study the effect of these interactions on polymer conformations. To understand why these interactions are often important, we need to estimate the number of monomer–monomer contacts within a single coil. This number depends on the probability for a given monomer to encounter any other monomer that is separated from it by many bonds along the polymer.

A mean-field estimate of this probability can be made for the general case of an ideal chain in d-dimensional space by replacing a chain with an ideal gas of N monomers in the pervaded volume of a coil $\sim R^d$. The probability of a given monomer to contact any other monomer within this mean-field approximation is simply the overlap volume fraction ϕ^*, of a chain inside its pervaded volume, determined as the product of the monomer 'volume' b^d and the number density of monomers in the pervaded volume of the coil N/R^d:

$$\phi^* \approx b^d \frac{N}{R^d}. \tag{3.1}$$

Ideal chains obey Gaussian statistics in any dimension with $R = bN^{1/2}$, leading to the overlap volume fraction:

$$\phi^* \approx b^d \frac{N}{\left(bN^{1/2}\right)^d} \approx N^{1-d/2}. \tag{3.2}$$

The overlap concentration of long ideal coils is very low in spaces with dimension d greater than 2:

$$\phi^* \approx N^{1-d/2} \ll 1 \quad \text{for } d > 2 \text{ and } N \gg 1. \tag{3.3}$$

In particular, in three-dimensional space the probability of a given monomer contacting another monomer on the same chain is $\phi^* \approx N^{-1/2} \ll 1$.

The number of monomer–monomer contacts between pairs of monomers that are far away from each other along the chain, but get close together in space, is the product of the number of monomers in the chain and the volume fraction of chains in the pervaded volume of the coil:

$$N\phi^* \approx N^{2-d/2}. \tag{3.4}$$

In spaces with dimension above 4, this number is small and monomer–monomer contacts are rare. Therefore, linear polymers are always ideal in spaces with dimension $d > 4$. In spaces with dimension less than 4 (in particular, in three-dimensional space relevant to most experiments), the number of monomer–monomer contacts for a long ideal chain is very large:

$$N\phi^* \approx N^{1/2} \gg 1 \quad \text{for } d = 3 \text{ and } N \gg 1. \tag{3.5}$$

It is important to understand how the energy arising from these numerous contacts affects the conformations of a real polymer chain. The effective interaction between a pair of monomers depends on the difference between a monomer's direct interaction with another monomer and with other surrounding molecules. An attractive effective interaction means that the direct monomer–monomer energy is lower and monomers would rather be near each other than in contact with surrounding molecules. In the opposite case of repulsive effective interactions, monomers 'do not like' to be near each other and prefer to be surrounded by other molecules. In the intermediate case, with zero net interaction, monomers 'do not care' whether they are in contact with other monomers or with surrounding molecules. In this case there is no energetic penalty for monomer–monomer contact and the chain conformation is nearly ideal. In the next section, this qualitative description of the monomer–monomer interaction is quantified.

3.1 Excluded volume and self-avoiding walks

3.1.1 Mayer f-function and excluded volume

Consider the energy cost $U(r)$ of bringing two monomers from ∞ to within distance r of each other in a solvent. A typical profile of this function is sketched in Fig. 3.1. It contains a repulsive hard-core barrier that corresponds to the energy cost of steric repulsion of two overlapping monomers. Typical monomers 'like' each other more than they 'like' solvent and therefore there is usually an attractive well corresponding to this energy difference. On the other hand, if monomers are chemically identical to the solvent and there is no energy difference between their interactions, the energy $U(r)$ will contain only the hard-core repulsion (see Fig. 3.2). For reasons that become clear later, in this case the solvent is called **athermal**. On the other hand, if monomers 'like' each other less than surrounding solvent (for example, similarly charged monomers) there is no attractive well in $U(r)$ but instead, extra repulsion.

The probability of finding two monomers separated by a distance r in a solvent at temperature T is proportional to the Boltzmann factor

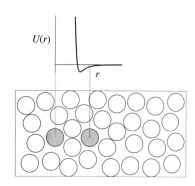

Fig. 3.1
Effective interaction potential between two monomers in a solution of other molecules.

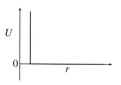

Fig. 3.2
The hard-core potential prevents monomers from overlapping.

Excluded volume and self-avoiding walks

exp $[-U(r)/(kT)]$ which is plotted in Fig. 3.3 for the potential of Fig. 3.1. The relative probability is zero at short distances, corresponding to the **hard-core repulsion** (it is impossible to find two overlapping monomers). The probability is large in the attractive well (it is energetically more favourable and therefore more likely to find the two monomers at these distances). The Boltzmann factor is equal to one at large distances if there are no long-range interactions.

The **Mayer f-function** is defined as the difference between the Boltzmann factor for two monomers at distance r and that for the case of no interaction (or at infinite distance):

$$f(r) = \exp[-U(r)/(kT)] - 1. \qquad (3.6)$$

At short distances, the energy $U(r)$ is large because of the hard-core repulsion, making the Mayer f-function negative. The probability of finding monomers at these distances is significantly reduced relative to the non-interacting case (see Fig. 3.3). The Mayer f-function is positive in the attractive well and the probability of finding a second monomer there is enhanced compared to the non-interacting case.

The **excluded volume** v is defined as minus the integral of the Mayer f-function over the whole space:

$$v = -\int f(r)\,d^3r = \int (1 - \exp[-U(r)/(kT)])\,d^3r. \qquad (3.7)$$

This single parameter summarizes the *net* two-body interaction between monomers. As shown in Fig. 3.4, the hard-core repulsion ($r < 1$) makes a negative contribution to the integration of the Mayer f-function and a positive contribution to excluded volume. The example in Fig. 3.4 also has an effective attraction between monomers ($r > 1$) that makes a positive contribution to the integration of the Mayer f-function and a negative contribution to excluded volume. The attraction and repulsion largely offset each other for this example, making the net excluded volume quite small. A net attraction has a negative excluded volume $v < 0$ and a net repulsion has $v > 0$.

3.1.1.1 Non-spherical monomers

The simple calculation of excluded volume in Eq. (3.7) is only valid for spherical monomers. Particularly because of the 'monomer' being defined as a Kuhn monomer, the monomer is better described as a cylinder of length equal to the Kuhn length b, but smaller diameter d, as depicted in Fig. 3.5. Polymers without bulky side groups, such as polyethylene and poly(ethylene oxide), have effective diameter $d \cong 5\,\text{Å}$. Polystyrene has $d \cong 8\,\text{Å}$, and the diameter of the cylindrical Kuhn monomer steadily increases as side groups increase in size. Most flexible polymers have aspect ratio b/d in the range 2–3, but this ratio is larger for stiffer polymers.

Excluded volume describes the two-body (pairwise) monomer–monomer interaction in solution. At low polymer concentrations, the interaction

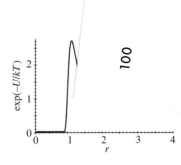

Fig. 3.3
The relative probability of finding a second monomer at distance r from a given monomer is given by the Boltzmann factor.

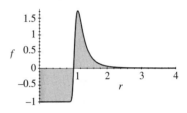

Fig. 3.4
The Mayer f-function and its integration (shaded regions) to determine excluded volume.

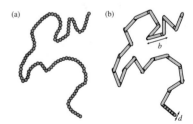

Fig. 3.5
(a) Chain with symmetric monomers.
(b) Chain with strongly asymmetric cylindrical Kuhn segments of length b and diameter d.

part of the free energy density F_{int}/V can be written as a virial expansion in powers of the monomer number density c_n. The coefficient of the c_n^2 term is proportional to the excluded volume v and the coefficient of the c_n^3 term is related to the **three-body interaction coefficient** w:

$$\frac{F_{int}}{V} = \frac{kT}{2}(vc_n^2 + wc_n^3 + \cdots) \approx kT\left(v\frac{N^2}{R^6} + w\frac{N^3}{R^9} + \cdots\right). \tag{3.8}$$

This virial expansion is analogous to that used for the osmotic pressure in Chapter 1 [Eq. (1.74)] and we will see in Section 3.3.4, how the excluded volume is related to the second virial coefficient.

For athermal spherical monomers of diameter d, $v \approx d^3$ and $w \approx d^6$. The interaction energy must not change if we redefine what is meant by a monomer. The chain in Fig. 3.5 can be thought of as a chain of n spheres of diameter d or a chain of $N = nd/b$ cylinders of length b and diameter d. Each term in the virial expansion must be unchanged by these choices, which requires:

$$v_s n^2 = v_c N^2 \quad w_s n^3 = w_c N^3. \tag{3.9}$$

Using the renormalization of $N = nd/b$ and the spherical results $v_s \approx d^3$ and $w_s \approx d^6$, the cylindrical Kuhn monomer has

$$v_c \approx v_s\left(\frac{n}{N}\right)^2 \approx v_s\left(\frac{b}{d}\right)^2 \approx b^2 d, \tag{3.10}$$

$$w_c \approx w_s\left(\frac{n}{N}\right)^3 \approx w_s\left(\frac{b}{d}\right)^3 \approx b^3 d^3 \tag{3.11}$$

as coefficients in Eq. (3.8). The excluded volume of strongly asymmetric objects (long rods) $v_c \approx b^2 d$ is much larger than their occupied volume $v_0 \approx bd^2$, since $b \gg d$. The ratio of excluded volume and occupied volume is the aspect ratio $v_c/v_0 = b/d$. If the aspect ratio of the rod polymer is large enough, excluded volume creates nematic liquid crystalline ordering in solutions of rods, originally derived by Onsager. Once the excluded volumes of neighbouring rods overlap, these rods strongly interact and prefer to orient predominantly parallel to neighbouring rods. Similar nematic ordering is seen with polymers as well if the rigidity of the monomers is large enough (making the aspect ratio b/d large). Further discussion of strongly asymmetric monomers is beyond the scope of this book. Here, we focus on flexible polymers, which typically have aspect ratios in the range 2–3. For this reason, we write results below in terms of cylindrical monomers, which reduce to the results for a spherical monomer when $b \approx d$. The spherical monomer results are often used in the rest of this book, owing to the simplicity of a single length scale to describe the monomer. The excluded volume discussed in this book always refers to the excluded volume of a Kuhn monomer. The transformation rules of

Eq. (3.9) can be used to recast the excluded volume in terms of any part of the chain, including the chemical monomer.

(A) *Athermal solvents.* In the high-temperature limit, the Mayer f-function has a contribution only from hard-core repulsion. The excluded volume becomes independent of temperature at high temperatures, making the solvent athermal. An example is polystyrene in ethyl benzene (essentially polystyrene's repeat unit). The excluded volume in athermal solvent was derived in Eq. (3.10):

$$v \approx b^2 d. \tag{3.12}$$

(B) *Good solvents.* In the athermal limit, the monomer makes no energetic distinction between other monomers and solvent. In a typical solvent, the monomer–monomer attraction is slightly stronger than the monomer–solvent attraction because dispersion forces usually favor identical species. Benzene is an example of a good solvent for polystyrene. The net attraction creates a small attractive well $U(r) < 0$ that leads to a lower excluded volume than the athermal value:

$$0 < v < b^2 d. \tag{3.13}$$

As temperature is lowered, the Mayer f-function increases in the region of the attractive well, reducing the excluded volume.

(C) *Theta solvents.* At some special temperature, called the θ-**temperature**, the contribution to the excluded volume from the attractive well exactly cancels the contribution from the hard-core repulsion, resulting in a net zero excluded volume:

$$v = 0. \tag{3.14}$$

The chains have nearly ideal conformations at the θ-temperature[1] because there is no net penalty for monomer–monomer contact. Polystyrene in cyclohexane at $\theta \cong 34.5\,^\circ\text{C}$ is an example of a polymer–solvent pair at the θ-temperature.

(D) *Poor solvents.* At temperatures below θ, the attractive well dominates the interactions and it is more likely to find monomers close together. In such poor solvents the excluded volume is negative signifying an effective attraction:

$$-b^2 d < v < 0. \tag{3.15}$$

Ethanol is a poor solvent for polystyrene.

(E) *Non-solvents.* The limiting case of the poor solvent is called non-solvent:

$$v \approx -b^2 d. \tag{3.16}$$

In this limit of strong attraction, the polymer's strong preference for its own monomers compared to solvent nearly excludes all solvent from being

---

[1] There are actually logarithmic corrections at the θ-temperature that make the chain conformation not quite ideal.

within the coil. Water is a non-solvent for polystyrene, which is why styrofoam coffee cups are made from polystyrene.

In a typical case of the Mayer f-function with an attractive well, repulsion dominates at higher temperatures and attraction dominates at lower temperatures. In athermal solvents with no attractive well there is no temperature dependence of the excluded volume. It is possible to have monomer–solvent attraction stronger than the monomer–monomer attraction. In this case, there is a soft barrier in addition to the hard-core repulsion and the excluded volume $v > b^2 d$ decreases to the athermal value $v = b^2 d$ at high temperatures.

3.1.2 Flory theory of a polymer in a good solvent

The conformations of a real chain in an athermal or good solvent are determined by the balance of the effective repulsion energy between monomers that tends to swell the chain and the entropy loss due to such deformation. One of the most successful simple models that captures the essence of this balance is the **Flory theory**, which makes rough estimates of both the energetic and the entropic contributions to the free energy.

Consider a polymer with N monomers, swollen to size $R > R_0 = bN^{1/2}$. Flory theory assumes that monomers are uniformly distributed within the volume R^3 with no correlations between them. The probability of a second monomer being within the excluded volume v of a given monomer is the product of excluded volume v and the number density of monomers in the pervaded volume of the chain N/R^3. The energetic cost of being excluded from this volume (the energy of excluded volume interaction) is kT per exclusion or $kTvN/R^3$ per monomer. For all N monomers in the chain, this energy is N times larger [see the first term of Eq. (3.8) with $V \approx R^3$]:

$$F_{int} \approx kTv\frac{N^2}{R^3}. \tag{3.17}$$

The Flory estimate of the entropic contribution to the free energy of a real chain is the energy required to stretch an ideal chain to end-to-end distance R [Eq. (2.101)]:

$$F_{ent} \approx kT\frac{R^2}{Nb^2}. \tag{3.18}$$

The total free energy of a real chain in the Flory approximation is the sum of the energetic interaction and the entropic contributions:

$$F = F_{int} + F_{ent} \approx kT\left(v\frac{N^2}{R^3} + \frac{R^2}{Nb^2}\right). \tag{3.19}$$

The minimum free energy of the chain (obtained by setting $\partial F/\partial R = 0$) gives the optimum size of the real chain in the Flory theory, $R = R_F$:

$$\frac{\partial F}{\partial R} = 0 = kT\left(-3v\frac{N^2}{R_F^4} + 2\frac{R_F}{Nb^2}\right),$$

$$R_F^5 \approx vb^2N^3,$$

$$R_F \approx v^{1/5}b^{2/5}N^{3/5}. \tag{3.20}$$

The size of long real chains is much larger than that of ideal chains with the same number of monomers, as reflected in the swelling ratio:

$$\frac{R_F}{bN^{1/2}} \approx \left(\frac{v}{b^3}N^{1/2}\right)^{1/5} \quad \text{for} \quad \frac{v}{b^3}N^{1/2} > 1. \tag{3.21}$$

If the total interaction energy of a chain in its ideal conformation $F_{int}(R_0)$ [Eq. (3.17) for $R = R_0 = bN^{1/2}$] is less than kT, the chain will not swell. In this case, $N^{1/2}v/b^3 < 1$ and the chain's conformation remains nearly ideal. Excluded volume interactions only swell the chain when the **chain interaction parameter,**

$$z \equiv \left(\frac{3}{2\pi}\right)^{3/2}\frac{v}{b^3}N^{1/2} \approx \frac{F_{int}(R_0)}{kT} \approx v\frac{N^2}{R_0^3} \approx \frac{v}{b^3}N^{1/2}, \tag{3.22}$$

becomes sufficiently large. Equation (3.20) is therefore only valid for chain interaction parameters that are larger than some number of order unity. The precise value of this number is discussed in Section 3.3.4.

The predictions of the Flory theory are in good agreement with both experiments and with more sophisticated theories (renormalization group theory, exact enumerations and computer simulations). However, the success of the Flory theory is due to a fortuitous cancellation of errors. The repulsion energy is overestimated because the correlations between monomers along the chain are omitted. The number of contacts per chain is estimated to be $b^3N^2/R^3 \approx N^{1/5}$. Computer simulations of random walks with excluded volume show that the number of contacts between monomers that are far apart along the chain does not grow with N. Hence, Flory overestimated the interaction energy. The elastic energy is also overestimated in the Flory theory because the ideal chain conformational entropy is assumed. The conformations of real chains are qualitatively different from the ideal chains as will be demonstrated in the remainder of this chapter. Simple modifications of the Flory theory that take into account only some of these effects usually fail. However, Flory theory is useful because it is simple and provides a reasonable answer. We will make calculations in a similar spirit throughout this book. Mean-field estimates of the energetic part of the free energy, ignoring correlations between monomers, are used with entropy estimates based on ideal chain statistics. We will refer to such simple calculations as 'Flory theory' and will hope that the errors will cancel again.

It is important to realize that Flory theory leads to a universal power law dependence of polymer size R on the number of monomers N:

$$R \sim N^\nu \tag{3.23}$$

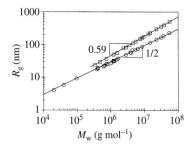

Fig. 3.6
Molar mass dependence of the radius of gyration from light scattering in dilute solutions for polystyrenes in a θ-solvent (cyclohexane at $\theta = 34.5\,°C$, circles) and in a good solvent (benzene at $25\,°C$, squares). Data are compiled in L. J. Fetters, *et al.*, *J. Phys. Chem. Ref. Data*, **23**, 619 (1994).

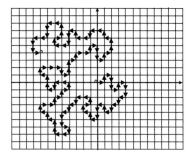

Fig. 3.7
A two-dimensional self-avoiding walk on a square lattice. The direction of each step is randomly chosen from four possible directions (up, down, right or left) with the requirement that previously visited sites cannot be visited again.

The quality of solvent, reflected in the excluded volume v, enters only in the prefactor, but does not change the value of the scaling exponent ν for any v > 0. The Flory approximation of the scaling exponent is $\nu = 3/5$ for a swollen linear polymer. For the ideal linear chain the exponent $\nu = 1/2$. In the language of fractal objects, the fractal dimension of an ideal polymer is $\mathcal{D} = 1/\nu = 2$, while for a swollen chain it is lower $\mathcal{D} = 1/\nu = 5/3$. More sophisticated theories lead to a more accurate estimate of the scaling exponent of the swollen linear chain in three dimensions:

$$\nu \cong 0.588. \tag{3.24}$$

Comparison of Eq. (3.23) with experimental data for polystyrenes in cyclohexane at the θ-temperature (a θ-solvent) and in toluene (a good solvent) is shown in Fig. 3.6. Both data sets obey Eq. (3.23), with $\nu = 1/2$ in θ-solvent and $\nu \cong 0.59$ in good solvent.

While the ideal chain discussed in Chapter 2 has a random walk conformation, the real chain has additional correlations because two monomers cannot occupy the same position in space. The real chain's conformation is similar to that of a **self-avoiding walk**, which is a random walk on a lattice that never visits the same site more than once. An example of a self-avoiding walk is shown in Fig. 3.7, on a two-dimensional square lattice.

3.2   Deforming real and ideal chains

3.2.1   Polymer under tension

In order to emphasize the difference between ideal and real chains, we compare their behaviour under tension. Consider a polymer containing N monomers of size b, under tension in two different solvents: a θ-solvent with nearly ideal chain statistics and an athermal solvent with excluded volume $v \approx b^3$. An ideal chain under tension was already discussed in Section 2.6.1 and is repeated for comparison with that of a swollen chain. The major difference between ideal and real chains is that in the latter there are excluded volume interactions between monomers that are far apart along the chain when they approach each other in space.

The end-to-end distances of the chains in the unperturbed state (with no applied external force) are given by Eqs (2.18) and (3.20) with $v \approx b^3$:

$$R_0 \approx bN^{1/2} \quad \text{ideal,} \tag{3.25}$$

$$R_F \approx bN^{3/5} \quad \text{real.} \tag{3.26}$$

Since both ideal and real chains are self-similar fractals, the same scaling applies to subsections of the chains of size r containing n monomers:

$$r \approx bn^{1/2} \quad \text{ideal,} \tag{3.27}$$

$$r \approx bn^{3/5} \quad \text{real.} \tag{3.28}$$

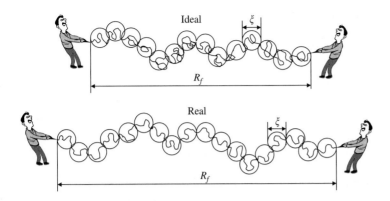

Fig. 3.8
Maxwell demons stretching ideal and real chains of the same contour length with the same force f.

Note that there are fewer monomers within the same distance r in the real chain case compared with the ideal chain because the real chain is swollen.

Let us now employ Maxwell demons to put both chains under tension with force of magnitude f applied at both ends of each chain, stretching them out as sketched in Fig. 3.8. As in Section 2.6.1, we subdivide each chain into tension blobs of size ξ containing g monomers each, such that on length scales smaller than these tension blobs the chain statistics are unperturbed,

$$\xi \approx bg^{1/2} \quad \text{ideal,} \tag{3.29}$$

$$\xi \approx bg^{3/5} \quad \text{real,} \tag{3.30}$$

while on larger length scales both chains are fully extended arrays of tension blobs.

Since each chain is a stretched array of tension blobs, their end-to-end distance R_f in an extended state is the product of the tension blob size ξ and the number of these blobs N/g per chain:

$$R_f \approx \xi \frac{N}{g} \approx \frac{Nb^2}{\xi} \approx \frac{R_0^2}{\xi} \quad \text{ideal,} \tag{3.31}$$

$$R_f \approx \xi \frac{N}{g} \approx \frac{Nb^{5/3}}{\xi^{2/3}} \approx \frac{R_F^{5/3}}{\xi^{2/3}} \quad \text{real.} \tag{3.32}$$

These equations can be solved for the size of the tension blobs in terms of the normal size (R_0 or R_F) and stretched size (R_f) of the chains:

$$\xi \approx \frac{R_0^2}{R_f} \quad \text{ideal,} \tag{3.33}$$

$$\xi \approx \frac{R_F^{5/2}}{R_f^{3/2}} \quad \text{real.} \tag{3.34}$$

Real chains

As discussed in Section 2.6.1, the free energy cost for stretching the chains is of the order kT per tension blob (we are neglecting coefficients of order unity):

$$F(N, R_f) \approx kT\frac{N}{g} \approx kT\frac{R_f}{\xi} \approx kT\left(\frac{R_f}{R_0}\right)^2 \quad \text{ideal,} \qquad (3.35)$$

$$F(N, R_f) \approx kT\frac{N}{g} \approx kT\frac{R_f}{\xi} \approx kT\left(\frac{R_f}{R_F}\right)^{5/2} \quad \text{real.} \qquad (3.36)$$

The force necessary to stretch the chain to end-to-end distance R_f is of the order of the thermal energy kT per tension blob of size ξ:

$$f \approx \frac{kT}{\xi} \approx \frac{kT}{R_0^2}R_f \approx \frac{kT}{R_0}\frac{R_f}{R_0} \quad \text{ideal,} \qquad (3.37)$$

$$f \approx \frac{kT}{\xi} \approx \frac{kT}{R_F^{5/2}}R_f^{3/2} \approx \frac{kT}{R_F}\left(\frac{R_f}{R_F}\right)^{3/2} \quad \text{real.} \qquad (3.38)$$

The same result (up to numerical prefactors of order unity) can be obtained by differentiation of the free energy with respect to end-to-end distance:

$$f = \frac{\partial F(N, R_f)}{\partial R_f}. \qquad (3.39)$$

It is very important to notice the difference between the results for ideal and real chains under tension. Ideal chains satisfy Hooke's law with force f linearly proportional to elongation R_f. For real chains the dependence of force f on chain elongation R_f is non-linear with the exponent equal to $3/2$ for the Flory value of $\nu = 3/5$. This non-linear dependence of force on elongation for real chains was first derived by Pincus and tension blobs are often called **Pincus blobs**. The differences between real and ideal chains can be clearly seen when we consider the dimensionless stretching force:

$$\frac{fb}{kT} \approx \frac{R_f}{Nb} \quad \text{ideal} \quad \text{for } R_f < Nb, \qquad (3.40)$$

$$\frac{fb}{kT} \approx \left(\frac{R_f}{Nb}\right)^{3/2} \quad \text{real} \quad \text{for } R_f < Nb. \qquad (3.41)$$

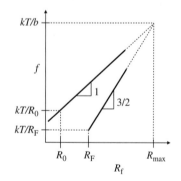

Fig. 3.9
Extensional force f as a function of end-to-end distance R_f on logarithmic scales. Comparison between ideal chains (upper line) and real chains (lower line).

The stretching energy is of order kT per monomer when either chain is nearly fully stretched ($R_f \approx Nb$) resulting in $f \approx kT/b$. The force required to stretch the real chain increases more rapidly with R_f, but is always *smaller* than the force required to stretch the ideal chain to the same end-to-end distance R_f, as shown in Fig. 3.9. Both chains have fewer possible conformations when they are stretched, but the real chain has fewer possible conformations to lose, resulting in a smaller stretching force.

A similar scaling calculation can be carried out for stretching a linear chain with fractal dimension $1/\nu$. The free energy cost of stretching a chain from its original size bN^{ν} to end-to-end distance R is (derived in Problem 3.15)

$$F \approx kT \left(\frac{R}{bN^{\nu}} \right)^{1/(1-\nu)}. \qquad (3.42)$$

The fractal dimension of an ideal chain is $1/\nu = 2$ and Eq. (3.42) reduces to free energy of stretching an ideal chain [Eq. (3.35)]. The Flory estimate of the fractal dimension of a real chain is $1/\nu = 5/3$ and Eq. (3.42) reduces to Eq. (3.36). A more accurate estimate of the fractal dimension of a real chain is $1/\nu \cong 1/0.588 \cong 1.7$ with corresponding free energy of stretching [Eq. (3.42)] $F \approx kT (R/R_F)^{2.4}$.

The divergence of the force near maximal extension ($f \to \infty$ as $R_f \to R_{\max}$) is not described by this scaling approach and is not shown in Fig. 3.9. This divergence is discussed in Section 2.6.2 for freely jointed and worm-like chain models.

3.2.2 Polymer under compression

Two simple examples comparing the properties of ideal and real chains are discussed in this section: uniaxial and biaxial compression. A related example of triaxial confinement shall be discussed in Section 3.3.2 for the case where polymers collapse into globules due to attraction between monomers.

3.2.2.1 Biaxial compression

We consider first biaxial compression corresponding to squeezing of a chain into a cylindrical pore of diameter D. The diameter of the pore defines a natural **compression blob** size. On length scales smaller than D, sections of the chain do not 'know' that it is compressed and their statistics are still the same as the statistics of an undeformed chain:

$$D \approx bg^{1/2} \quad \text{ideal}, \qquad (3.43)$$

$$D \approx bg^{3/5} \quad \text{real}. \qquad (3.44)$$

These equations can be solved for the number of monomers g in a compression blob of size D:

$$g \approx \left(\frac{D}{b} \right)^2 \quad \text{ideal}, \qquad (3.45)$$

$$g \approx \left(\frac{D}{b} \right)^{5/3} \quad \text{real}. \qquad (3.46)$$

The above relations are identical to the corresponding equations for tension blobs (Section 3.2.1) because in both examples the conformational statistics are unperturbed on the shortest scales.

The length of a tube $R_\parallel$ occupied by an ideal chain can be estimated as a random walk of N/g compression blobs along the contour of the tube:

$$R_\parallel \approx D\left(\frac{N}{g}\right)^{1/2} \approx bN^{1/2} \quad \text{ideal.} \tag{3.47}$$

As expected, the size of the ideal chain along the contour of the tube is not affected by the confinement. This is an important property of an ideal chain. Deformation of the ideal chain in one direction does not affect its properties in the other directions because each coordinate's random walk is *independent*.

In the case of confinement of a real chain, the compression blobs repel each other and fill the pore in a sequential array. Therefore, the length of the tube $R_\parallel$ occupied by a real chain is the size of one compression blob D times the number N/g of these blobs:

$$R_\parallel \approx D\left(\frac{N}{g}\right) \approx \left(\frac{b}{D}\right)^{2/3} Nb \quad \text{real in a cylinder.} \tag{3.48}$$

Note that in the case of a real chain confined to a tube, the occupied length of the tube $R_\parallel$ is linearly proportional to the number of monomers N in the chain. The occupied length increases as the tube diameter D decreases. Ideal and real chains of the same length, confined in a cylinder of diameter D, are shown schematically in Fig. 3.10. There is no penalty to overlap the compression blobs of an ideal chain, whereas the compression blobs of the real chain have strong excluded volume interactions that prevent overlap.

The free energy of confinement is of the order of kT per compression blob for either chain:

$$F_{\text{conf}} \approx kT\frac{N}{g} \approx kTN\left(\frac{b}{D}\right)^2 \approx kT\left(\frac{R_0}{D}\right)^2 \quad \text{ideal,} \tag{3.49}$$

$$F_{\text{conf}} \approx kT\frac{N}{g} \approx kTN\left(\frac{b}{D}\right)^{5/3} \approx kT\left(\frac{R_F}{D}\right)^{5/3} \quad \text{real.} \tag{3.50}$$

Ideal

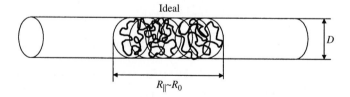

Real

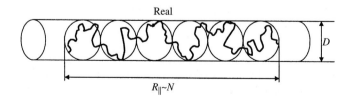

Fig. 3.10
Ideal and real chains of the same length, confined in a cylinder of diameter D.

R_0 and R_F are the end-to-end distances of unconfined ideal and real chains, respectively. These calculations can be generalized to confinement a polymer with fractal dimension $1/\nu$ from its original size bN^ν to a cylinder with diameter D. The confinement free energy in this case is (derived in Problem 3.16)

$$F_{conf} \approx kT \left(\frac{bN^\nu}{D} \right)^{1/\nu}, \tag{3.51}$$

with Eq. (3.49) corresponding to an ideal chain with $1/\nu = 2$ and Eq. (3.50) being the result for a real chain with the Flory estimate of fractal dimension $1/\nu = 5/3$. For a more accurate estimate of the fractal dimension of real chains $1/\nu = 1.70$ the confinement free energy is $F_{conf} \approx kT \, (R_F/D)^{1.70}$.

3.2.2.2 Uniaxial compression

The free energy of confinement of a chain between parallel plates in a slit of spacing D is the same as in the cylindrical pore (up to numerical prefactors of order unity [Eqs (3.49) and (3.50)]. The longitudinal size $R_\parallel$ of an ideal chain confined between parallel plates is still the same as for an unperturbed ideal chain [Eq. (3.47)] because the different x, y, z components of an ideal chain's random walk are not coupled. In the case of a real chain confined between parallel plates, the compression blobs repel each other, leading to a two-dimensional swollen conformation (see Fig. 3.11). The size of a two-dimensional swollen chain of compression blobs can be estimated from the Flory theory (Section 3.1.2). The 'excluded area' of each compression blob is $\approx D^2$, making the two-dimensional analogue of Eq. (3.17) for the repulsive interaction energy of the chain of N/g compression blobs $kTD^2(N/g)^2/R_\parallel^2$, where $R_\parallel^2$ is the area of the chain. The entropic part of the free energy that resists increasing the area of the chain of N/g compression blobs of size D is $kTR_\parallel^2/[(N/g)D^2]$ for a real chain confined between two parallel plates:

$$F \approx kT \left(D^2 \frac{(N/g)^2}{R_\parallel^2} + \frac{R_\parallel^2}{(N/g)D^2} \right). \tag{3.52}$$

Minimizing this free energy with respect to $R_\parallel$ gives the size of a real chain between plates of spacing D:

$$R_\parallel \approx D \left(\frac{N}{g} \right)^{3/4} \approx N^{3/4} b \left(\frac{b}{D} \right)^{1/4} \quad \text{real between plates} \tag{3.53}$$

The size of the real chain confined between plates is again much larger than that of an ideal chain (where $R_\parallel \approx bN^{1/2}$) because the compression blobs of the real chain repel each other. The maximum confinement corresponds to thickness D of the order of the Kuhn monomer size b. In this case the chain becomes effectively two-dimensional with size

$$R_\parallel \approx N^{3/4} b \quad \text{real two-dimensional.} \tag{3.54}$$

The exponent $\nu = 3/4$ is universal for two-dimensional linear chains with excluded volume repulsion.

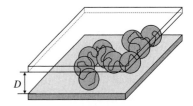

Fig. 3.11
Uniaxial compression—a real chain in a slit of spacing D between two parallel plates.

3.2.3 Adsorption of a single chain

For the final example comparing the properties of ideal and real chains, consider a polymer in dilute solution near a weakly adsorbing surface. Let the energy gain for a monomer in contact with the surface be $-\delta kT$, where we assume that $0 < \delta < 1$ (weak adsorption). The chain would like to increase the number of monomers in contact with the surface in order to gain adsorption energy. In order to do that, however, it would have to confine itself to a layer of thickness smaller than its unperturbed polymer size ($\xi_{ads} < R$), thereby losing conformational entropy.

3.2.3.1 Scaling calculation

The thickness ξ_{ads} of the adsorbed layer defines the **adsorption blob** size (see Fig. 3.12). This adsorption blob size is the length scale on which the cumulative interaction energy of a small section of the chain with the surface is of the order of the thermal energy kT. On smaller length scales, the interaction energy is weaker than the thermal energy and the chain remains in an unperturbed conformation, which is Gaussian for ideal chains [Eq. (3.45)] and swollen for real chains [Eq. (3.46)]. On scales larger than the adsorption blob, the interaction energy of the chain with the surface is larger than kT and the consecutive adsorption blobs are forced to be in contact with the surface. Therefore, the conformation of an adsorbed chain is a two-dimensional array of adsorption blobs and is similar to that for a chain confined between two parallel plates, discussed in the previous section.

In order to calculate the size of the adsorption blob ξ_{ads}, we need to calculate the number of monomers in contact with the surface for a chain section of size ξ_{ads}. The average volume fraction in a chain section of size ξ_{ads} containing g_{ads} monomers is ϕ:

$$\phi \approx \frac{b^3 g_{ads}}{\xi_{ads}^3} \approx \frac{b}{\xi_{ads}} \quad \text{ideal,} \tag{3.55}$$

$$\phi \approx \frac{b^3 g_{ads}}{\xi_{ads}^3} \approx \left(\frac{b}{\xi_{ads}}\right)^{4/3} \quad \text{real.} \tag{3.56}$$

The number of monomers in each adsorption blob that are in direct contact with the surface (within a layer of thickness b from it) is estimated as the product of the mean-field number density of monomers in the blob ϕ/b^3 and the volume of this layer within distance b of the surface, $\xi_{ads}^2 b$:

$$\frac{\phi}{b^3} \xi_{ads}^2 b \approx \frac{\xi_{ads}}{b} \quad \text{ideal,} \tag{3.57}$$

$$\frac{\phi}{b^3} \xi_{ads}^2 b \approx \left(\frac{\xi_{ads}}{b}\right)^{2/3} \quad \text{real.} \tag{3.58}$$

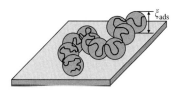

Fig. 3.12
A chain adsorbed to a weakly attractive surface.

The energy gain per monomer in contact with the surface is δkT. Therefore, the energy gain per adsorption blob is

$$\delta kT \frac{\xi_{ads}}{b} \approx kT \quad \text{ideal,} \tag{3.59}$$

$$\delta kT \left(\frac{\xi_{ads}}{b}\right)^{2/3} \approx kT \quad \text{real,} \tag{3.60}$$

leading to the adsorption blob size:

$$\xi_{ads} \approx \frac{b}{\delta} \quad \text{ideal,} \tag{3.61}$$

$$\xi_{ads} \approx \frac{b}{\delta^{3/2}} \quad \text{real.} \tag{3.62}$$

The free energy of an adsorbed chain can be estimated as the thermal energy kT per adsorption blob:

$$F_{ads} \approx -kT \frac{N}{g_{ads}} \approx -kTN\delta^2 \quad \text{ideal,} \tag{3.63}$$

$$F_{ads} \approx -kT \frac{N}{g_{ads}} \approx -kTN\delta^{5/2} \quad \text{real.} \tag{3.64}$$

The adsorbed layer is thicker and bound less strongly for the real chain (since for weak adsorption $0 < \delta < 1$) because it pays a higher confinement penalty than the ideal chain. The excluded volume interaction of real chains make them more difficult to compress or adsorb than ideal chains. These scaling calculations can be generalized to adsorption of a polymer with general fractal dimension $1/\nu$:

$$F_{ads} \approx -kTN\delta^{1/(1-\nu)}. \tag{3.65}$$

The same result can be obtained using the Flory theory, as demonstrated below.

3.2.3.2 Flory theory of an adsorbed chain

A mean-field estimate of the free energy of adsorption and the thickness of the adsorbed chain can be made by assuming the monomers are uniformly distributed at different distances from the surface up to thickness ξ_{ads}. Then the fraction of monomers in direct contact with the surface (within distance b from the surface) is b/ξ_{ads}. The number of adsorbed monomers Nb/ξ_{ads} is multiplied by the adsorption energy per monomer–surface contact $(-\delta kT)$ to calculate the energetic gain from the surface interaction:

$$F_{int} \approx -\delta kTN \frac{b}{\xi_{ads}}. \tag{3.66}$$

In order to gain this energy, the chain must pay the entropic confinement free energy F_{conf}, derived in the example above [Eqs (3.49) and (3.50)]. Therefore, the total free energy of a weakly adsorbing chain is

$$F = F_{conf} + F_{int} \approx kTN\left(\frac{b}{\xi_{ads}}\right)^2 - kTN\delta\frac{b}{\xi_{ads}} \quad \text{ideal,} \tag{3.67}$$

$$F = F_{conf} + F_{int} \approx kTN\left(\frac{b}{\xi_{ads}}\right)^{5/3} - kTN\delta\frac{b}{\xi_{ads}} \quad \text{real.} \tag{3.68}$$

The minimum of the free energy corresponds to the optimal thickness of the adsorbed layer, determined from $\partial F/\partial\xi_{ads} = 0$:

$$\xi_{ads} \approx \frac{b}{\delta} \quad \text{ideal,} \tag{3.69}$$

$$\xi_{ads} \approx \frac{b}{\delta^{3/2}} \quad \text{real.} \tag{3.70}$$

These estimates are identical to the scaling results [Eqs (3.61) and (3.62)]. Substituting Eqs (3.69) and (3.70) into each individual term[2] in Eqs (3.67) and (3.68) shows that each term is actually of order the free energy of adsorption [Eqs (3.63) and (3.64)].

The adsorbed layer thickness for a polymer with general fractal dimension $1/\nu$ is derived in Problem 3.18:

$$\xi_{ads} \approx b\delta^{-\nu/(1-\nu)}. \tag{3.71}$$

Substituting this adsorbed layer thickness into the confinement free energy [Eq. (3.51)] or into the interaction free energy [Eq. (3.66)] gives the expected result for the free energy of adsorption [Eq. (3.65)].

3.2.3.3 Proximity effects

Both theories of single-chain adsorption, described above, ignore a very important effect—the loss of conformational entropy of a strand due to its proximity to the impenetrable surface. Each adsorption blob has $1/\delta$ contacts with the surface and each strand of the chain near these contacts loses conformational entropy due to the proximity effect. In order to overcome this entropic penalty, the chain must gain finite energy E_{cr} per contact between a monomer and the surface. This critical energy E_{cr} corresponds to the adsorption transition. For ideal chains $E_{cr} \approx kT$. The small additional free energy gain per contact $kT\delta$ should be considered in excess of the critical value E_{cr},

$$E = E_{cr} + \delta kT. \tag{3.72}$$

Polymer adsorption is, therefore, a sharp transition with chain thickness changing rapidly in the small interval δkT of monomer–surface interaction energy E above E_{cr} (Fig. 3.13). Strictly speaking, this correction [Eq. (3.72)]

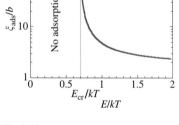

Fig. 3.13
The thickness ξ_{ads} of an adsorbed ideal chain decreases rapidly as the adsorption energy E is increased above the adsorption transition E_{cr}.

---

[2] Notice that if Eqs (3.69) and (3.70) are blindly substituted into Eqs (3.67) and (3.68), the conclusion would be that the adsorption free energy is zero for both ideal and real chains. This exemplifies the disadvantage of scaling calculations. There are unspecified prefactors of order unity in both terms of Eqs (3.67) and (3.68), which invalidates the blind substitution.

for the proximity effect is valid only for ideal chains. It is much harder to take into account the proximity effect for real chains due to strong correlation effects in these polymers. However, qualitatively there is still a threshold value of energy needed for the real chain to adsorb, as depicted in Fig. 3.13. For adsorption of real chains, the actual concentration inside each adsorption blob decays as a power law in distance from the surface. This power law decay modifies the exponent in Eqs (3.62) and (3.70) (see Problem 3.22).

3.3 Temperature effects on real chains

3.3.1 Scaling model of real chains

Several examples of scaling with different types of scaling blobs have already been introduced for tension, compression, and adsorption. The main idea in all scaling approaches is a separation of length scales. The blob in each case corresponds to the length scale at which the interaction energy is of the order of the thermal energy kT. On smaller scales the interaction is not important and smaller sections of the chain follow the unperturbed statistics (either ideal or swollen). On length scales larger than the blob size, the interaction energy is larger than kT and polymer conformations are controlled by interactions.

In this section, we will consider the excluded volume interaction following a similar scaling approach. The main idea is that of a thermal length scale (the **thermal blob**). On length scales smaller than the thermal blob size ξ_T, the excluded volume interactions are weaker than the thermal energy kT and the conformations of these small sections of the chain are nearly ideal. The thermal blob contains g_T monomers in a random walk conformation:

$$\xi_T \approx b g_T^{1/2} \tag{3.73}$$

The thermal blob size can be estimated by equating the Flory excluded volume interaction energy [Eq. (3.17)] for a single thermal blob and the thermal energy kT.

$$kT|v|\frac{g_T^2}{\xi_T^3} \approx kT. \tag{3.74}$$

In this section, we discuss both good ($v > 0$) and poor ($v < 0$) solvents and therefore, use $|v|$ in the definition of the thermal blob. The above two equations are combined to estimate the number of monomers in a thermal blob

$$g_T \approx \frac{b^6}{v^2}, \tag{3.75}$$

and the size of the thermal blob

$$\xi_T \approx \frac{b^4}{|v|}, \tag{3.76}$$

in terms of the monomer size b and the excluded volume v.

The thermal blob size is the length scale at which excluded volume becomes important. For $v \approx b^3$, the thermal blob is the size of a monomer ($\xi_T \approx b$) and the chain is fully swollen in an athermal solvent [Eq. (3.12)]. For $v \approx -b^3$, the thermal blob is again the size of a monomer ($\xi_T \approx b$) and the chain is fully collapsed in a non-solvent [Eq. (3.16)]. For $|v| < b^3 N^{-1/2}$, the thermal blob is larger than the chain size ($\xi_T > R_0$) and the chain is nearly ideal. For $b^3 N^{-1/2} < |v| < b^3$ the thermal blob is between the monomer size and the chain size, with either intermediate swelling in a good solvent [$v > 0$, Eq. (3.13)] or intermediate collapse in a poor solvent [$v < 0$, Eq. (3.15)].

3.3.1.1 Excluded volume repulsion ($v > 0$)

On length scales larger than the thermal blob size ξ_T, in athermal and good solvents, the excluded volume repulsion energy is larger than the thermal energy kT and the polymer is a swollen chain of N/g_T thermal blobs (Fig. 3.14). The end-to-end distance of this chain is determined as a self-avoiding walk of thermal blobs with fractal dimension $D = 1/\nu \cong 1.7$:

$$R \approx \xi_T \left(\frac{N}{g_T}\right)^\nu \approx b \left(\frac{v}{b^3}\right)^{2\nu-1} N^\nu. \tag{3.77}$$

For the swelling exponent $\nu \cong 0.588$ the expression for the chain size is $R \approx b \, (v/b^3)^{0.18} N^{0.588}$. Note that this scaling result reduces to the prediction of the Flory theory [Eq. (3.20)] for exponent $\nu = 3/5$.

3.3.1.2 Excluded volume attraction ($v < 0$)

In poor solvents on length scales larger than the thermal blob size ξ_T, the excluded volume attraction energy is larger than the thermal energy kT. This causes the thermal blobs to adhere to each other, forming a dense globule (Fig. 3.14). The size of the globule is calculated by assuming a dense packing of thermal blobs:

$$R_{gl} \approx \xi_T \left(\frac{N}{g_T}\right)^{1/3} \approx \frac{b^2}{|v|^{1/3}} N^{1/3}. \tag{3.78}$$

Thermal blobs in a poor solvent attract each other like molecules in a liquid droplet. The shape of the globule is roughly spherical to reduce the area of the unfavourable interface between it and the pure solvent. The volume fraction inside the globule is independent of the number of monomers N and is the same as inside a thermal blob:

$$\phi \approx \frac{Nb^3}{R_{gl}^3} \approx \frac{|v|}{b^3}. \tag{3.79}$$

The dependence of the size R of the chain on the number of monomers N, for solvents of different quality, is sketched in Fig. 3.15. In athermal solvent ($v = b^3$), in θ-solvent ($v = 0$) and in non-solvent ($v = -b^3$) the dependence of size R on number of monomers N is a single power law $R \approx bN^\nu$ for $N \gg 1$. The scaling exponent ν adopts three values: $\nu \cong 3/5$ in an athermal solvent, $\nu = 1/2$ in a θ-solvent, and $\nu = 1/3$ in a non-solvent. In good and

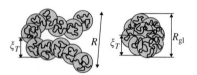

Fig. 3.14
The conformation of a single chain in a good solvent (left side) is a self-avoiding walk of thermal blobs while the conformation in a poor solvent (right side) is a collapsed globule of thermal blobs. Note that the conformation of a dilute globule in a poor solvent is a random walk of thermal blobs up to the globule size. There is almost no change in chain conformation at the length scale of a thermal blob. Density fluctuations in space saturate at this length scale and the density of a globule is uniform at length scales larger than the thermal blob.

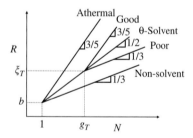

Fig. 3.15
End-to-end distance of dilute polymers in various types of solvents, sketched on logarithmic scales. In a θ-solvent the thermal blob size is infinite. For athermal solvent and non-solvent the thermal blob is the size of a single monomer. Good and poor solvents have intermediate thermal blob size (shown here for the specific example of equivalent thermal blobs in good and poor solvent.

poor solvents the dependence follows the ideal chain scaling for polymers (or sections of polymers) smaller than the thermal blob ξ_T. On larger scales, the chain follows the corresponding limiting scaling, with good solvent exponent $\nu \cong 3/5$ for $v > 0$ and with collapsed globule exponent $\nu = 1/3$ for $v < 0$. Fig. 3.15 shows that finite length chains have essentially ideal conformations for small values of the excluded volume. Chains have approximately ideal conformations as long as $N < g_T \approx b^6/v^2$ because the net excluded volume interaction in the whole chain is still of smaller magnitude than the thermal energy. Note that Fig. 3.15 suggests the crossover at the thermal blob size is abrupt, while in reality the crossover will be smooth with intermediate effective slopes observed over limited ranges of data.

3.3.2   Flory theory of a polymer in a poor solvent

The scaling result for a polymer in a poor solvent can also be found using Flory theory. The Flory free energy for a polymer chain is given by Eq. 3.19:

$$F \approx kT\left(\frac{R^2}{Nb^2} + v\frac{N^2}{R^3}\right). \tag{3.80}$$

In poor solvent, the excluded volume is negative, indicating a net attraction and the minimum of the free energy of Eq. (3.80) corresponds to $R = 0$. Both entropic and energetic contributions decrease with decreasing R. Such strong collapse of a polymer into a point is unphysical and we need to add a stabilizing term to this free energy.

3.3.2.1   Entropy of confinement
Earlier in this chapter, we have discussed the entropic cost due to confinement of an ideal chain into a cylindrical tube or in a slit between two parallel walls. A similar entropic penalty has to be paid if a chain is confined within a spherical cavity of size $R < bN^{1/2}$. Each compression blob corresponds to a random walk that fills the cavity. Thus, the number of monomers in each compression blob is determined by ideal chain statistics within the blob:

$$g \approx \left(\frac{R}{b}\right)^2. \tag{3.81}$$

The N/g compression blobs of the ideal chain fully overlap for a chain confined in a spherical pore. The free energy cost of confinement within the spherical cavity is of the order of the thermal energy kT per compression blob:

$$F_{conf} \approx kT\frac{N}{g} \approx kT\frac{Nb^2}{R^2}. \tag{3.82}$$

The entropic part of the free energy, that includes both the penalty for stretching and one for confinement, and is valid for both $R > bN^{1/2}$ and for

$R < bN^{1/2}$, is a simple sum of the stretching and confinement terms:

$$F_{ent} \approx kT \left(\frac{R^2}{Nb^2} + \frac{Nb^2}{R^2} \right). \tag{3.83}$$

Note that this entropic free energy alone has a minimum at $R^2 = Nb^2$, which is the conformation of an ideal chain.

Adding the excluded volume interaction term, we obtain a total free energy of the chain with three terms:

$$F \approx kT \left(\frac{R^2}{Nb^2} + \frac{Nb^2}{R^2} + v \frac{N^2}{R^3} \right). \tag{3.84}$$

This free energy still has a minimum at $R = 0$. The confinement entropy term is not strong enough to stabilize the collapse of the chain due to excluded volume attraction because $Nb^2/R^2 \ll |v|N^2/R^3$ for $R \to 0$.

3.3.2.2 Three-body repulsion

The stabilization of the collapsing coil comes from other terms of the interaction part of the free energy. The interaction energy per unit volume is an intrinsic property of any mixture, that is often expressed as a virial expansion in powers of the number density of monomers c_n [Eq. (3.8)]. The relevant volume of interest here is the pervaded coil volume R^3. The excluded volume term is the first term in the virial series and counts two-body interactions as vc_n^2. The next term in the expansion counts three-body interactions as wc_n^3, where w is the three-body interaction coefficient:

$$\frac{F_{int}}{R^3} \approx kT(vc_n^2 + wc_n^3 + \cdots). \tag{3.85}$$

At low concentration, the two-body term dominates the interaction. The three-body term becomes important at higher concentrations and can stabilize the collapse of the globule (since $w > 0$). The interaction free energy within the coil is estimated using the monomer concentration inside the coil $c_n = N/R^3$:

$$F_{int} \approx kT \left(v \frac{N^2}{R^3} + w \frac{N^3}{R^6} \right). \tag{3.86}$$

The total free energy of the chain is dominated by the interaction terms at higher densities (smaller chain sizes R):

$$F \approx kT \left(\frac{R^2}{Nb^2} + \frac{Nb^2}{R^2} + v \frac{N^2}{R^3} + w \frac{N^3}{R^6} \right)$$
$$\approx kT \left(v \frac{N^2}{R^3} + w \frac{N^3}{R^6} \right) \text{ for } R \ll R_0. \tag{3.87}$$

The globule seeks to minimize this free energy, by balancing the two-body attraction ($v < 0$) and three-body repulsion ($w > 0$) terms:

$$R_{gl} \approx \left(\frac{wN}{|v|} \right)^{1/3}. \tag{3.88}$$

A typical value of the three-body interaction coefficient for almost symmetric monomers is $w \approx b^6$ leading to the prediction of the globule size identical to that of the scaling approach [Eq. (3.78)]. For the cylindrical Kuhn monomer of length b and diameter d, Eq. (3.11) gives $w \approx (bd)^3$, making the globule size

$$R_{gl} \approx bd \left(\frac{N}{|v|} \right)^{1/3}, \tag{3.89}$$

and the volume fraction within this globule is proportional to the magnitude of the excluded volume:

$$\phi \approx \frac{Nbd^2}{R_{gl}^3} \approx \frac{|v|}{b^2 d}. \tag{3.90}$$

In a non-solvent, $v \approx -b^2 d$ [Eq. (3.16)] and the globule is fully collapsed with volume fraction $\phi \approx 1$ and size $R_{gl} \approx (bd^2 N)^{1/3}$. This state is the result of a dense packing of N monomers, since the volume of the cylindrical Kuhn monomer is bd^2.

3.3.3 Temperature dependence of the chain size

All results for chain size are now written in terms of the excluded volume. To understand how the chain size changes with temperature, we simply need the temperature dependence of the excluded volume. There are two important parts of the Mayer f-function, from which the excluded volume is calculated [Eq. (3.7)]. The first part is the hard-core repulsion, encountered when two monomers try to overlap each other (monomer separation $r < b$). In the hard-core repulsion, the interaction energy is enormous compared to the thermal energy, so the Mayer f-function for $r < b$ is -1:

$$f(r) = \exp \left[-\frac{U(r)}{kT} \right] - 1 \cong -1 \quad \text{for } r < b, \text{ where } U(r) \gg kT. \tag{3.91}$$

The second part is for monomer separations larger than their size ($r > b$), where the magnitude of the interaction potential is small compared to the thermal energy. In this regime, the exponential can be expanded and the Mayer f-function is approximated by the ratio of the interaction energy and the thermal energy:

$$f(r) = \exp \left[-\frac{U(r)}{kT} \right] - 1 \cong -\frac{U(r)}{kT} \quad \text{for } r > b, \text{ where } |U(r)| < kT. \tag{3.92}$$

The excluded volume v can be estimated using Eq. (3.7) with these two parts of the Mayer f-function:

$$v = -4\pi \int_0^\infty f(r) r^2 \, dr \approx 4\pi \int_0^b r^2 \, dr + \frac{4\pi}{kT} \int_b^\infty U(r) r^2 \, dr$$
$$\approx \left(1 - \frac{\theta}{T} \right) b^3. \tag{3.93}$$

The first term is the contribution of the hard-core repulsion, and is of the order of the monomer volume b^3. The second term contains the temperature dependence, and the coefficient of $1/T$ defines an effective temperature called the θ-temperature:

$$\theta \approx -\frac{1}{b^3 k} \int_b^\infty U(r) r^2 \, dr. \tag{3.94}$$

Since $U(r) < 0$ in the attractive well, the θ-temperature is positive. This results in a very simple approximate temperature dependence of the excluded volume [Eq. (3.93)]:

$$v \approx \frac{T - \theta}{T} b^3. \tag{3.95}$$

For $T < \theta$ the excluded volume is negative, indicating a net attraction between monomers (poor solvent). For temperatures far below θ, the chain collapses into a dry globule that excludes nearly all solvent (with $v \approx -b^3$) at $\theta - T \approx T$ and Eq. (3.95) does not apply below this temperature. For $T = \theta$ the net excluded volume is zero and the chain adopts a nearly ideal conformation (θ-solvent). $T > \theta$ has a positive excluded volume, resulting in swelling of the coil (good solvent). For $T \gg \theta$, excluded volume becomes independent of temperature ($v \approx b^3$) and such solvents are termed athermal.

The temperature dependence of the radius of gyration, reduced by the radius of gyration at the θ-temperature $R_\theta = bN^{1/2}$, is shown in Fig. 3.16 for both experimental data and Monte-Carlo simulations of chains made of N freely jointed monomers interacting via a **Lennard–Jones potential**:

$$U(r) = 4\epsilon \left[\left(\frac{\sigma}{r}\right)^{12} - \left(\frac{\sigma}{r}\right)^6 \right] \tag{3.96}$$

The abscissa of Fig. 3.16 is proportional to the chain interaction parameter [Eq. (3.22)]:

$$z \approx \frac{v}{b^3} N^{1/2} \approx \frac{T - \theta}{T} N^{1/2}. \tag{3.97}$$

Note that the square of the chain interaction parameter z is equal to the number of thermal blobs in a chain $z^2 \approx N/g_T$.

$$g_T \approx \frac{N}{z^2} \approx \left(\frac{T}{T - \theta}\right)^2. \tag{3.98}$$

The data reduction for R_g/R_θ as a function of chain interaction parameter z in Fig. 3.16 is remarkable for both simulation and experiment. Notice in Fig. 3.16 that the θ-temperature is a *compensation point* where the excluded volume happens to be zero. Below the θ-temperature, the chains are collapsed in poor solvent ($v < 0$), while above the θ-temperature, the coils are swollen in good solvent ($v > 0$).

The relative contraction of chains in poor solvents can be expressed in terms of the chain interaction parameter z [Eqs. (3.78) and (3.97)]:

$$\frac{R}{bN^{1/2}} \approx \frac{b}{|v|^{1/3} N^{1/6}} \approx |z|^{-1/3} \quad \text{for } T < \theta. \tag{3.99}$$

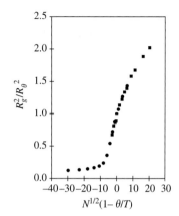

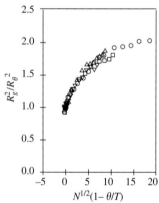

Fig. 3.16

Temperature dependence of radius of gyration in universal form. Upper plot shows Monte-Carlo simulation data on Lennard-Jones chains, with the filled squares from W. W. Graessley *et al.*, *Macromolecules* **32**, 3510 (1999) and the filled circles are courtesy of I. Withers. Lower plot shows experimental data on polystyrene in decalin: open circles have $M_w = 4\,400\,000\ \text{g mol}^{-1}$, open squares have $M_w = 1\,560\,000\ \text{g mol}^{-1}$, open triangles have $M_w = 1\,050\,000\ \text{g mol}^{-1}$ and open upside-down triangles have $M_w = 622\,000\ \text{g mol}^{-1}$, from G. C. Berry, *J. Chem. Phys.* **44**, 4550 (1966).

The relative swelling in good solvents can also be written as a function of the chain interaction parameter z [Eq. (3.21)].

$$\frac{R}{bN^{1/2}} \approx z^{2\nu-1} \quad \text{for } T > \theta. \tag{3.100}$$

The relative swelling is proportional to $z^{0.18}$ for $\nu \cong 0.588$.

3.3.4 Second virial coefficient

The second virial coefficient A_2 is determined from the concentration dependence of osmotic pressure [Eq. (1.76)] or scattered light intensity [Eq. (1.91)] from dilute polymer solutions. A_2 is a direct measure of excluded volume interactions between pairs of chains.

In solvents near the θ-temperature, the thermal blob is larger than the chain ($g_T > N$ meaning $|z| < 1$ or $|T - \theta|/T < N^{-1/2}$) and the excluded volume interactions are weak. The interaction energy of two overlapping chains is less than the thermal energy kT, so chains can easily interpenetrate each other. In this limit, monomers interact directly and A_2 is proportional to the excluded volume v of a Kuhn monomer. The second virial coefficient of Eqs (1.76) and (1.91) has units of $\text{m}^3\,\text{mol}\,\text{kg}^{-2}$, making the relation

$$v = \frac{2M_0^2}{\mathcal{N}_{Av}} A_2 \quad \text{for } \left|\frac{T-\theta}{T}\right| < N^{-1/2}, \tag{3.101}$$

as will be derived in Chapter 4 [see Eq. (4.72)]. Using Eq. (3.97), A_2 can be written in terms of the chain interaction parameter z:

$$A_2 \approx \frac{\mathcal{N}_{Av}v}{M_0^2} \approx \frac{\mathcal{N}_{Av}b^3}{M_0^{3/2}} \frac{z}{M^{1/2}} \quad \text{for } |z| < 1. \tag{3.102}$$

For good solvents ($z > 1$), chains repel each other strongly and do not interpenetrate. The volume excluded by a chain is of the order of its pervaded volume R^3 and the molar mass of the chain is M:

$$\frac{A_2 M^2}{\mathcal{N}_{Av}} \approx R^3 \quad \text{for } \frac{T-\theta}{T} > N^{-1/2}. \tag{3.103}$$

Using Eq. (3.100) for the chain size R allows the second virial coefficient in good solvent to be determined:

$$A_2 \approx \frac{\mathcal{N}_{Av}}{M^2} R^3 \approx \frac{\mathcal{N}_{Av}b^3}{M_0^{3/2}} \frac{z^{6\nu-3}}{M^{1/2}} \quad \text{for } z > 1. \tag{3.104}$$

The second virial coefficient is proportional to $z^{0.53}$ for exponent $\nu \cong 0.588$. Combining Eqs (3.102) and (3.104), we see that both θ-solvent and good solvent should have $A_2 M^{1/2}$ a function of chain interaction parameter z alone:

$$\frac{A_2 M^{1/2} M_0^{3/2}}{\mathcal{N}_{Av}b^3} = f(z) \approx \left\{ \begin{array}{ll} z & |z| < 1 \\ z^{6\nu-3} & z > 1 \end{array} \right\}. \tag{3.105}$$

The success of this functional form is demonstrated in Fig. 3.17 for polystyrene of various molar masses in decalin from slightly below θ to

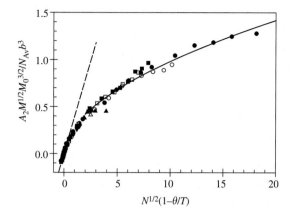

Fig. 3.17
Universal plot of second virial coefficient for linear polystyrenes in decalin (filled circles have
$M_w = 4\,400\,000$ g mol^{-1}, open circles have $M_w = 1\,560\,000$ g mol^{-1}, filled squares have
$M_w = 1\,050\,000$ g mol^{-1}, open squares have $M_w = 622\,000$ g mol^{-1}, filled triangles have
$M_w = 186\,000$ g mol^{-1}, open triangles have $M_w = 125\,000$ g mol^{-1} and filled inverted triangles
have $M_w = 48\,200$ g mol^{-1}. Data from G. C. Berry, *J. Chem. Phys.* **44**, 4550 (1966).

$\theta + 100\,K$. The collapse of the data is superb. The solid curve is the large
z branch of Eq. (3.105), $0.29\,[N^{1/2}(1 - \theta/T)]^{6\nu - 3}$. The slope of the dashed
line drawn in Fig. 3.17 is 0.39. The crossover between these two branches
occurs when there is a single thermal blob per chain ($N = g_T$). Using
Eq. (3.98) allows the second virial coefficient to be written in terms of the
number of thermal blobs per chain N/g_T.

$$\frac{A_2 M^{1/2} M_0^{3/2}}{\mathcal{N}_{Av} b^3} \cong 0.20 \left\{ \begin{array}{ll} (N/g_T)^{1/2} & N < g_T \\ (N/g_T)^{0.264} & N > g_T \end{array} \right\}. \tag{3.106}$$

This identifies the prefactors in Eq. (3.98)

$$g_T \cong 0.25 \left(\frac{T}{T - \theta} \right)^2, \tag{3.107}$$

and in Eq. (3.95)

$$\frac{v}{b^3} \cong 0.78 \frac{T - \theta}{T} \cong \frac{0.39}{\sqrt{g_T}}. \tag{3.108}$$

Examples of excluded volume and numbers of Kuhn monomers per
thermal blob are given in Table 3.1.

Indeed, data on the temperature dependence of second virial coefficient
for a variety of polymer–solvent combinations and from computer simulation show that Eq. (3.106) can be written as a simple crossover function:

$$\frac{A_2 M^{1/2} M_0^{3/2}}{\mathcal{N}_{Av} b^3} = 0.20 \left[\left(\frac{g_T}{N} \right)^{1.32} + \left(\frac{g_T}{N} \right)^{0.70} \right]^{-0.38}. \tag{3.109}$$

Table 3.1 Number of Kuhn monomers per thermal blob and excluded volume of a Kuhn monomer for polystyrene in various solvents

Polymer/solvent	T (°C)	$T - \theta$ (K)	g_T	v/b^3	v (Å^3)
Polystyrene/cyclohexane	50	15	120	0.036	210
Polystyrene/cyclohexane	70	35	24	0.079	460
Polystyrene/decalin	115	100	4	0.21	1200
Polystyrene/benzene	25	~200^a	0.6	0.5	3000

^{a} For benzene (and most other good solvents) the θ-temperature is far below all measurement temperatures and use of Eq. (3.107) to extrapolate to the θ-temperature has considerable error.

Measurement of the temperature dependence of second virial coefficient A_2 for polymers with known molar mass M and Kuhn length b allows estimation of the number of thermal blobs per chain N/g_T using Eq. (3.109).

3.4 Distribution of end-to-end distances

We have seen numerous examples of the qualitative difference in properties between ideal and real chains with excluded volume interactions. It is therefore not surprising that the distribution of end-to-end vectors of real chains is significantly different from the Gaussian distribution function of ideal chains [Eq. (2.86)].

Relative probabilities to find chain ends at distances much larger than the average end-to-end distance are related to the free energy penalty due to chain elongation [see Problem 3.15 and Eq. (3.42):

$$F \approx kT \left(\frac{R}{\sqrt{\langle R^2 \rangle}} \right)^\delta , \qquad (3.110)$$

where the exponent $\delta = 1/(1 - \nu)$ is related to the exponent ν of the root-mean-square end-to-end distance of the chain:

$$\sqrt{\langle R^2 \rangle} \approx bN^\nu. \qquad (3.111)$$

For ideal chains, $\nu = 1/2$ and $\delta = 2$ [see Eq. (3.35)], while for real chains in a good solvent $\nu \cong 0.588$ and $\delta \cong 2.43$ [see Eq. (3.36)]. The tail of the probability distribution function for end-to-end distances is determined by the Boltzmann factor arising from this free energy penalty [Eq. (3.110)]

$$P(N, R) \sim \exp\left(-\frac{F}{kT} \right) \sim \exp\left[-\alpha \left(\frac{R}{\sqrt{\langle R^2 \rangle}} \right)^\delta \right] \quad \text{for } R > \sqrt{\langle R^2 \rangle},$$

$$(3.112)$$

where α is a numerical coefficient of order unity. For ideal chains ($\delta = 2$) this leads to the Gaussian distribution function [Eq. (2.86)]. For real chains, a faster decay of the distribution function is expected due to the higher power $\delta \cong 2.43$ in the exponential:

$$P(N, R) \sim \exp\left[-\alpha \left(\frac{R}{\sqrt{\langle R^2 \rangle}} \right)^{2.43} \right] \quad \text{for } R > \sqrt{\langle R^2 \rangle}. \qquad (3.113)$$

Another major difference between ideal and real chains is the reduced probability of two ends of a real chain to be near each other due to excluded volume repulsion of these and neighboring monomers. Recall from Section 2.5 that the probability of finding one end of an ideal chain within a small spherical shell of volume $4\pi R^2\, dR$ around the other end is proportional to the volume of this shell [see Eq. (2.86) for $R \ll bN^{1/2}$]. This probability is significantly reduced for real chains by an additional factor

$$P(N, R) \sim \left(\frac{R}{\sqrt{\langle R^2\rangle}}\right)^g \quad \text{for } R \ll \sqrt{\langle R^2\rangle}, \tag{3.114}$$

due to excluded volume repulsion between sections of the polymer, as they approach each other. The exponent $g=0$ for ideal chains because there is no reduction of probabilities for small end-to-end distances. For real chains, the exponent $g \cong 0.28$ in three dimensions and $g=11/24$ in two dimensions.

By combining the two limits [Eqs. (3.113) and (3.114)], the distribution function of normalized end-to-end distances can be constructed:

$$P(x) \sim x^g \exp\left(-\alpha x^\delta\right), \tag{3.115}$$

where

$$x = \frac{R}{\sqrt{\langle R^2\rangle}}. \tag{3.116}$$

An approximate expression for the three-dimensional distribution function for real chains results:

$$P(x) \cong 0.278 x^{0.28} \exp\left(-1.206 x^{2.43}\right) \quad \text{real.} \tag{3.117}$$

For ideal chains, the corresponding function is Gaussian:

$$P(x) = \left(\frac{3}{2\pi}\right)^{3/2} \exp\left(-1.5 x^2\right) \quad \text{ideal.} \tag{3.118}$$

The two functions are compared in Fig. 3.18. Note the dramatic difference between them. Real chains in an athermal solvent rarely have ends in close proximity. The probability to find chain ends within relative distance dx of x is $4\pi x^2 P(x)\, dx$. The coefficients of the distributions of end-to-end distances are chosen so that they are normalized:

$$\int P(x)\, d^3x = \int_0^\infty P(x) 4\pi x^2\, dx = 1. \tag{3.119}$$

Their second moment is also equal to unity

$$\int x^2 P(x)\, d^3x = \int_0^\infty x^2 P(x) 4\pi x^2\, dx = 1, \tag{3.120}$$

due to the definition of the relative distance x [Eq. (3.116)].

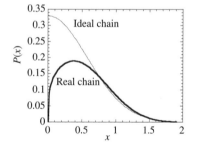

Fig. 3.18
Distribution function $P(x)$ of normalized end-to-end distances $x = R/\sqrt{\langle R^2\rangle}$. Thin curve, ideal chain; thick curve, real chain.

3.5 Scattering from dilute solutions

The size and shape of a polymer chain in dilute solution is best studied using scattering methods. Each monomer absorbs the incident radiation

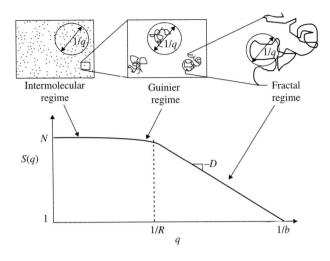

Fig. 3.19
Scattering function for a dilute solution on logarithmic scales.

and re-emits it in all directions. If there is contrast between monomers and solvent they can be distinguished. The scattered intensity at a given scattering wavevector is determined by this contrast and by the coherence of the re-emitted radiation from pairs of monomers. The **scattering function** $S(\vec{q})$ is defined as a sum over all pairs of n monomers in the scattering volume,

$$S(\vec{q}) = \frac{1}{n}\sum_{j=1}^{n}\sum_{k=1}^{n}\langle\exp[-i\vec{q}\cdot(\vec{r}_j - \vec{r}_k)]\rangle, \qquad (3.121)$$

where $\vec{q}$ is the scattering wavevector [Eq. (2.131)] and $\vec{r}_j$ is the position vector of jth monomer. This scattering function is simply a dimensionless version of the ratio of scattering intensity at wavevector $\vec{q}$ and concentration. The isotropic scattering function from dilute solution is sketched in Fig. 3.19. At large wavevectors $q \gg 1/R$ for each monomer j the sum over k has contribution of order unity from each of the n_q monomers within distance $1/q$ from monomer j since for $\vec{q}\cdot(\vec{r}_j - \vec{r}_k) < 1$ the exponential is close to one. On the other hand, the contribution from monomers further away from monomer j averages to zero. The scattering function at large wavevectors is

$$S(q) \approx \frac{1}{n}\sum_{j=1}^{n}n_q = n_q \quad \text{for } q \gg 1/R \qquad (3.122)$$

where n_q is the number of monomers in the volume $1/q^3$. The number of monomers n_q is related to the size of the chain segment $1/q$ through the fractal dimension of the chain $\mathcal{D}$:

$$S(q) \approx n_q \approx (qb)^{-\mathcal{D}} \quad \text{for } 1/R < q < 1/b. \qquad (3.123)$$

This power law extends from the monomer size b to the size of the chain R, provided that the entire chain has the same fractal dimension $\mathcal{D}$. Scattering on such small scales is dominated by **intra-molecular scattering** from monomers inside individual coils and is related to the pair correlation function within the coil [Eq. (2.123)]:

$$g(r) \approx \frac{m}{r^3} \sim r^{\mathcal{D}-3}. \qquad (3.124)$$

For real chains in an athermal solvent, $\mathcal{D} = 1/\nu \cong 1.7$, so $S(q) \sim q^{-1.7}$ and $g(r) \sim r^{-1.3}$ within the coil.

For smaller wavevectors $q < 1/R$, the number of monomers n_q within distance $1/q$ from monomer j saturates at the number of monomers in the chain N. The exact form of the scattering function in this Guinier regime enables calculation of the radius of gyration R_g [Eq. (2.152)]. In both the Guinier and fractal regimes, the scattering comes from pairs of monomers on the same chain and the scattering function is proportional to the form factor:

$$S(q) = NP(q) \quad \text{for } q > \left(\frac{c_n}{N}\right)^{1/3}. \qquad (3.125)$$

The **inter-molecular scattering** dominates the scattering function at wavevectors q smaller than the reciprocal distance between chains $(c_n/N)^{1/3}$, where c_n is the number density of monomers in solution. The inter-molecular regime is controlled by concentration fluctuations arising from the difference in the number of chains in volumes $1/q^3$. Assuming there are no interactions between chains (strictly valid only in very dilute solutions), the mean-square fluctuation in the number of chains in the volume $1/q^3$ is of the order of the average number $n_q/N \approx c_n/(Nq^3)$. The fluctuation in the number of monomers in volumes of size $1/q^3$ is $\sqrt{\langle (\delta n_q)^2 \rangle} = N\sqrt{c_n/(Nq^3)}$. The scattering function is the mean-square fluctuation in the number of monomers in the volume $1/q^3$ normalized by the number of monomers n_q in this volume:

$$S(q) = \frac{\langle (\delta n_q)^2 \rangle}{n_q} = \frac{N^2 c_n/(Nq^3)}{N c_n/(Nq^3)} = N \quad \text{for } q \ll \left(\frac{c_n}{N}\right)^{1/3}. \qquad (3.126)$$

Note that this value matches the low-q end of the fractal regime. It is hardly surprising that the scattering function contains information about the chain length, since in Chapter 1 we demonstrated how light scattering can be used to determine molar mass from the low concentration limit of $R_\theta/(Kc)$, where R_θ is Rayleigh ratio [Eq. (1.87)], K is the optical constant [Eq. (1.89)], and c is mass concentration. The scattering function for light scattering is related to the Rayleigh ratio as

$$S(q) = \frac{R_\theta}{KcM_0}, \qquad (3.127)$$

where M_0 is the molar mass of a monomer. We give this relation for light scattering for completeness, but scattering inside the polymer coils is usually measured using neutrons and X-rays, which extend the range of wavevectors to 1 nm^{-1}. The scattering function from all of these scattering experiments is the same, with the prefactor relating the scattering function to scattered intensity being specific to the type of radiation used.

3.6   Summary of real chains

Real chains have interactions between monomers. If the attraction between monomers just balances the effect of the hard core repulsion, the net excluded volume is zero ($v = 0$) and the chain will adopt a nearly ideal conformation (see Chapter 2):

$$R_0 = bN^{1/2} \quad \text{for } \theta\text{-solvent}. \tag{3.128}$$

Such a situation with zero net excluded volume is called the θ-condition, corresponding to a particular θ-temperature for a given solvent.

If the attraction between monomers is weaker than the hard-core repulsion, the excluded volume is positive and the chain swells. This corresponds to a good solvent at a temperature above the θ-temperature, and the coil size is larger than the ideal size:

$$R_F \approx b\left(\frac{v}{b^3}\right)^{2\nu-1} N^\nu \approx b\left(\frac{v}{b^3}\right)^{0.18} N^{0.588} \quad \text{for good solvent}. \tag{3.129}$$

The chain conformation is a self-avoiding walk of thermal blobs, whose size decreases as temperature is raised.

In an athermal solvent, the monomer–solvent energetic interaction is identical to the monomer–monomer interaction. This makes the net interaction between monomers zero, leaving only the hard core repulsion between monomers. The excluded volume is independent of temperature ($v \approx b^3$), and the chain is a self-avoiding walk of monomers:

$$R \approx bN^\nu \approx bN^{0.588} \quad \text{for athermal solvent}. \tag{3.130}$$

If the attraction between monomers is stronger than the hard-core repulsion, the excluded volume is negative and the chain collapses. This occurs below the θ-temperature, and corresponds to a poor solvent. In a poor solvent, the polymer is in a collapsed globular conformation corresponding to a dense packing of thermal blobs. The size of a globule is smaller than the ideal size:

$$R_{gl} \approx |v|^{-1/3} b^2 N^{1/3} \quad \text{for poor solvent}. \tag{3.131}$$

A chain in a poor solvent collapses into a globule with significant amounts of solvent inside. Most chains agglomerate with other chains and precipitate from solution. Only a very small number of polymers remain in the solvent-rich phase of a poor solvent in a globular conformation described by Eq. (3.131). Far below the θ-temperature, the attraction dominates completely and the excluded volume $v \approx -b^3$. This limit is called a non-solvent, and an individual chain in that solvent would have a fully collapsed conformation:

$$R \approx bN^{1/3} \quad \text{for non-solvent}. \tag{3.132}$$

In this case, most chains precipitate from solution into a melt excluding nearly all solvent, and the chains then adopt ideal conformations to maximize their entropy.

The good solvent and poor solvent results only apply to chains that are sufficiently long. Short chains with degree of polymerization less than the number of monomers in a thermal blob remain ideal, as depicted in Fig. 3.15.

Several examples were given of scaling models that utilize blobs to separate regimes of chain conformation. The common idea in these scaling models is that, on the smallest length scales (inside the blobs), there is not enough cumulative interaction to alter the chain conformation. On length scales larger than the blob size, the cumulative interactions become larger than the thermal energy, and can then modify the conformation of the chain of blobs. Since the cumulative interaction energy of each blob is roughly the thermal energy kT, the total interaction energy can be conveniently estimated as kT per blob.

The free energy of stretching a real linear chain in a good solvent has a stronger dependence on size R than the quadratic dependence of the ideal chain:

$$F \approx kT \left(\frac{R}{R_F} \right)^{1/(1-\nu)} \approx kT \left(\frac{R}{R_F} \right)^{2.43}. \tag{3.133}$$

The stretching force for a real chain increases non-linearly with elongation:

$$\frac{fb}{kT} = \frac{b}{kT} \frac{\partial F}{\partial R} \approx \left(\frac{R}{Nb} \right)^{\nu/(1-\nu)} \approx \left(\frac{R}{Nb} \right)^{1.43}. \tag{3.134}$$

The free energy of confining a real linear chain in a good solvent either into a slit of spacing D or to a cylindrical pore of diameter D is larger than for an ideal chain because the real chain has repulsive interactions:

$$F \approx kT \left(\frac{R_F}{D} \right)^{1/\nu} \approx kT \left(\frac{R_F}{D} \right)^{1.7}. \tag{3.135}$$

Excluded volume changes with temperature in the vicinity of the θ-temperature:

$$v \approx b^3 \left(\frac{T - \theta}{T} \right). \tag{3.136}$$

Good solvents typically have $T \gg \theta$, and their θ-temperature is not accessible because the solvent crystallizes at much higher temperatures. Similarly, θ-solvents cannot usually be heated far enough above the θ-temperature to reach the athermal limit because the solvent will boil at a lower temperature.

Problems

Section 3.1

3.1 (i) Taking the volume of a cylindrical Kuhn monomer to be bd^2, derive an expression for the cylindrical monomer diameter d in terms of the characteristic ratio, molar mass per backbone bond, melt density, Kuhn length b and bond angle θ.

   (ii) Using the melt density of polyethylene $\rho = 0.784$ g cm^{-3} and the melt density of polystyrene $\rho = 0.784$ g cm^{-3}, along with the data of Table 2.1 for C_∞ and b, calculate the diameter of the Kuhn cylindrical monomer for these two polymers.

3.2 Consider the excluded volume interaction between hard spheres of radius R.

   (i) What is the shortest possible distance between their centres?
   (ii) What is the interaction potential between these spheres?
   (iii) Demonstrate that the excluded volume of hard spheres is eight times larger than the volume v_0 a sphere:

$$v = \frac{32\pi}{3} R^3 = 8v_0.$$

3.3 Consider the excluded volume interaction between spherical particles with effective pairwise interaction potential

$$U(r) = \begin{cases} \infty & \text{for } r \leq 2R \\ -kT_0(2 - r/(2R)) & \text{for } 2R \leq r \leq 4R \\ 0 & \text{for } r > 4R \end{cases},$$

where kT_0 is the strength of the attractive potential with $T_0 = 100$ K.

   (i) Calculate the excluded volume of these particles.
   (ii) Plot the dimensionless excluded volume v/R^3 as a function of temperature and determine the θ-temperature of these particles.

3.4 Consider two cylindrical rods of length b and diameter d with $b \gg d$. Fix the centre of one of the rods at the origin of the coordinate system, pointing in the x direction.

   (i) Estimate the volume excluded for the second rod if it is fixed to always point in the y direction (perpendicular rods).
   (ii) Estimate the volume excluded for the second rod if it is fixed to always point in the x direction (parallel rods).
   (iii) How do you expect the excluded volume to change at different fixed angles between the two rods?

3.5 Consider a linear polymer chain with N monomers of length b, restricted to the air–water interface (two-dimensional conformations). Repeat the Flory theory calculation and demonstrate that the size R of the chain as a function of the 'excluded area' a per monomer (two-dimensional analogue of excluded volume v) is

$$R = a^{1/4}b^{1/2}N^{3/4}. \tag{3.137}$$

Compare the size of this chain at the interface to that in the bulk for parameters $N = 1000$, $b = 3$ Å, $a = 7.2$ Å^2, $v = 21.6$ Å^3.

3.6 Consider a randomly branched polymer in a dilute solution. Let us assume that the radius of gyration for this polymer in an ideal state (in the absence of excluded volume interactions) is

$$R_0 = bN^{1/4},$$

where b is the Kuhn monomer size and N is the number of Kuhn monomers. Use a Flory theory to determine the size R of this randomly branched polymer in a good solvent with excluded volume v. What is the size R of a randomly branched polymer with $N = 1000$, $b = 3$ Å, $v = 21.6$ Å^3? Compare this size to the size of a linear chain with the same degree of polymerization in the same good solvent and in θ-solvent.

3.7 Using the results of Problem 3.6, calculate the overlap volume fraction for the three cases:

   (i) randomly branched monodisperse polymer in good solvent.
   (ii) linear chain in good solvent.
   (iii) linear chain in θ-solvent.

3.8 Consider a randomly branched polymer with N monomers of length b. The polymer is restricted to the air–water interface and thus assumes a two-dimensional conformation. The ideal size of this polymer R_0 in the absence of excluded volume interactions is

$$R_0 = bN^{1/4}.$$

   (i) Repeat the Flory theory calculation to determine the size R of the branched polymer at the interface as a function of the 'excluded area' a per monomer (two-dimensional analogue of the excluded volume v), degree of polymerization N and monomer length b.
   (ii) Calculate the size of the branched polymer with $N = 1000$, $b = 3$ Å, $a = 7.2$ Å^2 at the air–water interface.
   (iii) Calculate the surface coverage (number of monomers per square Angstrom) at overlap for this randomly branched polymer at the air–water interface in good solvent.
   (iv) How much higher is the surface coverage at the overlap of randomly branched chains with $N = 100$, $b = 3$ Å, $a = 7.2$ Å^2 at the air–water interface, compared with $N = 1000$?

3.9 Consider a linear polymer chain with $N = 400$ Kuhn monomers of Kuhn length $b = 4$ Å in a solvent with θ-temperature of $27\,^\circ$C. The mean-field approximation of the interaction part of the free energy for a chain of size R is

$$F_{\text{int}} \approx kTR^3 \left[v\left(\frac{N}{R^3}\right)^2 + w\left(\frac{N}{R^3}\right)^3 + \cdots \right],$$

where the excluded volume of a monomer is

$$v \approx \left(1 - \frac{\theta}{T}\right) b^3,$$

and $w \approx b^6$ is the three-body interaction coefficient.

   (i) Use Flory theory to estimate the size of the chain swollen at the θ-temperature due to three-body repulsion.
   (ii) For what values of the excluded volume v does the two-body repulsion dominate over the three-body repulsion? Is the chain almost ideal or swollen if the two interactions are of the same order of magnitude?
   (iii) For what values of temperature T does the two-body repulsion dominate over the three-body repulsion?

(iv) Use Flory theory to estimate the size of the chain swollen at 60 °C due to excluded volume repulsion (ignore the three-body repulsion).

(v) Estimate the overlap volume fraction ϕ^* of the chain at 60 °C.

(vi) What is the number of Kuhn monomers in the largest chain that stays ideal at 60 °C?

3.10 Consider an oligomer with $N=3$ bonds occupying four lattice sites on a two-dimensional square lattice with lattice constant b. One end of the oligomer is fixed at the origin of the lattice.

(i) How many different conformations would such an oligomer have if it *can* occupy the same lattice site many times (simple random walk)?

(ii) How many different conformations would such an oligomer have if it *cannot* occupy the same lattice site (self-avoiding walk)?

(iii) Find the root-mean-square end-to-end distance of the oligomer for the first case.

(iv) Find the root-mean-square end-to-end distance of the oligomer for the second case.

3.11 Why is there no temperature dependence of the excluded volume in an athermal solvent?

3.12 If the monomer–solvent interaction potential is identical to the monomer–monomer interaction potential, the solvent is called:

(i) good,
(ii) θ,
(iii) athermal.

Explain your answer.

3.13 (i) Construct a Flory theory for the free energy of a polyelectrolyte chain consisting of N monomers of length b and net charge of the chain $Q=efN$, where f is the fraction of Kuhn monomers bearing a charge.

Hint: The electrostatic energy of the chain is $Q^2/(\epsilon R)$, where ϵ is the dielectric constant of the solvent and R is the size of the chain.

(ii) Show that the size of the chain at temperature T is

$$ R \approx Nbf^{2/3} \left(\frac{l_B}{b} \right)^{1/3}, $$

where the Bjerrum length is defined as $l_B \equiv e^2/(\epsilon k T)$.

3.14 (i) What is the relation of the fourth virial coefficients of spherical ($v_{4,s}$) and cylindrical ($v_{4,c}$) monomers if there are b/d spheres per cylinder?

(ii) What is the relation of the kth virial coefficients of spherical ($v_{k,s}$) and cylindrical ($v_{k,c}$) monomers?

Section 3.2

3.15 Calculate the free energy $F(N, R_f)$ and the force f for stretching a chain with an arbitrary scaling exponent ν in the dependence of the end-to-end distance on the number of monomers $R=bN^{\nu}$.

3.16 Calculate the free energy for compressing a real chain into a cylindrical tube with diameter D. Assume an arbitrary scaling exponent ν in the dependence of end-to-end distance of the chain on the number of monomers $R=bN^{\nu}$.

3.17 Calculate the free energy for squeezing a real chain between parallel plates into a slit of width D. Assume an arbitrary scaling exponent ν in the dependence of the end-to-end distance of the chain on the number of monomers $R = bN^\nu$.

3.18 Calculate the thickness ξ_{ads} of the adsorbed layer for a polymer with N monomers of size b. The interaction energy of a monomer in contact with the (planar) surface is $-\delta KT$. Assume an arbitrary scaling exponent ν in the dependence of end-to-end distance of the chain on the number of monomers $R = bN^\nu$.

3.19 Scaling theory of two-dimensional adsorption.

Consider a linear chain confined to an air–water interface. The attraction of each monomer at the contact line between the edge of the interface and the walls of the container is $-\delta kT$.

(i) Estimate the thickness and length of the ideal adsorbed chain of N Kuhn monomers with Kuhn length b.

(ii) Calculate the energy of adsorption of the ideal polymer of part (i) to the contact line.

(iii) Estimate the thickness and length of the real adsorbed chain of N Kuhn monomers with Kuhn length b. Recall that the unperturbed size of the real chain confined to the air–water interface in good solvent is $R \approx bN^{3/4}$ [Eq. (3.54)].

(iv) Calculate the energy of adsorption of the real polymer of part (iii) to the contact line.

3.20 Flory theory of adsorption from a two-dimensional interface onto a one-dimensional line.

Consider a real linear chain confined to an air–water interface. The attraction of a monomer at the contact line between the edge of the interface and the walls of the container is $-\delta kT$.

Calculate the thickness ξ_{ads} of the adsorbed real chain and the energy of adsorption using Flory theory. Recall that the unperturbed size of the real chain confined to the air–water interface in good solvent is given by Eq. (3.54).

3.21 Consider a polymer chain consisting of N Kuhn monomers of length b, adsorbed from a good solvent onto a solid substrate. A monomer in contact with the surface has interaction energy $-\delta kT$.

(i) What is the thickness ξ_{ads} of the adsorbed chain?

(ii) What is the size of the adsorption blob if $N = 1000$, $b = 3\ \text{Å}$, and $\delta = 0.4$?

Suppose that one of the ends of the adsorbed chain is attached to the tip of an atomic force microscope and is pulled away from the surface (still in a good solvent) with force f.

(iii) What is the minimal force f required to pull the chain away from the surface at room temperature?

(iv) What would be the minimal force f required to pull the chain away from the surface if the tip of the atomic force microscope were attached to a middle monomer rather than to the end monomer?

(v) Would the minimal force required to pull the chain away from the surface in a θ-solvent be smaller or larger (as compared to a good solvent) for the same attractive energy $-\delta kT$? Explain your answer.

3.22 Consider a real chain adsorbed at a surface with an excess free energy gain per monomer δkT. Assume that the monomer concentration decreases as a power law of the distance z from the surface

$$c(z) = c(0)\left(\frac{b}{z}\right)^a \quad \text{for } 0 < z < \xi_{ads},$$

where exponent $0 < a < 1$ and ξ_{ads} is the thickness of the adsorbed chain.

(i) Calculate the fraction of monomers within distance b of the surface. These are the monomers that lower their energy by favourable contacts with the surface.

(ii) Construct a modified Flory theory for the adsorption of a real chain and estimate the thickness ξ_{ads} of the adsorbed chain as a function of the excess free energy gain per monomer δkT. Ignore the effects of the density profile $c(z)$ on the confinement free energy penalty.

3.23 The effective interaction between each monomer and an adsorbing surface is

$$W(z) = -kT \frac{b^3}{z^3} A,$$

where A is a dimensionless parameter proportional to the **Hamaker constant** of the polymer–surface interaction. Consider an ideal chain adsorbed at the surface. Find the relation between the free energy gain per contact $-\delta kT$ and the effective polymer-surface interaction parameter A.

Section 3.3

3.24 Calculate the force needed to stretch a chain, of $N = 1000$ Kuhn monomers with Kuhn length $b = 5 \text{Å}$ in a good solvent with excluded volume $v = 37.5 \text{Å}^3$, by a factor of 4 from its unperturbed root-mean-square end-to-end distance at room temperature.

3.25 Consider a chain of N Kuhn monomers with Kuhn length b in a good solvent with excluded volume v confined between parallel plates in a slit of width D.

(i) What is the size of a thermal blob and the number of monomers in a thermal blob?

(ii) What is the size of a compression blob?

(iii) What is the number of monomers in a compression blob? Note: Be careful with respect to relative sizes of compression and thermal blobs.

(iv) What is the free energy of confinement of the chain in a slit?

(v) At what thickness of the slit D does the free energy change form between real and ideal chain expressions?

(vi) Estimate the value of this crossover thickness D for a chain with $N = 1000$ Kuhn monomers with Kuhn length $b = 5 \text{Å}$ in a good solvent with excluded volume $v = 20 \text{Å}^3$.

3.26 Consider a randomly branched polymer in a dilute solution. The ideal size of this polymer, R_0, in the absence of excluded volume interactions is

$$R_0 = bN^{1/4},$$

where b is the monomer size and N is the degree of polymerization.

(i) Use the Flory theory to determine the size R of this randomly branched polymer in a good solvent with excluded volume v.

(ii) What is the number of monomers in a thermal blob for this randomly branched polymer as a function of excluded volume v and monomer size b. How does it compare to the similar expression for the number of monomers in a thermal blob in linear polymers?

(iii) What is the size, ξ_T, of a thermal blob for a randomly branched polymer? How does it compare to the similar expression for the size of the thermal blob in linear polymers.

Hint: Recall that the cumulative energy of all excluded volume interactions inside a thermal blob is equal to kT.

(iv) What is the size of a randomly branched polymer with $N = 1000$, $b = 3$ Å, in a good solvent with $v = 2.7$ Å^3?

(v) What is the size of a linear chain for the same set of parameters?

(vi) What are the size and the number of monomers in the largest randomly branched polymer that stays ideal for monomer length $b = 3$ Å, and excluded volume $v = 2.7$ Å^3? (*Hint*: Thermal blob.)

(vii) What are the size and the number of monomers in the largest linear polymer that stays ideal for monomer length $b = 3$ Å, and excluded volume $v = 2.7$ Å^3? (*Hint*: Thermal blob.)

3.27 Assume the simple approximation for a temperature dependence of the excluded volume:

$$v = \left(1 - \frac{\theta}{T}\right)b^3.$$

Consider a chain with degree of polymerization $N = 1000$ and monomer size $b = 3$ Å in a solvent with θ-temperature of 30 °C.

(i) What would be the chain size at temperatures: $T = 10$, 30, and 60 °C?

(ii) Sketch the temperature dependence of the size of this polymer.

(iii) What is the degree of polymerization of the largest chain that stays ideal at $T = 60$ °C (*Hint*: Thermal blob)?

(iv) Estimate the degree of polymerization of the largest chain that dissolves in the solvent at $T = 10$ °C (*Hint*: Thermal blob)?

3.28 Use the following light scattering data for the temperature dependence of the second virial coefficient of a linear poly(methyl methacrylate) with $M_w = 2\,380\,000$ g mol^{-1} in a water/t-butyl alcohol mixture to determine the temperature dependence of excluded volume, assuming the Kuhn length of PMMA is 17 Å:

T (°C)	37.0	38.0	40.0	43.8	50.0	55.8
$10^5 A_2$ (cm^3 mol g^{-2})	−6.4	−3.4	−0.4	0.5	3.5	4.1

(i) Estimate the excluded volume at each of the six temperatures.

(ii) Estimate the θ-temperature from these data.

(iii) To measure the excluded volume of this polymer/solvent system at lower temperature, should a higher or lower molar mass sample be studied?

Data from M. Nakata, *Phys. Rev. E* **51**, 5770 (1995).

3.29 Derive an equation for the second virial coefficient in a solution of collapsed globules below their θ-temperature, in terms of the number of Kuhn monomers per chain N, the Kuhn monomer size b and the reduced temperature $(\theta - T)/T$. Can this second virial coefficient be related to the chain interaction parameter of Eq. (3.97)?

3.30 Determine the relation between the chain interaction parameter z [defined in Eq. (3.22)] and the number of thermal blobs per chain N/g_T.

Section 3.4

3.31 Polymerization of ring polymers.

Ring polymers are synthesized by linking two reactive ends of linear polymers in dilute solution. The cyclization probability can be defined as the probability of two ends of a chain being found within monomeric distance b of each other.

(i) What are the cyclization probabilities of N-mers in θ-solvent and in good solvent? What is the ratio of these probabilities for $N = 100$?
(ii) Are the resulting ring polymers obtained by cyclization in θ-solvent and good solvent statistically equivalent? In other words will these rings have the same size if they are placed in the same solvent? If they are different, which one is larger? Explain your answer.

Section 3.5

3.32 A pair correlation function $g(\vec{r})$ was defined in Section 2.7 as the probability of finding a monomer in a unit volume at distance $\vec{r}$ away from a given monomer (labeled by $j = 1$). Note that $j = 1$ is not necessarily the end monomer of any chain. The pair correlation function $g(\vec{r})$ can be written in terms of the delta function summed over all monomers except for the one at $\vec{r}_1$

$$g(\vec{r}) = \left\langle \sum_{j \neq 1} \delta(\vec{r} - (\vec{r}_j - \vec{r}_1)) \right\rangle \qquad (3.138)$$

(i) Show that the Fourier transform of the pair correlation function is

$$g(\vec{q}) = \int g(\vec{r}) \exp(-i\vec{q} \bullet \vec{r}) d^3r = \sum_{j=1}^{n} \langle \exp[-i\vec{q} \bullet (\vec{r}_j - \vec{r}_1)] \rangle - 1 \qquad (3.139)$$

(ii) Recognize that the choice of the $j = 1$ monomer was arbitrary and use the definition of the scattering function (Eq. 3.121) to show that

$$S(\vec{q}) = 1 + g(\vec{q}) = 1 + \int g(\vec{r}) \exp(-i\vec{q} \bullet \vec{r}) d^3r \qquad (3.140)$$

Bibliography

des Cloizeaux, J. and Jannink, G. *Polymers in Solution: Their Modelling and Structure* (Clarendon Press, Oxford, 1990).
Eisenriegler, E. *Polymers near Surfaces* (World Scientific, Singapore, 1993).
Flory, P. J. *Principles of Polymer Chemistry* (Cornell University Press, Ithaca, NY, 1953).
Freed, K. *Renormalization Group Theory of Macromolecules* (Wiley, New York, 1987).
de Gennes, P. G. *Scaling Concepts in Polymer Physics* (Cornell University Press, Ithaca, NY, 1979).
Grosberg, A. Yu. and Khokhlov, A. R. *Statistical Physics of Macro-molecules* (AIP Press, Woodbury, NY, 1994).
McQuarrie, D. A. *Statistical Mechanics* (University Science, 2000).
Yamakawa, H. *Modern Theory of Polymer Solutions* (Harper & Row, New York, 1971).

Thermodynamics of blends and solutions

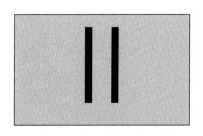

Thermodynamics of mixing

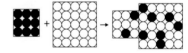

4

Mixtures are systems consisting of two or more different chemical species. **Binary mixtures** consist of only two different species. An example of a binary mixture is a blend of polystyrene and polybutadiene. Mixtures with three components are called **ternary**. An example of a ternary mixture is a solution of polystyrene and polybutadiene in toluene. If the mixture is uniform and all components of the mixture are intermixed on a molecular scale, the mixture is called **homogeneous**. An example of a homogeneous mixture is a polymer solution in a good solvent. If the mixture consists of several different phases (regions with different compositions), it is called **heterogeneous**. An example of a heterogeneous mixture is that of oil and water. Whether an equilibrium state of a given mixture is homogeneous or heterogeneous is determined by the composition dependence of the entropy and energy changes on mixing. Entropy always favours mixing, but energetic interactions between species can either promote or inhibit mixing.

4.1 Entropy of binary mixing

Consider the mixing of two species A and B. For the moment, assume that the two mix together to form a single-phase homogeneous liquid (criteria for such mixing will be determined later in this chapter). For purposes of illustration, the mixing is shown on a two-dimensional square lattice in Fig. 4.1. More generally, it is assumed that there is no volume change on mixing: volume V_A of species A is mixed with volume V_B of species B to make a mixture of volume $V_A + V_B$. The mixture is macroscopically uniform and the two components are randomly mixed to fill the entire lattice. The volume fractions of the two components in the binary mixture are ϕ_A and ϕ_B:

$$\phi_A = \frac{V_A}{V_A + V_B} \quad \text{and} \quad \phi_B = \frac{V_B}{V_A + V_B} = 1 - \phi_A. \tag{4.1}$$

While Fig. 4.1 shows the mixing of two small molecules of equal molecular volumes, similar mixing is possible if one or both of the species are polymers. In the more general case, the lattice site volume v_0 is defined by the smallest units (solvent molecules or monomers), and larger molecules

Fig. 4.1
Mixing two species with no volume change.

occupy multiple connected lattice sites. A molecule of species A has molecular volume

$$v_A = N_A v_0 \tag{4.2}$$

and a molecule of species B has molecular volume

$$v_B = N_B v_0, \tag{4.3}$$

where N_A and N_B are the numbers of lattice sites occupied by each respective molecule.[1] There are three cases of interest that are summarized in Table 4.1.

Regular solutions are mixtures of low molar mass species with $N_A = N_B = 1$. Polymer solutions are mixtures of macromolecules ($N_A = N \gg 1$) with the low molar mass solvent defining the lattice ($N_B = 1$). **Polymer blends** are mixtures of macromolecules of different chemical species ($N_A \gg 1$ and $N_B \gg 1$).

The combined system of volume $V_A + V_B$ occupies

$$n = \frac{V_A + V_B}{v_0} \tag{4.4}$$

lattice sites, while all molecules of species A occupy $V_A/v_0 = n\phi_A$ sites.

The entropy S is determined as the product of the Boltzmann constant k and the natural logarithm of the number of ways Ω to arrange molecules on the lattice (the number of states).

$$S = k \ln \Omega. \tag{4.5}$$

The number of translational states of a given single molecule is simply the number of independent positions that a molecule can have on the lattice, which is equal to the number of lattice sites. In a homogeneous mixture of A and B, each molecule has

$$\Omega_{AB} = n \tag{4.6}$$

possible states, where n is the total number of lattice sites of the combined system [Eq. (4.4)]. The number of states Ω_A of each molecule of species A before mixing (in a pure A state) is equal to the number of lattice sites occupied by species A.

$$\Omega_A = n\phi_A. \tag{4.7}$$

For a single molecule of species A, the entropy change on mixing is

$$\Delta S_A = k \ln \Omega_{AB} - k \ln \Omega_A = k \ln \left(\frac{\Omega_{AB}}{\Omega_A} \right)$$
$$= k \ln \left(\frac{1}{\phi_A} \right) = -k \ln \phi_A. \tag{4.8}$$

Table 4.1 The number of lattice sites occupied per molecule

	N_A	N_B
Regular solutions	1	1
Polymer solutions	N	1
Polymer blends	N_A	N_B

[1] The lattice sites are of the order of monomer sizes, but do not necessarily correspond precisely to either the chemical monomer or the Kuhn monomer.

Since the volume fraction is less than unity ($\phi_A < 1$), the entropy change upon mixing is always positive $\Delta S_A = -k \ln \phi_A > 0$. Equation (4.8) holds for the entropy contribution of each molecule of species A, with a similar relation for species B. To calculate the total entropy of mixing, the entropy contributions from each molecule in the system are summed:

$$\Delta S_{mix} = n_A \Delta S_A + n_B \Delta S_B = -k(n_A \ln \phi_A + n_B \ln \phi_B). \qquad (4.9)$$

There are $n_A = n\phi_A/N_A$ molecules of species A and $n_B = n\phi_B/N_B$ molecules of species B. The **entropy of mixing** per lattice site $\Delta \bar{S}_{mix} = \Delta S_{mix}/n$ is an intrinsic thermodynamic quantity:

$$\Delta \bar{S}_{mix} = -k \left[\frac{\phi_A}{N_A} \ln \phi_A + \frac{\phi_B}{N_B} \ln \phi_B \right]. \qquad (4.10)$$

The entropy of mixing per unit volume is $\Delta \bar{S}_{mix}/v_0$, where v_0 is the volume per lattice site.

A regular solution has $N_A = N_B = 1$ and a large entropy of mixing:

$$\Delta \bar{S}_{mix} = -k \left[\phi_A \ln \phi_A + \phi_B \ln \phi_B \right] \quad \text{for regular solutions.} \qquad (4.11)$$

A polymer solution has $N_A = N$ and $N_B = 1$:

$$\Delta \bar{S}_{mix} = -k \left[\frac{\phi_A}{N} \ln \phi_A + \phi_B \ln \phi_B \right] \quad \text{for polymer solutions.} \qquad (4.12)$$

Equations (4.10)–(4.12) predict enormous differences between the entropies of mixing for regular solutions, polymer solutions, and polymer blends. Consider the 10×10 square lattice of Fig. 4.2 with three different mixtures that each have $\phi_A = \phi_B = 0.5$. A regular solution of small molecules is shown in Fig. 4.2(a), using 50 black balls and 50 white balls. A polymer solution with five 10-ball black chains and 50 white balls is shown in Fig. 4.2(b) and a polymer blend with ten 10-ball chains (five black and five white) is shown in Fig. 4.2(c). The entropies of mixing per site for these mixtures are summarized in Table 4.2.

Typically N is large, making the first term in Eq. (4.12) negligible compared to the second term. For solutions with $\phi_A = \phi_B = 0.5$, as in Fig. 4.2 and Table 4.2, the entropy of mixing for the polymer solution is roughly half of that for the regular solution. For polymer blends, both N_A and N_B

Table 4.2 The mixing entropy per site for the three situations depicted in Fig. 4.2

Mixture	$\Delta \bar{S}_{mix}/k$
50 black balls and 50 white balls	0.69
Five 10-ball black chains and 50 white balls	0.38
Five 10-ball black chains and five 10-ball white chains	0.069

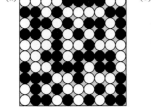 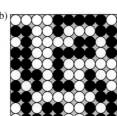 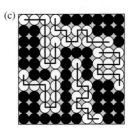

Fig. 4.2
Binary mixtures of (a) a regular solution of 50 white balls and 50 black balls, (b) a polymer solution of five black 10-ball chains, and (c) a polymer blend of five white 10-ball chains and five black 10-ball chains.

are typically large, making the entropy of mixing [Eq. (4.10)] very small. For this reason, polymers have *stymied entropy*. Connecting monomers into chains drastically reduces the number of possible states of the system. To illustrate this point, simply try to recreate Fig. 4.2(c) with molecules in a different state.

Despite the fact that the mixing entropy is small for polymer blends, it is always positive and hence promotes mixing. Mixtures with no difference in interaction energy between components are called **ideal mixtures**. Let us denote the volume fraction of component A by $\phi_A = \phi$ and the corresponding volume fraction of component B becomes $\phi_B = 1 - \phi$. The free energy of mixing per site for ideal mixtures is purely entropic:

$$\Delta \bar{F}_{\text{mix}} = -T\Delta \bar{S}_{\text{mix}} = kT\left[\frac{\phi}{N_A}\ln \phi + \frac{1-\phi}{N_B}\ln(1-\phi)\right]. \quad (4.13)$$

Ideal mixtures are always homogeneous as a result of the mixing entropy always being positive. Figure 4.3 shows the mixing free energy of an ideal regular solution, an ideal polymer solution, and an ideal polymer blend.

The mixing entropy calculated above includes only the translational entropy that results from the many possible locations for the centre of mass of each component. The calculation assumes that the conformational entropy of a polymer is identical in the mixed and pure states. This assumption is very good for polymer blends, where each chain is nearly ideal in the mixed and pure states. However, many polymer solutions have excluded volume that changes the conformation of the polymer in solution, as discussed in Chapter 3. Another important assumption in the entropy of mixing calculation is no volume change on mixing. Real polymer blends and solutions have very small, but measurable, volume changes when mixed.

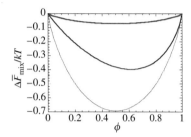

Fig. 4.3
The mixing free energy of an ideal mixture is always favourable and all compositions are stable. The bottom curve is a regular solution with $N_A = N_B = 1$. The middle curve is a polymer solution with $N_A = 10$ and $N_B = 1$. The top curve is a polymer blend with $N_A = N_B = 10$.

4.2 Energy of binary mixing

Interactions between species can be either attractive or repulsive. In most experimental situations, mixing occurs at constant pressure and the enthalpic interactions between species must be analysed to find a minimum of the Gibbs free energy of mixing. In the simplified lattice model (**Flory–Huggins theory**) discussed in the present chapter, components are mixed at constant volume and therefore we will be studying the energy of interactions between components and the change in the Helmholtz free energy of mixing.

The energy of mixing can be either negative (promoting mixing) or positive (opposing mixing). Regular solution theory allows for both possibilities, using the lattice model. To estimate the energy of mixing this theory places species into lattice sites randomly, ignoring any correlations. Thus, for all mixtures, favourable or unfavourable interactions between monomers are assumed to be small enough that they do not affect the random placement. Worse still, the regular solution approach effectively cuts the polymer chain into pieces that are the size of the solvent molecules (the lattice size) and distributes these pieces randomly. Such a mean-field approach ignores the correlations between monomers along the chain (the chain connectivity). Here, for simplicity, it is assumed that in polymer blends the monomer volumes of species A and B are identical.

Regular solution theory writes the energy of mixing in terms of three pairwise interaction energies (u_{AA}, u_{AB}, and u_{BB}) between adjacent lattice sites occupied by the two species. A mean field is used to determine the average pairwise interaction U_A of a monomer of species A occupying one lattice site with a neighbouring monomer on one of the adjacent sites. The probability of this neighbour being a monomer of species A is assumed to be the volume fraction ϕ_A of these molecules (ignoring the effect of interactions on this probability). The probability of this neighbour being a monomer of species B is $\phi_B = 1 - \phi_A$. The average pairwise interaction of an A-monomer with one of its neighbouring monomers is a volume fraction weighted sum of interaction energies:

$$U_A = u_{AA}\phi_A + u_{AB}\phi_B. \tag{4.14}$$

The corresponding energy of a B-monomer with one of its neighbours is similar to Eq. (4.14):

$$U_B = u_{AB}\phi_A + u_{BB}\phi_B. \tag{4.15}$$

Each lattice site of a regular lattice has z nearest neighbours, where z is the **coordination number** of the lattice. For example, $z = 4$ for a square lattice and $z = 6$ for a cubic lattice. Therefore, the average interaction energy of an A monomer with all of its z neighbours is zU_A. The average energy per monomer is half of this energy ($zU_A/2$) due to the fact that every pairwise interaction is counted twice (once for the monomer in question and once for its neighbour). The corresponding energy per site occupied by species B is $zU_B/2$. The number of sites occupied by species A (the number of monomers of species A) is $n\phi_A$, where n is the total number of sites in the combined system. The number of sites occupied by monomers of species B is $n\phi_B$. Summing all the interactions gives the total interaction energy of the mixture:

$$U = \frac{zn}{2}[U_A\phi_A + U_B\phi_B]. \tag{4.16}$$

Denoting the volume fraction of species A by $\phi = \phi_A = 1 - \phi_B$, Eqs (4.14)–(4.16) are combined to get the total interaction energy of a binary mixture with n lattice sites:

$$\begin{aligned} U &= \frac{zn}{2}\{[u_{AA}\phi + u_{AB}(1 - \phi)]\phi + [u_{AB}\phi + u_{BB}(1 - \phi)](1 - \phi)\} \\ &= \frac{zn}{2}[u_{AA}\phi^2 + 2u_{AB}\phi(1 - \phi) + u_{BB}(1 - \phi)^2]. \end{aligned} \tag{4.17}$$

The interaction energy per site in a pure A component before mixing is $zu_{AA}/2$, because each monomer of species A before mixing is only surrounded by species A. We ignore the boundary effects because of the very small surface-to-volume ratio for most macroscopic systems. The total number of monomers of species A is $n\phi$ and therefore the total energy of species A before mixing is

$$\frac{zn}{2}u_{AA}\phi$$

and the total energy of species B before mixing is

$$\frac{zn}{2}u_{BB}(1-\phi).$$

The total energy of both species before mixing is the sum of the energies of the two pure components:

$$U_0 = \frac{zn}{2}[u_{AA}\phi + u_{BB}(1-\phi)]. \tag{4.18}$$

The energy change on mixing is

$$\begin{aligned}
U - U_0 &= \frac{zn}{2}[u_{AA}\phi^2 + 2u_{AB}\phi(1-\phi) + u_{BB}(1-\phi)^2 - u_{AA}\phi - u_{BB}(1-\phi)] \\
&= \frac{zn}{2}[u_{AA}(\phi^2 - \phi) + 2u_{AB}\phi(1-\phi) + u_{BB}(1 - 2\phi + \phi^2 - 1 + \phi)] \\
&= \frac{zn}{2}[u_{AA}\phi(\phi - 1) + 2u_{AB}\phi(1-\phi) + u_{BB}\phi(\phi - 1)] \\
&= \frac{zn}{2}\phi(1-\phi)(2u_{AB} - u_{AA} - u_{BB}) \tag{4.19}
\end{aligned}$$

It is convenient to study the intensive property, which is the energy change on mixing per site:

$$\Delta \bar{U}_{mix} = \frac{U - U_0}{n} = \frac{z}{2}\phi(1-\phi)(2u_{AB} - u_{AA} - u_{BB}). \tag{4.20}$$

The **Flory interaction parameter** χ is defined to characterize the difference of interaction energies in the mixture:

$$\chi \equiv \frac{z}{2}\frac{(2u_{AB} - u_{AA} - u_{BB})}{kT}. \tag{4.21}$$

Defined in this fashion, χ is a dimensionless measure of the differences in the strength of pairwise interaction energies between species in a mixture (compared with the same species in their pure component states). Using this definition, we write the **energy of mixing** per lattice site as

$$\Delta \bar{U}_{mix} = \chi\phi(1-\phi)kT. \tag{4.22}$$

This energy equation is a mean-field description of all binary regular mixtures: regular solutions, polymer solutions, and polymer blends.

Combining with Eq. (4.10) for the entropy of mixing, we arrive at the **Helmholtz free energy of mixing** per lattice site:

$$\Delta \bar{F}_{\text{mix}} = \Delta \bar{U}_{\text{mix}} - T \Delta \bar{S}_{\text{mix}}$$
$$= kT \left[\frac{\phi}{N_A} \ln \phi + \frac{1 - \phi}{N_B} \ln(1 - \phi) + \chi \phi (1 - \phi) \right]. \tag{4.23}$$

The free energy of mixing per unit volume is $\Delta \bar{F}_{\text{mix}} / v_0$. Equation (4.23) was first calculated by Huggins and later independently derived by Flory, and is commonly referred to as the **Flory–Huggins equation**.

For non-polymeric mixtures with $N_A = N_B = 1$, this equation was developed earlier by Hildebrand and is called **regular solution theory**:

$$\Delta \bar{F}_{\text{mix}} = kT \left[\phi \ln \phi + (1 - \phi) \ln(1 - \phi) + \chi \phi (1 - \phi) \right]. \tag{4.24}$$

For polymer solutions, $N_A = N$ and $N_B = 1$, reducing Eq. (4.23) to the **Flory–Huggins equation for polymer solutions**:

$$\Delta \bar{F}_{\text{mix}} = kT \left[\frac{\phi}{N} \ln \phi + (1 - \phi) \ln(1 - \phi) + \chi \phi (1 - \phi) \right]. \tag{4.25}$$

The first two terms in the free energy of mixing [Eq. (4.23)] have entropic origin and always act to promote mixing, although with blends of long-chain polymers these terms are quite small. The last term has energetic origin, and can be positive (opposing mixing), zero [ideal mixtures—Eq. (4.13)], or negative (promoting mixing) depending on the sign of the interaction parameter χ.

If there is a net attraction between species (i.e. they like each other better than they like themselves), $\chi < 0$ and a single-phase mixture is favourable for all compositions. More often there is a net repulsion between species (they like themselves more than each other) and the Flory interaction parameter is positive $\chi > 0$. In Section 4.4, we will show that in this case the equilibrium state of the mixture depends not on the sign of the free energy of mixing $\Delta \bar{F}_{\text{mix}}$ at the particular composition of interest, but on the functional dependence of this free energy on the composition ϕ for the whole range of compositions. This functional dependence $\Delta \bar{F}_{\text{mix}}(\phi)$ depends on the value of the Flory interaction parameter χ as well as on the degrees of polymerization of both molecules N_A and N_B.

It is very important to know the value of the Flory interaction parameter χ for a given mixture. Methods of measuring this parameter are discussed in Section 4.6 and tables of χ parameters are listed in many reference books (see the 1996 review by Balsara).

For non-polar mixtures with species interacting mainly by dispersion forces, the interaction parameter χ can be estimated by the method developed by Hildebrand and Scott. It is based on the **solubility parameter** δ related to the energy of vapourization ΔE of a molecule. For example, for a molecule of species A the solubility parameter is defined as

$$\delta_A \equiv \sqrt{\frac{\Delta E_A}{v_A}}, \tag{4.26}$$

where v_A is the volume of molecule A [Eq. (4.2)]. The energy of vapourization ΔE_A of a molecule A is the energy of all the interactions between the molecule and its neighbours that have to be disrupted to remove the molecule from the pure A state. The ratio $\Delta E_A/v_A$ is called the cohesive energy density and is the interaction energy per unit volume between the molecules in the pure A state. The interaction energy per site in the pure A state $zu_{AA}/2$ [see the paragraph below Eq. (4.17)] is therefore related to the solubility parameter δ_A.

$$-\frac{zu_{AA}}{2} = v_0 \frac{\Delta E_A}{v_A} = v_0 \delta_A^2, \qquad (4.27)$$

where v_0 is the volume per site. Note that the minus sign is due to the fact that the interaction energy is negative $u_{AA} < 0$, while the energy of vapourization is defined to be positive. Similarly, the interaction energy per site in the pure B state is

$$-\frac{zu_{BB}}{2} = v_0 \frac{\Delta E_B}{v_B} = v_0 \delta_B^2, \qquad (4.28)$$

where v_B is the volume of molecule B [Eq. (4.3)]. The cohesive energy density of interaction between molecules A and B is estimated from the geometric mean approximation

$$-\frac{zu_{AB}}{2} = v_0 \delta_A \delta_B. \qquad (4.29)$$

Substituting Eqs (4.27)–(4.29) into the definition of the Flory interaction parameter [Eq. (4.21)] allows it to be written in terms of solubility parameter difference.[2]

$$\chi \approx v_0 \frac{[\delta_A^2 + \delta_B^2 - 2\delta_A\delta_B]}{kT} = \frac{v_0}{kT}(\delta_A - \delta_B)^2. \qquad (4.30)$$

Since χ is related to the *square* of the difference in solubility parameters it is clear why the Flory interaction parameter is usually positive $\chi > 0$. The above approach works reasonably well for non-polar interactions, which only have van der Waals forces between species, and does not work in mixtures with strong polar or specific interactions, such as hydrogen bonds.

One of the major assumptions of the Flory–Huggins theory is that there is no volume change on mixing and that monomers of both species can fit on the sites of the same lattice. In most real polymer blends, the volume per monomer changes upon mixing. Some monomers may pack together better with certain other monomers. The volume change on mixing and local packing effects lead to a temperature-independent additive constant in the expression of the Flory interaction parameter. In practice, these

---

[2] Note that since the Flory χ parameter is defined in terms of energies *per site*, it is proportional to the site volume v_0. The site volume, therefore, must be specified whenever χ is discussed.

effects are not fully understood and all deviations from the lattice model are lumped into the interaction parameter χ, which can display non-trivial dependences on composition, chain length, and temperature. Empirically, the temperature dependence of the Flory interaction parameter is often written as the sum of two terms:

$$\chi(T) \cong A + \frac{B}{T} \tag{4.31}$$

The temperature-independent term A is referred to as the 'entropic part' of χ, while B/T is called the 'enthalpic part'. The parameters A and B have been tabulated for many polymer blends and we list representative examples in Table 4.3. Isotopic blends typically have small positive χ parameters (deuterated polystyrene blended with ordinary polystyrene dPS/PS is an example) making them only phase separate at very high molar masses. PS/PMMA has four entries in Table 4.3, which reflect the differences encountered by labelling various species with deuterium. PS/PMMA is typical of many polymer pairs, for which the χ parameter is positive and of order 0.01, making only low molar mass polymers form miscible blends. PVME/PS, PS/PPO, and PS/TMPC have a strongly negative χ parameter over a wide range of temperatures (of order -0.01) but since $A > 0$ and $B < 0$, these blends phase separate on heating. PEO/PMMA, PP/hhPP and PIB/hhPP, all represent blends with very weak interactions between components ($\chi \cong 0$).

Additionally, the parameters A and B are often found to depend weakly on chain lengths and composition. Shortcomings of the Flory–Huggins theory are usually lumped into the interaction parameter χ. The Flory–Huggins equation (with all the corrections combined in χ) contains all of the thermodynamic information needed to decide the equilibrium

Table 4.3 Temperature dependence of the Flory interaction parameters of polymer blends [Eq. (4.31)] with $v_0 = 100$ Å^3

Polymer blend	A	B (K)	T range (°C)
dPS/PS	-0.00017	0.117	150–220
dPS/PMMA	0.0174	2.39	120–180
PS/dPMMA	0.0180	1.96	170–210
PS/PMMA	0.0129	1.96	100–200
dPS/dPMMA	0.0154	1.96	130–210
PVME/PS	0.103	-43.0	60–150
dPS/PPO	0.059	-32.5	180–330
dPS/TMPC	0.157	-81.3	190–250
PEO/dPMMA	-0.0021	–	80–160
PP/hhPP	-0.00364	1.84	30–130
PIB/$d$hhPP	0.0180	-7.74	30–170

dPS—deuterated polystyrene; PS—polystyrene; PMMA—poly(methyl methacrylate); dPMMA—deuterated poly(methyl methacrylate); PVME—poly(vinyl methyl ether); PPO—poly(2,6-dimethyl 1,4-phenylene oxide); TMPC—tetramethylpolycarbonate; PEO—poly(ethylene oxide); PP—polypropylene; hhPP—head-to-head polypropylene; PIB—polyisobutylene; $d$hhPP—deuterium labelled head-to-head polypropylene (after N. P. Balsara, *Physical Properties of Polymers Handbook*, AIP Press, 1996, Chapter 19).

state of a mixture and whether any metastable states are possible, as discussed next.

4.3   Equilibrium and stability

The definition of thermodynamic equilibrium is the state of the system with minimum free energy. Consider the states of a brick, shown in Fig. 4.4. The angle that the long side of the brick makes with the ground is θ, as defined in state A. The stable equilibrium state, or ground state of the brick, is shown as state B in Fig. 4.4, with the brick lying on the ground. This state is stable because any perturbations in the angle that the brick makes with the ground lead to its centre-of-mass being higher above the ground than the ground state, thereby increasing its potential energy. If the brick is balanced on its edge (state A in Fig. 4.4), any small fluctuations would lead to its fall and state A is called unstable. When standing on one end (state C), the brick has its centre-of-mass at half of the height of the brick. Any small change in θ from state C will increase the potential energy by raising the centre-of-mass of the brick. Thus, state C is **metastable**: small perturbations do not allow the brick to move from state C to state B, even though state B has lower energy and thus is the equilibrium state of the brick. Indeed, the brick in state C would stand until an earthquake causes it to move to state B. Hence, a long time duration of a given state is insufficient information to conclude that the state is the equilibrium state. The graph in Fig. 4.4 summarizes the free energy of the brick as a function of angle θ.

Consider the local stability of a homogeneous mixture of composition ϕ_0 with free energy $F_{mix}(\phi_0)$ that is either locally concave or convex, shown in Fig. 4.5. Stability is determined by whether the free energy of the mixed state $F_{mix}(\phi_0)$ is higher or lower than that of a phase separated state, $F_{\alpha\beta}(\phi_0)$. If the system with overall composition ϕ_0 is in a state with two phases, with volume fraction of A species in the α phase ϕ_α and the fraction of A component in the β phase ϕ_β (see Fig. 4.5), the relative amounts of each phase are determined from the **lever rule**. With the fraction f_α of the volume of the material having composition ϕ_α (and fraction $f_\beta = 1 - f_\alpha$ having composition ϕ_β), the total volume fraction of A component in the system is the sum of contributions from the two phases:

$$\phi_0 = f_\alpha \phi_\alpha + f_\beta \phi_\beta. \tag{4.32}$$

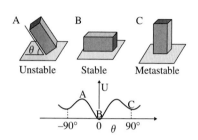

Fig. 4.4
The states of a brick.

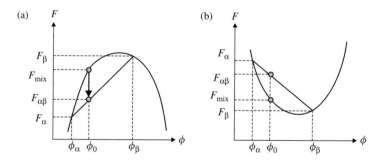

Fig. 4.5
Composition dependence of free energy, with examples of systems that are (a) unstable and (b) locally stable. Local stability is determined by the sign of the second derivative of free energy with respect to composition.

This equation can be solved for the fractions of the material that will have each composition (since $f_\beta = 1 - f_\alpha$):

$$f_\alpha = \frac{\phi_\beta - \phi_0}{\phi_\beta - \phi_\alpha} \quad \text{and} \quad f_\beta = 1 - f_\alpha = \frac{\phi_0 - \phi_\alpha}{\phi_\beta - \phi_\alpha}. \tag{4.33}$$

The free energy of the demixed state is the weighted average of the free energies of the material in each of the two states (F_α and F_β), neglecting the interfacial energy (surface tension) between the two phases:

$$F_{\alpha\beta}(\phi_0) = f_\alpha F_\alpha + f_\beta F_\beta = \frac{(\phi_\beta - \phi_0) F_\alpha + (\phi_0 - \phi_\alpha) F_\beta}{\phi_\beta - \phi_\alpha}. \tag{4.34}$$

This linear composition dependence of the free energy of the demixed state $F_{\alpha\beta}(\phi_0)$ results in the straight lines in Fig. 4.5 that connect the free energies F_α and F_β of the two compositions ϕ_α and ϕ_β. The local curvature of the free energy determines local stability, as demonstrated in Fig. 4.5. If the composition dependence of the free energy is concave [Fig. 4.5(a)], the system can spontaneously lower its free energy by phase separating into two phases, since $F_{\alpha\beta}(\phi_0) < F_{\text{mix}}(\phi_0)$.

On the other hand, when the composition dependence of the free energy is convex, as shown in Fig. 4.5(b), any mixed state has lower free energy than any state the blend could phase separate into $F_{\text{mix}}(\phi_0) < F_{\alpha\beta}(\phi_0)$, making the mixed state locally stable. The criterion for local stability is written in terms of the second derivative of the free energy:

$$\frac{\partial^2 F_{\text{mix}}}{\partial \phi^2} < 0 \quad \text{unstable}, \tag{4.35}$$

$$\frac{\partial^2 F_{\text{mix}}}{\partial \phi^2} > 0 \quad \text{locally stable}. \tag{4.36}$$

Ideal mixtures with $\Delta \bar{U}_{\text{mix}} = 0$ have their free energy of mixing [Eq. (4.13)] convex over the entire composition range, as can be seen in Fig. 4.3. To understand why it is convex, we differentiate Eq. (4.13) with respect to composition

$$\frac{\partial \Delta \bar{F}_{\text{mix}}}{\partial \phi} = -T \frac{\partial \Delta \bar{S}_{\text{mix}}}{\partial \phi} = kT \left[\frac{\ln \phi}{N_A} + \frac{1}{N_A} - \frac{\ln(1 - \phi)}{N_B} - \frac{1}{N_B} \right]. \tag{4.37}$$

Notice that this purely entropic contribution diverges at both extremes of composition ($\partial \Delta \bar{F}_{\text{mix}}/\partial \phi \to -\infty$ as $\phi \to 0$ and $\partial \Delta \bar{F}_{\text{mix}}/\partial \phi \to \infty$ as $\phi \to 1$). This divergence means that a small amount of either species will *always* dissolve even if there are strong unfavourable energetic interactions. Differentiating the free energy of mixing a second time determines the stability of the mixed state for ideal mixtures

$$\frac{\partial^2 \Delta \bar{F}_{\text{mix}}}{\partial \phi^2} = -T \frac{\partial^2 \Delta \bar{S}_{\text{mix}}}{\partial \phi^2} = kT \left[\frac{1}{N_A \phi} + \frac{1}{N_B(1 - \phi)} \right] > 0. \tag{4.38}$$

Homogeneous ideal mixtures are stable for all compositions because *entropy always acts to promote mixing*, and the ideal mixture does not have any energetic contribution to its free energy.

The opposite case where the energy dominates is found at $T=0$ K, because the entropic contribution vanishes. The free energy only has an energetic part given, for example, by Eq. (4.22) for regular mixtures. Differentiating Eq. (4.22) twice with respect to composition determines whether the blend is locally stable at 0 K

$$\frac{\partial^2 \Delta \bar{F}_{mix}}{\partial \phi^2} = \frac{\partial^2 \Delta \bar{U}_{mix}}{\partial \phi^2} = -2\chi kT \tag{4.39}$$

The stability criterion at $T=0$ K can be determined from either Eq. (4.21) or Eq. (4.31)

$$\frac{\partial^2 \Delta \bar{F}_{mix}}{\partial \phi^2} = -z(2u_{AB} - u_{AA} - u_{BB}) = -2kB \tag{4.40}$$

The parameter B describes the temperature dependence of χ in Eq. (4.31). If the components of the mixture like themselves more than each other

$$u_{AB} > \frac{u_{AA} + u_{BB}}{2} \quad \text{or} \quad B > 0$$

the free energy of mixing is concave (Fig. 4.6, top curve) and homogeneous mixtures are unstable for all compositions at $T=0$ K because the second derivative of the free energy of mixing is negative [Eq. (4.35)]. Any mixture phase separates into the two pure components at $T=0$ K since entropy makes no contribution at this special temperature. This case corresponds to positive Flory interaction parameter $\chi > 0$.

If the components like each other better than themselves

$$u_{AB} < \frac{u_{AA} + u_{BB}}{2} \quad \text{or} \quad B < 0$$

the free energy of mixing is convex (Fig. 4.6, bottom curve) and homogeneous mixtures of any composition are stable at $T=0$ K. This case corresponds to negative Flory interaction parameter $\chi < 0$.

Real mixtures have both energetic and entropic contributions to their free energy of mixing. The local stability of the mixture is determined by the sign of the second derivative of the free energy with respect to composition:

$$\frac{\partial^2 \Delta \bar{F}_{mix}}{\partial \phi^2} = \frac{\partial^2 \Delta \bar{U}_{mix}}{\partial \phi^2} - T\frac{\partial^2 \Delta \bar{S}_{mix}}{\partial \phi^2}$$

$$= kT\left[\frac{1}{N_A \phi} + \frac{1}{N_B(1-\phi)}\right] - 2\chi kT. \tag{4.41}$$

At finite temperatures, $\Delta \bar{F}_{mix}$ is convex at both ends of the composition range because its second derivative is positive due to the diverging slope of the entropy of mixing $\Delta \bar{S}_{mix}$.

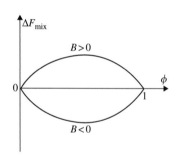

Fig. 4.6
At $T=0$ K, the mixing free energy is determined by the energy of mixing. If $B > 0$, mixing is unfavourable and all blend compositions are unstable. If $B < 0$, mixing is favourable and all blend compositions are stable.

For example, consider a polymer blend with $N_A = 200$ and $N_B = 100$, for which $\chi T = 5$ K. At high temperatures the entropic term of the mixing free energy dominates, and all blend compositions are stable, as shown in Fig. 4.7 at 350 K.

As temperature is lowered the entropic term diminishes, allowing the repulsive energetic term to start to be important at intermediate compositions. Entropy always dominates the extremes of composition (due to the divergent first derivative) making those extremes stable. Below some critical temperature T_c (defined in detail in Section 4.4), a composition range with concave free energy appears, which makes intermediate compositions unstable. Below T_c there is a range of compositions for which there are phase separated states with lower free energy than the homogeneous state. Many demixed states have lower free energy than the homogeneous state, but the lowest free energy state defines the equilibrium state. Straight lines connecting the two phase compositions determine the free energy of the phase separated state. In order to minimize the free energy, the system chooses the compositions that have the lowest possible straight line, which is a common tangent. The phases present are thus determined by the **common tangent rule**. This common tangent minimization of the free energy of mixing effectively requires that the chemical potential of each species in both phases are balanced at equilibrium. The two equilibrium compositions ϕ' and ϕ'' at 250 K correspond to a common tangent line in Fig. 4.7. For any overall composition in the **miscibility gap** between ϕ' and ϕ'', the system can minimize its free energy by phase separating into two phases of composition ϕ' and ϕ''. The amounts of each phase are determined by the lever rule outlined above [Eq. (4.33)]. The composition ranges $0 < \phi < \phi'$ or $\phi'' < \phi < 1$ are outside the miscibility gap and the homogeneously mixed state is the stable equilibrium state for these blend compositions.

Within the miscibility gap there are unstable and metastable regions, separated by inflection points at which the second derivative of the free energy is zero ($\partial^2 \Delta \bar{F}_{mix} / \partial \phi^2 = 0$). Between the inflection points, the second derivative of the free energy is negative and the homogeneously mixed state is unstable. Even the smallest fluctuations in composition lower the free energy, leading to spontaneous phase separation (called **spinodal decomposition**). Between the infection points and the equilibrium phase separated compositions, there are two regions that have positive second derivative of the free energy of mixing. Even though the free energy of the homogeneous state is larger than that of the phase-separated state (on the common tangent line) the mixed state is locally stable to small composition fluctuations. Such states are metastable because large fluctuations are required for the system to reach thermodynamic equilibrium. Phase separation in this metastable regime occurs by **nucleation and growth**. The nuclei of the more stable phase must be larger than some critical size in order to grow in the metastable region because of the surface tension between phases (see Problem 4.15). The new phase can grow only when a sufficiently large fluctuation creates a domain larger than the critical size.

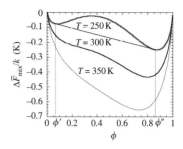

Fig. 4.7
Composition dependence of the free energy of mixing at three temperatures for a hypothetical blend with $N_A = 200$ and $N_B = 100$, for which $\chi = (5$ K$)/T$.

4.4 Phase diagrams

By considering the temperature dependence of the free energy of mixing, a phase diagram can be constructed to summarize the phase behaviour of the mixture, showing regions of stability, instability, and metastability. Recall the free energy of mixing for a polymer blend

$$\Delta \bar{F}_{\text{mix}} = kT \left[\frac{\phi}{N_A} \ln \phi + \frac{1 - \phi}{N_B} \ln(1 - \phi) + \chi \phi (1 - \phi) \right]. \quad (4.42)$$

The phase boundary is determined by the common tangent of the free energy at the compositions ϕ' and ϕ'' corresponding to the two equilibrium phases

$$\left(\frac{\partial \Delta \bar{F}_{\text{mix}}}{\partial \phi} \right)_{\phi = \phi'} = \left(\frac{\partial \Delta \bar{F}_{\text{mix}}}{\partial \psi} \right)_{\phi = \phi''}. \quad (4.43)$$

This derivative of the free energy of mixing per site with respect to volume fraction of component A is

$$\frac{\partial \Delta \bar{F}_{\text{mix}}}{\partial \phi} = kT \left[\frac{\ln \phi}{N_A} + \frac{1}{N_A} - \frac{\ln(1 - \phi)}{N_B} - \frac{1}{N_B} + \chi(1 - 2\phi) \right]. \quad (4.44)$$

For the simple example of a symmetric polymer blend with $N_A = N_B = N$, the common tangent line is horizontal.

$$\left(\frac{\partial \Delta \bar{F}_{\text{mix}}}{\partial \phi} \right)_{\phi = \phi'} = \left(\frac{\partial \Delta \bar{F}_{\text{mix}}}{\partial \phi} \right)_{\phi = \phi''}$$

$$= kT \left[\frac{\ln \phi}{N} - \frac{\ln(1 - \phi)}{N} + \chi(1 - 2\phi) \right] = 0. \quad (4.45)$$

The above equation can be solved for the interaction parameter corresponding to the phase boundary—the **binodal** (solid line in the bottom part of Fig. 4.8) of a symmetric blend:

$$\chi_b = \frac{1}{2\phi - 1} \left[\frac{\ln \phi}{N} - \frac{\ln(1 - \phi)}{N} \right] = \frac{\ln(\phi/(1 - \phi))}{(2\phi - 1)N}. \quad (4.46)$$

Using the phenomenological temperature dependence of the interaction parameter [Eq. (4.31)], this relation can be transformed to the binodal of the phase diagram in the space of temperature and composition:

$$T_b = \frac{B}{\ln[\phi/(1 - \phi)]/[(2\phi - 1)N] - A}. \quad (4.47)$$

The binodal for binary mixtures coincides with the **coexistence curve**, since for a given temperature (or $N\chi$) with overall composition in the two-phase region, the two compositions that coexist at equilibrium can be read off the binodal. Any overall composition at temperature T within the miscibility gap defined by the binodal has its minimum free energy in a

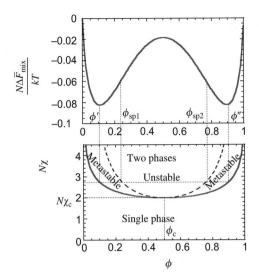

Fig. 4.8
Composition dependence of the free energy of mixing for a symmetric polymer blend with the product $\chi N = 2.7$ (top figure) and the corresponding phase diagram (bottom figure). Binodal (solid curve) and spinodal (dashed curve) are shown on the phase diagram.

phase-separated state with the compositions given by the two coexistence curve compositions ϕ' and ϕ''.

Returning to the general case of an asymmetric blend, the inflection points in $\Delta \bar{F}_{\text{mix}}(\phi)$ can be found by equating the second derivative of the free energy [Eq. (4.41)] to zero:

$$\frac{\partial^2 \Delta \bar{F}_{\text{mix}}}{\partial \phi^2} = kT\left[\frac{1}{N_A \phi} + \frac{1}{N_B(1-\phi)} - 2\chi\right] = 0. \qquad (4.48)$$

The curve corresponding to the inflection point is the boundary between unstable and metastable regions and is called the **spinodal** (the dashed line in the bottom part of Fig. 4.8):

$$\chi_s = \frac{1}{2}\left[\frac{1}{N_A \phi} + \frac{1}{N_B(1-\phi)}\right]. \qquad (4.49)$$

This spinodal can also be transformed to a phase diagram in the temperature–composition plane by using the experimentally determined $\chi(T)$ via Eq. (4.31):

$$T_s = \frac{B}{\frac{1}{2}[1/(N_A \phi) + 1/(N_B(1-\phi))] - A}. \qquad (4.50)$$

In a binary blend the lowest point on the spinodal curve corresponds to the **critical point**:

$$\frac{\partial \chi_s}{\partial \phi} = \frac{1}{2}\left[-\frac{1}{N_A \phi^2} + \frac{1}{N_B(1-\phi)^2}\right] = 0. \qquad (4.51)$$

The solution of this equation gives the **critical composition**:

$$\phi_c = \frac{\sqrt{N_B}}{\sqrt{N_A} + \sqrt{N_B}}. \tag{4.52}$$

Substituting this critical composition back into the equation of the spinodal [Eq. (4.49)] determines the critical interaction parameter:

$$\chi_c = \frac{1}{2}\frac{(\sqrt{N_A} + \sqrt{N_B})^2}{N_A N_B} = \frac{1}{2}\left(\frac{1}{\sqrt{N_A}} + \frac{1}{\sqrt{N_B}}\right)^2. \tag{4.53}$$

Equation (4.31) can again be utilized to determine the **critical temperature** from χ_c:

$$T_c = \frac{B}{\chi_c - A} = \frac{B}{\frac{1}{2}(1/\sqrt{N_A} + 1/\sqrt{N_B})^2 - A}. \tag{4.54}$$

For a symmetric polymer blend ($N_A = N_B = N$), the whole phase diagram is symmetric (see Fig. 4.8) with the critical composition

$$\phi_c = \frac{1}{2} \tag{4.55}$$

and very small critical interaction parameter

$$\chi_c = \frac{2}{N}. \tag{4.56}$$

Since this critical interaction parameter is very small for blends of long chains, most polymer blends have $\chi > \chi_c$ and thus are phase separated over some composition range (within the miscibility gap). Only blends with either very weak repulsion ($0 < \chi < \chi_c$), or a net attraction between components of the mixture ($\chi < 0$) form homogeneous (single-phase) blends over the whole composition range.

In polymer solutions ($N_A = N$ and $N_B = 1$), the phase diagram is strongly asymmetric with low critical composition

$$\phi_c = \frac{1}{\sqrt{N} + 1} \simeq \frac{1}{\sqrt{N}} \tag{4.57}$$

and critical interaction parameter close to 1/2

$$\chi_c = \frac{1}{2} + \frac{1}{\sqrt{N}} + \frac{1}{2N} \simeq \frac{1}{2} + \frac{1}{\sqrt{N}}. \tag{4.58}$$

Note that the spinodal and binodal for any binary mixture meet at the critical point (Fig. 4.8). For interaction parameters χ below the critical one (for $\chi < \chi_c$) the homogeneous mixture is stable at any composition $0 \le \phi \le 1$. For higher values of the interaction parameter (for $\chi > \chi_c$) there is a miscibility gap between the two branches of the binodal in Fig. 4.8. For any composition in a miscibility gap, the equilibrium state corresponds to two phases with compositions ϕ' and ϕ'' located on the two branches of the coexistence curve at the same value of χ.

Experimentally, the interaction parameter is most conveniently changed by varying temperature T [see Eq. (4.31)]. Phase diagrams are typically plotted in the temperature – composition plane. Examples of phase diagrams for a polymer blend and a polymer solution are shown in Fig. 4.9. The binodal line separates the phase diagrams into a single-phase region and a two-phase region.

If $B > 0$ in Eq. (4.31), then χ decreases as temperature is raised. This situation is depicted in Fig. 4.10. The highest temperature of the two-phase region is the **upper critical solution temperature (UCST)** T_c. For all $T > T_c$, the homogeneous mixtures are stable. On the other hand, if $B < 0$ in Eq. (4.31), then χ decreases as temperature is lowered. The lowest temperature of the two-phase region is the **lower critical solution temperature (LCST)**, and this case is shown in Fig. 4.11. While the case of $B > 0$ is more

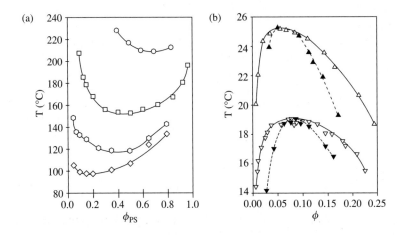

Fig. 4.9
Phase diagrams of polymer blends and solutions (open symbols are binodals and filled symbols are spinodals). (a) Polymer blends of poly(vinyl methyl ether) ($M = 51\,500$ g mol^{-1}) and various molar masses of polystyrene (circles have $M = 10\,000$ g mol^{-1}, squares have $M = 20\,400$ g mol^{-1}, hexagons have $M = 51\,000$ g mol^{-1}, diamonds have $M = 200\,000$ g mol^{-1}), data from T. K. Kwei and T. T. Wang, in: *Polymer Blends*, Vol. 1 (D. R. Paul and S. Newman, editors), Academic Press, 1978. (b) Polyisoprene solutions in dioxane (upside-down triangles have $M = 53\,300$ g mol^{-1}, triangles have $M = 133\,000$ g mol^{-1}), data from N. Takano *et al.*, *Polym. J.* **17**, 1123 (1985).

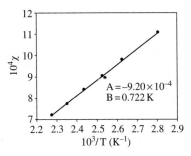

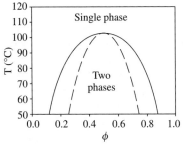

Fig. 4.10
Temperature dependence of χ for mixtures of hydrogenated polybutadiene (88% vinyl) and deuterated polybutadiene (78% vinyl) and the calculated phase diagram from Flory–Huggins theory with $N_A = N_B = 2000$ and $v_0 = 100$ Å^3. The binodal is the solid curve and the spinodal is dashed. Adapted from N. P. Balsara, *Physical Properties of Polymers Handbook* (J. E. Mark, editor), AIP Press, 1996, Chapter 19.

Fig. 4.11

Temperature dependence of χ for mixtures of polyisobutylene and deuterated head-to-head polypropylene and the calculated phase diagram from Flory–Huggins theory with $N_A = N_B = 6000$ and $v_0 = 100$ Å^3. The binodal is the solid curve and the spinodal is dashed. Adapted from N. P. Balsara, *Physical Properties of Polymers Handbook* (J. E. Mark, editor), AIP Press, 1996, Chapter 19.

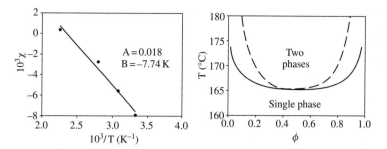

common and better understood, there are many examples of polymer blends that phase separate when temperature is raised, such as polystyrene/poly(vinyl methyl ether). There are also examples where B varies with temperature, changing sign as temperature is changed and resulting in both UCST and LCST, as seen for the polymer solution polystyrene/cyclopentane.

Consider a sudden temperature jump that brings a homogeneous mixture at the critical composition ϕ_c into the two-phase region. The system will spontaneously phase separate into two phases with compositions given by the values on the coexistence curve at that new temperature. This spontaneous phase separation, called spinodal decomposition, occurs because the mixture is locally unstable. Any small composition fluctuation is sufficient to initiate the phase separation process. At any point inside the spinodal curve, the mixture is locally unstable and spontaneously phase separates by the spinodal decomposition process.

The points of the phase diagram between the spinodal and binodal curves correspond to metastable mixtures. The metastable homogeneous state is stable against small composition fluctuations and requires a larger nucleation event to initiate phase separation into the equilibrium phases given by the coexistence curve. This phase separation process is called nucleation and growth.

4.5   Mixtures at low compositions

Consider adding a small amount of A molecules to a liquid of B molecules ($\phi \ll 1$). The free energy of mixing per site

$$\Delta \bar{F}_{\text{mix}} = kT \left[\frac{\phi}{N_A} \ln \phi + \frac{1-\phi}{N_B} \ln (1 - \phi) + \chi \phi (1 - \phi) \right] \qquad (4.59)$$

can be expanded into a power series in composition ϕ of the A-molecules. For small values of composition $\phi \ll 1$, the expansion of the logarithm is $\ln(1 - \phi) \cong -\phi - \phi^2/2 - \phi^3/3 - \cdots$ The second term in the free energy of mixing [Eq. (4.59)] becomes a power series for small ϕ (written here up to the third order in ϕ):

$$\frac{(1-\phi)}{N_B} \ln(1 - \phi) = \frac{1}{N_B} \left(-\phi + \frac{\phi^2}{2} + \frac{\phi^3}{6} + \cdots \right). \qquad (4.60)$$

The free energy of mixing per site can then be rewritten for small ϕ:

$$\Delta\bar{F}_{\text{mix}} = kT\left[\frac{\phi}{N_A}\ln\phi + \phi\left(\chi - \frac{1}{N_B}\right) + \frac{\phi^2}{2}\left(\frac{1}{N_B} - 2\chi\right) + \frac{\phi^3}{6N_B} + \cdots\right].$$

(4.61)

4.5.1 Osmotic pressure

Imagine a semipermeable membrane that prevents passage of A molecules, but allows passage of B molecules. The difference of pressure across this membrane is called the osmotic pressure of A molecules (see Section 1.7.1). The osmotic pressure is defined as the rate of change of the total free energy of the system $\Delta F_{\text{mix}} = n\Delta\bar{F}_{\text{mix}}$ with respect to volume at constant number of A molecules:

$$\Pi \equiv -\left.\frac{\partial\Delta F_{\text{mix}}}{\partial V}\right|_{n_A}$$

(4.62)

The volume fraction ϕ of n_A molecules each with N_A monomers is the ratio of their volume to the volume V of the system:

$$\phi = \frac{b^3 n_A N_A}{V}.$$

(4.63)

The derivative with respect to volume V can be expressed in terms of the derivative with respect to composition ϕ at constant number of A–molecules n_A:

$$\partial V = (b^3 n_A N_A)\partial\left(\frac{1}{\phi}\right) = -\frac{b^3 n_A N_A}{\phi^2}\partial\phi.$$

(4.64)

Note that the number of lattice sites n can be expressed in terms of the number of A molecules n_A as $n = n_A N_A/\phi$. The osmotic pressure is then calculated from the derivative of $\Delta\bar{F}_{\text{mix}}/\phi$ with respect to composition:

$$\Pi = -\left.\frac{\partial(n\Delta\bar{F}_{\text{mix}})}{\partial V}\right|_{n_A} = \frac{\phi^2}{b^3 n_A N_A}\left.\frac{\partial(n_A N_A \Delta\bar{F}_{\text{mix}}/\phi)}{\partial\phi}\right|_{n_A}$$

$$= \frac{\phi^2}{b^3}\left.\frac{\partial(\Delta\bar{F}_{\text{mix}}/\phi)}{\partial\phi}\right|_{n_A}.$$

(4.65)

Differentiating the ratio of free energy of mixing $\Delta\bar{F}_{\text{mix}}$ and composition ϕ with respect to composition gives the mean-field expression for osmotic pressure, valid for small ϕ:

$$\Pi = \frac{kT}{b^3}\left[\frac{\phi}{N_A} + \frac{\phi^2}{2}\left(\frac{1}{N_B} - 2\chi\right) + \frac{\phi^3}{3N_B} + \cdots\right].$$

(4.66)

This expression of osmotic pressure can be written in the form of the virial expansion in terms of number density of A monomers $c_n = \phi/b^3$ [see Eq. (3.8)]

$$\Pi = kT\left[\frac{c_n}{N_A} + \left(\frac{1}{N_B} - 2\chi\right)b^3\frac{c_n^2}{2} + \frac{b^6}{3N_B}c_n^3 + \cdots\right]$$
$$= kT\left[\frac{c_n}{N_A} + \frac{v}{2}c_n^2 + wc_n^3 + \cdots\right], \tag{4.67}$$

where

$$v = \left(\frac{1}{N_B} - 2\chi\right)b^3 \tag{4.68}$$

is the measure of two-body interactions called excluded volume [see Eq. (3.8)] and

$$w = \frac{b^6}{3N_B} \tag{4.69}$$

is the three-body interaction coefficient (see Section 3.3.2.2).

The first term of this virial expansion [Eq. (4.67)] is linear in composition and is called the van't Hoff Law [Eq. (1.72)], which is valid for very dilute solutions:

$$\Pi = \frac{kT}{b^3}\frac{\phi}{N_A} = kT\frac{c_n}{N_A} = kTv. \tag{4.70}$$

The concentration $c_n = \phi/b^3$ is the number density of A monomers and $v = c_n/N_A$ is the number density of A molecules. The last relation of the above equation is a general statement of the van't Hoff Law, as each solute molecule contributes kT to the osmotic pressure in very dilute solutions. The membrane allows the B molecules to pass freely, but restricts all A molecules to stay on one side. This restriction leads to a pressure which is analogous to the ideal gas law (the osmotic pressure is kT per restricted molecule $\Pi = kTv$). This pressure is due to the translational entropy loss caused by the confinement of the A molecules.

In polymer solutions $N_A = N$ and $N_B = 1$, so the osmotic pressure [Eq. (4.66)] at low polymer concentrations has the virial expansion form

$$\Pi = \frac{kT}{b^3}\left[\frac{\phi}{N} + (1 - 2\chi)\frac{\phi^2}{2} + \frac{\phi^3}{3} + \cdots\right]. \tag{4.71}$$

At the θ-temperature, the interaction parameter $\chi = 1/2$ and *the energetic part of two-body interactions exactly cancels the entropic part, making the net two-body interaction zero* $(v = (1 - 2\chi)\ b^3 = 0)$. For $\chi < 1/2$, the two-body interactions increase the osmotic pressure of dilute polymer solutions. Hence, measurement of the osmotic pressure in dilute solutions provides a direct way of determining the Flory interaction parameter χ.

Near the θ-temperature, the second virial coefficient A_2 is related to χ and v by comparing Eqs (1.74) and (4.71), remembering that mass concentration $c = M_0\phi/(b^3\mathcal{N}_{Av})$ and molar mass $M = M_0N$:

$$\frac{v}{b^3} = \frac{2M_0^2}{b^3\mathcal{N}_{Av}}A_2 \approx 1 - 2\chi \approx \frac{T - \theta}{T}. \tag{4.72}$$

As χ is lowered, the polymer likes the solvent more, increasing the osmotic pressure. However, the mean-field theory that is the basis of Eq. (4.71) is only valid close to the θ-temperature, where chains interpenetrate each other freely [Eq. (3.102)]. Far above the θ-temperature (in good solvent), the second virial coefficient A_2 is related to chain volume [Eq. (3.104)] rather than monomer excluded volume v. Recall that the second virial coefficient can also be determined from the concentration dependence of scattering intensity [Eq. (1.91)].

4.5.2 Polymer melts

Consider a binary blend of chemically identical chains with a small concentration of chains with N_A monomers in a melt of chains with N_B monomers. For such a blend there is no energetic contribution to mixing ($\chi = 0$) and the excluded volume contains only a small entropic part:

$$v = \frac{b^3}{N_B}. \tag{4.73}$$

This parameter describes the excluded volume interactions of an A molecule with itself, mediated by the melt of B molecules. This excluded volume is small for polymer melts because each chain has difficulty distinguishing contacts with itself from contacts with surrounding chains. This very important result was first pointed out by Flory: *melts of long polymers have* $v \approx 0$ *and adopt nearly ideal chain conformations.*

The effect of this interaction on the conformations of an A chain can be analysed using the scaling approach described in detail in Chapter 3. On small length scales (smaller than the thermal blob size ξ_T), the excluded volume interactions barely affect the Gaussian statistics of the chain $\xi_T \approx bg_T^{1/2}$, where g_T is the number of monomers in a thermal blob. The thermal blob is defined as the section of the chain with excluded volume interactions of order of the thermal energy:

$$kTv\frac{g_T^2}{\xi_T^3} \approx kT\frac{v}{b^3}\frac{g_T^2}{g_T^{3/2}} \approx kT. \tag{4.74}$$

The number of monomers in a thermal blob is very large when the excluded volume is small:

$$g_T \approx \frac{b^6}{v^2} \approx N_B^2. \tag{4.75}$$

The thermal blob has random walk statistics:

$$\xi_T \approx b g_T^{1/2} \approx b N_B. \qquad (4.76)$$

If an A chain is smaller than the thermal blob ($N_A < N_B^2$), its conformation is almost ideal. In a monodisperse melt with $N_A = N_B$, or in a weakly polydisperse melt, all chains have ideal statistics. On the other hand, strongly asymmetric binary blends of dilute long chains in a melt of short chains with $N_A > N_B^2$ have swollen long chains. The size of these swollen long chains can be estimated as a self-avoiding walk of thermal blobs (as described in Chapter 3):

$$R_A \approx \xi_T \left(\frac{N_A}{g_T}\right)^\nu \approx b N_B \left(\frac{N_A}{N_B^2}\right)^\nu$$

$$= \frac{b}{N_B^{2\nu-1}} N_A^\nu = b N_A^{1/2} \left(\frac{N_A}{N_B^2}\right)^{\nu-1/2}. \qquad (4.77)$$

The swelling coefficient $(N_A/N_B^2)^{0.088} \approx (N_A/g_T)^{0.088}$ for $\nu \cong 0.588$ increases as the number of monomers in the long chains N_A increases beyond that in a thermal blob $g_T \approx N_B^2$. The size of the A chain R_A is plotted as a function of N_A in Fig. 4.12(a) and as a function of N_B in Fig. 4.12(b). When $N_B = 1$, the thermal blob is one monomer and the athermal solvent chain size is recovered [Eq. (3.130)]. Figure 4.12(b) shows how the long chain deswells and eventually crosses over to the ideal chain size at $N_B \approx \sqrt{N_A}$ as the length of the short chains increases.

Three-dimensional polymer melts are strongly interpenetrating. To demonstrate this point, consider a monodisperse melt with $N_A = N_B = N$. The average number of other chains that are inside the pervaded volume of a given polymer (the overlap parameter P, defined in Chapter 1) is the product of this volume $R^3 = N^{3/2} b^3$ and polymer number density $1/Nb^3$ and is equal to $P = \sqrt{N}$. Since N is large, chains are strongly interpenetrated in three-dimensional polymer melts. The presence of so many other chains means that each chain has difficulty distinguishing the intramolecular contacts, that give rise to the excluded volume interaction, from intermolecular contacts. The surrounding chains in the melt have effectively screened the excluded volume interaction, with $v = b^3/N \cong 0$. For this reason, Flory's insightful conjecture that chains in the melt are nearly ideal is correct.

Two-dimensional melts are quite different. The thermal blob for a dilute A polymer in a two-dimensional melt of chemically identical B polymers with excluded area $a = b^2/N_B$ can be estimated in a similar way. The excluded area interactions of a thermal blob with g_T monomers and size $\xi_T \approx b g_T^{1/2}$ is written by analogy with Eq. (3.74):

$$kTa \frac{g_T^2}{\xi_T^2} \approx kTa \frac{g_T^2}{b^2 g_T} \approx kT \quad \text{for two-dimensional melts.} \qquad (4.78)$$

(a)

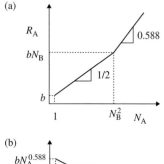

(b)

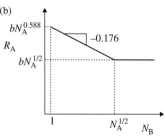

Fig. 4.12
The size of the long A chain as functions of (a) the number of monomers in the A chain and (b) the number of monomers in the B chain, on logarithmic scales.

The number of monomers in a two-dimensional thermal blob is smaller than in the three-dimensional thermal blob.

$$g_T \approx \frac{b^2}{a} \approx N_B \quad \text{for two-dimensional melts.} \qquad (4.79)$$

The chains in a monodisperse two-dimensional melt are roughly the size of a thermal blob and are therefore barely ideal. The number of the other chains in a pervaded area of a given chain is the product of this area $R^2 \approx Nb^2$ and the two-dimensional number density of chains $1/Nb^2$ and is of the order of unity ($P \approx 1$). Thus, chains do not significantly interpenetrate each other in two dimensions. This is the expected result whenever the fractal dimension of the object and the dimension of space are the same.

4.6 Experimental investigations of binary mixtures

The Flory interaction parameter χ can be determined in homogeneous single-phase blends by measuring composition fluctuations using scattering. Consider a homogeneous blend at equilibrium with average composition of A monomers $\bar{\phi}$. In a small volume containing n total monomers with $n_A = \phi n$ A monomers, a small fluctuation in composition $\delta\phi$ can occur spontaneously at equilibrium:

$$\delta\phi \equiv \phi - \bar{\phi}. \qquad (4.80)$$

This fluctuation corresponds to a transfer of δn_A A monomers from the rest of the blend into the small volume with a concurrent transfer of the same number of B monomers out of the small volume (an effective exchange of A and B monomers):

$$\delta n_A = n\delta\phi. \qquad (4.81)$$

The free energy of mixing in this small volume ΔF_{mix} can be expanded in powers of this fluctuation $\delta\phi$:

$$\Delta F_{\text{mix}}(\phi) = \Delta F_{\text{mix}}(\bar{\phi}) + \frac{\partial \Delta F_{\text{mix}}}{\partial \phi}\delta\phi + \frac{1}{2}\frac{\partial^2 \Delta F_{\text{mix}}}{\partial \phi^2}(\delta\phi)^2 + \cdots \qquad (4.82)$$

The term linear in $\delta\phi$ can be rewritten in terms of the number of monomers exchanged:

$$\frac{\partial \Delta F_{\text{mix}}}{\partial \phi}\delta\phi = \frac{\partial \Delta F_{\text{mix}}}{\partial (n\phi)}n\delta\phi = \frac{\partial \Delta F_{\text{mix}}}{\partial n_A}\delta n_A. \qquad (4.83)$$

The derivative of the free energy of mixing with respect to the number of A monomers is the **exchange chemical potential**, the change in free energy of mixing arising from the exchange of one A monomer for one B monomer. The exchange changes the free energy in the rest of the blend by the exchange chemical potential multiplied by the change in number of

A monomers $(-\delta n_A)$ in the rest of the blend:

$$\frac{\partial \Delta F_{\text{mix}}}{\partial n_A}(-\delta n_A).$$

The free energy change in the system δF arising from this fluctuation is the sum of the free energy change in the small volume containing n monomers and the free energy change in the rest of the blend:

$$\delta F = \Delta F_{\text{mix}}(\phi) - \Delta F_{\text{mix}}(\bar{\phi}) - \frac{\partial \Delta F_{\text{mix}}}{\partial n_A} \delta n_A$$

$$= \frac{1}{2} \frac{\partial^2 \Delta F_{\text{mix}}}{\partial \phi^2}(\delta \phi)^2 + \cdots \qquad (4.84)$$

Note that the rest of the blend is considered very large and exchange of δn_A A monomers does not change its composition significantly. The typical free energy change is of the order of the thermal energy, $\delta F \approx kT$, giving a simple relation for the mean-square composition fluctuation:[3]

$$\left\langle (\delta \phi)^2 \right\rangle \approx kT \left(\frac{\partial^2 \Delta F_{\text{mix}}}{\partial \phi^2} \right)^{-1} = \frac{kT}{n} \left(\frac{\partial^2 \Delta \bar{F}_{\text{mix}}}{\partial \phi^2} \right)^{-1}. \qquad (4.85)$$

The final relation involves the free energy of mixing per site, $\Delta \bar{F}_{\text{mix}}$, an intensive quantity. Hence, Eq. (4.85) clearly shows that thermally-driven composition fluctuations diminish as the volume considered (reflected in the number of sites n) increases. Small volumes with only a few monomers (small n) can have large fluctuations, but any macroscopic volume (large n) has a composition that is indistinguishable from the mean blend composition. The mean-square fluctuation is related to the low wavevector limit of the scattering function [Eq. (3.126)]

$$S(q) = \frac{\left\langle (\delta n_A)^2 \right\rangle}{n} = n \left\langle (\delta \phi)^2 \right\rangle, \qquad (4.86)$$

where the number of monomers in the small volume is $n = (qb)^{-3}$. Since $\left\langle (\delta \phi)^2 \right\rangle \sim 1/n$, $S(q)$ saturates at small values of the wavevector. The scattering function at zero wavevector $S(0)$ is thus related to the second derivative of the free energy of mixing:

$$S(0) = n \left\langle (\delta \phi)^2 \right\rangle = kT \left(\frac{\partial^2 \Delta \bar{F}_{\text{mix}}}{\partial \phi^2} \right)^{-1}. \qquad (4.87)$$

This is an example of a much more general thermodynamic relationship between $S(0)$ and osmotic compressibility [see Eq. (1.91)].

---

[3] The real derivation of the mean-square fluctuation is obtained from an average over all magnitudes of the composition fluctuation with the corresponding Boltzmann factor $\exp(-\delta F/kT)$:

$$\left\langle (\delta \phi)^2 \right\rangle = \frac{\int_{-\infty}^{\infty} (\delta \phi)^2 \exp(-\delta F/kT) \, \mathrm{d}(\delta \phi)}{\int_{-\infty}^{\infty} \exp(-\delta F/kT) \, \mathrm{d}(\delta \phi)} = kT \left(\frac{\partial^2 \Delta F_{\text{mix}}}{\partial \phi^2} \right)^{-1}.$$

Using Eq. (4.48), the Flory–Huggins theory predicts

$$\frac{1}{S(0)} = \frac{1}{kT}\frac{\partial^2 \Delta \bar{F}_{\text{mix}}}{\partial \phi^2} = \frac{1}{N_A \phi} + \frac{1}{N_B(1-\phi)} - 2\chi. \qquad (4.88)$$

This means that the Flory interaction parameter χ can be determined from the low wavevector limit of the scattering function of a single-phase blend of A chains (with N_A monomers) and B chains (with N_B monomers), where ϕ is the volume fraction of A chains. In practice, the concentration fluctuations in the blend provide sufficient scattering contrast for neutron scattering, as long as one of the components is at least partially labelled with deuterium.

The standard assumption, called the **random-phase approximation**, extends Eq. (4.88) to non-zero wavevectors q using the form factor of an ideal chain $P(q, N)$:

$$\frac{1}{S(q)} = \frac{1}{N_A \phi P(q, N_A)} + \frac{1}{N_B(1-\phi)P(q, N_B)} - 2\chi. \qquad (4.89)$$

Recall from Section 2.8.4 that the form factor for an ideal chain is the Debye function [Eq. (2.160)]. The high q limit of the Debye function is

$$P(q, N) \cong \frac{2}{q^2 \langle R_g^2 \rangle} = \frac{12}{q^2 N b^2} \quad \text{for } q \gg 1/R_g, \qquad (4.90)$$

where we used the standard radius of gyration for an ideal chain $\langle R_g^2 \rangle = Nb^2/6$ [Eq. (2.54)]. The low q limit of any form factor is $P(q, N) = 1$, and a simple crossover expression emerges for the reciprocal of the Debye function:[4]

$$\frac{1}{P(q, N)} = 1 + \frac{q^2 N b^2}{12}. \qquad (4.91)$$

Substituting this result into Eq. (4.89) (twice) gives a simple result for the reciprocal scattering function:

$$\begin{aligned}
\frac{1}{S(q)} &= \frac{1}{N_A \phi} + \frac{q^2 b^2}{12\phi} + \frac{1}{N_B(1-\phi)} + \frac{q^2 b^2}{12(1-\phi)} - 2\chi \\
&= \frac{1}{N_A \phi} + \frac{1}{N_B(1-\phi)} - 2\chi + \frac{q^2 b^2}{12}\left(\frac{1}{\phi} + \frac{1}{1-\phi}\right) \\
&= \frac{1}{S(0)} + \frac{q^2 b^2}{12\phi(1-\phi)}.
\end{aligned} \qquad (4.92)$$

The final result made use of Eq. (4.88).

This form for scattering is actually far more general, valid for many systems with scattering arising from random fluctuations. Small angle

---

[4] This crossover expression is never more than 15% different from the Debye function over all q. For small qR_g a better expression is Eq. (2.146) or (2.161).

neutron scattering data on miscible polymer blends are customarily fit to the **Ornstein–Zernike scattering function**:

$$S(q) = \frac{S(0)}{1 + (q\xi)^2} \tag{4.93}$$

Comparing Eqs (4.92) and (4.93) reveals the **correlation length** for the mean-field theory of binary mixtures:

$$\xi = \sqrt{\frac{b^2 S(0)}{12\phi(1 - \phi)}}. \tag{4.94}$$

The correlation length effectively divides the form of the scattering into two regions. For $q \ll 1/\xi$, the scattering function approaches its zero wavevector limit $S(0)$. For $q \gg 1/\xi$, the scattering function decreases as a power law $S(q) \sim q^{-2}$. Ideally, experiments would extend to sufficiently low q at which the zero wavevector limit would be nearly realized. However, in practice this is often not the case, with q of order $1/\xi$ for most small-angle neutron scattering (SANS) data on single-phase blends. SANS data at three temperatures for a blend of a polyisobutylene with a deuterium-labelled ethylene–butene random copolymer are shown in Fig. 4.13. This blend phase separates on heating at 95 ± 5 °C. The scattering intensity increases as temperature is raised, which means that concentration fluctuations are getting stronger. The data at all three temperatures are reasonably fit by the Ornstein–Zernike equation [curves are fits to Eq. (4.93)]. The scattering intensity is independent of temperature at high $q \gg 1/\xi$ because

$$S(q) = \frac{S(0)}{(q\xi)^2} = \frac{12\phi(1 - \phi)}{(qb)^2} \quad \text{for } q \gg 1/\xi \tag{4.95}$$

in this limit.

Like $S(0)$, the correlation length ξ has important physical significance and is related to concentration fluctuations. On length scales smaller than the correlation length, correlated chain sections of $(qb)^{-2}$ monomers fluctuate in and out of the volume q^{-3}. Mean-square fluctuations in the number of A and B chain sections is proportional to $(qb)^{-1}$, the number of these sections in the volume q^{-3}. Mean-square fluctuations in the number of A and B *monomers* in this small volume is the product of the mean-square fluctuations in the number of chain sections and the square of the number of monomers in each chain section $\langle (\delta n_A)^2 \rangle \sim (qb)^{-5}$. From Eq. (4.86) we find that coherent fluctuations of chain sections on length scales smaller than the correlation length ($q^{-1} < \xi$) lead to $S(q) \sim \langle (\delta n_A)^2 \rangle / n \sim (qb)^{-2}$ [Eq. (4.95)] independent of ξ or χ. This makes the scattered intensity at high q in Fig. 4.13 independent of temperature [Eq. (4.95)]. Concentration fluctuations in different correlation volumes are incoherent, so on length scales larger than the correlation length, mean-square fluctuations in the number of A and B monomers in the large volume q^{-3} is proportional to the product of the mean-square fluctuations within a correlation volume $(\xi/b)^5$ and $(\xi q)^{-3}$, the number of correlation volumes

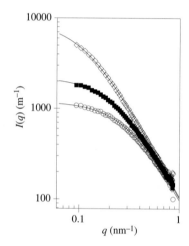

Fig. 4.13

SANS intensity for a $\phi = 0.5$ miscible blend of polyisobutylene ($M_w = 160\,000$ g mol^{-1}) and a random copolymer of ethylene and butene (66 wt% butene, $M_w = 114\,000$ g mol^{-1}) at three temperatures with fits to Eq. (4.93) (curves). Open circles are at 27 °C, where $I(0) = 1180$ m^{-1} and $\xi = 3.3$ nm. Filled squares are at 51 °C, where $I(0) = 2180$ m^{-1} and $\xi = 4.5$ nm. Open diamonds are at 83 °C, where $I(0) = 8850$ m^{-1} and $\xi = 9.1$ nm. Data from R. Krishnamoorti *et al.*, *Macromolecules* **28**, 1252 (1995).

in the volume q^{-3}. From Eq. (4.86), it follows that the structure factor saturates at low $q < \xi^{-1}$ with $S(0) \sim (\xi/b)^2$ [Eq. (4.94)]. Both $S(0)$ and ξ^2 contain the same information about the Flory interaction parameter χ and it is important to realize that this information is only obtained at low q. Since SANS has a limited q-range, in practice $S(0)$ and ξ are determined by fitting data to Eq. (4.93), as shown in Fig. 4.13.

The astute reader will recognize that the scattering function for polymer blends [Eq. (4.86)] is defined in a subtly different manner than for polymer solutions [Eq. (3.126)]. In both cases, the scattering function is normalized by the number of monomers in the system. In Section 3.5, monomers occupy volume fraction ϕ of the total volume, while in the blend the combined volume fraction of monomers of type A and B is unity. The scattering function of Section 3.5 is related to that of the present section as $S(q)/\phi$. To facilitate comparison, we rewrite Eq. (4.88):

$$\frac{\phi}{S(0)} = \frac{1}{N_A} + \frac{\phi}{N_B(1-\phi)} - 2\chi\phi. \tag{4.96}$$

For a polymer solution $N_B = 1$ and $\phi \ll 1$

$$\frac{\phi}{S(0)} \cong \frac{1}{N_A} + (1 - 2\chi)\phi, \tag{4.97}$$

which is the usual virial expansion in dilute solutions [Eq. (1.91)].

The equilibrium concentrations of two-phase blends are calculated from the phase diagram—a coexistence curve in the temperature–composition plane. The phase diagram can be conveniently determined by monitoring light scattering as a function of temperature for various overall compositions, as long as sufficient time is allowed to reach equilibrium at each temperature. Starting at a temperature in the single-phase state, the blend is transparent and the scattering is low. When the temperature reaches the binodal curve, the scattering increases, as phase separation creates domains with different refractive indices.[5] Simple thermodynamic considerations link the phase boundary to the interaction parameter, as described in this chapter.

4.7 Summary of thermodynamics

In this chapter, the thermodynamics of binary mixtures was discussed in the framework of a lattice model. For simplicity, polymers were divided into 'monomers' that fit onto this lattice and the free energy of mixing was written *per lattice site* ($\Delta \bar{F}_{mix}$). Rescaling the monomers to more conventional definitions (such as either the chemical monomer of Chapter 1 or the Kuhn monomer of Chapter 2) is trivial because the volume of an A chain $v_A = N_A v_0$ and the volume of a B chain $v_B = N_B v_0$ must be independent of the choice of lattice site volume v_0. In practice, fitting the monomers onto the lattice is inconvenient and in this summary results are given for the more usual case of mixing two polymers with

---

[5] For this reason, the binodal is often referred to as the cloud point.

different monomer volumes, in terms of the free energy *per unit volume* (called the free energy density $\Delta \bar{F}_{\text{mix}}/v_0$).

The free energy of mixing has two parts: entropic and energetic. The entropic part per unit volume,

$$-\frac{T\Delta S_{\text{mix}}}{V} = -\frac{T\Delta \bar{S}_{\text{mix}}}{v_0} = kT\left[\frac{\phi}{v_A}\ln\phi + \frac{1-\phi}{v_B}\ln(1-\phi)\right], \qquad (4.98)$$

simply counts translational entropy of the mixed state compared with the pure component states. Entropy always promotes mixing. Mixtures of two small molecules have large entropy of mixing, solutions of polymers in small molecule solvents have less entropy of mixing, and blends of two polymers have very little mixing entropy.

The energetic part of the free energy density

$$\frac{\Delta U_{\text{mix}}}{V} = \frac{\Delta \bar{U}_{\text{mix}}}{v_0} = kT\frac{\chi}{v_0}\phi(1-\phi) = \frac{k}{v_0}(AT+B)\,\phi\,(1-\phi) \qquad (4.99)$$

is the difference of intermolecular interaction energies in the mixed and pure states, and is reflected in the Flory interaction parameter χ. Equation (4.99) clearly points out the importance of specifying the reference volume v_0 when stating the value of the Flory χ parameter.[6] The energy of mixing can be either positive (meaning that the different species prefer to be next to themselves) favouring segregation, or negative (meaning that the different species prefer each other) promoting mixing. Interactions between components are often of the van der Waals type, meaning that they are weak and repulsive. Despite this repulsion, many simple liquid pairs form regular solutions that have entropically driven mixing. It is somewhat less likely for a polymer to dissolve in a solvent simply because of lower entropy of mixing for larger molecules. However, most polymers will dissolve in a number of common solvents. On the other hand, miscible polymer blends are very unlikely because the entropy of mixing two long-chain polymers is extremely small. The rule of thumb is that polymers never mix, but there are many exceptions to this rule because interactions between components are not always repulsive.

The shape of the free energy density of mixing as a function of composition

$$\frac{\Delta F_{\text{mix}}}{V} = \frac{\Delta \bar{F}_{\text{mix}}}{v_0} = \frac{\Delta \bar{U}_{\text{mix}}}{v_0} - \frac{T\Delta \bar{S}_{\text{mix}}}{v_0}$$

$$= kT\left[\frac{\phi}{v_A}\ln\phi + \frac{1-\phi}{v_B}\ln(1-\phi) + \frac{\chi}{v_0}\phi(1-\phi)\right]$$

$$= kT\left[\frac{\phi}{v_A}\ln\phi + \frac{1-\phi}{v_B}\ln(1-\phi) + \frac{A}{v_0}\phi(1-\phi)\right]$$

$$+ \frac{kB}{v_0}\phi(1-\phi). \qquad (4.100)$$

---

[6] All numbers for the Flory χ parameter in this book use a reference volume of $v_0 = 100$ Å^3.

determines the stability of a homogeneously mixed state. This function is always convex near the boundaries of the composition range (for small ϕ and for ϕ near unity) because the entropic part always dominates there at any practical (non-zero) temperature. If the composition dependence of the free energy of mixing is convex over the whole composition range, the mixture is homogeneous at all compositions. If the free energy is concave in some part of the composition range, the line of common tangent to the free energy curve determines the range of the miscibility gap (see Fig. 4.7).

The mean-field lattice model of Flory and Huggins predicts that $A \equiv 0$ but in practice this is not observed. If $A < 0$ and $B < 0$, then all four terms in Eq. (4.100) are negative and miscible mixtures are stable at all temperatures. If $A > 0$ and $B < 0$, the blend has a LCST and phase separates at high temperatures. If $B > 0$, the blend has an UCST and phase separation occurs as temperature is lowered.

The Flory interaction parameter in miscible polymer blends is measured using small-angle neutron scattering, usually involving deuterium labelling of one blend component. The χ parameter is determined from the zero wavevector limit of the scattering function $S(0)$:

$$\chi = \frac{v_0}{2} \left[\frac{1}{v_A \phi} + \frac{1}{v_B(1 - \phi)} \right] - \frac{1}{2S(0)}. \qquad (4.101)$$

The binodal separates the homogeneous (single phase) and heterogeneous (two phase) regions in the phase diagram (see Figs 4.10 and 4.11). For binary mixtures, the binodal line is also the coexistence curve, defined by the common tangent line to the composition dependence of the free energy of mixing curve, and gives the equilibrium compositions of the two phases obtained when the overall composition is inside the miscibility gap. The spinodal curve, determined by the inflection points of the composition dependence of the free energy of mixing curve, separates unstable and metastable regions within the miscibility gap.

Melts of long chains have nearly ideal conformations because the excluded volume is screened by the presence of other chains ($v \approx 0$). The excluded volume in a melt is $v \approx b^3/N$. Excluded volume therefore gradually increases as the short chains in a polymer blend are shortened. The short B chains make the A chains swell when $N_A > N_B^2$. Hence, miscible blends of high molar mass polymers with $N_A < N_B^2$ have nearly ideal conformations.

Problems

Section 4.1

4.1   (i) Calculate the number of ways to arrange 10 identical solute molecules on a lattice of 100 sites. Each molecule occupies one lattice site.

  (ii) Calculate the number of ways of arranging an oligomer consisting of 10 repeat units on a cubic lattice of 100 sites. Each repeat unit occupies one lattice site. Ignore long-range (along the chain) excluded volume interactions. Assume that each site has coordination number $z = 6$ (ignore the boundary effects).

4.2 Calculate the entropy of mixing per site $\Delta \bar{S}_{mix}$ on a three-dimensional cubic lattice of:

  (i) 100 black 50-ball chains with 100 white 50-ball chains on a lattice with 10 000 sites (one ball per site).
  (ii) 100 black 50-ball chains with 100 identical black 50-ball chains on a lattice with 10 000 sites (one ball per site).

Explain the difference between cases (i) and (ii).

Section 4.2

4.3 Estimate the Flory interaction parameter χ between polystyrene and polybutadiene at room temperature if the solubility parameter of polystyrene is $\delta_{PS} = 1.87 \times 10^4$ (J m^{-3})$^{1/2}$ and the solubility parameter of polybutadiene is $\delta_{PB} = 1.62 \times 10^4$ (J m^{-3})$^{1/2}$. For simplicity assume $v_0 \cong 100$ Å^3.

4.4 What is the free energy of mixing 1 mol of polystyrene of molar mass $M = 2 \times 10^5$ g mol, with 1×10^4 L of toluene, at 25 °C (Flory interaction parameter $\chi = 0.37$). The density of polystyrene is 1.06 g cm^{-3}, the density of toluene is 0.87 g cm^{-3}. Assume no volume change upon mixing.

4.5 Compare the magnitudes of the two terms in Eq. (4.31) for χ using the data for the 11 polymer blends in Table 4.3 at the lowest temperature studied (corresponding to the largest value of B/T). Is the Flory–Huggins assumption that $|B/T| \gg |A|$ correct?

4.6 (i) Derive the relation between A and B in Eq. (4.31) and the Hildebrand–Scott solubility parameter difference.

  (ii) What values of A and B are possible in the solubility parameter approach?

Section 4.3

4.7 At $T = 0$ K, the entropic contributions to the free energy of mixing disappear, and only the energetic contributions remain. Substitute Eq. (4.31) into the Flory–Huggins equation to write the free energy of mixing in terms of the parameters A and B. Sketch the composition dependence of the free energy for cases where $B < 0$, $B = 0$, and $B > 0$, and discuss whether any of those situations lead to a stable mixture at $T = 0$ K. Does your answer depend on whether regular solutions, polymer solutions, or polymer blends are considered?

4.8 Plot on a single graph, the composition dependence of the free energy of mixing per site (normalized by the thermal energy) $\Delta \bar{F}_{mix}/kT$ of a symmetric polymer blend with $N_A = N_B = 100$ using five different choices for the parameter $\chi = 0, 0.01, 0.02, 0.03, 0.04$. Which choices of χ make the blends miscible in all proportions (i.e. over the whole composition range $0 \leq \phi \leq 1$) and why?

Section 4.4

4.9 The free energy of mixing (per mole of lattice sites) for the regular solution theory can be written as

$$\mathcal{R}T[\phi \ \ln \phi + (1 - \phi) \ \ln (1 - \phi) + \chi \phi(1 - \phi)],$$

where $\mathcal{R}$ is the gas constant and the interaction parameter is $\chi = B/T$, where $B = 600$ K. Construct the binodal and spinodal curves in the temperature–composition phase diagram.

4.10 The free energy of mixing (per mole of lattice sites) of a polymer solution (according to the Flory–Huggins model) is

$$\mathcal{R}T\left[\frac{\phi}{N}\ln\phi + (1-\phi)\ln(1-\phi) + \chi\phi(1-\phi)\right],$$

where $\mathcal{R}$ is the gas constant and the interaction parameter is $\chi = B/T$ where $B = 300$ K. Plot the critical parameters (ϕ_c, χ_c, and T_c) for the solution as a function of the degree of polymerization N.

4.11 Calculate the free energy density $\Delta F_{mix}/V$ of mixing polystyrene of molar mass 20 000 g mol^{-1} with cyclohexane at 34 °C, to make up a 5% by volume solution? Assume no volume change upon mixing.

Note that 34 °C is the θ-temperature for a polystyrene solution in cyclohexane (Flory interaction parameter $\chi = 1/2$):

the density of polystyrene is 1.06 g cm^{-3};
the density of cyclohexane is 0.78 g cm^{-3};
the molar mass of cyclohexane (C_6H_{12}) is 84 g mol^{-1}.

4.12 (i) What is the free energy of mixing 1 g of polystyrene of molar mass $M = 10^5$ g mol^{-1}, with 1 mol of cyclohexane at 34 °C? Note that 34 °C is the θ-temperature for a polystyrene solution in cyclohexane (Flory interaction parameter $\chi = 1/2$). The molar volume of polystyrene is $v_{ps} = 9.5 \times 10^4$ cm^3 mol^{-1}, the molar volume of cyclohexane is $v_{cyc} = 108$ cm^3 mol^{-1}. Assume no volume change upon mixing and assume that the volume of one solvent molecule is the lattice site volume v_0.

(ii) What does the sign of the free energy of mixing imply about the stability of a homogeneous solution?

(iii) Under what conditions does the homogeneous solution spontaneously phase separate by spinodal decomposition?

(iv) When is the homogeneous solution metastable?

4.13 Since the mean-field Flory–Huggins theory puts everything that is not understood about thermodynamics into the χ parameter, this parameter is experimentally found to vary with composition and temperature. For solutions of linear polystyrene in cyclohexane, the interaction parameter

$$\chi = 0.2035 + \frac{90.65\,\text{K}}{T} + 0.3092\phi + 0.1554\phi^2 \qquad (4.102)$$

was determined by R. Koningsveld *et al.*, *J. Polym. Sci. A-2* **8**, 1261 (1970).

(i) What is the critical temperature for a very high molar mass polystyrene in cyclohexane with polymer volume fraction $\phi = 0.01$?

(ii) Does the polystyrene/cyclohexane system have a UCST or an LCST?

(iii) Determine the θ-temperature at a volume fraction $\phi = 0.1$.

4.14 (i) Derive a general expression for the critical temperature of a mixture in terms of the solubility parameter difference $\delta_A - \delta_B$ and the number of monomers in each component N_A and N_B.

(ii) What is the criterion for miscibility in this approach?

(iii) What is the largest solubility parameter difference that allows small molecule mixtures (with $N_A = N_B = 1$) to be miscible?

(iv) What is the largest solubility parameter difference that allows polymer solutions (with $N_A = 10^4$ and $N_B = 1$) to be miscible?

(v) What is the largest solubility parameter difference that allows polymer blends (with $N_A = N_B = 10^4$) to be miscible?

4.15 Consider a nucleation process from a uniform metastable state of a polymer solution. Denote by $\Delta\mu = \mu_1 - \mu_2$ the chemical potential difference between N-mers in a uniform solution μ_1 and in a phase separated solution μ_2. If $\Delta\mu$ is positive, the phase separated solution is the equilibrium state. However, a small drop of a phase with higher concentration of molecules c/N formed in the homogeneous phase could be unstable due its positive surface energy with surface tension γ. Calculate the Gibbs free energy of a spherical drop of concentrated phase of radius R and determine the critical radius R_c for nucleation in terms of γ, $\Delta\mu$ and the number density of chains. Nuclei smaller than R_c shrink and disappear, while larger ones grow into domains of the dense phase.

4.16 (i) What is the critical value of χ required for high molar mass polymers to dissolve in a solvent in all proportions?

(ii) In Chapter 5, we will learn that polymer solutions are not described well by the mean-field theory because the connectivity of the chains keeps monomers from being uniformly distributed in solution (particularly at low polymer concentrations). An empirical form that better relates χ to the Hildebrand solubility parameters in polymer solutions is widely used with an entropic part of χ of 0.34:

$$\chi = 0.34 + \frac{v_0}{kT}(\delta_A - \delta_B)^2. \tag{4.103}$$

Use the following table to decide which solvents will dissolve poly (dimethyl siloxane) ($\delta_{PDMS} = 14.9 \ (MPa)^{1/2}$) and which will dissolve polystyrene ($\delta_{PS} = 18.7 \ (MPa)^{1/2}$) at room temperature.

Solvent	n-Heptane	Cyclohexane	Benzene	Chloroform	Acetone
Molar volume $(cm^3 \ mol^{-1})$	195.9	108.5	29.4	80.7	74.0
Solubility parameter $\delta \ (MPa)^{1/2}$	15.1	16.8	18.6	19.0	20.3

(iii) Which solvent is closest to the athermal limit for each polymer?

Section 4.5

4.17 Consider a melt of B chains with degree of polymerization $N_B = 100$ and Kuhn length $b = 3$ Å. What is the root-mean-square end-to-end distance of isolated chemically identical A-chains in this melt with degree of polymerization:

(i) $N_A = 10^2$? (ii) $N_A = 10^4$? (iii) $N_A = 10^6$?

4.18 Consider a monodisperse melt of randomly branched polymers with N Kuhn monomers of length b. Randomly branched polymers in an ideal state (in the absence of excluded volume interactions) have fractal dimension $D = 4$. Do these randomly branched polymers overlap in a three-dimensional monodisperse melt?

Hint: What would be the N-dependence of density if monodisperse randomly branched polymers overlapped in the melt?

4.19* Demonstrate that the excluded volume in a polydisperse melt is $v = b^3/N_w$, where N_w is the weight-average molar mass of the melt.

Section 4.6

4.20 Ginzburg criterion for polymer blends

Estimate the size of the critical region near the critical point in a symmetric polymer blend by comparing the mean-square composition fluctuations $\langle(\delta\phi_A)^2\rangle$ with the square of the difference in volume fractions of the two phases $(\phi'' - \phi')^2$ in the miscibility gap. Such considerations determine the point where mean-field theory (which assumes fluctuations are small) fails, known as the Ginzburg criterion.

(i) Expand the equation for the binodal (Eq. 4.47 with $A = 0$) near the critical composition $\phi_c = 1/2$ to derive the dependence of the order parameter $\phi'' - \phi'$ on the relative temperature difference from the critical temperature $(T - T_c)/T_c$.

$$\phi'' - \phi' \approx \sqrt{\frac{T_c - T}{T_c}} \quad \text{for } T < T_c$$

(ii) Demonstrate that the mean-square composition fluctuations on the scale of the correlation length are of order

$$\langle(\delta\phi)^2\rangle \approx \sqrt{\frac{1}{N}\frac{T_c - T}{T_c}} \quad \text{for } T < T_c$$

(iii) Estimate the size of the critical region (Ginzburg criterion) by comparing the square of the order parameter $(\phi'' - \phi')^2$ with the mean-square composition fluctuations on the scale of correlation length $\langle(\delta\phi)^2\rangle$.

$$\frac{T_c - T}{T_c} \approx \frac{1}{N} \quad \text{for } T < T_c$$

For blends of long chain polymers (large N) the critical region is very small and the mean-field theory applies at nearly all temperatures.

4.21 Use the data in Table 4.3 to calculate the zero wavevector limit of the scattering function $S(0)$ and the mean-square concentration fluctuation $\langle(\delta\phi)^2\rangle$ at a 50 Å scale (at $q = 2\pi/50$ Å$^{-1} = 0.126$ Å^{-1}) assuming the Ornstein–Zernike form for $S(q)$, for the blends listed below (with $N_A = N_B = 100$ in each case). For each blend, plot $S(0)$ and $\langle(\delta\phi)^2\rangle$ as functions of temperature over the temperature range of Table 4.3.

(i) 50% by volume poly(vinyl methyl ether) mixed with polystyrene;

(ii) 50% by volume polyisobutylene mixed with deuterated head-to-head polypropylene;

(iii) 30% by volume poly(ethylene oxide) mixed with deuterated polymethyl (methacrylate).

Identify the blends that phase separate and state whether they have an LCST or a UCST.

4.22 Use the fitting results in the caption of Fig. 4.13 to analyze the temperature dependence of thermodynamics for blends of polyisobutylene and a random ethylene-butene copolymer (66% butene).

(i) Plot the extrapolated zero wavevector intensity against the square of the correlation length for concentration fluctuations. What is the physical significance of $I(0) \sim \xi^2$?

(ii) The correlation length diverges at the spinodal temperature T_s as a power law:

$$\xi \sim \left(\frac{T_s - T}{T_s} \right)^{-1/2}. \tag{4.105}$$

Use the fitting results in the caption of Fig. 4.13 for the three different temperatures to estimate the spinodal temperature T_s for this blend. How does this spinodal temperature compare with the observed cloud point for this blend of $95 \pm 5\,^{\circ}\text{C}$?

(iii) $S(0)$ can be determined from the measured $I(0)$ using the scattering lengths of the polymers (which determine the contrast in SANS) which can be calculated from their chemical structures. However, in practice the background subtractions required for absolute intensity are fraught with difficulties. Derive the following relation between the Flory interaction parameter χ and the correlation length ξ for concentration fluctuations that involves the volume fraction ϕ, the reference volume v_0 and information about *both* components, regarding the mean-square end-to-end distance R_i^2, and the occupied volume v_i of each chain.

$$\chi = \frac{v_0}{2 v_A \phi} + \frac{v_0}{2 v_B (1 - \phi)} - \frac{1}{24 \xi^2} \left[\frac{v_0 R_A^2}{v_A \phi} + \frac{v_0 R_B^2}{v_B (1 - \phi)} \right] \tag{4.106}$$

(iv) Determine the Flory interaction parameter χ, at each of the three temperatures in Fig. 4.13, from the correlation length for concentration fluctuations and fit those determinations to Eq. (4.31) to determine the temperature dependence of χ. Use a reference volume of $v_0 = 100\,\text{Å}^3$ and the following data for the random ethylene-butene copolymer (PEB) and polyisobutylene (PIB).

polymer	$b(\text{Å})$	M (g/mol)	M_0 (g/mol)	ρ (g/cm^3)
PEB	12.1	114 000	211	0.802
PIB	12.5	160 000	274	0.918

Bibliography

Balsara, N. P. Thermodynamics of polymer blends In: *Physical Properties of Polymers Handbook*, ed. J. E. Mark (AIP Press, 1996).

Flory, P. J. *Principles of Polymer Chemistry* (Cornell University Press, Ithaca, NY, 1953).

Higgins, J. S. and Benoit, H. C. *Polymers and Neutron Scattering* (Clarendon Press, Oxford, 1994).

Hildebrand, J. H. *Regular Solutions* (Prentice-Hall, New York, 1962).

Koningsveld, R. Stockmayer, W. H. and Nies, E., *Polymer Phase Diagrams*, Oxford University Press (2001).

Kurata, M. *Thermodynamics of Polymer Solutions* (Harwood Academic, Chichester, 1982).

Paul, D. R. and Bucknall, C. B. eds. *Polymer Blends*, Vols 1 and 2 (Wiley, New York, 2000).

Polymer solutions

5

In this chapter, our understanding of ideal chain conformations (Chapter 2), real chain conformations (Chapter 3), and thermodynamics (Chapter 4) will be combined to describe the conformations of polymer solutions at all concentrations and temperatures. In this chapter, the focus is on semidilute and concentrated solutions that span the large range of concentrations between dilute solutions and melts.

5.1 Theta solvent

The phase diagram for a polymer solution is shown in Fig. 5.1. Here the attention is focused on the case where $A = 0$ and $B > 0$ in Eq. (4.31), which is the most common case for polymer solutions. The θ-temperature separates the poor solvent (bottom) half of the diagram from the good solvent (top) half. At this special temperature ($T = \theta$) the interaction parameter $\chi = 1/2$ and the excluded volume is zero [see Eq. (4.72)]:

$$v = (1 - 2\chi)b^3 = \frac{T - \theta}{T}b^3 = 0. \tag{5.1}$$

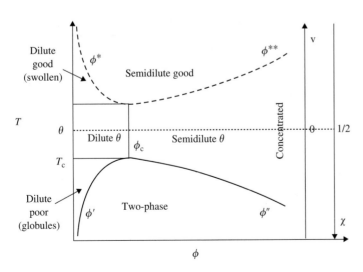

Fig. 5.1
Phase diagram for polymer solutions with a UCST. The solid curve denotes the binodal and phase separation occurs for polymer solutions with T and ϕ below the binodal. The dashed curve is the low temperature boundary of the semidilute good solvent regime.

The net excluded volume at the θ-temperature is zero because of the exact cancelation between the constant steric repulsion between monomers (the b^3) and the solvent-mediated attraction between monomers (the $-2\chi b^3$). At the θ-temperature, the chains have nearly ideal conformations at *all* concentrations:

$$R = b\sqrt{N}. \tag{5.2}$$

Subtle deviations from ideal chain statistics are caused by three-body monomer interactions (see Section 3.3.2.2) at the θ-temperature, leading to logarithmic corrections to scaling.

At very low concentrations, the polymers exist as isolated coils that are very far apart. The concentration increases moving from left to right in Fig. 5.1. At $T = \theta$, there is a special concentration that equals the concentration inside the pervaded volume of the coil. This is the **overlap concentration** for θ-solvent [see Eq. (1.21)]:

$$\phi_\theta^* \approx \frac{Nb^3}{R^3} = \frac{1}{\sqrt{N}}. \tag{5.3}$$

Polymer solutions with volume fractions $\phi < \phi_\theta^*$ are called **dilute** θ**-solutions**. Above ϕ_θ^*, linear chains interpenetrate each other. At volume fractions above overlap $(\phi > \phi_\theta^*)$ at $T = \theta$, polymer solutions are called **semidilute** θ. This name originates from the fact that at low volume fractions $\phi_\theta^* < \phi \ll 1$ the solution consists of mostly solvent, but many solution properties are dictated by overlapping chains.

Chains are nearly ideal not only at the θ-temperature, but also at temperatures sufficiently close to θ. Recall that for real chains with $T \neq \theta$, the conformations deviate from ideal statistics only on length scales larger than the thermal blob size [Eq. (3.76)].

$$\xi_T \approx \frac{b^4}{|v|}. \tag{5.4}$$

The entire chain is nearly ideal if its size $R \approx bN^{1/2}$ is smaller than the thermal blob size ξ_T. This condition defines the two temperature boundaries of the dilute θ-regime.

$$|v| = |1 - 2\chi|b^3 = \left|1 - \frac{\theta}{T}\right|b^3 \approx \frac{b^3}{\sqrt{N}}. \tag{5.5}$$

Solving Eq. (5.5) for T gives the temperature at which chains begin to either swell (above θ) or collapse (below θ):

$$T \approx \theta\left(1 \pm \frac{1}{\sqrt{N}}\right). \tag{5.6}$$

Note that the ideal dilute regime is restricted to being quite close to the θ-temperature for long chains.

5.2 Poor solvent

Solvent quality decreases as temperature is lowered, leading to polymer collapse and possible phase separation (the lower part of Fig. 5.1). In Section 4.4, the binodal curve that describes the phase boundary was defined. The highest point on the binodal line is the critical point with critical composition [Eq. (4.57)]:

$$\phi_c \cong \frac{1}{\sqrt{N}}, \tag{5.7}$$

and critical interaction parameter [Eq. (4.58)],

$$\chi_c = \frac{1}{2} + \frac{1}{\sqrt{N}} + \frac{1}{2N}. \tag{5.8}$$

The critical composition ϕ_c and the overlap concentration ϕ^* at the θ-temperature nearly coincide for monodisperse polymer solutions. The critical temperature for a polymer solution [Eq. (4.54) with $N_A = N$ and $N_B = 1$] can be written in terms of the θ-temperature, since $1/2 = A + B/\theta$ from Eq. (4.31):

$$\frac{1}{T_c} = \frac{\chi_c - A}{B} = \frac{1}{B}\left(\frac{1}{2} + \frac{1}{\sqrt{N}} + \frac{1}{2N} - A\right)$$

$$= \frac{1}{\theta} + \frac{1}{B}\left(\frac{1}{\sqrt{N}} + \frac{1}{2N}\right). \tag{5.9}$$

This critical temperature is close to the boundary of the ideal dilute regime [Eq. (5.6)]. Longer chains phase separate at higher temperatures (closer to the θ-temperature). Phase diagrams of different molar mass polystyrenes in cyclohexane are shown in Fig. 5.2(a). The Flory–Huggins prediction [Eq. (5.9)] for the dependence of the critical temperature T_c on the degree of polymerization N is in good agreement with experiments as seen in Fig. 5.2(b), whose intercept is $1/\theta$ and slope is $1/B$.

Below the binodal, homogeneous solutions phase separate into a dilute supernatant of isolated globules and a concentrated sediment (assuming that the polymer has higher density than the solvent). The overall composition of the dilute phase of isolated globules is the lower volume fraction branch ϕ' of the coexistence curve for monodisperse polymer solutions. The branch of the coexistence curve at higher compositions is the volume fraction of polymer ϕ'' of the coexisting sediment for binary mixtures.

The shape of the binodal near the critical point is not predicted correctly by the mean-field (Flory–Huggins) theory as demonstrated in Fig. 5.3(a). The difference in the two concentrations coexisting at equilibrium in the two-phase region is called the **order parameter**. This order parameter is analogous to the order parameter of van der Waals for the liquid–vapour phase transition, that is proportional to the density difference between the two coexisting phases. This order parameter is predicted to vary as a power law of the proximity to the critical point:

$$\phi'' - \phi' \sim (\chi - \chi_c)^\beta \sim (T_c - T)^\beta. \tag{5.10}$$

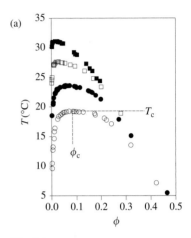

(a)

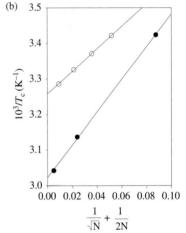

(b)

Fig. 5.2
(a) Phase diagrams for polystyrenes in cyclohexane, $M = 43\,600\ \mathrm{g\,mol^{-1}}$ (open circles), $M = 89\,000\ \mathrm{g\,mol^{-1}}$ (filled circles), $M = 250\,000\ \mathrm{g\,mol^{-1}}$ (open squares), $M = 1\,270\,000\ \mathrm{g\,mol^{-1}}$ (filled squares). (b) Chain length dependence of the critical temperature for polystyrene in cyclohexane (open circles) from part (a) and for polyisobutylene in diisobutyl ketone (filled circles). All data from A. R. Shultz and P. J. Flory, *J. Am. Chem. Soc.* **74**, 4760 (1952).

(a)

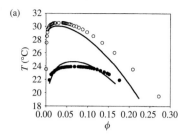

(b)

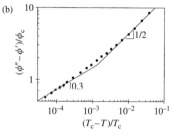

Fig. 5.3

(a) Phase diagrams for polystyrene in methylcyclohexane with $M_w = 200\,000\ \mathrm{g\,mol^{-1}}$ (solid circles) and $M_w = 1\,560\,000\ \mathrm{g\,mol^{-1}}$ (open circles) are compared with the mean-field predictions (curves). (b) Dependence of the order parameter on $\varepsilon \equiv (T_c - T)/T_c$. Close to the critical point (at small ε) the order parameter is larger than expected by mean-field theory (line with slope $= 1/2$) and the slope is $\beta \approx 0.3$. Data are from T. Dobashi *et al.*, *J. Chem. Phys.* **72**, 6692 (1980).

The mean-field prediction for the exponent is $\beta = 1/2$. As shown in Fig. 5.3(a), the experimentally measured binodal is wider than the mean-field prediction. Mean-field theories of phase transitions always fail sufficiently close to the critical point. There is a critical region near T_c where mean-field theories do not work because they ignore concentration fluctuations. The **Ginzburg criterion** determines the width of the critical region by checking the self-consistency of the mean-field theory. The mean-field theory assumes that monomers are uniformly distributed with no concentration fluctuations. However, in dilute solutions the monomer concentration fluctuates enormously, from zero between coils to a higher value within a coil. It fluctuates even within each coil (as discussed in Section 2.7). In semidilute solutions, concentration fluctuations are considerably smaller. A quantitative measure of concentration fluctuations is provided by the overlap parameter P—the average number of other chains inside the pervaded volume of a given chain [Eq. (1.22)]. At the critical point for a general A–B mixture, the overlap parameter of the A-chains is the ratio of the overall volume fraction ϕ_c and the volume fraction of a single A-chain within its pervaded volume:

$$P_c = \frac{\phi_c R_A^3}{N_A b^3} = \phi_c \sqrt{N_A} = \frac{\sqrt{N_A N_B}}{\sqrt{N_A} + \sqrt{N_B}}. \tag{5.11}$$

To obtain this result, Eq. (4.52) has been used for the critical composition of an A–B mixture and the ideal coil size is $R_A = b\sqrt{N_A}$. For symmetric blends, $N_A = N_B = N \gg 1$, the overlap parameter at the critical point with $\phi_c = 1/2$ is very large $P_c = \sqrt{N}/2 \gg 1$ and mean-field theory works well for symmetric blends. For asymmetric blends with $N_A \gg N_B$, the overlap parameter at the critical point is $P_c = \sqrt{N_B}$. Mean-field theory is still applicable as long as $N_B \gg 1$. In polymer solutions $N_B = 1$ and the mean-field theory does not work near the critical point.

Critical theories take into account concentration fluctuations and describe phase transitions near the critical point. The critical theory for this phase transition is the Ising model which also describes liquid–vapour transitions near their critical points. The prediction of the three-dimensional Ising model for the critical exponent $\beta \cong 0.3$ is in good agreement with experiments as shown in Fig. 5.3(b). Mean-field and critical theories and the Ginzburg criterion for the boundary between them for a qualitatively different transition, called percolation, will be described in detail in Chapter 6.

Further from the critical point, the high concentration branch of the coexistence curve ϕ'' (the volume fraction of the coexisting sediment) is well described by the mean-field theory because the overlap parameter $P \gg 1$. This concentration is determined by the balance of the second and the third terms of the virial expansion. The second term $kT\phi^2(1 - 2\chi)/2 < 0$ is the two-body attraction that causes phase separation. The three-body repulsion term $kT\phi^3/6 > 0$ stabilizes the concentration. Minimizing the sum of these two terms of the free energy gives the concentration of the sediment:

$$\phi'' \approx 2\chi - 1 = -\frac{v}{b^3}. \tag{5.12}$$

Recall from Section 3.3 that the balance of the same two terms of the free energy determines the concentration inside the globules in the dilute coexisting phase:

$$\frac{Nb^3}{R_{gl}^3} \approx 2\chi - 1 = -\frac{v}{b^3} \approx \phi''. \tag{5.13}$$

The size of the globules is proportional to the one-third power of the number of monomers in them [Eq. (3.78)]:

$$R_{gl} \approx \frac{bN^{1/3}}{(2\chi - 1)^{1/3}} \approx \frac{b^2 N^{1/3}}{|v|^{1/3}}. \tag{5.14}$$

The globules in dilute solution behave as little droplets. The **surface tension** (the energy per unit area of the surface) of the droplets ensures that they are roughly spherical. Monomers within the droplets attract each other, but monomers at the surface of the droplet also contact the pure solvent. The missing attraction energy for the monomers at the surface of the droplet is the origin of the surface tension. In polymeric globules, neighbouring thermal blobs attract each other with energy of order of the thermal energy kT. Blobs along the surface of the globules have a deficit of this attraction energy due to the absence of neighbouring thermal blobs outside the globule. Therefore, the surface tension γ of the globules is of the order of kT per thermal blob at the surface:

$$\gamma \approx \frac{kT}{\xi_T^2} \approx \frac{kT}{b^8} v^2 \approx \frac{kT}{b^2} (2\chi - 1)^2. \tag{5.15}$$

The total surface energy of a globule is the product of the surface tension γ and the globule surface area R_{gl}^2:

$$\gamma R_{gl}^2 \approx kT \frac{R_{gl}^2}{\xi_T^2} \approx kT \frac{|v|^{4/3}}{b^4} N^{2/3} \approx kT(2\chi - 1)^{4/3} N^{2/3}. \tag{5.16}$$

The thickness of the interface between the globules and the pure solvent is of the order of the size of a thermal blob $\xi_T \approx b/(2\chi - 1)$.

Owing to the high cost of their surface energy, globules would like to stick together, forming larger clusters with lower surface energy per molecule. This tendency results in the formation of the second phase—the sediment. The equilibrium concentration of globules in the supernatant phase is very low. Their high surface energy [Eq. (5.16)] can only be balanced by their translational entropy $kT \ln \phi$ per molecule. Therefore, the concentration of the dilute supernatant phase is lower than that in the concentrated sediment phase by the exponential of the globule surface energy:

$$\phi' = \phi'' \exp\left[-\frac{\gamma R_{gl}^2}{kT}\right] \approx \frac{|v|}{b^3} \exp\left[-\frac{|v|^{4/3}}{b^4} N^{2/3}\right]. \tag{5.17}$$

This dependence differs significantly from the prediction of the mean-field (Flory–Huggins) theory. This is not surprising because mean-field theory is expected to fail in dilute solutions, where the small overlap parameter ($P \ll 1$) makes concentration fluctuations large.

Equation (5.17) allows the surface tension of globules to be determined by measuring the very low concentration ϕ' of globules, their size R_{gl} and the concentration of the sediment. There is a detailed balance and exchange of the chains between the sediment and supernatant phases. As in any equilibrium between coexisting phases, the chemical potential of polymer chains is the same in both phases. Chains increase the enthalpic part of their free energy by the increase in surface energy when they leave the sediment, but they gain translational entropy that lowers the entropic part of their free energy.

There is also a major difference in the conformation of chains in the supernatant globules and the condensed sediment. The chains in globules are collapsed, with size given by Eq. (5.14), arising from a dense packing of thermal blobs [see Fig. 3.14]. The same chains in the sediment are in their ideal conformations with size $R = bN^{1/2}$. As globules stick together forming a sediment, they effectively organize into a melt of thermal blobs. Even though thermal blobs of each chain still attract each other with the same energy kT, they also attract blobs of other chains, just as monomers in a melt attract each other. The surrounding chains screen this intramolecular interaction and each one opens up to maximize its conformational entropy at the same interaction energy (kT per thermal blob). Thus, the excluded volume attraction in the sediment is screened by overlapping chains in a similar way to the screening of excluded volume repulsion in a polymer melt.

5.3 Good solvent

5.3.1 Correlation length and chain size

The upper part of the phase diagram (Fig. 5.1) corresponds to good solvents. At low concentrations, polymer coils are far from each other and behave as isolated real chains (see Section 3.3.1.1). At temperatures for which the excluded volume interaction within each chain exceeds the thermal energy kT, they begin to swell. The Flory theory prediction for the size of swollen real chains with excluded volume $v > b^3/\sqrt{N}$ is the same as the result for a self-avoiding walk of thermal blobs [Eq. (3.77)]:

$$R \approx b \left(\frac{v}{b^3} \right)^{2\nu - 1} N^\nu. \tag{5.18}$$

For swelling exponent $\nu \cong 0.588$ the expression for chain size is $R \approx b(v/b^3)^{0.18} N^{0.588}$. Chains begin to overlap when their volume fraction ϕ exceeds the volume fraction of monomers inside each isolated coil, called the overlap concentration ϕ^*.

$$\phi^* \approx \frac{Nb^3}{R^3} \approx \left(\frac{b^3}{v} \right)^{6\nu - 3} N^{1 - 3\nu}. \tag{5.19}$$

The overlap concentration decreases with chain length more rapidly than in θ-solvent, with $\phi^* \sim N^{-0.76}$ since the exponent $\nu \cong 0.588$. The quality of solvent and the excluded volume increase with temperature and therefore, the overlap concentration decreases with increasing temperature.

At higher concentrations, chains interpenetrate and the solution is called semidilute. The polymer volume fraction in semidilute solution is still very low $\phi^* < \phi \ll 1$. Therefore at small distances each monomer is surrounded by mostly solvent and a few monomers belonging to the same chain. If a monomer tries to reach monomers on other chains by a CB radio, the radio must have a sufficiently long range ξ (see Fig. 5.4). This length scale ξ, called the correlation length, is one of the most important concepts in semidilute solutions. On length scales smaller than the correlation length, each monomer is surrounded by mostly solvent and other monomers belonging to the same chain. If the range of the CB radio $r < \xi$, it is unlikely that the monomer will be able to call any monomer on any other chain. The conformations of the section of the chain of size ξ containing g monomers are very similar to those for a chain in a dilute solution with a solvent of the same quality [see Eq. (5.18)].

$$\xi \approx b \left(\frac{v}{b^3} \right)^{2\nu - 1} g^{\nu}. \tag{5.20}$$

This small section of a chain is hardly aware that the solution is semidilute because it interacts mostly with solvent and with monomers from the same section of its own chain. The *correlation volumes are space-filling*. They are at overlap with each other and a semidilute solution can be subdivided into a set of densely packed correlation volumes, each with the overall solution concentration of monomers inside them:

$$\phi \approx \frac{gb^3}{\xi^3}. \tag{5.21}$$

Combining Eqs (5.20) and (5.21), provides a relation between the volume fraction ϕ and the number of monomers g in a correlation blob of size ξ:

$$\phi \approx \left(\frac{\xi}{b} \right)^{1/\nu} \left(\frac{b^3}{v} \right)^{(2\nu-1)/\nu} \left(\frac{b}{\xi} \right)^{3} \approx \left(\frac{b^3}{v} \right)^{(2\nu-1)/\nu} \left(\frac{\xi}{b} \right)^{-(3\nu-1)/\nu}. \tag{5.22}$$

The correlation length in a semidilute solution decreases with increasing concentration as a power law:

$$\xi \approx b \left(\frac{b^3}{v} \right)^{(2\nu-1)/(3\nu-1)} \phi^{-\nu/(3\nu-1)}. \tag{5.23}$$

The concentration dependence of the correlation length is $\xi \sim \phi^{-0.76}$ since the exponent $\nu \cong 0.588$. The number of monomers in each correlation blob

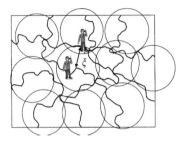

Fig. 5.4
If the range of a monomer's CB radio is shorter than the correlation length ξ, the monomer can only talk with monomers on the same chain, whereas if the range extends beyond the correlation length, the monomer can talk with other monomers on many different chains.

decreases with increasing concentration as another power law:

$$g \approx \left(\frac{b^3}{v}\right)^{3(2\nu-1)/(3\nu-1)} \phi^{-1/(3\nu-1)}. \tag{5.24}$$

The concentration dependence of the number of monomers in a correlation blob is $g \sim \phi^{-1.3}$ since the exponent $\nu \cong 0.588$.

On length scales larger than the correlation length, the excluded volume interactions are screened by the overlapping chains. The semidilute solution on these length scales behaves as a melt of chains made of correlation blobs and the polymer conformation is a random walk of correlation blobs:

$$R \approx \xi \left(\frac{N}{g}\right)^{1/2}. \tag{5.25}$$

The concentration dependence of polymer size in semidilute solution is determined by substituting the expressions for the concentration dependence of the correlation length ξ [Eq. (5.23)] and the number of monomers in a correlation blob g [Eq. (5.24)] into Eq. (5.25):

$$R \approx \xi \left(\frac{N}{g}\right)^{1/2} \approx b \left(\frac{v}{b^3 \phi}\right)^{(\nu-1/2)/(3\nu-1)} N^{1/2}. \tag{5.26}$$

The size of polymer chains decreases weakly with concentration in semidilute solution as $R \sim \phi^{-0.12}$ since the exponent $\nu \cong 0.588$.

There are three scaling regimes for a chain in a semidilute solution, summarized in Fig. 5.5.

Regime (i). On scales up to the thermal blob size ξ_T, the chain is nearly ideal because excluded volume interactions are weaker than the thermal energy. The size of a polymer subsection grows as the square root of the number of monomers in it up to the thermal blob size ξ_T.

Regime (ii). On length scales larger than a thermal blob ξ_T, but smaller than the correlation blob ξ, the excluded volume interactions are strong enough to swell the chain, but are not yet screened by the surrounding chains. The size of polymer sections on this intermediate length scale grows as the 0.588 power of the number of monomers in them.

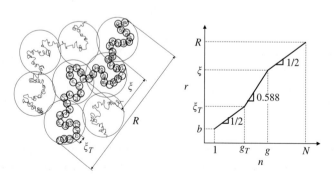

Fig. 5.5
Schematic representation of a chain in semidilute solution in a good solvent. The plot shows the scaling of the end-to-end distance r with the number of monomers n in a subsection of a chain in a good solvent on logarithmic scales.

Regime (iii). On length scales larger than the correlation length ξ, the excluded volume interactions are screened and the chain is a random walk of correlation blobs.

The concentration dependence of polymer size [Eq. (5.26)] will now be derived using an alternative **de Gennes scaling theory**. The main assumption of any scaling theory is that each quantity (such as polymer size R) changes as a power of another quantity (such as the volume fraction ϕ). At the overlap concentration [Eq. (5.19)],

$$\phi^* \approx \left(\frac{b^3}{v}\right)^{6\nu-3} N^{1-3\nu}, \tag{5.27}$$

the chain has its dilute solution size [Eq. (5.18)]:

$$R_F \approx b\left(\frac{v}{b^3}\right)^{2\nu-1} N^{\nu}. \tag{5.28}$$

The size of a chain in a semidilute solution can be written as some power x of concentration that matches the dilute size at ϕ^*:

$$R \approx R_F \left(\frac{\phi}{\phi^*}\right)^x \approx b\left(\frac{v}{b^3}\right)^{2\nu-1+6\nu x-3x} N^{\nu+3\nu x-x} \phi^x. \tag{5.29}$$

Chains in semidilute solutions are random walks on their largest length scales and their size is thus proportional to the square root of the number of monomers $R \sim N^{1/2}$. Therefore

$$\nu + 3\nu x - x = \frac{1}{2} \implies x = -\frac{\nu - 1/2}{3\nu - 1}, \tag{5.30}$$

leading to

$$R \approx R_F \left(\frac{\phi}{\phi^*}\right)^{-(\nu-1/2)/(3\nu-1)} \approx b\left(\frac{\phi b^3}{v}\right)^{-(\nu-1/2)/(3\nu-1)} N^{1/2}. \tag{5.31}$$

The same de Gennes scaling approach can be applied to the correlation length:

$$\xi \approx R_F \left(\frac{\phi}{\phi^*}\right)^y \approx b\left(\frac{v}{b^3}\right)^{2\nu-1+6\nu y-3y} N^{\nu+3\nu y-y} \phi^y. \tag{5.32}$$

At the overlap concentration, the correlation length is equal to the dilute coil size because the coils are space-filling at ϕ^* and the correlation blobs are always space-filling. Above the overlap concentration, the *correlation length does not depend on the number of monomers in a chain*. Correlation blobs behave as shorter chains with g monomers at overlap [compare Eqs (5.19) and (5.21)]. A solution of longer chains (with $N_1 > N > g$) at the same concentration ϕ can be thought of as a melt of correlation blobs with more of these blobs per chain (with $N_1/g > N/g$). The correlation blobs are the same in these two solutions. The number of monomers g in each correlation blob and the size ξ of each correlation blob depend only on the volume fraction ϕ and excluded volume v, but not on the chain length N. From the simple fact that the correlation length is independent of chain length, the exponent y can be determined:

$$\nu + 3\nu y - y = 0 \quad \Rightarrow \quad y = -\frac{\nu}{3\nu - 1}. \tag{5.33}$$

The final expression for the correlation length is identical to Eq. (5.23):

$$\xi \approx R_{\mathrm{F}}\left(\frac{\phi}{\phi^*}\right)^{-\nu/(3\nu-1)} \approx b\left(\frac{b^3}{\mathrm{v}}\right)^{(2\nu-1)/(3\nu-1)} \phi^{-\nu/(3\nu-1)}. \tag{5.34}$$

The correlation length ξ decreases with increasing concentration, but the size of the thermal blob ξ_T is independent of concentration [Eq. (5.4)]. Therefore at some concentration, ϕ^{**} they are equal to each other and the intermediate swollen regime (ii) disappears. This concentration can be determined from the relation $\xi \approx \xi_T$ using Eqs (5.4) and (5.34):

$$b\left(\frac{b^3}{\mathrm{v}}\right)^{(2\nu-1)/(3\nu-1)} (\phi^{**})^{-\nu/(3\nu-1)} \approx \frac{b^4}{\mathrm{v}}, \tag{5.35}$$

$$\phi^{**} \approx \frac{\mathrm{v}}{b^3}. \tag{5.36}$$

Note that this concentration is analogous to ϕ'' in poor solvent [Eq. (5.12)], at which point the two- and three-body interactions are balanced (see Fig. 5.1), but in good solvent, both interactions are repulsive. Notice that in the athermal limit ($\mathrm{v} \approx b^3$) the semidilute solution persists to high concentrations, since $\phi^{**} \approx 1$.

At and above the concentration ϕ^{**}, chains are ideal on all length scales. On length scales smaller than the correlation length, the chains are ideal because the excluded volume interactions are weaker than the thermal energy (since $\xi < \xi_T$). On length scales larger than the correlation length, the chains are ideal because excluded volume interactions are screened by overlapping chains. The ϕ^{**} concentration is the crossover boundary between the semidilute regime with some swelling on intermediate length scales and the **concentrated solution** regime (Fig. 5.1) with ideal chain statistics on all scales $R \approx bN^{1/2}$. In the semidilute regime ($\phi^* < \phi < \phi^{**}$) the coils shrink from their size [Eq. (5.18)] in dilute solutions ($\phi < \phi^*$) to their ideal size in the concentrated regime ($\phi > \phi^{**}$) as shown in Fig. 5.6. The concentration dependence of the polymer size in semidilute solution $R \sim \phi^{-0.12}$ [see Eq. (5.31)] can be rewritten with ϕ^{**} as the reference point:

$$R \approx R_0\left(\frac{\phi}{\phi^{**}}\right)^{-(\nu-1/2)/(3\nu-1)} \qquad \phi^* < \phi < \phi^{**}. \tag{5.37}$$

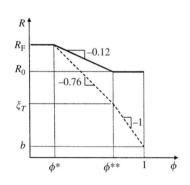

Fig. 5.6
Concentration dependence of coil size (solid line) and correlation length (dashed line) in an intermediate good solvent on logarithmic scales.

In an athermal solvent, the thermal blob is of the order of one monomer $\xi_T \approx b$ and regime (iii) disappears. The relations (5.23) and (5.26) are simplified because $v \approx b^3$.

$$\xi \approx b\phi^{-\nu/(3\nu-1)} \approx b\phi^{-0.76}, \tag{5.38}$$

$$R \approx bN^{1/2}\phi^{-(\nu-1/2)/(3\nu-1)} \approx bN^{1/2}\phi^{-0.12}. \tag{5.39}$$

The crossover volume fraction is of order unity ($\phi^{**} \approx 1$) in an athermal solvent, meaning that the chains are partially swollen at all concentrations. The excluded volume in an athermal solvent is fully screened only in the melt state ($\xi \approx b \approx \xi_T$ at $\phi = 1$).

5.3.2 Osmotic pressure

5.3.2.1 Mean-field theory
Recall the mean-field virial expansion for the osmotic pressure of polymer solutions discussed in Section 4.5.1 [Eq. (4.67)].

$$\Pi = kT\left[\frac{c_n}{N} + \frac{v}{2}c_n^2 + wc_n^3 + \cdots\right] = \frac{kT}{b^3}\left[\frac{\phi}{N} + \frac{v}{2b^3}\phi^2 + \frac{w}{b^6}\phi^3 + \cdots\right]. \tag{5.40}$$

The Flory–Huggins mean-field theory recovers the van't Hoff Law [Eq. (1.72)] in the dilute limit (as $\phi \to 0$). At higher concentrations, the mean-field theory predicts that two-body excluded volume interactions make osmotic pressure proportional to the mean-field probability of monomer–monomer contact (ϕ^2):

$$\Pi \approx kTvc_n^2 \approx \frac{kT}{b^6}v\phi^2 \quad \text{mean-field.} \tag{5.41}$$

The mean-field theory correctly predicts that the osmotic pressure is independent of molar mass in semidilute solution. However, the mean-field theory does not take into account the correlations between monomers along the chain, but instead assumes that they are distributed uniformly as in a solution of monomers. This is the reason why the two-body interactions [Eq. (5.41)] become important at volume fraction:

$$\phi_{2body} \approx \frac{b^3}{vN} \quad \text{mean-field,} \tag{5.42}$$

estimated by equating the first two terms in the virial series [Eq. (5.40)]. The two-body interaction concentration ϕ_{2body} is far below the overlap concentration ($\phi_{2body} < \phi^*$ for $v/b^3 > 1/\sqrt{N}$) predicting that two-body interactions are important in dilute solutions. This prediction of the mean-field theory overestimates the two-body interactions in dilute solution by distributing monomers uniformly everywhere. The two-body interactions between chains do not dominate the osmotic pressure until the overlap concentration ϕ^* is reached. In order to properly take into account chain connectivity, de Gennes developed a scaling model of osmotic pressure.

5.3.2.2 de Gennes scaling theory

Up to the overlap concentration ϕ^*, the van't Hoff Law should approximately describe the osmotic pressure. Above ϕ^*, the osmotic pressure should increase as a stronger function of concentration. This function must have the following form, to match the van't Hoff Law when $\phi = \phi^*$:

$$\Pi = \frac{kT}{b^3} \frac{\phi}{N} f\left(\frac{\phi}{\phi^*}\right). \tag{5.43}$$

In dilute solutions, the van't Hoff Law should be valid and the function f approaches unity at low values of the argument:

$$f\left(\frac{\phi}{\phi^*}\right) \approx 1 \quad \text{for } \phi < \phi^*. \tag{5.44}$$

Indeed, measurement of osmotic pressure in dilute solution can determine the chain length N, with results for a polydisperse sample providing the number-average chain length (see Section 1.7.1). In the semidilute regime, a power law form for the function f is assumed:

$$f\left(\frac{\phi}{\phi^*}\right) \approx \left(\frac{\phi}{\phi^*}\right)^z \quad \text{for } \phi > \phi^*. \tag{5.45}$$

The exponent z can be determined from the scaling form of the osmotic pressure:

$$\Pi \approx \frac{kT}{b^3} \frac{\phi}{N} \left(\frac{\phi}{\phi^*}\right)^z \approx \frac{kT}{b^3} \phi^{1+z} \left(\frac{v}{b^3}\right)^{3z(2\nu-1)} N^{(3\nu-1)z-1}. \tag{5.46}$$

The final relation was obtained using Eq. (5.27) for ϕ^*. *The osmotic pressure in semidilute solution is independent of chain length.* Therefore the exponent of N in Eq. (5.46) must be zero

$$(3\nu - 1)z - 1 = 0 \quad \Rightarrow \quad z = \frac{1}{3\nu - 1}. \tag{5.47}$$

The semidilute osmotic pressure has a stronger concentration dependence than predicted by the mean-field virial expansion [Eq. (5.41)]:

$$\Pi \approx \frac{kT}{b^3} \left(\frac{v}{b^3}\right)^{3(2\nu-1)/(3\nu-1)} \phi^{3\nu/(3\nu-1)} \approx \frac{kT}{b^3} \left(\frac{v}{b^3}\right)^{0.69} \phi^{2.3}. \tag{5.48}$$

The scaling approach to semidilute solutions, described above, takes into account the correlations between monomers along the chain. Chains begin to interact at length scales of order of the correlation length ξ. In good and athermal solvents, *neighbouring blobs repel each other with energy of order kT.* Therefore the scaling model prediction for the osmotic pressure in semidilute solutions is of order kT per correlation blob.

$$\Pi \approx \frac{kT}{b^3} \left(\frac{v}{b^3}\right)^{3(2\nu-1)/(3\nu-1)} \phi^{3\nu/(3\nu-1)} \approx \frac{kT}{\xi^3}. \tag{5.49}$$

In the final expression, Eq. (5.23) was used for the correlation length ξ. The scaling prediction for osmotic pressure is significantly different from the mean-field prediction because the exponents for the concentration dependence differ (2.3 instead of 2). The scaling prediction is in excellent agreement with experiments, as demonstrated in Fig. 5.7 (the high concentrations are described by $\Pi/c \sim c^{1.3}$). Equation (5.49) demonstrates that the osmotic pressure provides a direct measure of the correlation length in semidilute solutions.

Figure 5.7(b) demonstrates that the functional form of Eq. (5.43) reduces osmotic pressure data at various M and ϕ (or c) to a universal curve. The limiting scaling laws of $\Pi \sim \phi$ or $\Pi \sim \phi^{2.3}$ are only valid sufficiently far from the overlap concentration. Near ϕ^* (and more generally near any crossover point), a more complicated functional form than a simple power law is needed. For osmotic pressure in a good solvent (and many other examples) the full functional form of Eq. (5.43) is well described by a simple sum of the two limiting behaviours:

$$\frac{\Pi b^3 N}{\phi k T} = f\left(\frac{\phi}{\phi^*}\right) \approx 1 + \left(\frac{\phi}{\phi^*}\right)^{1.3}. \qquad (5.50)$$

5.4 Semidilute theta solutions

5.4.1 Correlation length

In semidilute θ-solutions and in concentrated solutions ($\phi > \phi^* = \phi_c$ and between ϕ'' and ϕ^{**} curves in Fig. 5.1), the chains are almost ideal with $R \approx bN^{1/2}$. The correlation blob of size ξ is still defined as the volume in which most of the monomers belong to the same chain. As in good solvents, the volume of the semidilute θ-solution is divided into space-filling correlation blobs (Fig. 5.4) and the correlation length ξ is the scale at which monomers from other chains are seen. However, unlike good solvents, *there is no change in chain conformation at the correlation length in θ-solvent*. In good solvents, the excluded volume interactions are screened by overlapping chains at the correlation length and the chain statistics change to that of an ideal coil at larger length scales. In θ-solvents, chain statistics are nearly ideal on all length scales at all concentrations, and no change in polymer conformation occurs at the correlation length ξ. In Chapter 8, we will demonstrate that this correlation length is important for polymer dynamics in semidilute θ-solutions. In Section 5.4.2, this correlation length will be related to osmotic pressure.

The correlation length is determined by recognizing that only g monomers from a single chain are inside the correlation volume ξ^3 and that these correlation volumes are space-filling. The volume fraction of polymer in solution is then the ratio of the occupied volume of the g monomers in the strand $b^3 g$ and the correlation volume ξ^3:

$$\phi \approx \frac{b^3 g}{\xi^3} \approx \frac{b^3 (\xi/b)^2}{\xi^3} \approx \frac{b}{\xi}. \qquad (5.51)$$

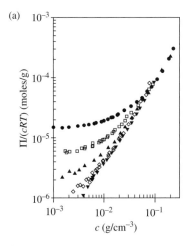

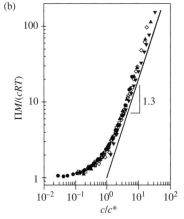

Fig. 5.7
Concentration dependence of osmotic pressure data for five poly(α–methyl styrene)s in the good solvent toluene at 25 °C. (a) Raw data—the data below c^* for the lowest three molar masses are plotted on linear scales in Fig. 1.22. (b) Data reduced in the scaling form expected by Eq. (5.43). Filled circles are data for $M_n = 70\,800\,\text{g mol}^{-1}$, open squares are data for $M_n = 200\,000\,\text{g mol}^{-1}$, filled triangles are data for $M_n = 506\,000\,\text{g mol}^{-1}$, open diamonds are data for $M_n = 1\,190\,000\,\text{g mol}^{-1}$, and filled upside-down triangles are data for $M_w = 1\,820\,000\,\text{g mol}^{-1}$. (after I. Noda, et al. Macromolecules, **14**, 668, 1981).

The concentration dependence of the correlation length in semidilute θ-solutions is stronger than that in good solvent:

$$\xi \approx \frac{b}{\phi}. \tag{5.52}$$

This concentration dependence can also be determined by a scaling argument similar to the one used in semidilute good solvent. The correlation length ξ is equal to the ideal chain size R_0 at the overlap concentration ϕ^*:

$$\xi \approx R_0 = b\sqrt{N} \quad \text{at } \phi^* \approx \frac{Nb^3}{R_0^3} \approx \frac{1}{\sqrt{N}}. \tag{5.53}$$

In the semidilute regime, the scaling assumption is that the correlation length decreases as a power law in concentration:

$$\xi \approx bN^{1/2}\left(\frac{\phi}{\phi^*}\right)^x \approx b\phi^x N^{(1+x)/2}. \tag{5.54}$$

The exponent x is once again determined from the condition that the correlation length in semidilute solution should be independent of chain length $\xi \sim N^0$:

$$\frac{1+x}{2} = 0 \quad \Rightarrow \quad x = -1. \tag{5.55}$$

Substituting $x = -1$ in Eq. (5.54) also yields Eq. (5.52) for the concentration dependence of the correlation length in θ-solvent.

The correlation length has been measured using scattering experiments, as discussed in Section 5.7. Small-angle neutron scattering (SANS) data on polystyrene in deuterated cyclohexane at $\theta = 38.0\,^\circ$C are summarized in Fig. 5.8. These data are in good agreement with the power law expected by Eq. (5.52), shown as the line.

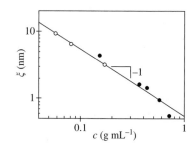

Fig. 5.8
Correlation length from SANS data fit to Eq. (5.52), for polystyrene in perdeuterated cyclohexane at the θ-temperature (38 °C). Open circles are from J. P. Cotton *et al.*, *J. Chem. Phys.* **65**, 1101 (1976) and filled circles are from E. Geissler *et al.*, *Macromolecules* **23**, 5270 (1990).

5.4.2 Osmotic pressure

The mean-field prediction for the osmotic pressure [Eq. (5.40)] in θ-solvents is the virial expansion with vanishing excluded volume (v = 0):

$$\Pi = \frac{kT}{b^3}\left[\frac{\phi}{N} + \frac{w}{b^6}\phi^3 + \cdots\right]. \tag{5.56}$$

The first term is proportional to the number density of chains [the van't Hoff law Eq. (1.72)] and is important in dilute solutions. The three-body term is larger than the linear term $(w\phi^3/b^6 > \phi/N)$ at concentrations above overlap $\phi > b^3/\sqrt{wN} \approx \phi^*$, since the three-body interaction coefficient $w \approx b^6$. In semidilute θ-solutions, the osmotic pressure is determined by the third virial term:

$$\Pi \approx \frac{kT}{b^3} \phi^3. \tag{5.57}$$

This mean-field prediction can be substantiated by constructing a simple de Gennes scaling theory. A similar scaling form for the osmotic pressure is assumed in θ-solutions as was used in good solvent [Eq. (5.43)]:

$$\Pi = \frac{kT}{b^3} \frac{\phi}{N} h\left(\frac{\phi}{\phi^*}\right). \tag{5.58}$$

In dilute solutions $\phi < \phi^*$, this scaling function approaches unity $h(\phi/\phi^*) \approx 1$ and the osmotic pressure obeys the van't Hoff law. In semi-dilute θ-solutions, $h(\phi/\phi^*)$ is again assumed to be a power law:

$$\Pi \approx \frac{kT}{b^3} \frac{\phi}{N} \left(\frac{\phi}{\phi^*}\right)^y \approx \frac{kT}{b^3} \phi^{1+y} N^{y/2-1}. \tag{5.59}$$

The exponent y is determined from the condition that the osmotic pressure in semidilute solutions is independent of chain length, $\Pi \sim N^0$:

$$\frac{y}{2} - 1 = 0 \implies y = 2. \tag{5.60}$$

The scaling argument leads to the same prediction in semidilute θ-solutions as the mean-field theory [Eq. (5.57)]. The osmotic pressure in semidilute solutions is again of the order of the thermal energy kT per correlation volume:

$$\Pi \approx \frac{kT}{\xi^3} \approx \frac{kT}{b^3} \phi^3. \tag{5.61}$$

The correlation length in θ-solution is of the order of the distance between three-body contacts (because the excluded volume of two-body interactions is $v = 0$ in θ-solvents). The density of n-body contacts is proportional to the probability of n monomers being in the same small volume of space. The mean-field prediction for this probability and the corresponding contribution to osmotic pressure is proportional to ϕ^n. The number density of such contacts is ϕ^n/b^3. The distance between n-body contacts r_n in three-dimensional space is of the order of

$$r_n \approx b\phi^{-n/3} \quad \text{three dimensions,} \tag{5.62}$$

within the mean-field theory (ignoring correlations). This distance defines the corresponding correlation length. The mean-field predictions for the correlation length [Eq. (5.62) with $n = 3$] and the osmotic pressure [Eq. (5.61)] in semidilute θ-solutions are in agreement with the scaling theory prediction and with experiments (Figs. 5.8 and 5.9).

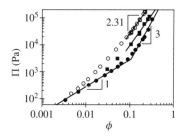

Fig. 5.9
Osmotic pressure of
$M_n = 90\,000\,\mathrm{g\,mol^{-1}}$ polyisobutylene
solutions from P. J. Flory and H.
Daoust, *J. Polym. Sci.* **25**, 429 (1957).
Filled circles are in benzene at
$\theta = 24.5\,^{\circ}\mathrm{C}$, filled squares are in benzene
at 50 °C, open circles are in cyclohexane
at 30 °C, and open squares are in
cyclohexane at 8 °C. The lines are the
power laws expected from scaling
theory.

Figure 5.9 compares osmotic pressure in a θ-solvent (filled circles) with osmotic pressure in the same solvent 25 K above θ (filled squares) and in an athermal solvent (open symbols). The θ-solvent data clearly exhibit the slopes of 1 and 3 expected from Eq. (5.56). At $T = \theta + 25\,\mathrm{K}$, the solvent is a good solvent, with the slope of 2.3 expected by Eq. (5.48) observed at high concentrations. The fact that osmotic pressure data for polyisobutylene solutions in cyclohexane are the same at two temperatures indicates that cyclohexane is an athermal solvent. The factor of 2.75 relating the coefficients of the 2.3 power laws for athermal solvent and benzene at 50 °C indicates that polyisobutylene in benzene at 50 °C has excluded volume smaller than in cyclohexane by a factor of $2.75^{-1/0.69} \cong 0.23$ [see Eq. (5.48)].

The mean-field theory does not work in good solvent because excluded volume interactions strongly affect chain statistics and reduce the probability of inter-chain contacts. In θ-solvents, the chain statistics and the probability of the inter-chain contacts are almost unaffected by interactions and well approximated by the mean-field theory.

5.5   The Alexander – de Gennes brush

The concepts of correlation length and scaling can be used to understand a wide variety of topics in polymer physics. As an application of these ideas, consider a simple scaling estimate of the height H of a grafted layer (called a brush). The layer consists of σ chains per unit surface area grafted to the repulsive surface in an athermal solvent. If the grafting density σ is high enough, the attached chains form an overlapping layer that behaves like a semidilute solution. The distance between the grafting points defines the distance between the chains which is the correlation length in this layer:

$$\xi \approx \frac{1}{\sqrt{\sigma}}. \tag{5.63}$$

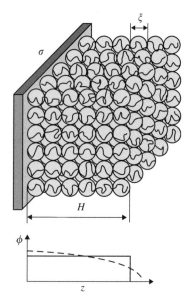

Fig. 5.10
The Alexander – de Gennes brush of
chains grafted to a surface in a good
solvent and corresponding density
profiles. The solid line is the
Alexander – de Gennes step function
profile. The dashed line is the
Semenov–Milner–Witten parabolic
profile.

The correlation blobs of overlapping grafted chains repel each other with energy of order kT, forcing them into an array perpendicular to the surface (see Fig. 5.10). The number of monomers g in a correlation volume in an athermal solvent is determined by the self-avoiding statistics of the chains inside each correlation volume:

$$g \approx \left(\frac{\xi}{b}\right)^{1/\nu} \approx \sigma^{-1/(2\nu)} b^{-1/\nu}. \tag{5.64}$$

The number of correlation blobs per chain is N/g:

$$\frac{N}{g} \approx N\sigma^{1/(2\nu)} b^{1/\nu}. \tag{5.65}$$

The Alexander – de Gennes approximation for a brush is that it is a stretched array of correlation blobs. The height of the brush is the size of a correlation blob times the number of these blobs per chain:

$$H \approx \xi \frac{N}{g} \approx N\sigma^{(1-\nu)/(2\nu)} b^{1/\nu}. \tag{5.66}$$

The height H increases linearly with the number of monomers N per chain at constant grafting density. The stretching energy per chain E_{chain} in the grafting layer is kT times the number of correlation blobs per chain:

$$E_{chain} \approx kT \frac{N}{g} \approx kTN\sigma^{1/(2\nu)} b^{1/\nu}. \tag{5.67}$$

The stretching energy per unit volume in the brush is balanced by the interchain repulsive energy per unit volume, which is proportional to the osmotic pressure in the layer:

$$\frac{E_{chain}}{V_{chain}} \approx E_{chain} \frac{\sigma}{H} \approx kT \frac{N}{g} \frac{\sigma}{H} \approx kT \frac{\sigma}{\xi} \approx \frac{kT}{\xi^3} \approx \Pi. \tag{5.68}$$

The Alexander – de Gennes approximation assumes that the chains are uniformly stretched, with the free end of each chain located at the top of the brush at height H from the surface. The density profile of the Alexander – de Gennes brush is a step function:

$$\phi \approx \begin{cases} b^3 g/\xi^3 \approx (\sigma b^2)^{(3\nu-1)/(2\nu)} & \text{for } z < H \\ 0 & \text{for } z > H. \end{cases} \tag{5.69}$$

A more accurate solution of this problem was developed by Semenov and by Witten and Milner using a self-consistent field theory. It allows the free ends of the chains to be distributed throughout the whole grafted layer and predicts a parabolic density profile of the brush (see Fig. 5.10).

5.6 Multichain adsorption

In Section 3.2.3, a scaling picture of single-chain adsorption was presented. A chain forms a flat pancake at an adsorbing surface in order to balance the number of contacts with the surface against the loss of entropy due to confinement. Even though each monomer usually gains energy less than kT upon contact with the surface, the whole chain can gain energy much larger than kT because of many contact points. Thus, the chains in solution are attracted to the surface and try to maximize the number of contacts with it. This leads to a high concentration of monomers near the surface in the realistic case of multichain adsorption.

Crowded chains near the surface repel each other, limiting other chains from getting into this concentrated layer. The monomer concentration in the first layer near the surface is determined by an energy balance between attraction to the surface and repulsion from surrounding chains. The correlation length ξ_{ads} in the first layer defines the adsorption blob size with g_{ads} monomers. The chain sections of the adsorption blob size are attracted to the surface with energy of order kT and are repelled from each other with similar energy. The size of the adsorption blob was calculated in Section 3.2.3 [Eq. (3.62)] where δkT is the energy gain per monomer adsorbed:

$$\xi_{ads} \approx \frac{b}{\delta^{\nu/(1-\nu)}} \quad \text{good solvent.} \tag{5.70}$$

The adsorption blob size decreases with increasing surface attraction δ as $\xi_{ads} \approx b\delta^{-1.4}$ for exponent[1] $\nu \cong 0.588$.

Only the sections of the chains in this first layer gain energy due to contact with the surface. Sections further away from the surface do not gain attraction energy, but are relaxing the concentration from the high value near the surface to a low value in the bulk of the solution in the most optimal way. The correlation length $\xi \approx b\phi^{-\nu/(3\nu-1)}$ in a semidilute athermal solution defines the distance from the surface $z \approx \xi$ at which the concentration decays to a value ϕ:

$$\phi \approx (\xi/b)^{-(3\nu-1)/\nu} \approx (z/b)^{-(3\nu-1)/\nu}. \tag{5.71}$$

The concentration in the adsorbed layer decreases away from the surface as $\phi \approx (z/b)^{-1.3}$ for exponent $\nu \cong 0.588$. This power law concentration profile in an adsorbed polymer layer was proposed by de Gennes and is called the **de Gennes self-similar carpet**. This profile of adsorbed polymer can be described by a set of layers of correlation blobs with their size ξ of order of their distance to the surface z (see Fig. 5.11). The self-similar concentration profile starts at the adsorption blob ξ_{ads} in the first layer. The self-similar profile ends either at the correlation length of the surrounding solution if it is semidilute or at the chain size $R_F \approx bN^\nu$ if the surrounding solution is dilute.

The adsorbed amount Γ is the number of monomers adsorbed per unit area of the surface, and is controlled by the densest layer with thickness of the order of adsorption blob size ξ_{ads}:

$$\Gamma \approx \int \frac{\phi(z)}{b^3} \, dz \approx b^{-3} \int_{\xi_{ads}}^{R_F} (z/b)^{-(3\nu-1)/\nu} \, dz$$

$$\approx b^{-2} \left(\frac{b}{\xi_{ads}} \right)^{(2\nu-1)/\nu} \approx \frac{\delta^{(2\nu-1)/(1-\nu)}}{b^2} \tag{5.72}$$

The adsorbed amount increases with the adsorption energy per monomer as $\Gamma \approx \delta^{0.43}/b^2$ for exponent $\nu \cong 0.588$. Notice that the integral in Eq. (5.72) for $\nu > 1/2$ is dominated by the lower limit $z = \xi_{ads}$ and therefore the adsorbed amount Γ is almost independent of the chain length of the adsorbed polymers. It is proportional to the number of monomers in the first layer with thickness of the order of the size of the adsorption blob [Eq. (5.70)]:

$$\frac{g_{ads}}{\xi_{ads}^2} \approx \left(\frac{\xi_{ads}}{b} \right)^{1/\nu} \frac{1}{\xi_{ads}^2} \approx \frac{1}{b^2} \left(\frac{b}{\xi_{ads}} \right)^{(2\nu-1)/\nu} \approx \frac{\delta^{(2\nu-1)/(1-\nu)}}{b^2}. \tag{5.73}$$

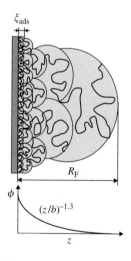

Fig. 5.11
The de Gennes self-similar carpet and corresponding concentration profile for an adsorbed polymer layer in a good solvent.

---

[1] The exponent was $-3/2$ in Eq. (3.62) because in Section 3.2.3 we used $\nu = 3/5$.

There is considerable difference between the power law profile of the adsorbed polymer layer [Eq. (5.71) and Fig. 5.11] and the step function (or parabolic) profile of a grafted polymer brush (Fig. 5.10).

5.7 Measuring semidilute chain conformations

Semidilute solutions have an important length scale called the correlation length ξ, at which neighboring chains start to interact. The correlation length can be experimentally measured using SANS from a polymer solution in a perdeuterated solvent (all solvent hydrogen atoms are substituted with deuterium). The scattering function for a semidilute solution of ideal chains is given by the Ornstein–Zernike scattering function [Eq. (4.93)].

$$S(q) = \frac{S(0)}{1 + (q\xi)^2}. \tag{5.74}$$

This equation allows the correlation length ξ to be measured from SANS on semidilute θ-solutions.

For polymers dissolved in a good solvent, the coils are swollen on length scales smaller than ξ, and SANS data from a solution in a perdeuterated good solvent fit a slightly different function that utilizes the swollen fractal dimension $\mathcal{D} = 1/\nu \cong 1.7$:

$$S(q) \cong \frac{S(0)}{1 + (q\xi)^{1/\nu}}. \tag{5.75}$$

Figure 5.12 compares Eq. (5.75) ($1/\nu \cong 1.7$) with SANS data on a semidilute solution of polystyrene in carbon disulphide. Since the solvent contains no protons, no deuterium labelling is necessary for good scattering contrast.[2] Fitting Eq. (5.75) with the good solvent value of $1/\nu \cong 1.7$ to the data yields $\xi = 5.5$ nm and provides an excellent description of the data.

The coil size in semidilute solution can also be determined using SANS, by utilizing a mixture of deuterium-labelled and unlabelled chains in an appropriate mixture of deuterium-labelled and unlabelled solvent with the same average contrast as the polymer mixture (contrast matched, see Problem 5.26). If the fraction x of the chains are labelled, the q-dependent scattering intensity $I(q)$ is proportional to the form factor $P(q)$ (see Problem 5.26)

$$I(q) \sim x(1-x)N^2 KP(q), \tag{5.76}$$

where K is the number of chains in the scattering volume. In practice, the proportionality constant is known and the radius of gyration is obtained from $P(q)$ by fitting to the Debye function [Eq. (2.160)] for $q < 1/\xi$, where random walk statistics apply.

A common measure of the thermodynamics of interchain interactions in polymer solutions is the osmotic pressure. Osmotic pressure is the pressure

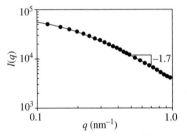

Fig. 5.12
Small-angle neutron scattering data for a semidilute solution of polystyrene ($M_w = 1\,100\,000$ g mol^{-1}) in CS$_2$ with $c = 0.025$ g/ml$^{-1} \cong 2.5c*$. The curve is Eq. (5.75) with good solvent fractal dimension $\mathcal{D} = 1/\nu \cong 1.7$. Data from M. Daoud *et al.*, *Macromolecules* **8**, 804 (1975).

---

[2] In most situations of interest in polymer science, neutron scattering from protons dominates the intensity.

difference across a membrane that separates the polymer solution from pure solvent. The membrane allows solvent to pass freely but prevents polymer from crossing. In Chapters 1 and 4, we learned that osmotic pressure measurements in dilute solution determine polymer molar mass, since osmotic pressure is kT per chain. In this chapter, we learned that the osmotic pressure measurement in semidilute solution provides another means of determining the correlation length because the osmotic pressure is of the order of kT per correlation volume. The correlation length can also be determined from the osmotic compressibility measured by scattering at low wavevector [Eq. (1.91)].

5.8   Summary of polymer solutions

The phase diagram of polymer solutions is shown in Fig. 5.1, assuming the usual case of $B > 0$ in Eq. (4.31) (with $\chi = A + B/T$ a decreasing function of temperature). In the poor solvent half of the diagram (at temperatures below θ) the binodal separates the two-phase region from the two single-phase regions.

There are dilute globules with size

$$R_{gl} \approx \frac{b^2 N^{1/3}}{|v|^{1/3}}, \tag{5.77}$$

at very low concentrations ($\phi < \phi'$) and concentrated solutions with overlapping ideal chains for $\phi > \phi''$.

At temperatures near θ (for $|T - \theta|/\theta < 1/\sqrt{N}$) there are two regions. Dilute θ-solutions with non-overlapping chains for $\phi < \phi_\theta^* \approx 1/\sqrt{N}$ and semidilute θ-solutions with overlapping chains for $\phi > \phi_\theta^*$. Chains in both θ-regions have nearly ideal coil size:

$$R \approx R_0 = bN^{1/2}. \tag{5.78}$$

At sufficiently high temperatures, the solvent is good, with three regimes. There is a dilute good solvent regime at concentrations $\phi < \phi^* \approx (b^3/v)^{6\nu - 3} N^{1 - 3\nu}$, with non-overlapping swollen chains whose size was determined in Chapter 3:

$$R_F \approx b\left(\frac{v}{b^3}\right)^{2\nu - 1} N^\nu \approx b\left(\frac{v}{b^3}\right)^{0.18} N^{0.588}. \tag{5.79}$$

At concentrations $\phi^* < \phi < \phi^{**} \approx v/b^3$, there is a semidilute good solvent regime. In semidilute solution, the chain conformation is similar to dilute solutions on small length scales, while the conformation is analogous to polymer melts on large length scales. The overlapping chains in semidilute solution are swollen at intermediate length scales between the thermal blob size and the correlation length $\xi_T < r < \xi$ and ideal at smaller ($r < \xi_T$) and larger ($r > \xi$) length scales. The chain size in semidilute solutions in a good solvent decreases weakly as the concentration is increased:

$$R \approx R_F \left(\frac{\phi}{\phi^*}\right)^{-(\nu - 1/2)/(3\nu - 1)} \approx R_0 \left(\frac{\phi}{\phi^{**}}\right)^{-0.12}. \tag{5.80}$$

Concentrated solutions occur above the concentration ϕ^{**} at which the thermal blob size and the correlation length coincide (at $\phi = \phi^{**} \approx v/b^3$). Chains have nearly ideal statistics on all length scales in concentrated solution. This regime is simply an extension of the semidilute θ-solution region to higher temperatures (see Fig. 5.1).

Semidilute and concentrated solutions are characterized by a correlation length ξ, the scale at which a given chain starts to find out about other chains. This correlation length is

$$\xi \approx b \left(\frac{b^3}{v}\right)^{(2\nu-1)/(3\nu-1)} \phi^{-\nu/(3\nu-1)} \approx b \left(\frac{b^3}{v}\right)^{0.23} \phi^{-0.76} \qquad (5.81)$$

in good solvents, and

$$\xi \approx \frac{b}{\phi} \qquad (5.82)$$

in semidilute θ and concentrated solutions. The correlation length is the average distance between segments on neighbouring chains and is independent of the degree of polymerization. Inside the correlation blob, dilute chain statistics apply, whereas the large-scale conformation of the chain is that of a melt of correlation blobs. Hence, the chain size in semidilute solution is always determined as a random walk of correlation blobs.

The semidilute good solvent predictions have been tested using SANS on polystyrene solutions in carbon disulphide in Fig. 5.13, showing remarkable agreement. Carbon disulphide was chosen for the solvent because no deuterium labelling is needed since this solvent has no protons. Apparently, CS_2 is an athermal solvent for polystyrene, since the radius of gyration continues to decrease all the way to the melt.

The correlation length also determines the osmotic pressure to be of order kT per blob:

$$\Pi \approx \frac{kT}{\xi^3}. \qquad (5.83)$$

This equation holds for theta, good, and athermal solvents. Hence, osmotic pressure or osmotic compressibility measurements provide a convenient means of measuring the correlation length in semidilute solutions.

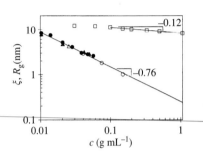

Fig. 5.13
Correlation length ξ and radius of gyration R_g for polystyrene in carbon disulfide from M. Daoud *et al.*, *Macromolecules* **8**, 804 (1975). Radius of gyration data (open squares $M = 1.14 \times 10^5 \, g \, mol^{-1}$) above $c = 0.1 \, g \, ml^{-1}$ were fit to a power law with slope -0.12 that agrees well with the scaling prediction [Eq. (5.80)]. The correlation length is independent of molar mass (filled circles $M = 2.1 \times 10^6 \, g \, mol^{-1}$, filled triangles $M = 6.5 \times 10^5 \, g \, mol^{-1}$, open circles $M = 5 \times 10^5 \, g \, mol^{-1}$) and the power law slope of -0.76 for all data agrees well with scaling prediction [Eq. (5.81)]. Note that $c = 1.06 \, g \, mL^{-1}$ corresponds to a polystyrene melt.

Problems

Section 5.2

5.1 Consider a dilute polymer solution with excluded volume $v = -12.5 \, \text{Å}^3$ and Kuhn monomer length $b = 5 \, \text{Å}$. What is the root-mean-square end-to-end distance of this polymer if the number of Kuhn monomers is
(i) $N = 50$? (ii) $N = 10^2$? (iii) $N = 10^3$?

5.2 A solution is prepared using chains having $N = 10^4$ Kuhn monomers of length $b = 4 \, \text{Å}$, with Flory interaction parameter $\chi = 0.55$ at volume fraction $\phi = 0.01$.

(i) Will this solution remain homogeneous or phase separate?
(ii) What is the polymer volume fraction ϕ'' in the sediment?

   (iii) What is the root-mean-square end-to-end distance of the polymer in the sediment?

   (iv) What is the polymer volume fraction ϕ' in the supernatant?

   (v) What is the root-mean-square end-to-end distance of the polymer in the supernatant?

   (vi) What is the volume fraction within the globule in the supernatant?

Section 5.3

5.3 Consider a polymer solution with $N = 10^3$ Kuhn monomers of length $b = 7$ Å, and Flory interaction parameter $\chi = 0.45$ at volume fraction $\phi = 0.025$.

   (i) Do you expect this solution to stay homogeneous or to phase separate? Explain.

   (ii) What is the polymer volume fraction ϕ in the sediment?

   (iii) What is the root-mean-square end-to-end distance of the polymer in the sediment?

   (iv) What is the polymer volume fraction ϕ in this solution?

   (v) What is the root-mean-square end-to-end distance of the polymer in this solution?

5.4 Consider three polymer solutions with Flory interaction parameter $\chi = 0.49$, Kuhn length $b = 7$ Å, and the number of Kuhn monomers: (a) $N = 100$; (b) $N = 1000$; and (c) $N = 10,000$.

   (i) How many monomers are in a thermal blob?

   (ii) Find the root-mean-square size of these polymers in dilute solutions.

   (iii) Find the overlap volume fraction for each of these three polymers.

   (iv) What is the size of each of these polymers in solutions with volume fraction $\phi = 0.015$?

5.5 Consider a polymer solution with Flory interaction parameter $\chi = 0.4$, consisting of chains with $N = 10^3$ Kuhn monomers of length $b = 3$ Å.

   (i) What is the overlap volume fraction ϕ^* of these chains?

   (ii) What is the ϕ^{**} volume fraction for this solution?

What is the root-mean-square end-to-end distance of the polymer at volume fraction

   (iii) $\phi = 0.005$?

   (iv) $\phi = 0.05$?

   (v) $\phi = 0.1$?

   (vi) $\phi = 0.2$?

   (vii) $\phi = 0.4$?

5.6 Recall the two-dimensional size of an isolated real chain at the air–water interface $R = a^{1/4} b^{1/2} N^{3/4}$, where a is the excluded area parameter, b is the monomer size, and N is the degree of polymerization [Eq. (3.137)].

   (i) What is the overlap surface coverage, σ^*?

   (ii) What is the thermal blob size, ξ_T?

   (iii) What is the correlation length ξ at surface coverage σ?

   (iv) What is the size of the polymer R? Describe the conformation of the chain at surface coverages σ above the overlap coverage σ^*.

   (v) What is the surface pressure (the two-dimensional analog of osmotic pressure)?

5.7 Consider an athermal semidilute polymer solution (with excluded volume $v = 125$ Å^3, and the Kuhn length $b = 5$ Å) at volume fraction $\phi = 0.01$. The degree of polymerization of chains is $N = 10^4$. Estimate the osmotic pressure of this solution at room temperature using

(i) mean-field theory;
(ii) scaling theory.

5.8 Consider a semidilute polymer solution of chains with N_B monomers, volume fraction ϕ and excluded volume v. A trace amount of longer chemically identical chains with N_A monomers is added to the solution. What is the size R_A of these A-chains, if they are assumed not to overlap with each other and not to change the overall volume fraction ϕ?

5.9 Derive Eqs (5.23) and (5.26) for good solvents with $0 < v < b^3$ from Eqs (5.38) and (5.39) for athermal solvents. *Hint*: Renormalize the monomer to the thermal blob.

5.10 Plot the size r of a labelled section of n consecutive Kuhn monomers of a chain for different regions of the diagram in Fig. 5.1. (i) Dilute θ-solvent; (ii) semidilute θ-solvent; (iii) dilute poor solvent; (iv) two-phase region; (v) concentrated poor solvent; (vi) dilute good solvent; (vii) semidilute good solvent; (viii) concentrated good solvent.

5.11 Plot the total number of Kuhn monomers belonging to all chains within a small sphere of radius r with the centre at one monomer for different regions of the diagram in Fig. 5.1. (i) Dilute θ-solvent; (ii) semidilute θ-solvent; (iii) dilute poor solvent; (iv) two-phase region; (v) concentrated poor solvent; (vi) dilute good solvent; (vii) semidilute good solvent; (viii) concentrated good solvent.

5.12 Stretching a chain in semidilute solution.

Consider a semidilute solution with volume fraction ϕ of chains with N Kuhn monomers of length b and excluded volume v. Calculate the free energy cost to stretch a chain to end-to-end distance R for the following cases:

(i) Consider the case of relatively weak stretching, with the Pincus blob larger than the correlation length.
(ii) Consider the case of intermediate stretching, with the Pincus blob smaller than the correlation length but larger than the thermal blob.
(iii) Consider the case of strong stretching, with the Pincus blob smaller than the thermal blob.
(iv) Over what range of end-to-end distance does each case apply?

Section 5.4

5.13 Consider a semidilute polymer solution at room temperature with Flory interaction parameter $\chi = 0.4$, having $N = 10^3$ Kuhn monomers of length $b = 3\,\text{Å}$.

(i) Calculate the size of a thermal blob ξ_T.
(ii) Calculate the size of a correlation blob ξ as a function of polymer volume fraction ϕ. Note: separately consider two cases: $\xi > \xi_T$ and $\xi < \xi_T$.
(iii) What is the concentration dependence of osmotic pressure $\Pi(\phi)$ at room temperature?

5.14 In order to better understand why the distance between three-body contacts is of the order of the distance between monomers on neighbouring chains, the problem can be generalized to ideal chains in d dimensions.

(i) Calculate the distance r_n between n-body contacts in d dimensions.

(ii) Calculate the average distance ξ between monomers on neighbouring chains in d dimensions at 'volume' fraction ϕ.

(iii) What is the relation between n and d that is needed for the distance r_n between n-body contacts to be proportional to the average distance ξ between monomers in d dimensions at any volume fraction ϕ?

(iv) Does this condition work for three-body contacts in three dimensions? For which interactions does this condition work in four dimensions?

Section 5.5

5.15 Consider a wet Alexander – de Gennes brush formed by chains with $N = 10^4$ Kuhn monomers of length $b = 3\,\text{Å}$, attached to a surface with density $\sigma = 1 \times 10^{-4}\,\text{Å}^{-2}$ (chains per unit area) in a solvent with Flory interaction parameter $\chi = 0.43$.

(i) What is the correlation length ξ for this brush?

(ii) What is the size of a thermal blob ξ_T within the brush?

(iii) How many monomers of a chain are there inside a thermal blob (g_T) and inside a correlation blob (g)?

(iv) How many correlation blobs does each chain have?

(v) What is the thickness of the brush?

5.16 Calculate the thickness of a wet Alexander – de Gennes brush formed by chains with $N = 10^3$ Kuhn monomers of length $b = 5\,\text{Å}$, attached to the surface with density $\sigma = 1\,\text{nm}^{-2}$ in a good solvent with Flory interaction parameter $\chi = 0.3$.

5.17 Calculate the density profile in an f-arm star polymer in a θ-solvent as a function of the distance from the branch point. How does the size of the polymer depend on the number of arms f, the degree of polymerization N, and monomer size b.

5.18 Calculate the density profile in an f-arm star polymer in an athermal solvent as a function of the distance from the branch point. How does the size of the polymer depend on the number of arms f, the degree of polymerization N, and monomer size b.

5.19* Calculate the density profile in an f-arm star polymer in a good solvent as a function of the distance from the branch point. How does the size of the polymer depend on the number of arms f, the degree of polymerization N, monomer size b, and the excluded volume $v > 0$?

5.20 Calculate the density profile in an f-arm star polymer restricted to the air-water interface in an athermal good solvent (two-dimensional stars) as a function of the distance from the branch point. How does the size of the polymer depend on the number of arms f, the degree of polymerization N, and monomer size b.

5.21 Calculate the density profile of a cylindrical brush in an athermal solvent as a function of the distance from the axis of a cylinder. How does the diameter of the brush R depend on the line density of arms σ, the degree of polymerization N, and monomer size b.

5.22* Consider a spherical micelle made out of f diblock copolymers. The insoluble block consists of N_A monomers of size b_A and its Flory interaction parameter with the solvent is χ_A.

(i) Calculate the size R_{core} of the core of the micelle.

(ii) Calculate the surface energy of the core (per chain).

(iii) Calculate the size of the micelle $R_{micelle}$ in an athermal good solvent for the outer block $(\chi_B = 0)$. Calculate the free energy of the corona (per chain).

(iv) Calculate the size of the micelle R_{micelle} in a θ-solvent for the outer block ($\chi_B = 1/2$). Calculate the free energy of the corona (per chain).

(v) Calculate the size of the micelle R_{micelle} in a good solvent for the outer block ($\chi_B = 0.1$). Calculate the free energy of the corona (per chain).

(vi) Optimize the aggregation number f by balancing the core and corona parts of the free energy per chain.

Section 5.6

5.23 Consider multichain adsorption in a θ-solvent, of dilute chains with N monomers of size b, and with monomer–surface interaction of δkT. Calculate the density profile of the de Gennes self-similar carpet. Calculate the thickness ξ_{ads} of the adsorbed layer and the coverage Γ.

5.24 Consider multichain adsorption in a good solvent, of dilute chains with N monomers of size b, excluded volume $v = 0.01b^3$ and with monomer-surface interaction of $\delta kT = 0.1\ kT$. Calculate the density profile of the de Gennes self-similar carpet. Calculate the thickness ξ_{ads} of the adsorbed layer and the coverage Γ.

5.25 Consider dilute chains with N monomers of size b restricted to the air–water interface in an athermal good solvent (two-dimensional polymers). These chains are adsorbed to a contact line with monomer-surface interaction of $\delta kT = 0.1\ kT$. Calculate the density profile of the de Gennes self-similar carpet. Calculate the thickness ξ_{ads} of the adsorbed layer and the linear coverage Γ (the total number of monomers in the adsorbed layer per unit length of the contact line).

Section 5.7

5.26 Contrast matching in small-angle neutron scattering

(i) Separate the scattering function (Eq. 3.121) for a system of K polymers each containing N monomers into intramolecular and intermolecular contributions. Describe each scattering unit (monomer) by two indices. The first index (labeled by n or p) ranging from 1 to K indicates which molecule the monomer belongs to. The second index (labeled by j or k) varying from 1 to N describes the monomer number along the polymer. Express the sums over all monomers as sums over all monomers on each polymer summed over the number of polymers.

$$S(\vec{q}) = \frac{1}{NK}\sum_{n=1}^{K}\sum_{p=1}^{K}\sum_{j=1}^{N}\sum_{k=1}^{N}\langle\exp[-i\vec{q}(\vec{r}_{n,j} - \vec{r}_{p,k})]\rangle$$

Demonstrate that the scattering function $S(\vec{q})$ can be separated into a form factor $P(\vec{q})$ and an intermolecular contribution $Q(\vec{q})$

$$S(\vec{q}) = NP(\vec{q}) + NKQ(\vec{q})$$

where the form factor (Eq. 2.139) is the intramolecular contribution

$$P(\vec{q}) = \frac{1}{N^2}\sum_{j=1}^{N}\sum_{k=1}^{N}\langle\exp[-i\vec{q}(\vec{r}_{1,j} - \vec{r}_{1,k})]\rangle$$

The intermolecular contribution to the scattering function is defined as

$$Q(\vec{q}) = \frac{1}{N^2}\sum_{j=1}^{N}\sum_{k=1}^{N}\langle\exp[-i\vec{q}(\vec{r}_{1,j} - \vec{r}_{2,k})]\rangle$$

(ii) Consider a solution of N-mers with fraction x of chains having each monomer labelled with deuterium. This solution has xK deuterium-labelled and $(1 - x)K$ unlabelled chains. The scattering intensity consists of contributions from unlabelled pairs of monomers $S_{HH}(\vec{q})$, labelled pairs of monomers $S_{DD}(\vec{q})$ and the cross-term $S_{HD}(\vec{q})$ arising from pairs consisting of one labelled and one unlabelled monomer. The contrast in neutron scattering arises from scattering length differences. Let l_H and l_D be the coherent scattering lengths of hydrogenated and deuterated monomers, while l_0 is the scattering length of the solvent. The scattering intensity per monomer is the sum of three contributions

$$\frac{I(\vec{q})}{KN} = (l_H - l_0)^2 S_{HH}(\vec{q}) + (l_D - l_0)^2 S_{DD}(\vec{q})$$
$$+ 2(l_H - l_0)(l_D - l_0)S_{HD}(\vec{q})$$

The two polymers (labelled and unlabelled) are identical, except for their coherent scattering lengths. Therefore they are characterized by the same intramolecular ($P(\vec{q})$) and intermolecular ($Q(\vec{q})$) parts of the scattering function. The corresponding scattering functions are

$$S_{HH}(\vec{q}) = (1 - x)NP(\vec{q}) + (1 - x)^2 NKQ(\vec{q})$$

$$S_{DD}(\vec{q}) = xNP(\vec{q}) + x^2 NKQ(\vec{q})$$

$$S_{HD}(\vec{q}) = x(1 - x)NKQ(\vec{q})$$

Show that the intensity per monomer can be rewritten as

$$\frac{I(\vec{q})}{KN} = (l_H - l_D)^2 x(1 - x)NP(\vec{q})$$
$$+ [xl_D + (1 - x)l_H - l_0]^2 [NP(\vec{q}) + NKQ(\vec{q})]$$

(iii) Define the average scattering contrast between solvent and polymers as

$$\langle l \rangle = xl_D + (1 - x)l_H - l_0$$

and find the fraction x of labelled chains at which the intensity is directly proportional to the single chain form factor

$$I(\vec{q}) = (l_H - l_D)^2 x(1 - x)KN^2 P(\vec{q}) \tag{5.84}$$

Bibliography

Chaikin, P. M. and Lubensky, T. C. *Principles of Condensed Matter Physics* (Cambridge University Press, Cambridge, 1995).

des Cloizeaux, J. and Jannink, G. *Polymers in Solution: Their Modelling and Structure* (Clarendon Press, Oxford, 1990).

Flory, P. J. *Principles of Polymer Chemistry* (Cornell University Press, Ithaca, New York, 1953).

Fujita, H. *Polymer Solutions* (Elsevier, Amsterdam, 1990).

de Gennes, P. G. *Scaling Concepts in Polymer Physics* (Cornell University Press, Ithaca, New York, 1979).

Jones, R. A. L. and Richards, R. W. *Polymers at Surfaces and Interfaces* (Cambridge University Press, Cambridge, 1999).

Teraoka, I. *Polymer Solutions: An Introduction to Physical Properties* (Wiley, New York, 2002).

Networks and gelation

Random branching and gelation

6

6.1 Introduction

Everyday life encounters many materials in transition from liquid to solid, examples are preparing Jello® gelatin or mixing Epoxy glue. This fascinating phenomenon is called gelation and it is caused by the formation of crosslinks between polymer chains. In the above examples, these crosslinks are induced either by microcrystallization upon cooling (Jello®) or by covalent bonds formed on mixing (Epoxy). The final state after crosslinking consists of linear polymer **strands** connected by crosslinks.

Linking chains together leads to progressively larger branched polymers [see Fig. 6.1(a)]. The polydisperse mixture of branched polymers obtained as the result of such a process is called the **sol** (since the molecules are soluble). As the linking process continues, still larger branched polymers are obtained [Fig. 6.1(b)]. At a certain extent of reaction a molecule spanning the whole system appears. Such a huge molecule will not dissolve in a solvent, but may only swell in it. This 'infinite polymer' is called the **gel** or **network** and is permeated with finite branched polymers [Fig. 6.1(c)]. The transition from a system with only finite branched polymers (exclusively sol) to a system containing also an infinite molecule (gel) is called the **sol–gel transition** (or **gelation**) and the critical point where gel first appears is called the **gel point**.

The early studies of the sol–gel transition date back to the dawn of polymer science. The first quantitative theories of gelation—the mean-field theories—were formulated in the 1940s by Flory and Stockmayer. Critical percolation theory was successfully applied to gelation in the 1970s. A number of growth models (diffusion limited aggregation, cluster–cluster aggregation, kinetic gelation) have been developed in the 1980s to describe the kinetic aspects of aggregation and gelation.

Different types of gelation transitions are summarized in Fig. 6.2. Gelation can occur either by physical linking (as in the Jello® gelatin example above) or by chemical crosslinking (as in the Epoxy glue example). The first type is called **physical gelation**, while the second type is called **chemical gelation**.

It is convenient to distinguish between strong and weak physical gels. **Strong physical gels** have **strong physical bonds** between polymer chains that are effectively permanent at a given set of experimental conditions.

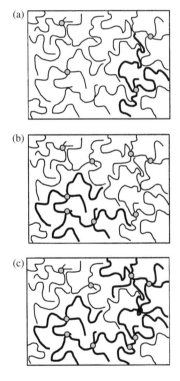

(a)

(b)

(c)

Fig. 6.1
Crosslinking of linear chains: (a) four crosslinks; (b) eight crosslinks; (c) 10 crosslinks. The largest branched polymer is highlighted and the 10th crosslink (dark) formed an incipient gel.

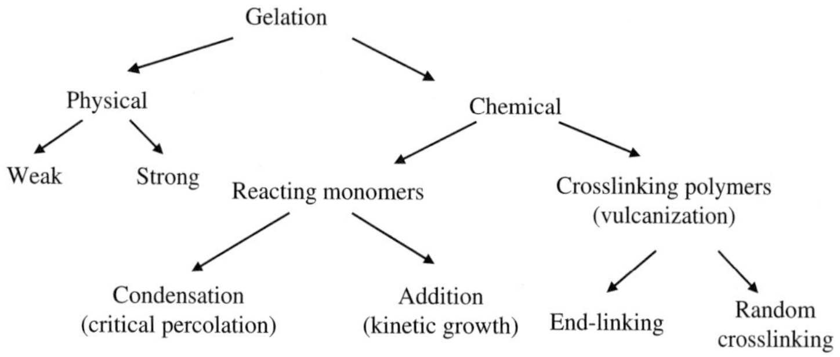

Gelation

Physical

Weak   Strong

Chemical

Reacting monomers

Crosslinking polymers
(vulcanization)

Fig. 6.2
Classification of gelation transitions
(and the models describing them).

Condensation
(critical percolation)

Addition
(kinetic growth)   End-linking

Random
crosslinking

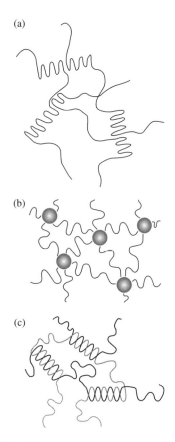

(a)

(b)

(c)

Fig. 6.3
Examples of strong physical gels
with crosslinks that are: (a) lamellar
microcrystals; (b) glassy nodules;
(c) double helices.

Examples of strong physical bonds are glassy and microcrystalline nodules, or double and triple helixes (see Fig. 6.3). The strong physical gels formed by these bonds are solids and can only melt and flow when the external conditions change (such as changing temperature for **thermoreversible gels** like Jello®). Hence, strong physical gels are analogous to chemical gels (discussed below).

The difference between strong physical gels and chemical gels has important industrial applications. **Thermoplastic elastomers** are examples of strong physical gels. Triblock copolymers of styrene–isoprene–styrene flow like liquids at high temperature (far above the polystyrene glass transition temperature of 100 °C). However, at room temperature the polystyrene blocks are immobilized in glassy nodules [Fig. 6.3(b)] that act as effective crosslinks for polyisoprene, whose glass transition temperature is −60 °C. These elastomers are reformable by simply heating above 100 °C, and hence are called thermoplastic.

Weak physical gels have reversible links formed from temporary associations between chains. These associations have finite lifetimes, breaking and reforming continuously. Examples of weak physical bonds are hydrogen bonds, block copolymer micelles above their glass transition, and ionic associations (Fig. 6.4). Such **reversible gels** are never truly solids but if the association lifetime is sufficiently long they can appear to be solids on certain time scales. Hence, whether a reversible gel is weak or strong depends on the time scale over which it is observed. Paint is a good example of a reversible gel. The weak hydrophobic associations in a water-based paint give it properties of a weak network at short time scales, but these associations have short lifetime and allow paint to flow at long times. These weak associations can also be easily broken by stirring, spraying, or brushing.

In contrast, chemical gelation involves formation of covalent bonds and always results in a strong gel. There are three main chemical gelation processes: condensation, vulcanization, and addition polymerization.

Condensation reactions typically start from a melt or solution of monomers that are capable of reacting with each other. If all monomers are bifunctional (able to react with at most two other monomers) then only linear chains result and no network can be formed. Condensation of bifunctional monomers and the molar mass distribution of the resulting linear chains are discussed in Section 1.6.2. If at least some of the starting monomers have functionality three or higher, so that they can form bonds with three or more other monomers, then branched polymers are formed and a sol–gel transition is possible. An example is a condensation reaction of a difunctional acid (any molecule with two acid groups, A_2) with a trifunctional alcohol (any molecule with three alcohol groups, B_3).[1] Each trifunctional alcohol can become a branch point leading to randomly branched polyesters and ultimately networks. Condensation of monomers with functionality greater than two, like the $A_2 + B_3$ system, in the melt (meaning that there is no non-reactive diluent present) is well described by the **critical percolation** model. This model is discussed in Section 6.5. Condensation polymerization of monomers with one acid group and $f-1$ alcohol groups, AB_{f-1} with functionality f higher than 2 produces randomly branched polymers that are **hyperbranched**, but does not lead to gelation, as will be discussed in Section 6.2.1.

Vulcanization refers to crosslinking of long linear chains that start out strongly overlapping each other. This process is well described by the **mean-field percolation** model, discussed in Section 6.4. Condensation and vulcanization are two closely related processes. By increasing the fraction of bifunctional alcohol monomers in the first process (i.e., increasing the fraction of B_2 in an $A_2 + B_2 + B_3$ system) or by decreasing the molar mass of linear chains in vulcanization it is possible to study the crossover between them. In vulcanization, one usually distinguishes end-linking [multifunctional units crosslinking the ends of the chains—see Fig. 6.5(a)] from random crosslinking with bonds formed between monomers along the chains [see Fig. 6.5(b)]. This random crosslinking vulcanization process was invented by Goodyear in 1839 to crosslink natural rubber (cis-polyisoprene) using sulphur. In Section 6.5.4, we will learn that the overlap of the branched polymers diminishes as the vulcanization proceeds, and consequently vulcanization crosses over to critical percolation sufficiently close to the gel point.

In **addition polymerization**, a free radical transfers from one vinyl monomer to another, leaving behind a trail of chemical bonds (Fig. 6.6). The distinction of addition polymerization, compared with condensation polymerization, is the high correlation of the formed bonds along the path of a free radical. Certain monomers (with two double bonds, such as divinylbenzene) can be visited twice by free radicals and become crosslinks (black circles in Fig. 6.6). As the neighbouring trails of formed bonds begin to overlap, the system approaches its gel point. The model describing branching and network formation via addition polymerization is called **kinetic gelation**.

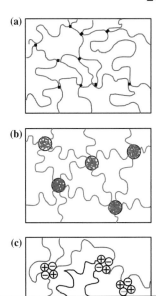

Fig. 6.4
Weak physical bonds: (a) hydrogen bonds; (b) block copolymer micelles; (c) ionic associations.

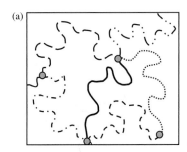

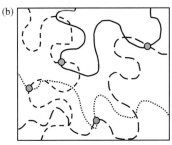

Fig. 6.5
Examples of vulcanization: (a) end-linking; (b) random crosslinking. Different lines represent different chains.

---

[1] In this notation, A groups only react with B groups.

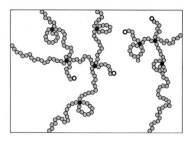

Fig. 6.6
Addition polymerization with branching. Free radicals are denoted by open circles and crosslinks by black circles.

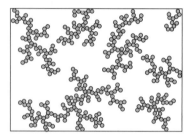

Fig. 6.7
Cluster–cluster aggregation leading to gelation.

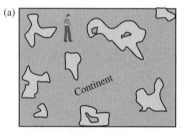

(a)

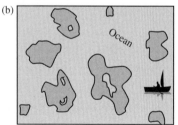

(b)

Fig. 6.8
Percolation transition from a continent with lakes to an ocean with islands.

The condensation reaction in solution can, at early stages, be **diffusion-controlled**, meaning that the time it takes for molecules to diffuse towards each other in order to react controls the reaction time. This diffusion-limited process is described by kinetic models such as **diffusion–limited aggregation** and **cluster–cluster aggregation**. During aggregation, highly ramified clusters are formed that are fractals with fractal dimension $\mathcal{D} < 3$, as depicted in Fig. 6.7. At first, only small clusters are formed and the monomer density inside each cluster is higher than the overall monomer density in solution. Since $\mathcal{D} < 3$, as the clusters grow by coalescing with other clusters, the monomer density inside them decreases and finally reaches the overall solution density. At this point the clusters overlap and their growth at later stages and the sol–gel transition itself is described by the critical percolation model. Examples are aggregation of colloidal gold and silica, the latter being a low-temperature method of making glasses.

It is very important to distinguish the equilibrium gelation models (such as mean-field gelation and critical percolation) where kinetics only affect reaction rates but not the structures formed, from the growth models (such as cluster–cluster aggregation and kinetic gelation) in which rate processes strongly influence the structures formed. All of these processes have at least a small region near the gel point (called the critical region) where the reacting structures are just at overlap, which is described by the critical percolation model. However, in many cases this region is too small to be experimentally accessible. In Section 6.5.4, the size of this small critical region in vulcanization is estimated using the Ginzburg criterion.

6.1.1 Percolation around us

Before considering the details of the gelation process, it is useful to mention the broad spectrum of other systems undergoing similar connectivity transitions and therefore described by mathematically similar percolation models. Familar examples are water percolating through sand and coffee percolating through ground coffee beans.

6.1.1.1 Deluge

The first written reference to a percolation-related process can be found in the Old Testament (Genesis 7–10). It describes a continent, different parts of which are at different heights above the sea level. Rain begins to fall and the sea level rises, submerging the low-lying parts of the continent under water. At the early stages of the deluge it is still possible to walk across the continent, travelling around numerous puddles, lakes, and perturbing bays [Fig. 6.8(a)]. As the water level rises above some critical point called the **percolation threshold**, the continent ceases to exist as it is broken into an archipelago of islands [Fig. 6.8(b)]. At the later stages of the deluge it is no longer possible to travel large distances over land and it becomes necessary to sail between the islands by boat. This transition from a continent containing lakes and bays [Fig. 6.8(a)] to an ocean containing islands [Fig. 6.8(b)] is an example of a percolation transition. When the deluge stops and the sea level lowers, an opposite transition takes place from an ocean with islands to a continent with lakes.

6.1.1.2   Forest fires

Another example of a percolation-related process is the spreading of a fire in a forest without wind. Consider a forest represented by a set of trees, as sketched in Fig. 6.9. Smokey the Bear goes into hibernation, leaving an unextinguished campfire under one of the trees and the tree catches fire. A burning tree can ignite a neighbouring tree with some probability p. A very important question is whether the fire will propagate and burn most of the forest or stop after burning only a small copse of trees.

The ignition probability p depends on the distance between the trees, the size of the trees, and on how dry the season is. Depending on these conditions, there are two possible outcomes resulting from Smokey's carelessness:

(1) An accident—If the trees are small and grow far apart and the season is relatively wet, the ignition probability p is low and the fire would stop after burning only a small cluster of trees.

(2) A major disaster—If the trees are large, grow close together and the season is dry, the ignition probability p is large and the fire would spread across the forest burning a significant fraction of the trees.

The transition between these two possible outcomes—an 'island' and a 'continent' of burned trees—is another example of a percolation transition. In the present example of a forest fire we have described percolation on a discrete set of trees, while in the example of a deluge we dealt with a continuously varying water level on a constant terrain. The two problems are quite different, but both are described by percolation models.

6.1.1.3   Spreading of a contagious disease

Another problem related to percolation and analogous to forest fires is the spreading of a disease in an apple orchard. In an orchard, the trees grow on a regular lattice [Fig. 6.10(a)]. A sick tree can contaminate its neighbour with probability p. Just as in the above example, for small values of p the disease remains contained and will only strike a finite cluster (island) of trees. For higher values of the disease transmission probability p above the percolation threshold, p_c, the disease will spread through the whole orchard, affecting a large fraction of its trees [Fig. 6.10(b)]. This problem is an example of bond percolation that was described by Hammersley in 1956. In a **bond percolation** model, all sites are occupied (in the present example, all lattice sites are occupied by trees). The percolation process is determined by the presence or absence of bonds between two neighbouring sites (in the present example, a bond corresponds to the spreading of the disease between two neighbouring trees). The island of sick trees corresponds to a cluster of sites connected by bonds. The spreading of a disease in an orchard is an example of bond percolation on a regular two-dimensional lattice. For a triangular lattice, the percolation threshold is $p_c \cong 0.34729$ and for the

Fig. 6.9
Spreading of a fire in a forest.

Fig. 6.10
Spreading of a disease in an orchard.

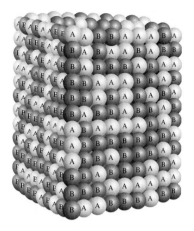

Fig. 6.11
A random three-dimensional substitutional alloy $A_{1-q}B_q$.

square lattice it is $p_c = 1/2$. In contrast, the forest fire is an example of two-dimensional bond percolation on a random set of points (trees in a forest).

6.1.1.4 Substitutional alloy

Consider a substitutional alloy $A_{1-q}B_q$ with A and B atoms randomly occupying sites of a regular lattice (see Fig. 6.11). A B cluster is defined as a set of neighbouring B atoms. For low values of the fraction q of B atoms in the $A_{1-q}B_q$ alloy, the B clusters are small. With increasing fraction q of B atoms, B clusters become progressively larger and above some critical value q_c a macroscopic B cluster spanning the whole crystal appears. This transition from a set of finite B clusters (for $q < q_c$) to a crystal with a macroscopic B cluster in addition to finite clusters (for $q > q_c$) is an example of **site percolation** on a lattice. In site percolation, only some sites of the lattice are occupied (in the present example—by B atoms). The neighbouring occupied sites are always connected into a cluster (in the language of the bond percolation model, the bond between neighbouring occupied sites is always present with probability $p = 1$). If B atoms are ferromagnetic, the B clusters would form magnetic domains. In this case, the percolation model gives a necessary condition for a transition from a paramagnet with only small magnetic domains to a ferromagnet with a macroscopic domain and finite spontaneous magnetization.

Note that for $q < q_c$ only A percolates, for $q_c < q < 1 - q_c$ both A and B percolate, and for $q > 1 - q_c$ only B percolates. Copper and gold form substitutional alloys on a face-centred cubic lattice ($Cu_{1-q}Au_q$). For $q < 0.2$, there are isolated gold islands in an ocean of copper. Conversely, for $q > 0.8$, there are isolated copper islands in an ocean of gold. For $0.2 < q < 0.8$ both copper and gold are co-continuous. There are still isolated islands of each metal but there are also continuous pathways across the macroscopic material consisting only of each individual metal.

6.1.1.5 Classes of percolation models

We have given four examples of connectivity transitions described by percolation models. Other examples include the conductor–insulator transition for a random resistor network, oil recovery through a porous rock, and communication networks (the world wide web).

The percolation transition can be defined on a continuous manifold (deluge) or on a discrete set of objects. This discrete set can be either random (such as trees in a forest) or regular (such as rows of trees in an orchard). Two types of percolation models have been distinguished: bond percolation (spreading of a disease) and site percolation (substitutional alloy). These can also be combined into a **site-bond percolation** model, where only some sites are occupied and bonds may only be placed between occupied adjacent sites. For example, if a fraction $1 - q$ of the trees in the orchard are randomly cut down, the spreading of a disease between the trees of the remaining fraction q with the spreading probability p could be studied using site-bond percolation.

The value of the percolation threshold depends on many details of the model, including the type of lattice. Table 6.1 compares site and bond

Table 6.1 Values of site and bond percolation thresholds on selected lattices

Lattice	d	f	q_c (site)	p_c (bond)
Honeycomb	2	3	0.696	0.652
Square	2	4	0.592	0.5
Triangular	2	6	0.5	0.347
Diamond	3	4	0.43	0.388
Simple cubic	3	6	0.312	0.249
BCC	3	8	0.246	0.180
FCC	3	12	0.198	0.119

The dimension of space is d and the coordination number of the lattice is f. BCC is body-centred cubic and FCC is face-centred cubic.

percolation thresholds for three two-dimensional lattices and four three-dimensional lattices. In each dimension, as the coordination number of the lattice increases, the percolation threshold decreases.

6.1.2   Percolation in one dimension

The percolation transition can be described in space of any dimension. Examples of two-dimensional percolation are deluge, forest fire, spreading of a contagious disease in an orchard, and gelation of a polymer at an air–water interface. Examples of three-dimensional percolation are substitutional alloys and bulk polymer gelation. A problem analogous to one-dimensional percolation is the condensation polymerization of bifunctional monomers described in Section 1.6.2.

Consider the condensation polymerization of bifunctional monomers, each with two different reactive groups A and B, where A is only allowed to react with B in an intermolecular fashion. For example, if group A were –OH and group B were –COOH, the resulting polymers would be polyesters with the link B–A denoting –COO–. For simplicity of notation, we call the unreacted monomer AB. The bonds allowed between groups A and B can be formulated as bond percolation on a one-dimensional lattice by simply placing the unreacted monomers on every site of the lattice and randomly connecting (or reacting) them with probability p.

AB–AB AB–AB–AB AB AB–AB–AB–AB AB–AB AB

There is exactly one unreacted A group (and one unreacted B group) per molecule. The number density of molecules $n_{\text{tot}}(p)$ (number of molecules per monomer) is therefore equal to the fraction $1 - p$ of unreacted groups:

$$n_{\text{tot}}(p) = 1 - p. \tag{6.1}$$

The number-average degree of polymerization (the number of monomers per molecule) is [see Eq. (1.55)] the reciprocal of the number density of molecules:

$$N_{\text{n}}(p) = \frac{1}{n_{\text{tot}}(p)} = \frac{1}{1 - p}. \tag{6.2}$$

A linear polymer (N-mer) is a cluster of N monomers (sites) connected by $N - 1$ bonds and containing one unreacted A group and one unreacted B group. *The number fraction distribution (mole fraction of N-mers) is given by the probability that a chosen unreacted A group is part of an N-mer.* This number fraction of N-mers is the probability of $N - 1$ formed bonds (p^{N-1}) and one unreacted B group ($1 - p$) [Eq. (1.52)]:

$$n_N(p) = p^{N-1}(1 - p). \tag{6.3}$$

For one-dimensional percolation, the system can either be below the percolation threshold ($p < p_{\text{c}} = 1$) with finite polymers or at the threshold ($p = p_{\text{c}} = 1$) with one infinite polymer, but not above the threshold. States above the threshold exist only for percolation in dimensions higher than one.

In a real polymerization reaction, monomers are distributed in three-dimensional space rather than on a one-dimensional lattice. An important consequence is that cyclic structures of various numbers of monomers can be created. Unlike their linear counterparts, these ring polymers have no reactive ends and never grow longer as reaction proceeds. They are not accounted for in the simple theory presented here.

6.2   Branching without gelation

6.2.1   Hyperbranched polymers

Another reaction that does not lead to gelation is the polymerization of hyperbranched polymers from AB_{f-1} monomers (each with a single functional group of type A and $f-1$ functional groups of type B) in which A can only react with B. For functionality $f=2$, this reduces to the condensation polymerization of linear chains, described above. For $f > 2$, such a process leads to formation of highly branched molecules.

Let p be the fraction of reacted B groups. The fraction of reacted A groups is $p(f-1)$ (the same number of reacted A and B groups, but $f-1$ times smaller total number of A groups in AB_{f-1} molecules). Therefore, the fraction of unreacted A groups is $1 - p(f-1)$. There is one unreacted A group per molecule, giving the total number of molecules per monomer:

$$n_{\text{tot}}(p) = 1 - p(f-1). \tag{6.4}$$

The number-average degree of polymerization (the average number of monomers per molecule) is the reciprocal of $n_{\text{tot}}(p)$:

$$N_{\text{n}}(p) = \frac{1}{1 - p(f-1)}. \tag{6.5}$$

This number-average degree of polymerization diverges at $p_{\text{c}} = 1/(f-1)$. This maximum possible fraction of reacted B groups corresponds to complete reaction of all A groups $(f-1)p_{\text{c}} = 1$.

Each N-mer in a condensation polymerization of AB_{f-1} monomers consists of $N-1$ reacted A–B bonds and exactly one unreacted A group, as shown in Fig. 6.12. There are $(f-1)N$ total B groups, of which $N-1$ have reacted, so there are $(f-2)N+1$ unreacted B groups in each N-mer. The probability that an unreacted A group is a part of an N-mer is proportional to the probability that $N-1$ B groups have reacted (p^{N-1}), while $(f-2)N+1$ B groups have not $(1-p)^{(f-2)N+1}$. The number fraction of N-mers is given by a product of probabilities:

$$n_N(p) = a_N p^{N-1}(1-p)^{(f-2)N+1}. \tag{6.6}$$

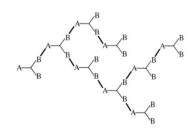

Fig. 6.12

A hyperbranched polymer. Darker lines correspond to bonds between monomers.

The **degeneracy** a_N is the number of unique ways of arranging N monomers AB_{f-1} into an N-mer. The number of ways of selecting the first B to form a bond out of $(f-1)N$ possible B groups of the N-mer is $(f-1)N$. The number of ways of selecting the second B group for the second bond out of the remaining B groups is $(f-1)N-1$. Each successive reaction has one less B group to choose from, all the way up to the last bond of an N-mer, leading to the total number of structurally different arrangements for all $N-1$ bonds:

$$(f-1)N[(f-1)N-1]\cdots[(f-1)N-(N-2)] = \frac{[(f-1)N]!}{[(f-2)N+1]!}. \quad (6.7)$$

Since all monomers of the N-mer are indistinguishable, the number of structurally different arrangements has to be reduced by $N!$ permutations of monomers leading to the number of unique N-mers:

$$a_N = \frac{[(f-1)N]!}{N![(f-2)N+1]!}. \quad (6.8)$$

Using this degeneracy in Eq. (6.6) gives the number fraction of N-mers at any extent of reaction p for hyperbranched polymers:

$$n_N(p) = \frac{[(f-1)N]!}{N![(f-2)N+1]!}p^{N-1}(1-p)^{(f-2)N+1}. \quad (6.9)$$

Comparing Eqs (6.3) and (6.9), we see that for $f \geq 3$, hyperbranched polymers have a quite different form for their number fraction distribution than linear polymers. The large number of unreacted B groups (Fig. 6.12) on large hyperbranched N-mers broadens the distribution because the larger molecules have a higher probability of growing than smaller molecules do.

In order to evaluate different moments, this number fraction distribution is rewritten as

$$n_N(p) = a_N\frac{1-p}{p}x^N \quad (6.10)$$

using the following definition of x:

$$x \equiv p(1-p)^{f-2}. \quad (6.11)$$

The number fraction is a normalized distribution:

$$\sum_{N=1}^{\infty} n_N(p) = \frac{1-p}{p}\sum_{N=1}^{\infty} a_N x^N = 1. \quad (6.12)$$

The k-moment of the number fraction distribution is related to the k-moment of the sum Σ_k:

$$m_k = \sum_{N=1}^{\infty} N^k n_N(p) = \frac{1-p}{p}\sum_{N=1}^{\infty} N^k a_N x^N = \frac{1-p}{p}\Sigma_k. \quad (6.13)$$

The consecutive moments of the sum

$$\Sigma_k \equiv \sum_{N=1}^{\infty} N^k a_N x^N \tag{6.14}$$

are related to each other through the derivative with respect to the variable x:

$$\Sigma_k = x \frac{\partial \Sigma_{k-1}}{\partial x} = x \frac{\partial \Sigma_{k-1}}{\partial p} \frac{\partial p}{\partial x}. \tag{6.15}$$

The rate of change of extent of reaction p with variable x can be evaluated from Eq. (6.11):

$$\begin{aligned}
\frac{\partial p}{\partial x} &= \frac{1}{\partial x / \partial p} = \frac{1}{(1-p)^{f-2} - p(f-2)(1-p)^{f-3}} \\
&= \frac{p(1-p)}{p(1-p)^{f-1} - p^2(f-2)(1-p)^{f-2}} = \frac{p(1-p)}{x[(1-p) - p(f-2)]} \\
&= \frac{p(1-p)}{x[1 - p(f-1)]}.
\end{aligned} \tag{6.16}$$

From Eq. (6.12), we find the zeroth moment of the sum:

$$\Sigma_0 = \frac{p}{1-p}. \tag{6.17}$$

All higher moments of the sum, Σ_k, can be evaluated from the zeroth moment, using the recurrence relation [Eq. (6.15)]. The first moment of the sum is

$$\begin{aligned}
\Sigma_1 &= x \frac{\partial \Sigma_0}{\partial p} \frac{\partial p}{\partial x} = x \frac{1}{(1-p)^2} \frac{p(1-p)}{x[1 - p(f-1)]} \\
&= \frac{p}{(1-p)[1 - p(f-1)]},
\end{aligned} \tag{6.18}$$

leading to the number-average degree of polymerization in agreement with Eq. (6.5):

$$N_n(p) = \sum_{N=1}^{\infty} N n_N(p) = \frac{1-p}{p} \Sigma_1 = \frac{\Sigma_1}{\Sigma_0} = \frac{1}{1 - p(f-1)}$$

$$\text{for } p < p_c = \frac{1}{f-1}. \tag{6.19}$$

The second moment of the sum is determined similarly:

$$\begin{aligned}
\Sigma_2 &= x \frac{\partial \Sigma_1}{\partial p} \frac{\partial p}{\partial x} = x \frac{1 - p^2(f-1)}{(1-p)^2[1 - p(f-1)]^2} \frac{p(1-p)}{x[1 - p(f-1)]} \\
&= \frac{(1 - p^2(f-1))p}{(1-p)[1 - p(f-1)]^3}.
\end{aligned} \tag{6.20}$$

The weight-average degree of polymerization is the ratio of the second and the first moments

$$N_{\mathrm{w}}(p) = \frac{\sum\limits_{N=1}^{\infty} N^2 n_N(p)}{\sum\limits_{N=1}^{\infty} N n_N(p)} = \frac{\Sigma_2}{\Sigma_1} = \frac{1 - p^2(f-1)}{[1 - p(f-1)]^2} \quad \text{for } p < p_{\mathrm{c}} = \frac{1}{f-1}.$$

$$(6.21)$$

This weight-average degree of polymerization $N_{\mathrm{w}}(p)$ diverges at p_{c} more rapidly than the number-average $N_{\mathrm{n}}(p)$. The polydispersity index of this distribution

$$\frac{N_{\mathrm{w}}(p)}{N_{\mathrm{n}}(p)} = \frac{1 - p^2(f-1)}{1 - p(f-1)} \quad \text{for } p < p_{\mathrm{c}} = \frac{1}{f-1} \qquad (6.22)$$

also diverges at $p_{\mathrm{c}} = 1/(f-1)$. All of the above results reduce for $f = 2$ to those for condensation polymerization of linear chains (Section 1.6.2) and for $f > 2$ correspond to the random branching that creates hyperbranched polymers. Like the linear condensation of Section 1.6.2, this branched condensation can, in principle, reach the gel point (where all monomers are connected into a single enormous hyperbranched polymer) but the reaction cannot proceed beyond the gel point, since there are no more unreacted A groups. In practice, it is difficult to achieve complete reaction.

Of particular interest is the molar mass distribution of high molar mass hyperbranched polymers that are produced when the reaction of AB_{f-1} monomers is driven close to completion. The number fraction of molecules [Eq. (6.9)] can be approximated for large N, using Stirling's formula:

$$N! \cong \sqrt{2\pi N} N^N \exp(-N). \qquad (6.23)$$

The degeneracy a_N [Eq. (6.8)] is approximated for $f > 2$ as

$$\begin{aligned}
a_N &= \frac{1}{(f-2)N + 1} \frac{[(f-1)N]!}{N![(f-2)N]!} \\
&\cong \frac{1}{(f-2)N + 1} \frac{\sqrt{2\pi(f-1)N}[(f-1)N]^{(f-1)N}}{\sqrt{2\pi N} N^N \sqrt{2\pi(f-2)N}[(f-2)N]^{(f-2)N}} \\
&\cong \frac{1}{(f-2)N} \sqrt{\frac{f-1}{2\pi(f-2)N}} \frac{(f-1)^{(f-1)N}}{(f-2)^{(f-2)N}}.
\end{aligned} \qquad (6.24)$$

Let us define the **relative extent of reaction**:

$$\varepsilon \equiv \frac{p - p_{\mathrm{c}}}{p_{\mathrm{c}}} = (f-1)p - 1. \qquad (6.25)$$

The limit of interest to us here is small negative values of ε. The fraction p of reacted B groups can be expressed in terms of this relative extent of reaction ε

$$p = \frac{1+\varepsilon}{f-1}, \tag{6.26}$$

as can the fraction of unreacted B groups:

$$1 - p = \frac{f-2-\varepsilon}{f-1} = \frac{f-2}{f-1}\left(1 - \frac{\varepsilon}{f-2}\right). \tag{6.27}$$

Consider the p-dependence of Eq. (6.9):

$$p^{N-1}(1-p)^{(f-2)N+1} = \left(\frac{1+\varepsilon}{f-1}\right)^{N-1}\left(\frac{f-2}{f-1}\right)^{(f-2)N+1}\left(1-\frac{\varepsilon}{f-2}\right)^{(f-2)N+1}$$

$$= \frac{f-2}{1+\varepsilon}\left(1 - \frac{\varepsilon}{f-2}\right)$$

$$\times \left[\left(\frac{1+\varepsilon}{f-1}\right)\frac{(f-2)^{f-2}}{(f-1)^{f-2}}\left(1-\frac{\varepsilon}{f-2}\right)^{f-2}\right]^{N}. \tag{6.28}$$

The small negative values of ε can be neglected outside the square bracket in Eq. (6.28), $1+\varepsilon \cong 1$ and $1-\varepsilon/(f-2) \cong 1$. Since the square bracket in Eq. (6.28) is taken to a large power N it dominates the ε-dependence. Expanding $[1-\varepsilon/(f-2)]^{f-2} \cong 1-\varepsilon$ to the lowest power in ε we find a simple form:

$$p^{N-1}(1-p)^{(f-2)N+1} \cong (f-2)\left[\frac{(f-2)^{f-2}}{(f-1)^{f-1}}(1+\varepsilon)\left(1-\frac{\varepsilon}{f-2}\right)^{f-2}\right]^{N}$$

$$\cong (f-2)\frac{(f-2)^{(f-2)N}}{(f-1)^{(f-1)N}}\left[(1+\varepsilon)\left(1-\varepsilon+\frac{1}{2}\frac{(f-3)}{(f-2)}\varepsilon^2\right)\right]^{N}$$

$$\cong \frac{(f-2)^{(f-2)N+1}}{(f-1)^{(f-1)N}}\left(1-\frac{1}{2}\frac{(f-1)}{(f-2)}\varepsilon^2\right)^{N}$$

$$\cong \frac{(f-2)^{(f-2)N+1}}{(f-1)^{(f-1)N}}\exp\left(-\frac{1}{2}\frac{(f-1)}{(f-2)}\varepsilon^2 N\right). \tag{6.29}$$

The number fraction of N-mers [Eq. (6.6)] can be approximated by combining Eqs (6.24) and (6.29):

$$n_N(p) = a_N p^{N-1}(1-p)^{(f-2)N+1}$$

$$\cong \sqrt{\frac{f-1}{2\pi(f-2)}}N^{-3/2}\exp\left(-\frac{1}{2}\frac{(f-1)}{(f-2)}\varepsilon^2 N\right) \quad \text{for } |\varepsilon| = 1-p(f-1) \ll 1. \tag{6.30}$$

This number fraction distribution has a form of a power law with an exponential cutoff at the **characteristic degree of polymerization** $N^* = \frac{2(f-2)}{f-1}\varepsilon^{-2}$:

$$n_N(p) \approx N^{-3/2}\exp(-N/N^*). \tag{6.31}$$

It is interesting to note that both in linear condensation polymerization (AB) and in condensation polymerization of AB_{f-1}, a similar asymptotic behaviour is predicted for the number-average degree of polymerization

$$N_n = -\varepsilon^{-1} \tag{6.32}$$

and the weight-average degree of polymerization

$$N_w \approx \varepsilon^{-2} \tag{6.33}$$

for small negative relative extents of reaction $\varepsilon = (f-1)p - 1$. The weight-average, z-average, and all higher-order average degrees of polymerization are proportional to the characteristic degree of polymerization N^*:

$$N_w \approx N_z \approx N_{z+1} \approx \cdots \approx N^* = \frac{2(f-2)}{f-1}\varepsilon^{-2}. \tag{6.34}$$

Just as for linear condensation polymers, the simple statistics presented above assume there are no intramolecular reactions. This assumption is never really correct. For linear polymers, the assumption gets progressively better for longer chains. However, for hyperbranched polymers it gets worse for species with large degrees of polymerization, since most of the unreacted B groups near an unreacted A will be on the same polymer. Properly including intramolecular reactions is a difficult and important problem. The molar mass distributions of hyperbranched polymers do not follow the simple statistics presented here because of intramolecular reactions.

6.2.2 Regular dendrimers

Another means of preparing branched polymers using condensation chemistry is to systematically grow a very regular structure, called a dendrimer. Dendrimers are typically grown from a B_n monomer core by reacting this core with AB_{f-1} monomers that temporarily have their B groups protected from further reaction. If done properly, this reaction will link each B_n core with n AB_{f-1} monomers, resulting in a first-generation dendrimer. By deprotecting the B groups, this first-generation dendrimer becomes a $B_{n(f-1)}$ core that can react with $n(f-1)$ AB_{f-1} monomers to create a second-generation dendrimer. In principle, this process can be repeated indefinitely, with each subsequent generation having another layer of monomers incorporated, which are connected to the central B_n core by the intermediate generations. For the special case of $f = n$, the dendrimer is a **Cayley tree**, shown in Fig. 6.13 for $f = n = 3$. Since the dendrimer is created one generation at a time, in principle each new generation could utilize different monomers with different functionalities and functional groups. However, in practice if $f \geq 3$ for all generations, real dendrimers cannot exceed a certain largest possible size and remain perfect because they eventually become too congested. To understand this congestion, we count the number of monomers in each generation of the dendrimer in Table 6.2.

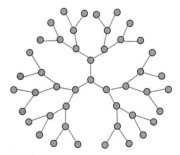

Fig. 6.13
Cayley tree or Bethe lattice with functionality $f = 3$.

Table 6.2 The number of monomers in each generation layer of a dendrimer and the total number of monomers in the dendrimer at each generation

Generation	Number of monomers	Cumulative number of monomers
0	1	1
1	n	$1+n$
2	$n(f-1)$	$1+n+n(f-1)$
3	$n(f-1)^2$	$1+n+n(f-1)+n(f-1)^2$
.	.	
.		
g	$n(f-1)^{g-1}$	$1+n+n(f-1)+n(f-1)^2$ $+\cdots+n(f-1)^{g-1}$

The number of monomers in generation g (the last generation) of the dendrimer is $n(f-1)^{g-1}$, while the total number of monomers in the dendrimer (from the core to generation g) is

$$N_g = 1 + n + n(f-1) + n(f-1)^2 + \cdots + n(f-1)^{g-1}$$
$$= 1 + n[1 + (f-1) + (f-1)^2 + \cdots + (f-1)^{g-1}]$$
$$= 1 + n\left[\frac{(f-1)^g - 1}{f-2}\right], \tag{6.35}$$

where the last step in the derivation used the sum of a geometric series $1 + x + x^2 + \cdots + x^{g-1} = (x^g - 1)/(x - 1)$. The number of monomers in the rest of the polymer (from the core to generation $g - 1$) is

$$N_{g-1} = 1 + n\left[\frac{(f-1)^{g-1} - 1}{f-2}\right]. \tag{6.36}$$

The ratio of the number of monomers in the last generation of a very large dendrimer $n(f-1)^{g-1}$ to the number of monomers in the rest of the dendrimer N_{g-1} is

$$\frac{n(f-1)^{g-1}}{1 + n[(f-1)^{g-1} - 1]/(f-2)} \cong f-2 \quad \text{for } g \gg 1. \tag{6.37}$$

Thus, for $f=3$ approximately half of the monomers are in the last generation of the dendrimer! For higher functionalities ($f>3$), an even larger fraction of monomers are in the last generation of the dendrimer.

The volume occupied by the polymer is $v_0 N_g$, where v_0 is the monomer volume and the maximum accessible volume for a fully stretched dendrimer is $(4\pi/3)(gl)^3$, where l is the monomer size. The occupied volume cannot exceed the maximum accessible volume, leading to the maximum possible generation g_{max} for a perfect dendrimer:

$$v_0 N_{g_{max}} \cong \frac{4\pi}{3}(g_{max}l)^3. \tag{6.38}$$

Using Eq. (6.35) relates the maximum generation for a perfect dendrimer to geometric properties of a monomer:

$$\frac{l^3}{v_0} \cong \frac{n}{f-2} \frac{(f-1)^{g_{max}}}{(4\pi/3)g_{max}^3}.$$

To make a large generation dendrimer, monomers with large aspect ratio are needed,[2] so that $l^3 \gg v_0$. For example, to make a perfect tetrafunctional ($n = f = 4$) seventh generation dendrimer requires $l^3/v_0 > 3$.

The regular lattice constructed in this way is called a **Bethe lattice** (see Fig. 6.13). The mean-field model of gelation corresponds to percolation on a Bethe lattice (see Section 6.4). The infinite Bethe lattice does not fit into the space of *any* finite dimension. Construction of progressively larger randomly branched polymers on such a lattice would eventually lead to a congestion crisis in three-dimensional space similar to the one encountered here for dendrimers.

6.3   Gelation: concepts and definitions

Gelation is a **connectivity transition** that can be described by a bond percolation model. Imagine that we start with a container full of monomers, which occupy the sites of a lattice (as sketched in Fig. 6.14). In a simple bond percolation model, all sites of the lattice are assumed to be occupied by monomers. The chemical reaction between monomers is modelled by randomly connecting monomers on neighbouring sites by bonds. The fraction of all possible bonds that are formed at any point in the reaction is called the **extent of reaction** p, which increases from zero to unity as the reaction proceeds. A polymer in this model is represented by a cluster of monomers (sites) connected by bonds. When all possible bonds are formed (all monomers are connected into one macroscopic polymer) the reaction is completed ($p = 1$) and the polymer is a fully developed network. Such fully developed networks will be the subject of Chapter 7, while in this chapter we focus on the gelation transition.

At the percolation threshold or gel point p_c, the system undergoes a connectivity transition. Slightly below the gel point, the system is a polydisperse mixture of branched polymers shown in Fig. 6.14(a). Slightly beyond the gel point, the system is still mostly a polydisperse mixture of branched polymers, but one structure percolates through the entire system [Fig. 6.14(b)]. This structure is called the **incipient gel**, which is a tenuous structure quite different from the fully developed network that exists far above p_c. This connectivity transition from a sol below p_c to a gel permeated with sol above p_c is called the gelation transition.

At any specified extent of reaction p, the dimensionless number density of molecules with N monomers is $n(p, N)$, defined as the number of N-mers divided by the total number of monomers. This number density is proportional to the probability that a randomly selected polymer has N

[2] Alternatively, short linear chains can be introduced between the f-functional branch points.

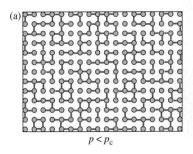
(a)

$p < p_c$

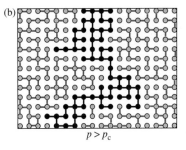
(b)

$p > p_c$

Fig. 6.14
Gelation is a bond percolation transition. The percolation cluster is indicated by darker shading.

monomers, which is the number fraction of N-mers $n_N(p)$ (the number fraction distribution function of Sections 1.6.2 and 6.1.2). The two functions are simply normalized differently [Eq. (1.45)]. The number density $n(p, N)$ is the number of N-mers per monomer. In contrast, the number fraction distribution $n_N(p)$ is normalized by the total number of molecules.

The **sol fraction** is defined as the fraction of all monomers that are either unreacted or belong to finite-size polymers (the sol):

$$P_{sol}(p) = \sum_{N=1}^{\infty} Nn(p, N) = \sum_{N=1}^{\infty} w(p, N). \qquad (6.39)$$

The sum in Eq. (6.39) is only made over the finite-size species, meaning that above the gel point the gel is excluded. The last equality employs the definition of the dimensionless weight density of N-mers:

$$w(p, N) = Nn(p, N). \qquad (6.40)$$

Analogous to the weight fraction distribution function $w_N(p)$ of Section 1.6.2, the weight density $w(p, N)$ is the probability that a randomly chosen monomer is part of a polymer with N monomers, but the two functions are normalized differently. The weight density $w(p, N)$ is normalized by the total number of monomers in the system, while the weight fraction $w_N(p)$ is normalized only by the monomers belonging to finite-size polymers [see Eq. (1.23)]. The sum of weight densities $w(p, N)$ of all finite size polymers is the sol fraction [Eq. (6.39)], while the sum of weight fractions $w_N(p)$ is always unity. Therefore, the weight density $w(p, N)$ is equal to the weight fraction $w_N(p)$ times the sol fraction $P_{sol}(p)$.

The **gel fraction** is defined as the fraction of all monomers belonging to the gel. Every monomer must be either part of the sol or part of the gel, so the sum of the sol and gel fractions is unity:

$$P_{gel}(p) + P_{sol}(p) = 1. \qquad (6.41)$$

Below the gel point, all monomers are either unreacted or belong to finite sized polymers and therefore the sol fraction is unity and the gel fraction is zero:

$$P_{sol}(p) = 1, \quad P_{gel}(p) = 0 \qquad \text{for } p \leq p_c; \qquad (6.42)$$

above the gel point, the gel fraction is non-zero and the sol fraction is less than unity.

$$P_{sol}(p) < 1, \quad P_{gel}(p) > 0 \qquad \text{for } p > p_c. \qquad (6.43)$$

The gel fraction is the probability that a randomly selected monomer belongs to the gel. The gel fraction is the order parameter for gelation. The order parameter tells us whether the reaction has passed the gel point and if above the gel point, it indicates the gel fraction. The growth of the gel fraction is accompanied by a simultaneous decay of the sol fraction, beyond the gel point p_c as shown in Fig. 6.15. The fact that the order parameter is continuous through the transition means that gelation is analogous to a **continuous phase transition**. Familiar examples of continuous phase transitions are the paramagnetic–ferromagnetic transition of iron at the Curie temperature (where the magnetization changes continuously) and the vapor-liquid transition with increasing pressure at the critical temperature (where the density changes continuously). A great deal is known about continuous phase transitions and the application of those ideas to gelation is discussed in Section 6.5.

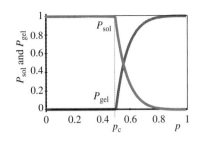

Fig. 6.15
Mean-field prediction of sol and gel fractions for functionality $f = 3$.

The total number density of molecules is the sum of number densities of all finite-size polymers and unreacted monomers. The number density of all finite molecules $n_{tot}(p)$ is the zeroth moment of the number density distribution function:

$$n_{tot}(p) = \sum_{N=1}^{\infty} n(p, N). \qquad (6.44)$$

The number-average degree of polymerization [see Eq. (1.27)] is the average number of monomers per finite-size polymer:

$$N_n(p) = \frac{\sum_{N=1}^{\infty} Nn(p, N)}{\sum_{N=1}^{\infty} n(p, N)} = \frac{P_{sol}(p)}{n_{tot}(p)}. \qquad (6.45)$$

The weight-average degree of polymerization is the ratio of second and first moments [recall Eq. (1.31)]:

$$N_w(p) = \frac{\sum_{N=1}^{\infty} N^2 n(p, N)}{\sum_{N=1}^{\infty} Nn(p, N)} = \frac{\sum_{N=1}^{\infty} Nw(p, N)}{\sum_{N=1}^{\infty} w(p, N)} = \frac{\sum_{N=1}^{\infty} Nw(p, N)}{P_{sol}(p)}. \qquad (6.46)$$

Note that the sums are understood to run only over finite-size molecules. Beyond the gel point, the gel fraction is excluded from such sums.

6.4 Mean-field model of gelation

A convenient way of presenting the mean-field model (though it is not the way it was originally defined) is by placing monomers at the sites of an infinite Bethe lattice, a small part of which is shown in Fig. 6.13. The Bethe lattice has the advantage of directly taking into account the functionality of the monomers f by adopting this functionality for the lattice. For trifunctional monomers (the $f = 3$ Bethe lattice of Fig. 6.13) there are three possible bonds emanating from each lattice site.

In a bond percolation model on a Bethe lattice, we assume that all lattice sites are occupied by monomers and the possible bonds between neighbouring monomers are either formed with probability p or left unreacted with probability $1 - p$. In the simplest version, called the random bond percolation model, the probability p of forming each bond is assumed to be independent of any other bonds in the system.

The basic assumptions of the mean-field model are implicit in the topology of the Bethe lattice. The most apparent assumption is the absence of closed loops. No intramolecular crosslinking is allowed—bonds can only be formed between monomers belonging to different polymers. This assumption significantly simplifies the analytical treatment of the model, but limits its practical utility.

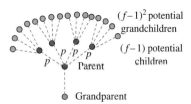

Fig. 6.16
Gel point calculation on a Bethe lattice. Each site that is already connected to the gel from its grandparent has $f-1$ possible additional connections (potential children).

6.4.1   Gel point

To calculate the gel point for bond percolation on an f-functional Bethe lattice, let us choose a single site and work progressively outwards from it, analysing the structure of the polymer obtained by forming bonds. Which site is chosen as the starting point of this procedure makes no difference because they are all statistically identical for an infinite Bethe lattice.

Let us assume that our starting ('parent') site has already formed a bond with one of its neighbours (the 'grandparent' site) as sketched in Fig. 6.16. We would like to calculate the average number of additional bonds the 'parent' site forms with its $f-1$ remaining neighbours (potential 'children'). The probability of each of these bonds being formed is p and is independent of other bonds. Therefore, the average number of bonds between a 'parent' site and its 'children' is $p(f-1)$. Similarly, the average number of bonds between each of its 'children' and their corresponding 'grandchildren' is also $p(f-1)$.

If this average number of bonds $p(f-1)$ is less than unity ($p < 1/(f-1)$), each new generation has, on average, fewer members and the dynasty does not survive for long (only finite-size branched polymers exist in the system). If this average number of bonds $p(f-1)$ is greater than unity ($p > 1/(f-1)$), each new generation has, on average, more members and the descendants of some parents multiply indefinitely, forming an infinite genealogical branched family tree.

The transition between these two cases is the gel point:

$$p_c = \frac{1}{f-1}. \tag{6.47}$$

Below the gel point (for $p < p_c$) there are only finite-size branched polymers, while above the gel point (for $p > p_c$) there is also at least one infinite polymer (the gel) in addition to many finite-size branched polymers. The distribution of polymer sizes changes with the fraction of formed bonds p. For small extents of reaction $p \ll p_c$ there are only small polymers, while near the gel point some large branched polymers are present.

A unique feature of percolation on a Bethe lattice is that there are many infinite polymers present in the same system above the gel point. The easiest way to understand this result is to start with a single infinite network polymer on a fully reacted Bethe lattice at $p = 1$. If a single bond is cut, the infinite polymer breaks into two infinite polymers. As bonds are randomly cut, progressively more infinite polymers are created, as well as many finite-size branched polymers. The reason for the existence of many infinite polymers on the Bethe lattice above the gel point is the absence of intra-molecular bonds. This is due to the fact that loops are prohibited for a Bethe lattice. In contrast, on a regular lattice (such as a simple cubic lattice) loops are allowed and there is only one infinite polymer above the gel point. Large overlapping polymers on a regular three-dimensional lattice have many possibilities for potential bonding and therefore high probability of being connected into one larger polymer.

6.4.2 Sol and gel fractions

Each site of the f-functional Bethe lattice has f possible paths to other sites (see Fig. 6.13). We define Q as the probability that a randomly selected site (A in Fig. 6.17) is *not* connected to the gel through a certain randomly selected path (the path from site A to neighbouring site B in Fig. 6.17). There are immediately two possibilities. The bond between sites A and B could be unreacted with probability $1 - p$, and if so this path cannot con-nect site A to the gel. The second possibility is that the bond between A and B is reacted with probability p. If this bond is formed, there are $f - 1$ remaining paths from site B that could connect to the gel. The probability that none of the $f - 1$ paths connect to the gel is Q^{f-1} and the probability that the bond between A and B is reacted and does not lead to the gel is pQ^{f-1}. Hence, there is a **recurrence relation** for the probability that a ran-domly selected site is not connected to the gel through a randomly selected potential bond (see Fig. 6.17):

$$Q = 1 - p + pQ^{f-1}. \tag{6.48}$$

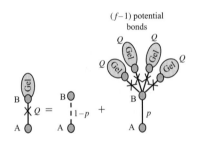

Fig. 6.17
A recurrence diagram for the probability of not being connected to the gel through a given bond.

The fraction of monomers in the sol (sol fraction) P_{sol} is the probability that a randomly selected site is not connected to the gel along any of its f paths (Fig. 6.18):

$$P_{\text{sol}} = Q^f. \tag{6.49}$$

Solving Eq. (6.49) for Q,

$$Q = P_{\text{sol}}^{1/f} \tag{6.50}$$

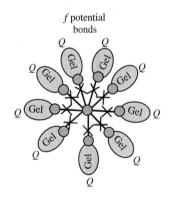

Fig. 6.18
Diagram for the sol fraction.

and substituting into the recurrence relation for Q [Eq. (6.48)] gives a relation between sol fraction and extent of reaction that is valid for any functionality f:

$$P_{\text{sol}}^{1/f} = 1 - p + pP_{\text{sol}}^{(f-1)/f}.\tag{6.51}$$

There is always one solution of this equation [Eq. (6.42)] with zero gel fraction below gel point ($P_{\text{sol}} = 1$). A second solution exists above the gel point for any $f \geq 3$. For $f = 3$ this second solution of Eq. (6.51) can be easily found because it becomes a quadratic equation for $P_{\text{sol}}^{1/3}$:

$$P_{\text{sol}} = \left(\frac{1-p}{p}\right)^3 \quad \text{for } f = 3 \text{ and } p > p_c.\tag{6.52}$$

The gel fraction above the gel point for $f = 3$ is calculated from the sol fraction:

$$P_{\text{gel}} = 1 - P_{\text{sol}} = 1 - \left(\frac{1-p}{p}\right)^3 \quad \text{for } f = 3 \text{ and } p > p_c.\tag{6.53}$$

The sol and gel fractions for functionality $f = 3$ are plotted in Fig. 6.15 as functions of extent of reaction p. For $p > p_c = 1/2$, the $f = 3$ system is beyond the gel point and the gel fraction grows steadily as the reaction proceeds.

6.4.3 Number-average molar mass below the gel point

Before any bonds are formed (at extent of reaction $p = 0$) there are only monomers present and the total number of molecules per site is $n_{\text{tot}}(0) = 1$ (every site is occupied by an unreacted monomer). Each bond reduces the number of molecules by one because there are no intra-molecular bonds. The maximum number of bonds per site on a Bethe lattice with functionality f is $f/2$. The total number of formed bonds per site at extent of reaction p is $pf/2$:

$$n_{\text{tot}}(p) = 1 - \frac{pf}{2} \quad \text{for } p < p_c.\tag{6.54}$$

The number-average degree of polymerization (the average number of sites per molecule) is the reciprocal of the average number of molecules per site.

$$N_n(p) = \frac{1}{n_{\text{tot}}(p)} = \frac{1}{1 - pf/2} \quad \text{for } p < p_c.\tag{6.55}$$

At the gel point $p_c = 1/(f-1)$, the number-average degree of polymerization exhibits no unusual behaviour for $f > 2$ (it is just a finite number):

$$N_n(p_c) = \frac{1}{1 - f/2(f-1)} = \frac{2(f-1)}{f-2}.\tag{6.56}$$

For $f = 3$, $N_n(p_c) = 4$ meaning that a randomly selected molecule at the gel point has an average degree of polymerization of four. The fact that N_n does not diverge at p_c is our first evidence that random branching in gelation leads to very different molar mass distribution than hyperbranched polymers (for which N_n does diverge at p_c), see Eq. (6.5).

6.4.4 Weight-average molar mass below the gel point

The weight-average molar mass at any extent of reaction below the gel point $p < p_c$ can be calculated in a fashion similar to the sol and gel fraction calculations of Section 6.4.2 (see Fig. 6.19). Let Υ be the mean number of monomers connected to a randomly selected site A through one of its f randomly chosen possible bonds (with neighbour B). For $\Upsilon > 0$, the bond connecting A and B must be reacted with probability p and site B contributes one monomer to Υ. Then, each of the remaining paths through $f - 1$ remaining potential bonds of B contributes a mean number of monomers Υ, giving the following recurrence relation:

$$\Upsilon = p[1 + (f - 1)\Upsilon]. \tag{6.57}$$

This recurrence relation can be easily solved for Υ:

$$\Upsilon = \frac{p}{1 - (f - 1)p}. \tag{6.58}$$

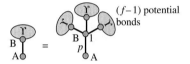

Fig. 6.19
Diagram of the recurrence relation for the average number of monomers connected to site A through a given bond.

The weight-average number of monomers per polymer N_w below the gel point is the average number of monomers in a polymer belonging to a randomly selected site. This number can be written as 1 for the site itself and Υ for each of the f paths emanating from that site (Fig. 6.20):

$$N_w = 1 + f\Upsilon = 1 + \frac{fp}{1 - (f - 1)p} = \frac{1 + p}{1 - (f - 1)p}$$

$$= \frac{1 + p}{1 - p/p_c} \quad \text{for } p < p_c. \tag{6.59}$$

The weight-average degree of polymerization diverges at the gel point, since the denominator of Eq. (6.59) is zero when $p = p_c$. Since the number-average degree of polymerization stays finite at p_c [Eq. (6.56)], the polydispersity index

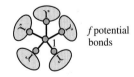

Fig. 6.20
Diagram for the calculation of N_w.

$$\frac{N_w}{N_n} = \frac{(1 + p)(1 - pf/2)}{1 - (f - 1)p} \quad \text{for } p < p_c \tag{6.60}$$

diverges at the gel point $p_c = 1/(f - 1)$ for $f > 2$.

6.4.5   Molar mass distribution

The structure of the branched polymers produced by any random branching process is the same. Any individual hyperbranched polymer structure made from reacting AB_{f-1} monomers can also be made by reacting A_f monomers. The difference between these branching processes is the molar mass *distribution*—the relative amounts of each structure produced.

The molar mass distributions resulting from AB_{f-1} and A_f reactions are related in a very simple way. Table 6.3 shows the unique relation between the number of monomers in an N-mer and the number of unreacted A groups in the N-mer. The property of the Bethe lattice not allowing any loops makes this unique relation possible.

Regardless of how they are connected, each new monomer brings f more A groups to the molecule, but also reacts two A groups in the process.[3] Each N-mer has exactly $(f-2)N+2$ unreacted A groups.

If any branched structure made from A_f condensation is held by one of its unreacted A groups, it is identical to a structure produced by AB_{f-1} condensation. For the AB_{f-1} reaction, the probability that an unreacted A group is part of an N-mer is given by the number fraction $n_N(p)$ of N-mers because there is only a *single* unreacted A group per molecule for AB_{f-1}. For A_f condensation, the probability $u_N(p)$ that a randomly selected unreacted A group is part of an N-mer is identical to the number fraction $n_N(p)$ of N-mers in the condensation polymerization of AB_{f-1} [Eq. (6.9)]:

$$u_N(p) = \frac{[(f-1)N]!}{N![(f-2)N+1]!} p^{N-1}(1-p)^{(f-2)N+1}. \tag{6.61}$$

The difference between A_f and AB_{f-1} arises from the degeneracy of A_f condensation, since there are $N(f-2)+2$ unreacted A groups on each N-mer (see Table 6.3). The number of N-mers per unreacted A group is obtained by dividing $u_N(p)$ by this degeneracy, since there are $N(f-2)+2$ different ways to select the same N-mer by randomly choosing an unreacted A:

$$\frac{u_N(p)}{(f-2)N+2}.$$

To obtain the number of N-mers per monomer, $n(p, N)$, we simply multiply this number of N-mers per unreacted A group by the average number of unreacted A groups per monomer $f(1-p)$:

$$n(p, N) = \frac{f(1-p)}{(f-2)N+2} u_N(p), \tag{6.62}$$

$$n(p, N) = f \frac{[(f-1)N]!}{N![(f-2)N+2]!} p^{N-1}(1-p)^{(f-2)N+2}. \tag{6.63}$$

Table 6.3 The number of unreacted A groups in an N-mer for A_f condensation on a Bethe lattice

N	Number of unreacted A groups
1	f
2	$2f-2$
3	$3f-4$
4	$4f-6$
...	...
N	$Nf-2N+2=(f-2)N+2$

---

[3] Formation of a bond converts two unreacted A groups to two reacted A groups. Since each site has fA groups, there are $f/2$ potential bonds per site.

This number density distribution function is more convenient than the previously used number fraction, when dealing with systems like A_f condensation that can form gels. The reason is that number fraction applies to the randomly branched polymers present but not to the gel, which makes this quantity awkwardly normalized beyond the gel point. The number density $n(p, N)$ avoids this complication since the total number of monomers is independent of extent of reaction.

Simply multiplying $n(p, N)$ by N gives the weight density distribution function, which is the probability of a randomly chosen monomer belonging to an N-mer:

$$w(p, N) = Nn(p, N)$$
$$= f \frac{[(f-1)N]!}{(N-1)![(f-2)N+2]!} p^{N-1}(1-p)^{(f-2)N+2}. \qquad (6.64)$$

The weight density distribution functions are plotted in Fig. 6.21 for functionality $f=3$ and extents of reaction $p=0.4$, 0.48, and 0.49.

From these distribution functions, their moments related to sol and gel fractions and to various averages of the degree of polymerization [Eqs (6.45) and (6.46)] may be calculated. Results for functionality $f=3$ are presented here. The sol fraction is defined as the fraction of all sites belonging to finite molecules

$$P_{\text{sol}}(p) = \sum_{N=1}^{\infty} w(p, N) = \left(\frac{1 - |1 - p/p_{\text{c}}|}{2p}\right)^3 \quad \text{for } f = 3, \qquad (6.65)$$

where $|x|$ denotes the absolute value of x. As expected, the sol fraction is equal to unity below the gel point [Eq. (6.42)]. Above the gel point the sol fraction decreases as the reaction proceeds (see Section 6.4.2 and Fig. 6.15) and Eq. (6.65) reduces to Eq. (6.52) for $p > p_{\text{c}} = 1/2$.

The number of molecules per site can be calculated from Eq. (6.63):

$$n_{\text{tot}}(p) = \sum_{N=1}^{\infty} n(p, N) = P_{\text{sol}}(p)\left(\frac{1}{4} + \frac{3}{4}|1 - p/p_{\text{c}}|\right) \quad \text{for } f = 3. \qquad (6.66)$$

It decreases linearly below the gel point [recall Eq. (6.54)]:

$$n_{\text{tot}}(p) = 1 - \frac{3p}{2} \quad \text{for } p < p_{\text{c}} = \frac{1}{2} \text{ for } f = 3 \qquad (6.67)$$

and non-linearly above the gel point

$$n_{\text{tot}}(p) = \left(\frac{1-p}{p}\right)^3 \frac{3p-1}{2} \quad \text{for } p \geq p_{\text{c}} = \frac{1}{2} \text{ for } f = 3 \qquad (6.68)$$

as plotted in Fig. 6.22.

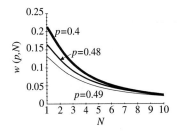

Fig. 6.21
Weight density distribution functions for functionality $f=3$ at three different extents of reaction.

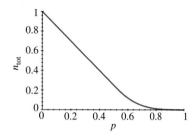

Fig. 6.22
Total number of molecules per site for functionality $f = 3$.

The relative extent of reaction ε is defined as a measure of the proximity to the gel point [Eq. (6.25)]. In the mean-field theory, the critical extent of the reaction is given by Eq. (6.47), $p_c = 1/(f-1)$, and the relative extent of the reaction is determined by f and p:

$$\varepsilon = \frac{p}{p_c} - 1 = (f-1)p - 1. \tag{6.69}$$

Using $f = 3$ gives the relative extent of reaction for the trifunctional Bethe lattice.

$$\varepsilon = 2p - 1 \quad \text{for } f = 3. \tag{6.70}$$

The number-average degree of polymerization

$$N_n(p) = \frac{\sum_{N=1}^{\infty} N n(p, N)}{\sum_{N=1}^{\infty} n(p, N)} = \frac{P_{sol}(p)}{n_{tot}(p)} = \frac{4}{1 + 3|\varepsilon|} \quad \text{for } f = 3 \tag{6.71}$$

grows below the gel point from $N_n(0) = 1$ to $N_n(1/2) = 4$ [see Eq. (6.55) for a more general expression] and decreases above the gel point, with $N_n(1) = 1$ (see Fig. 6.23).

The second moment of the number density distribution [Eq. (6.63)] is

$$\sum_{N=1}^{\infty} N^2 n(p, N) = P_{sol}(p) \frac{3 - |\varepsilon|}{2|\varepsilon|} \quad \text{for } f = 3. \tag{6.72}$$

The weight-average degree of polymerization is the ratio of the second and the first moments and diverges at the gel point (see Fig. 6.23):

$$N_w(p) = \frac{\sum_{N=1}^{\infty} N^2 n(p, N)}{\sum_{N=1}^{\infty} N n(p, N)} = \frac{3}{2|\varepsilon|} - \frac{1}{2} \quad \text{for } f = 3. \tag{6.73}$$

The third moment of the number density distribution

$$\sum_{N=1}^{\infty} N^3 n(p, N) = P_{sol}(p) \frac{3 + 9|\varepsilon| - 9\varepsilon^2 + |\varepsilon|^3}{4|\varepsilon|^3} \quad \text{for } f = 3 \tag{6.74}$$

gives the z-average degree of polymerization at any extent of reaction p:

$$N_z = \frac{\sum_{N=1}^{\infty} N^3 n(p, N)}{\sum_{N=1}^{\infty} N^2 n(p, N)} = \frac{3 + 9|\varepsilon| - 9\varepsilon^2 + |\varepsilon|^3}{2\varepsilon^2(3 - |\varepsilon|)} \quad \text{for } f = 3. \tag{6.75}$$

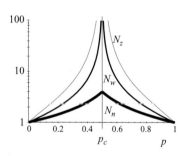

Fig. 6.23
Degree of polymerization averages for functionality $f = 3$. N_z—top line, N_w—middle line, and N_n—bottom line.

It diverges at the gel point faster than the weight-average degree of polymerization (see Fig. 6.23).

Close to the gel point (for $|\varepsilon| \ll 1$) the number density of polymers reaches an asymptotic form for large polymers that can be obtained from the similar expression for condensation polymerization of AB_{f-1} [Eqs (6.30) and (6.61)]:

$$u_N(p) \cong \sqrt{\frac{f-1}{2\pi(f-2)}} N^{-3/2} \exp\left(-\frac{f-1}{2(f-2)}\varepsilon^2 N\right) \cong n_N(p) \qquad (6.76)$$

for AB_{f-1}

using the relation between the two distribution functions [Eq. (6.62)]:

$$n(p, N) = \frac{f(1-p)}{(f-2)N+2}u_N(p) \cong \frac{f}{(f-2)N}\frac{f-2}{f-1}u_N(p)$$

$$\cong \frac{f}{\sqrt{2\pi(f-1)(f-2)}} N^{-5/2} \exp\left(-\frac{f-1}{2(f-2)}\varepsilon^2 N\right). \qquad (6.77)$$

In this conversion we made use of the fact that $p \cong p_c = 1/(f-1)$ near the gel point, so $(1-p)$ was replaced with $(f-2)/(f-1)$. The polymer number density near the gel point on the Bethe lattice decays as a power of the number of monomers in a polymer:

$$n(p, N) \approx N^{-5/2} f_\pm(N/N^*). \qquad (6.78)$$

The **cutoff function** $f_\pm(N/N^*)$ is an exponential

$$f_\pm(N/N^*) \approx \exp(-N/N^*) \qquad (6.79)$$

with the cutoff at the characteristic degree of polymerization:

$$N^* = \frac{2(f-2)}{f-1}\varepsilon^{-2}. \qquad (6.80)$$

The number density of polymers is plotted in Fig. 6.24 for several different extents of reaction p.

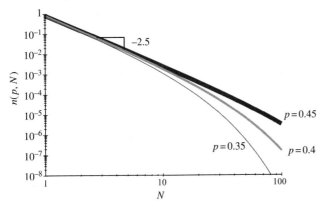

Fig. 6.24
Number density distribution function $n(p, N)$ for trifunctional randomly branched polymers at three different extents of reaction below the gel point ($p_c = 0.5$).

It is important to point out that this characteristic degree of polymerization is proportional to the z-average degree of polymerization near the gel point for any functionality f:

$$N^* \cong 2N_z \quad \text{for } |\varepsilon| \ll 1 \text{ and } f = 3. \tag{6.81}$$

The characteristic degree of polymerization diverges at the gel point, so the number density distribution becomes a power law with no cutoff:

$$n(p_c, N) \sim N^{-5/2}. \tag{6.82}$$

Similar relations with different coefficients and different exponents hold in all percolating systems near the transition point, not just in the mean-field case of percolation on a Bethe lattice, as will be shown in Section 6.5.

6.4.6   Size of ideal randomly branched polymers

So far we have discussed the distribution of degrees of polymerization of molecules both below and above the gel point. In the present section, we will describe their spacial sizes in the polymerization reactor. The Bethe lattice introduced above for the mean-field gelation model properly describes the connectivity of monomers into tree-like branched molecules, but is not designed to describe the location of these monomers in space. Indeed, the Bethe lattice has *infinite* fractal dimension because the number of sites grows exponentially with the number of generations [see Eq. (6.35) and Table 6.2] and would not fit into space with any finite dimension without significant overlap between different sites.

Here we calculate the size of ideal randomly branched polymers, ignoring excluded volume interactions and allowing each molecule to achieve the state of maximum entropy (recall the discussion of ideal chains in Chapter 2). Since branched molecules have many ends, the mean-square end-to-end distance used to characterize the size of linear chains is not appropriate for them. The simplest quantity describing the size of branched molecules is their mean-square radius of gyration R_g [see Eq. (2.44) for the definition].

In Chapter 2 we have presented a proof of the Kramers theorem for branched molecules containing N monomers of size b, but no loops (Eq. 2.65). The mean-square radius of gyration of these molecules is

$$\langle R_g^2 \rangle = b^2 \frac{\langle N_1(N - N_1) \rangle}{N}, \tag{6.83}$$

where the average is taken over all possible ways of cutting these molecules into two parts containing N_1 and $N - N_1$ monomers, respectively (see Fig. 2.7).

The Kramers theorem relates the ideal size of molecules to a purely structural property—the number of ways of dividing a molecule into two branches. The probability that a given bond is connected to a branch of N_1 monomers through a chosen one of the two ends of the bond is the same as the probability $u_{N_1}(p)$ that a given unreacted A group is a part of an N_1-mer [Eq. (6.61)]:

$$u_{N_1}(p) = \frac{[(f-1)N_1]!}{N_1![(f-2)N_1+1]!} p^{N_1-1}(1-p)^{(f-2)N_1+1}. \qquad (6.84)$$

The probability that a chosen bond divides a molecule into two branches—the first one with N_1 monomers and the second one with $N - N_1$ monomers is $u_{N_1}(p)u_{N-N_1}(p)$. Therefore, the Kramers theorem can be rewritten:

$$\langle R_g^2 \rangle = \frac{b^2}{N} \frac{\sum\limits_{N_1=1}^{N-1} N_1(N - N_1)u_{N_1}(p)u_{N-N_1}(p)}{\sum\limits_{N_1=1}^{N-1} u_{N_1}(p)u_{N-N_1}(p)}. \qquad (6.85)$$

The denominator of this equation is the probability that an occupied bond is a part of an N-mer. It is related to the number density of N-mers $n(p, N)$ by the ratio of sites to bonds on the Bethe lattice $2/f$, the number of occupied bonds per N-mer $(N-1)$ and the probability of the chosen bond to be formed p. The number of N-mers per bond is $(2/f)n(p, N)$, while the number of N-mers per occupied bond is larger by the factor $1/p$ and is equal to $(2/f)n(p, N)/p$. The probability that an occupied bond is a part of an N-mer is the number of bonds of all N-mers per occupied bond, which is larger than the number of N-mers per occupied bond by the number of bonds per N-mer $(N-1)$:

$$\sum_{N_1=1}^{N-1} u_{N_1}(p)u_{N-N_1}(p) = \frac{2}{f} \frac{N-1}{p} n(p, N)$$

$$= \frac{2(N-1)[(f-1)N]!}{N![(f-2)N+2]!} p^{N-2}(1-p)^{(f-2)N+2}. \qquad (6.86)$$

In the second relation, we used Eq. (6.63) for $n(p, N)$. Close to the gel point [for $\varepsilon = (f-1)p - 1 \ll 1$] this probability can be rewritten for large degree of polymerization $(N \gg 1)$ using Stirling's approximation [Eq. (6.23)]:

$$\sum_{N_1=1}^{N-1} u_{N_1}(p)u_{N-N_1}(p) \cong \sqrt{\frac{2(f-1)}{\pi(f-2)}} N^{-3/2} \exp\left(-\frac{f-1}{2(f-2)}\varepsilon^2 N\right). \qquad (6.87)$$

Each individual branch probability can be approximated in a similar way [recall Eq. (6.76)].

The summation in the numerator of Eq. (6.85) is dominated by the middle of the interval and can be evaluated by replacing the summation by integration using the asymptotic expression for $u_N(p)$ [Eq. (6.76)]:

$$
\sum_{N_1=1}^{N-1} N_1(N-N_1)u_{N_1}(p)u_{N-N_1}(p)
$$

$$
\cong \int_1^{N-1} N_1(N-N_1)u_{N_1}(p)u_{N-N_1}(p)\,dN_1
$$

$$
\cong \frac{(f-1)}{2\pi(f-2)}\exp\left(-\frac{f-1}{2(f-2)}\varepsilon^2 N\right)\int_1^{N-1}\frac{N_1(N-N_1)\,dN_1}{N_1^{3/2}(N-N_1)^{3/2}}
$$

$$
\cong \frac{f-1}{2\pi(f-2)}\exp\left(-\frac{f-1}{2(f-2)}\varepsilon^2 N\right)\int_{-1}^{1}\frac{dx}{\sqrt{1-x^2}}
$$

$$
\cong \frac{f-1}{2(f-2)}\exp\left(-\frac{f-1}{2(f-2)}\varepsilon^2 N\right), \tag{6.88}
$$

where we have introduced a new variable of integration $x = -1 + 2N_1/N$, and the integral over x is π.

Substituting Eqs (6.87) and (6.88) into the Kramers theorem expression [Eq. (6.85)] gives the mean-square radius of gyration for an ideal randomly branched N-mer:

$$
\langle R_g^2 \rangle \cong \sqrt{\frac{\pi(f-1)}{8(f-2)}}b^2 N^{1/2}. \tag{6.89}
$$

A similar relation was derived by Zimm and Stockmayer in 1949.[4]

As expected, the radius of gyration of an ideal randomly branched polymer is much smaller than that of an ideal linear chain with the same number and size of monomers (see Fig. 6.25). It is important to note that the dependence of the size on the degree of polymerization for randomly branched polymers

$$
R_g \approx bN^{1/4} \quad \text{for ideal randomly branched} \tag{6.90}
$$

is weaker than for ideal linear polymers [Eq. (2.54)]:

$$
R_g \approx bN^{1/2} \quad \text{for ideal linear.} \tag{6.91}
$$

The fractal dimension of an ideal randomly branched polymer is $\mathcal{D} = 4$ (because its degree of polymerization is proportional to its size to the fourth power $N \sim R_g^4$). In spaces with dimension $d < 4$ (in two-dimensional and three-dimensional spaces), ideal randomly branched polymers become too

$R_g \sim N^{1/4}$ $R_g \sim N^{1/2}$

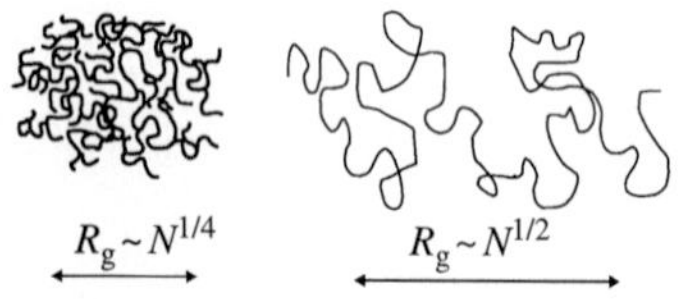

Fig. 6.25
The branched polymer is smaller than the linear polymer with the same number of monomers.

---

[4] The difference is that in Eq. (6.89) N is the number of monomers (occupied sites of an N-mer), while in the treatment of Zimm and Stockmayer N_{lin} is the number of linear segments per N-mer. Their asymptotic values are related to each other by $N = N_{\text{lin}}/(f-1)$. Therefore, the Zimm–Stockmayer prediction is $\langle R_g^2 \rangle \cong b^2 N_{\text{lin}}\sqrt{\pi/[8(f-1)(f-2)N]}$.

dense as reflected in the overlap parameter *decreasing* with degree of polymerization:

$$P = \frac{R_g^d}{Nb^d} \approx \frac{N^{d/4}}{N} \approx N^{-(4-d)/4}. \tag{6.92}$$

An important difference between randomly branched and linear polymers is that the fractal dimension of branched polymers is larger than the dimension of space ($d=3$). This severely limits the applicability of the mean-field theory to the crosslinking of long linear chains, called vulcanization. Long chains in the melt have a fractal dimension of $\mathcal{D} = 2$, which leaves lots of room inside the pervaded volume of the chain (i.e., filled by other chains in a polymer melt). The extra room created by the linear sections between crosslinks allows the fractal dimension of $\mathcal{D} = 4$ to exist in three-dimensional space *on a certain range of length scales* (see Section 6.5.4).

6.5 Scaling model of gelation

6.5.1 Molar mass distribution and gel fraction

It is possible to generalize the results, derived in Section 6.4 for the Bethe lattice, to any percolation problem. Of particular interest is gelation (percolation) in two-dimensional and three-dimensional spaces. Unlike mean-field percolation (on a Bethe lattice) there is no simple relation between the number of unreacted groups on a branched polymer and its degree of polymerization. Consequently, general percolation problems have no simple analytical solution, but have been solved by computer simulations and by analytical approximations. For example, polymerization of multifunctional monomers can be modeled on a three-dimensional lattice, while polymerization at a surface can be simulated on a two-dimensional lattice. In this section, we follow the scaling approach developed for treating continuous phase transitions, first applied to gelation by Stauffer and de Gennes.

Near the gel point, the system consists of a highly polydisperse distribution of polymers. One of the most important features of gelation is that the number density of polymers near the gel point has a power law dependence on the degree of polymerization, as was found for the Bethe lattice [Eq. (6.78)]. However, the cutoff function is more complicated than the simple exponential of the mean-field theory. In fact, it is asymmetric, having different form below and above the gel point. Therefore, we define the cutoff function $f_+(N/N^*)$ above the gel point and the cutoff function $f_-(N/N^*)$ below the gel point. These cutoff functions have the property of truncating the power law at the characteristic branched polymer with N^* monomers. In addition, $f_-(N/N^*)$ assures that the first moment of the distribution [the sol fraction, Eq. (6.39)] is unity below the gel point:

$$n(p, N) = N^{-\tau}f_-(N/N^*) \quad \text{for } p < p_c, \tag{6.93}$$

$$n(p, N) = N^{-\tau}f_+(N/N^*) \quad \text{for } p > p_c. \tag{6.94}$$

The critical exponent τ is the same above and below the gel point and is called the **Fisher exponent**. The number of monomers N^* in the characteristic branched polymer increases as the gel point is approached (from either side) and diverges as a power of the distance from the gel point, characterized by the relative extent of reaction ε [Eq. (6.25)]:

$$N^* \approx |\varepsilon|^{-1/\sigma}. \tag{6.95}$$

The values of the critical exponents τ and σ and the cutoff functions $f_+(N/N^*)$ and $f_-(N/N^*)$ depend only on the dimension of space in which gelation takes place. The percolation model has been solved analytically in one dimension ($d=1$, see Sections 1.6.2 and 6.1.2) and critical exponents have been derived for two dimensions ($d=2$). The mean-field model of gelation corresponds to percolation in spaces with dimension above the upper critical dimension ($d>6$). The cutoff function in the mean-field model [see Eq. (6.77)] is approximately a simple exponential function [Eq. (6.79)]. The exponents characterizing mean-field gelation are $\sigma=1/2$ and $\tau=5/2$ (see Section 6.4). The exact values of the critical exponents and the exact functional form of the cutoff functions for three-dimensional percolation are not known, but good estimates exist from computer simulations (see Table 6.4 for the values of critical exponents). In three-dimensional percolation, $\sigma \cong 0.45$ and $\tau \cong 2.18$. Both computer simulations and experiments have also verified that the cutoff function is sharp.

Together, Eqs (6.93)–(6.95) predict that plotting $n(p, N) (N^*)^\tau$ against N/N^* constructs a **universal scaling curve** that reduces all molar mass distributions at different extents of reaction below the gel point to a single curve. Two such scaling curves are shown in Fig. 6.26.

The moments of the molar mass distribution $n(p, N)$ were defined in Section 1.6:

$$m_k = \sum_{N=1}^{\infty} N^k n(p, N) \approx \int_1^{\infty} N^{k-\tau} f_\pm(N/N^*) \, dN, \tag{6.96}$$

where $f_\pm(N/N^*)$ refers to $f_-(N/N^*)$ below the gel point and $f_+(N/N^*)$ above the gel point. It is important to note that, while the sum and integral in Eq. (6.96) extend to infinity, they *only include the sol molecules*. Above the gel point, the gel is not included in $n(p, N)$ and hence does not affect the calculation of the moments.

Table 6.4 Summary of exponents for percolation in d-dimensions

Exponent	$d=1$	$d=2$	$d=3$	$d=4$	$d=5$	$d\geq 6$
β	0	5/36	0.41	0.64	0.84	1
γ	1	43/18	1.82	1.44	1.18	1
σ	1	36/91	0.45	0.48	0.49	1/2
τ	2	187/91	2.18	2.31	2.41	5/2
ν	1	4/3	0.88	0.68	0.57	1/2
$\mathcal{D}$	1	91/48	2.53	3.06	3.54	4

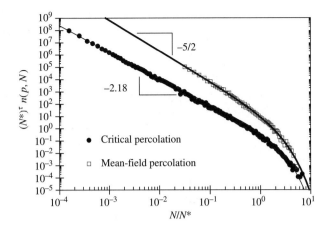

Fig. 6.26
Universal scaling curves for the molar
mass distribution of randomly branched
polyesters with many different samples
in each class. The filled symbols have
$N_0 = 2$ monomers between branch
points and correspond to critical
percolation in three dimensions. The
open symbols have $N_0 = 900$ monomers
between branch points and obey the
mean-field percolation model. Data of
C. P. Lusignan *et al.*, *Phys. Rev. E* **52**,
6271 (1995); **60**, 5657 (1999).

Since the Fisher exponent τ for percolation in any dimension is limited to
the interval $2 \leq \tau \leq 5/2$, the first two moments of the distribution m_0 and m_1
do *not* diverge at the gel point. These two moments are dominated by the
smaller polymers with a small contribution from the larger ones. Below
the gel point, the sol fraction is unity $[P_{sol}(p) = m_1 = 1$, see Eq. (6.42)] and
the gel fraction is zero $[P_{gel}(p) = 1 - P_{sol}(p) = 0]$. Above the gel point, we can
approximate the cutoff function as a step function that goes to zero
abruptly at N^*:

$$m_1 \approx \int_1^\infty N^{1-\tau} f_+(N/N^*)\, dN \approx \int_1^{N^*} N^{1-\tau}\, dN. \qquad (6.97)$$

The integral of a power law is easily evaluated,

$$P_{sol} = m_1 \approx \int_1^{N^*} N^{1-\tau}\, dN = 1 - C(N^*)^{2-\tau}, \qquad (6.98)$$

where C is a constant. The gel fraction increases with extent of reaction
beyond the gel point:

$$P_{gel}(p) = 1 - P_{sol}(p) = 1 - m_1 \approx C(N^*)^{2-\tau}.$$

Since $2 - \tau < 0$, the gel fraction is zero right at the gel point (where N^*
diverges) and grows steadily with extent of reaction above the gel point:

$$P_{gel} \sim (N^*)^{2-\tau} \sim \varepsilon^\beta \quad \text{for } p > p_c. \qquad (6.99)$$

The exponent β, defined for the growth of the gel fraction, is related to the
two exponents τ and σ by a **scaling relation**, obtained by combining
Eqs (6.95) and (6.99):

$$\beta = \frac{\tau - 2}{\sigma}. \qquad (6.100)$$

The absolute value of ε is not needed when discussing the gel fraction because the growth of P_{gel} only occurs above the gel point, where $\varepsilon > 0$. The mean-field value of this exponent is $\beta = 1$, since $\tau = 5/2$ and $\sigma = 1/2$. However, in three-dimensional percolation this exponent is $\beta \cong 0.41$.

The number-average degree of polymerization is given by the ratio of the first to the zeroth moments of the number density distribution ($N_n = m_1/m_0$) and does *not* diverge (stays finite) at the gel point.

Making the change of variables $x = N/N^*$ in the integral for the moments of the distribution function [Eq. (6.96)] allows us to extract the diverging part from the integral:

$$m_k \approx (N^*)^{k-\tau+1} \int_{1/N^*}^{\infty} x^{k-\tau} f_{\pm}(x)\, dx. \qquad (6.101)$$

The cutoff function $f_{\pm}(x)$ assures that the integral in the above equation converges to a number. This approach therefore determines the way in which the moments of the distribution function scale with the proximity to the gel point *without specifying the details of the cutoff function $f_{\pm}(x)$*:

$$m_k \sim (N^*)^{k-\tau+1} \sim |\varepsilon|^{-(k-\tau+1)/\sigma} \quad \text{for } k > \tau - 1. \qquad (6.102)$$

The weight-average degree of polymerization of the finite-size branched polymers is the ratio of the second and first moments. Since the first moment is finite and almost constant near the gel point, the weight-average degree of polymerization is proportional to the second moment of the distribution [Eq. (6.102) with $k = 2$]:

$$N_w = \frac{m_2}{m_1} \sim (N^*)^{3-\tau} \sim |\varepsilon|^{-\gamma}. \qquad (6.103)$$

Combining Eqs (6.95) and (6.103) leads to a scaling relation for the exponent γ, defined from the divergence of the weight-average degree of polymerization:

$$\gamma = \frac{3 - \tau}{\sigma}. \qquad (6.104)$$

The mean-field value of this exponent is $\gamma = 1$ [see Eq. (6.59) or substitute $\tau = 5/2$ and $\sigma = 1/2$ in the above equation] and $\gamma \cong 1.82$ in three dimensions. Experimental data on weight-average molar mass of branched polyurethanes in Fig. 6.27 clearly demonstrate that the experimental value of $\gamma = 1.7 \pm 0.1$ is in reasonable agreement with the expectation of three-dimensional percolation.

Moments of the distribution with $k \geq 2$ diverge at the gel point (because $2 \leq \tau \leq 5/2$ and $\sigma > 0$ always). The z-average degree of polymerization of the finite-size branched polymers is the ratio of the third and second moments:

$$N_z = \frac{m_3}{m_2} \sim N^* \sim |\varepsilon|^{-1/\sigma}. \qquad (6.105)$$

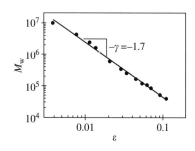

Fig. 6.27
Correlation of weight-average molar mass and relative extent of reaction for trifunctionally branched polyurethanes. Data from M. Adam *et al.*, *J. Phys. France* **48**, 1809 (1987).

Equation (6.80) gives the identical relation for the mean-field model (with $\sigma = 1/2$). Higher-order averages of the distribution are ratios of higher moments (m_{k+1}/m_k with $k \geq 3$). These higher-order averages are all proportional to the degree of polymerization of the characteristic branched polymer N^* and diverge at the gel point with the same exponent $-1/\sigma$.

Note that whenever a new exponent is defined, there is also a scaling relation that calculates this exponent from τ and σ. There are only two independent exponents that describe the distribution of molar masses near the gelation transition, with the other exponents determined from scaling relations. Table 6.4 summarizes the exponents in different dimensions that have been determined numerically, along with the exact results for $d=1$, $d=2$, and $d \geq 6$. It turns out that $d=6$ is the upper critical dimension for percolation, and the mean-field theory applies for all dimensions $d > 6$.

Each set of exponents corresponds to the **universality class** for percolation in space of a particular dimension. These exponents do not depend on the type of lattice (square vs. hexagonal) type of percolation (site vs. bond) and even whether it is on a lattice or not, as long as there are no long-range correlations. Of course, the value of the gel point p_c depends on all the above-mentioned details.

The cutoff functions $f_+(N/N^*)$ and $f_-(N/N^*)$ are expected to only depend on space dimension. In dimensions $1 < d < 6$, the cutoff functions are different above and below the gel point. This asymmetry makes the values of diverging quantities above and below the gel point differ. The ratio of a given quantity below and above the gel point defines an amplitude ratio. For example, the ratio of the weight-average degree of polymerization below and above the gel point by the same small extent x,

$$\lim_{x \to 0} \frac{N_w(p_c - x)}{N_w(p_c + x)}, \tag{6.106}$$

is believed to be universal. In the mean-field theory (and dimensions $d \geq 6$), the transition is symmetric and the amplitude ratios are equal to unity. In general, the percolation transition is not symmetric and the amplitude ratio defined by Eq. (6.106) is of order 10 in three dimensions and even larger (of order 200) in two dimensions.

6.5.2 Cutoff functions

The cutoff functions $f_-(N/N^*)$ and $f_+(N/N^*)$ defined in Eqs (6.93) and (6.94) are both simple exponentials in the mean-field gelation theory [Eq. (6.79)]:

$$f_-(N/N^*) = f_+(N/N^*) \approx \exp[-N/N^*] \approx \exp\left[-\frac{1}{2}\frac{f-1}{f-2}\varepsilon^2 N\right]$$

$$\approx \exp\left[-\frac{1}{2}\frac{f-1}{f-2}(\varepsilon N^{1/2})^2\right]. \tag{6.107}$$

Let us define a scaling parameter z,

$$z \equiv \varepsilon N^\sigma, \tag{6.108}$$

that combines the two cutoff functions $f_-(N/N^*)$ and $f_+(N/N^*)$ into a single general cutoff function $\hat{f}(z)$. Positive values of z correspond to the system above the gel point, while negative values apply to the system below the gel point. At the gel point, $z = 0$. The absolute value of the scaling parameter z is proportional to $(N/N^*)^\sigma$. The cutoff function in the mean-field theory is a Gaussian function of z with its maximum at the gel point:

$$\hat{f}(z) \approx \exp\left(-\frac{f-1}{2(f-2)}z^2\right). \tag{6.109}$$

The scaling variable z can be used to construct universal plots of molar mass distributions. There are two approaches to making these universal plots. To study the cutoff function itself, multiply both sides of Eqs (6.93) and (6.94) by N^τ.

$$N^\tau n(p, N) = \hat{f}(z) = \begin{cases} f_-(N/N^*) & \text{for } p < p_c, \\ f_+(N/N^*) & \text{for } p > p_c. \end{cases} \tag{6.110}$$

To study the universal properties of the distribution of molar masses, multiply both sides of Eqs (6.93) and (6.94) by $(N^*)^\tau$:

$$(N^*)^\tau n(p, N) = z^{-\tau/\sigma}\hat{f}(z) = \left(\frac{N}{N^*}\right)^{-\tau}\begin{cases} f_-(N/N^*) & \text{for } p < p_c, \\ f_+(N/N^*) & \text{for } p > p_c. \end{cases} \tag{6.111}$$

The analytical form of the cutoff function in three-dimensional critical percolation is not known. However, experimental and numerical universal plots have been constructed using the methods described above and constitute a convincing proof of the validity of the scaling ansatz (see Fig. 6.26).

Similar methods can be used to construct universal plots for molar mass distributions of linear and hyperbranched condensation polymers. The number distribution function $n(p, N)$ for linear condensation polymers is obtained from the number fraction distribution [Eq. (1.66)]:

$$n(p, N) = \frac{n_N(p)}{N_n} = \frac{1}{N_n^2}\exp\left(-\frac{N}{N_n}\right). \tag{6.112}$$

Identifying the characteristic degree of polymerization for linear condensation (with $\sigma = 1$ and $p_c = 1$)

$$N^* = N_n = \frac{p_c}{p_c - p} = \frac{1}{\varepsilon} \tag{6.113}$$

the distribution function $n(p, N)$ can be rewritten in the scaling ansatz form:

$$n(p, N) = N^{-2}\left(\frac{N}{N^*}\right)^2\exp\left(-\frac{N}{N^*}\right). \tag{6.114}$$

Comparing this expression with Eq. (6.93) identifies $\tau = 2$ and the one-dimensional percolation cutoff function

$$\hat{f}(z) = z^2 \exp(z) \tag{6.115}$$

for scaling variable

$$z = -\frac{N}{N^*} \tag{6.116}$$

that is never positive for linear condensation because $p \leq p_c = 1$. The scaled molar mass distribution for linear condensation is a simple exponential

$$n(p, N)N_n^2 = \exp\left(-\frac{N}{N_n}\right) \tag{6.117}$$

that does not have any power law features, even though the Fisher exponent for this case is $\tau = 2$. This is different from the scaled molar mass distribution in gelation that decays as a power law with exponent equal to the Fisher exponent. The fact that the cutoff function in linear condensation [Eq. (6.115)] is a power law itself at small z accounts for this difference.

   An even more striking example is the universal molar mass distribution of hyperbranched polymers. This distribution [Eq. (6.31)] can also be written in the scaling ansatz form with $\tau = 2$:

$$n(p, N) = \frac{n_N(p)}{N_n} \approx \frac{N^{-3/2}}{N_n} \exp(-N/N^*) \approx \frac{N^{-3/2}}{(N^*)^{1/2}} \exp(-N/N^*)$$

$$\approx N^{-2}\left(\frac{N}{N^*}\right)^{1/2} \exp(-N/N^*). \tag{6.118}$$

The characteristic degree of polymerization N^* is related to N_n by Eqs (6.32) and (6.34):

$$N^* = \frac{2(f-2)}{f-1}N_n^2 = \frac{2(f-2)}{(f-1)(1-p/p_c)^2}. \tag{6.119}$$

The cutoff function for hyperbranched polymers is

$$\hat{f}(z) = -z \exp\left(-\frac{f-1}{2(f-2)}z^2\right), \tag{6.120}$$

where the scaling parameter

$$z = -(1 - p/p_c)N^{1/2}. \tag{6.121}$$

Alternatively, the universal molar mass distribution [Eq. (6.111)] for hyperbranched polymers can be constructed:

$$n(p, N)(N^*)^2 \approx \left(\frac{N}{N^*}\right)^{-3/2} \exp(-N/N^*). \tag{6.122}$$

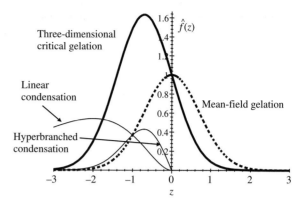

Fig. 6.28
Cutoff functions for critical gelation
(thick curve), mean-field gelation
(dotted curve), and linear and
hyperbranched condensation (thin
curves) for functionality of cross-links
$f = 3$.

Using the proportionality between the characteristic degree of poly-merization N^* and N_w, a universal distribution function for hyperbranched polymers can be constructed by plotting $n(p, N)N_w^2$ against N/N_w. The fact that the cutoff function contains the power law in z makes the apparent power law exponent $3/2$ in the molar mass distribution different from $\tau = 2$. The cutoff function for gelation has no power law so the apparent exponent is equal to τ.

The cutoff functions are plotted in Fig. 6.28. For critical gelation in three dimensions (and in the more general case of $1 < d < 6$) the cutoff function is asymmetric with the maximum shifted to the negative values of z. In contrast, the cutoff function for mean-field gelation is perfectly symmetric. Linear and hyperbranched condensation are defined only for negative scaling parameter z (for $p < p_c$). The cutoff function for both of these reactions vanishes at $z = 0$ because there is only one macromolecule at complete reaction ($p = p_c = 1$). For small negative values of the parameter z, the cutoff function grows as a power law of z ($\sim -z$ for hyperbranched and $\sim z^2$ for linear condensation). Both of these cutoff functions have a maximum and decay at large negative values of z.

A very important caveat with the universal molar mass distributions and cutoff functions is that calculated molar mass distributions of linear con-densation, hyperbranched condensation and mean-field percolation all assume no intramolecular reactions occur. Intramolecular reactions are hard to avoid in real polymerization experiments.

6.5.3   Size and overlap of randomly branched polymers

The correlation length ξ for percolation (and gelation) is the size of the characteristic branched polymer with N^* monomers. Randomly branched polymers are fractals, so the size R and the number of monomers N in a polymer are related by the fractal dimension $\mathcal{D}$:

$$N \sim R^{\mathcal{D}}. \tag{6.123}$$

The same relation is valid for the characteristic branched polymer:

$$N^* \sim \xi^{\mathcal{D}}. \tag{6.124}$$

Since N^* diverges at the gel point with exponent $1/\sigma$ [see Eq. (6.95)], the correlation length must also diverge:

$$\xi \sim (N^*)^{1/\mathcal{D}} \sim |\varepsilon|^{-\nu}. \qquad (6.125)$$

The exponentt ν describing the divergence of the correlation length is related to σ and the fractal dimension $\mathcal{D}$ by combining Eqs (6.95) and (6.125):

$$\nu = \frac{1}{\mathcal{D}\sigma}. \qquad (6.126)$$

In mean field, $\sigma = 1/2$ and $\mathcal{D} = 4$ (see Section 6.4) so $\nu = 1/2$. For three-dimensional percolation, this critical exponent is $\nu = 0.88$ and the fractal dimension of randomly branched polymers in the polymerization reactor is $\mathcal{D} = 2.53$.

6.5.3.1 Hyperscaling

For critical percolation, the exponent ν and the fractal dimension $\mathcal{D}$ are related to the critical exponents discussed in Section 6.5.2 by the **hyperscaling** relation in dimensions $1 < d < 6$. The idea of hyperscaling is that polymers with a given number of monomers N are at the overlap concentration with themselves and with similar sized parts of larger polymers (see Fig. 6.29). Polymers of a given size cannot strongly overlap because they would have many contact points and would have already reacted to make a larger polymer. The pervaded volume for a randomly branched polymer with N monomers in d-dimensional space is proportional to $R^d \sim N^{d/\mathcal{D}}$ [see Eq. (6.123)]. Molecules with K monomers (for $K > N$) can be divided into K/N parts, each containing N monomers. Since the molecules are fractal, these smaller parts have the same size as individual molecules with N monomers. The number of molecules with K monomers is proportional to their number density $n(p, K)$. Therefore, the total volume pervaded by molecules with N monomers as well as smaller parts (containing N monomers) of larger molecules is proportional to

$$\int_N^\infty R^d \frac{K}{N} n(p, K)\, \mathrm{d}K \sim \frac{R^d}{N} \int_N^\infty K n(p, K)\, \mathrm{d}K. \qquad (6.127)$$

The integral on the right-hand side of the above equation is the weight fraction of all molecules larger than N. This integral is dominated by the lower limit for critical exponent $\tau > 2$ and is proportional to $N^{2-\tau}$.

$$\int_N^\infty K n(p, K)\, \mathrm{d}K \sim N^{2-\tau}.$$

In order for the fragments of molecules containing N monomers to be at overlap (for all N), the combined pervaded volume has to be independent of N (and equal to the total volume of the system):

$$\int_N^\infty R^d \frac{K}{N} n(p, K)\, \mathrm{d}K \sim R^d N^{1-\tau} \sim N^{1-\tau+d/\mathcal{D}} \sim N^0. \qquad (6.128)$$

Fig. 6.29
Hyperscaling—polymers are at overlap with other chains of the same size and with similar sized parts of larger chains. Polymers of different size are denoted by lines of different thicknesses.

This condition leads to the hyperscaling relation between critical exponents:

$$\frac{d}{\mathcal{D}} = \tau - 1. \tag{6.129}$$

Hyperscaling relates the fractal dimension of randomly branched molecules in the gelation reaction with the Fisher exponent and the space dimension (see Table 6.4):

$$\mathcal{D} = \frac{d}{\tau - 1} \quad \text{for } 1 \leq d \leq 6. \tag{6.130}$$

The mean-field value of the Fisher exponent is $\tau = 5/2$. The fractal dimension of ideal randomly branched polymers is $\mathcal{D} = 4$ [see Eq. (6.90)]. Equation (6.130) gives $\mathcal{D} = 4$ for $\tau = 5/2$ in dimension $d = 6$, which is the upper critical dimension for percolation. For dimensions larger than 6, the mean-field theory works and the hyperscaling relation does not (polymers strongly overlap). For three-dimensional percolation, $\tau \cong 2.18$ and $\mathcal{D} \cong 2.53$. Experimental measures of fractal dimension by Adam and coworkers using small-angle neutron scattering experiments report $\mathcal{D} = 2.50 \pm 0.06$.

6.5.3.2 Flory–de Gennes theory

The fractal dimension of randomly branched polymers in the polymerization reaction can be estimated using Flory theory (see Section 3.1.2). The free energy of the characteristic branched polymer consists of entropic and interaction parts:

$$F \approx kT \left[\frac{\xi^2}{[b(N^*)^{1/4}]^2} + \frac{\text{v}}{N_\text{w}} \frac{(N^*)^2}{\xi^d} \right]. \tag{6.131}$$

The first term is the entropic penalty for swelling the branched polymer from its ideal size $b(N^*)^{1/4}$ [Eq. (6.90)] to its size in the reactor ξ. The second term is the estimate of the excluded volume repulsion between $(N^*)^2$ pairs of monomers spread over the pervaded volume of the polymer ξ^d with excluded volume interactions screened by the polydisperse mix of smaller branched polymers inside the pervaded volume of this characteristic branched polymer. In Section 4.5.2, the screening of excluded volume interactions by overlapping chains was discussed for the case of a melt of linear chains [Eq. (4.73)]. Analogous screening occurs in the case of a polydisperse sample with the excluded volume reduced by the weight-average degree of polymerization of the polymers inside the pervaded volume of a given polymer (see Problem 4.19). In the case of the characteristic branched polymer, the excluded volume is reduced by the weight-average degree of polymerization of all finite-size branched polymers v/N_w.

The size of the characteristic branched polymer ξ is obtained by minimizing the free energy [Eq. (6.131)] with respect to its size:

$$\xi \approx b\left(\frac{v}{b^d}\frac{(N^*)^{5/2}}{N_w}\right)^{1/(d+2)}. \tag{6.132}$$

Substituting the mean-field relation between the weight-average and characteristic degrees of polymerization,

$$N_w \sim \varepsilon^{-\gamma} \sim (N^*)^{\gamma\sigma} \sim (N^*)^{1/2} \tag{6.133}$$

provides the de Gennes prediction for the correlation length in gelation based on Flory theory.

$$\xi \approx b\left(\frac{v}{b^d}\right)^{1/(d+2)}(N^*)^{2/(d+2)}. \tag{6.134}$$

Hence, the fractal dimension of the branched polymers is predicted in every dimension of space:

$$\mathcal{D} \cong \frac{d+2}{2} \quad \text{for } 2 \leq d \leq 6. \tag{6.135}$$

Note that this prediction is quite close to the values reported in Table 6.4.

6.5.4   Vulcanization universality class

Consider the crosslinking of long linear precursor chains with degree of polymerization N_0 and functionality f in the melt. This class of gelation is called vulcanization, named after Goodyear's famous process to crosslink natural rubber using sulphur. Vulcanization is one type of gelation for which the mean-field theory, described in Section 6.4, works well in a wide range of relative extents of reaction $\varepsilon = (p - p_c)/p_c$ [Eq. (6.25)]. The effective functionality of a long chain with N_0 crosslinkable monomers is $f \approx N_0$. The mean-field prediction for the gel point is [Eq. (6.47)]

$$p_c = \frac{1}{f-1} \approx \frac{1}{N_0} \quad \text{for } N_0 \gg 1 \tag{6.136}$$

and corresponds to an average of *one crosslink per chain*. The gel fraction is proportional to this relative extent of reaction [see Eq. (6.99) with mean-field value of exponent $\beta = 1$]:

$$P_{gel} \approx \varepsilon \quad \text{for } 0 < \varepsilon \ll 1. \tag{6.137}$$

At relative extent of reaction $\varepsilon = 1$, the gel fraction is of order unity $P_{gel} \approx 1$, and most of the chains are attached to the gel. The percolation transition is almost complete, with very small sol fraction left $P_{sol} \ll 1$ at extent of reaction $p \approx 2p_c$ (where $\varepsilon \approx 1$). Since the gel point corresponds to an average of one crosslink per chain, the end of the gelation regime corresponds to an average of two crosslinks per chain. In some cases of vulcanization, the functionality is very large ($f \approx N_0 \gg 1$) leading to a small percolation threshold ($p_c \ll 1$). An example of such vulcanization is the Goodyear original process, where large values of the relative extent of reaction $\varepsilon \gg 1$ are possible. Such highly vulcanized networks (with many crosslinks per chain) are considered in the next chapter.

The correlation length ξ and the number of monomers in a characteristic branched polymer N^* have simple predictions for vulcanization. These predictions can be easily obtained from the mean-field percolation theory [Eqs (6.105) and (6.125) with exponents $\sigma = \nu = 1/2$] by replacing the monomer in the previous treatment by the precursor linear chain[5] of size $bN_0^{1/2}$ containing N_0 monomers.

$$\xi \approx bN_0^{1/2}|\varepsilon|^{-1/2}, \tag{6.138}$$

$$N^* \approx N_0\varepsilon^{-2}. \tag{6.139}$$

The number of monomers in the characteristic branched polymer N^* and its size ξ are symmetric around the gel point ($\varepsilon = 0$). The molar mass distribution is in fact similar for the same ε above and below the gel point (within the framework of mean-field scaling).

Beyond the gel point, the correlation length ξ also describes the size of the 'holes' in the network and N^* is the average number of monomers in a network strand. Each network strand is a branched polymer. The number of overlapping network strands P plays a vital role in our understanding of gelation. The volume fraction of a single network strand inside its pervaded volume is N^*b^3/ξ^3. The number of network strands with N^* monomers, sharing the same volume ξ^3 (which is the overlap parameter for the network strands) is given by the ratio of the volume fraction of all network strands (the gel fraction P_{gel}) and the volume fraction of a single strand:

$$P \approx \frac{P_{gel}}{N^*b^3/\xi^3} \approx \frac{P_{gel}}{N^*}\left(\frac{\xi}{b}\right)^3 \approx N_0^{1/2}|\varepsilon|^{3/2}. \tag{6.140}$$

The final result was obtained using Eqs (6.137)–(6.139). For long chains between branch points (large N_0), P can be large, meaning that network strands overlap each other extensively. This is hardly surprising, since we know that before any crosslinking takes place (at $\varepsilon = -1$) the long precursor chains will have $N_0^{1/2}$ other chains within their pervaded volume [see Eq. (5.11)]. The highly overlapping and interpenetrating precursor chains guarantee that there will be considerable overlap over most of the crosslinking reaction.

However, Eq. (6.140) shows that the overlap of the largest polymers diminishes as the gel point is approached. For any precursor chain length

---

[5] Although it decreases gradually as the reaction proceeds, the chain length between branch points *in the gelation regime* is of order N_0.

there is always a region very close to the gel point where the largest finite polymers (and network strands above the gel point) no longer overlap and the mean-field theory breaks down. As with other transitions that we have discussed, the mean-field theory only applies sufficiently far from the critical point. Closer to the critical point (near $\varepsilon = 0$) the mean-field theory must be replaced by the critical theory (discussed in detail in Section 6.5.1). The crossover point between the regions where each theory applies is given by the Ginzburg criterion. For percolation, the relative extent of reaction ε_G corresponding to the Ginzburg criterion is the point where the characteristic polymers (and network strands above the gel point) no longer overlap. Setting $P = 1$ in Eq. (6.140) determines the Ginzburg relative extent of reaction:

$$\varepsilon_G \approx N_0^{-1/3}. \tag{6.141}$$

For $|\varepsilon| > \varepsilon_G$, the mean-field theory describes the crosslinking reaction. For $|\varepsilon| < \varepsilon_G$, (sufficiently close to the gel point), critical percolation describes the actual transition. For very long precursor chains, ε_G is extremely small and the entire experimentally accessible region of crosslinking is described by mean-field percolation (in practice $N_0 \gtrsim 100$ is sufficient). The limit of $N_0 = 1$ is polymerization of monomers with functionality greater than 2, and all extents of reaction in the range $-1 < \varepsilon < 1$ are described by critical percolation.

The dependence of the overlap parameter on relative extent of reaction is sketched in Fig. 6.30. Equation (6.140) describes the decrease in overlap as the gel point is approached until $P = 1$ at the Ginzburg point. The characteristic polymers and the network strands above the gel point are just at their overlap concentration inside the critical region (where $P = 1$). Beyond the Ginzburg point on the gel side of the gel point, the overlap of network strands starts to increase again at ε_G and builds as the correlation length (and hence the network strand length) decreases. Notice that the overlap parameter is symmetric around the gel point.

Before the vulcanization reaction, long linear chains with degree of polymerization N_0 and size $bN_0^{1/2}$ overlap with each other strongly (recall from Problem 1.22 and Section 4.5.2 that there are $N_0^{1/2}$ other chains in the pervaded volume of any linear chain in the melt). Crosslinking these chains creates branched polymers with fractal dimension $\mathcal{D} = 4$ on scales larger than the linear chains. The size of these branched polymers with degree of polymerization N is

$$R_g \approx bN_0^{1/2}\left(\frac{N}{N_0}\right)^{1/4} \approx b(N_0 N)^{1/4}. \tag{6.142}$$

The intramolecular pair correlation function $g(r)$ is the number of monomers per unit volume of a section of chain with section size r inside its pervaded volume r^3, plotted in Fig. 6.31. The linear subsections of the randomly branched polymer have approximately N_0 monomers connected in a linear chain ($\mathcal{D} = 2$). The intramolecular pair correlation function $g(r)$

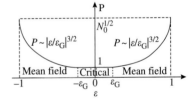

Fig. 6.30
Dependence of the overlap parameter P on relative extent of reaction ε.

Fig. 6.31
Intramolecular pair correlation function
$g(r)$ of directly connected monomers
within a sphere of radius r of a given
monomer, for randomly branched
polymers with N monomers made from
vulcanizing linear chains with degree of
polymerization N_0. Both axes have
logarithmic scales.

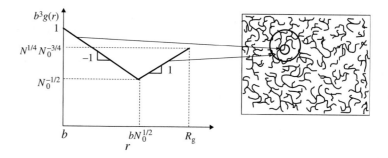

decreases for r smaller than $bN_0^{1/2}$ (the size of the linear chain sections), Eq. (2.121):

$$g(r) \approx \frac{n}{r^3} \approx \frac{1}{rb^2} \quad \text{for } r < bN_0^{1/2}. \tag{6.143}$$

This simple scaling results because a volume of size r contains $n = (r/b)^2$ monomers from that chain section. Equation (6.143) describes the minus one slope region of Fig. 6.31 on scales smaller than the size of the linear chains (with $r < bN_0^{1/2}$). When r reaches the size scale of the linear chain sections $bN_0^{1/2}$, Eq. (6.143) gives the expected result for the volume fraction of a linear chain inside its own pervaded volume in a melt, $\phi \approx N_0^{-1/2}$. On scales larger than the linear chains, the fractal dimension $\mathcal{D} = 4$ applies to chain sections of size r that are directly connected within the volume r^3 [see Eq. (6.142) and Fig. 6.31]:

$$r \approx b(N_0 n)^{1/4}. \tag{6.144}$$

The intramolecular pair correlation function grows with increasing r because the fractal dimension is larger than the space dimension:

$$g(r) \approx \frac{n}{r^3} \approx \frac{r}{b^4 N_0} \quad \text{for } bN_0^{1/2} < r < R_g. \tag{6.145}$$

The volume fraction of monomers inside the pervaded volume of the randomly branched polymer is $b^3 g(R_g) \approx N^{1/4} N_0^{-3/4}$, obtained by substituting R_g from Eq. (6.142) for r in Eq. (6.145). This approach is only valid in the vulcanization regime, with $\varepsilon > \varepsilon_G$. The number of monomers in a characteristic polymer at the Ginzburg point is $N^* \approx N_0 \varepsilon_G^{-2} \approx N_0^{5/3}$ [see Eqs (6.139) and (6.141)]. The volume fraction of monomers $b^3 g(R_g) \approx N_0^{-1/3} \ll 1$, so the polymers with fractal dimension $\mathcal{D} = 4$ fit in three-dimensional space.

The polymerization reactor contains a highly polydisperse distribution of molecules. The size of the characteristic polymer (with N^* monomers) defines the correlation length ξ [see Eq. (6.142)]:

$$\xi \approx b(N_0 N^*)^{1/4} \approx bN_0^{1/2}|\varepsilon|^{-1/2}. \tag{6.146}$$

This equation used Eq. (6.139) for the degree of polymerization of the characteristic polymer in the mean-field theory. The correlation length diverges as the gel point is approached for $|\varepsilon| > \varepsilon_G$ with mean-field exponent $\nu = 1/2$.

Below the gel point, the system is self-similar on length scales smaller than the correlation length ξ, with a power law distribution of molar masses with Fisher exponent $\tau = 5/2$ [Eq. (6.78)]. Each branched molecule is a self-similar fractal with fractal dimension $\mathcal{D} = 4$ for ideal branched molecules in the mean-field theory. The lower limit of this critical behaviour is the average distance between branch points ($\cong bN_0^{1/2}$). There are very few finite molecules larger than the characteristic branched polymer (with $N > N^*$) with size larger than the correlation length, because the distribution [Eq. (6.78)] has a sharp cutoff.

Above the gel point there is a macroscopic molecule (the gel). The structure of the gel is also self-similar (fractal) in the same range of length scales between the average distance between crosslinks (the linear chain size) and the correlation length:

$$bN_0^{1/2} < r < \xi. \tag{6.147}$$

The correlation length ξ is the average distance between branch points that are connected to several branches leading to 'infinity' (the boundary of the gel). At $p \approx 2p_c$ (at $\varepsilon \approx 1$) the gelation regime ends and most of the network strands are then simply linear chains. Chapter 7 shall discuss the properties of such well-developed networks.

6.6 Characterization of branching and gelation

Below the gel point, all species are soluble (in the appropriate solvent) allowing the standard dilute solution characterization methods to be utilized. Above the gel point, there is an insoluble gel fraction and a soluble sol fraction. When immersed in an excess of the appropriate solvent, the gel fraction will swell, and the sol fraction will slowly diffuse out of the swollen gel into the excess solvent. A convenient technique for such separation is **Soxhlet extraction**, shown in Fig. 6.32. Solvent is boiled in the bottom flask and condenses at the very top, dripping down onto the swollen gel. The gel is inside a carefully weighed glass thimble with a fritted filter at the bottom. As the thimble fills with solvent, the solvent flows through the filter, carrying with it the sol fraction. When the solvent outside the thimble reaches a certain level, it automatically siphons down into the lower boiling solvent. The Soxhlet extractor is designed to run continuously for many days virtually unattended. However, in practice a solvent with a very low boiling point is used to minimize degradation of the sol, and the high vapor pressure means that pure solvent must be added periodically. If the filter does not plug, the Soxhlet extractor eventually has all of the sol fraction in the boiling solvent and all of the gel fraction in the thimble. The gel fraction is characterized by its swelling (discussed in detail in Chapter 7). After allowing the excess solvent to flow through the filter, the thimble and swollen gel are weighed to determine the swollen mass. Then, the solvent is removed under vacuum and the thimble is weighed again to determine the gel fraction. The sol fraction can be characterized with the same dilute

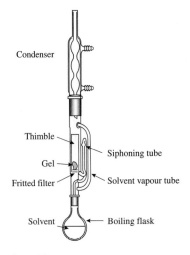

Fig. 6.32
Soxhlet extraction apparatus for separating sol and gel fractions.

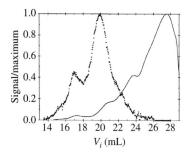

Fig. 6.33
Refractive index (line) and 15° light scattering (points) detector outputs as functions of elution volume for size exclusion chromatography on a randomly branched polyester sample dissolved in tetrahydrofuran. Data from C. P. Lusignan, Doctoral Dissertation, University of Rochester, 1996.

solution techniques used below the gel point. The fractal dimension of the swollen branched polymers is smaller than in their melt state.

Owing to the very broad molar mass distributions associated with random branching, the primary characterization tool is size exclusion chromatography, introduced in Section 1.7.4. For branched polymers, a low-angle light scattering detector is used together with a concentration-sensitive detector to determine the weight-average molar mass of each small volume of solution eluting from the columns. This experiment provides a wealth of information about the molecular characteristics of randomly branched polymers.

Depending on the chemical details of the polymer, a variety of concentration detectors can be used. The most common ones measure either refractive index or ultraviolet absorption. Figure 6.33 shows the measurements of refractive index and light scattering intensity at a scattering angle of 15° as functions of elution volume for a randomly branched polyester. Recall from Section 1.7.4 that the SEC separates polymers by size. The largest species access the smallest amount of column volume and thereby elute first (smaller elution volume V_i). Notice in Fig. 6.33 that there is a very small amount of the largest species, but they dominate the light scattering. This is because at low concentrations the intensity of scattered light is proportional to the product $c_i M_i$ [see Eq. (1.90)]. The scattering intensity has two local maxima. Experimentally, it is verified that the maximum intensity (at $V_i = 19.9$ mL in Fig. 6.33) corresponds to branched polymers with degree of polymerization proportional to N^*. The second peak at lower elution volume is an artefact of the exclusion limit of the columns. Close to the gel point, some branched polymers are so large that they do not fit into any of the pores in the columns. Such very large molecules are not separated by the SEC, and elute together (at $V_i \cong 17$ mL in Fig. 6.33). These imperfectly separated species must be removed before constructing universal molar mass distribution plots, such as Fig. 6.26.

Knowledge of the concentration c_i and weight-average molar mass M_{wi} of each elution volume enables calculation of the weight-average and z-average molar masses:

$$M_w = \frac{\sum_i c_i M_{wi}}{\sum_i c_i},\tag{6.148}$$

$$M_z \cong \frac{\sum_i c_i (M_{wi})^2}{\sum_i c_i M_{wi}}.\tag{6.149}$$

Comparison with Eqs (1.31) and (1.32) shows that Eq. (6.148) is exact, but Eq. (6.149) is exact only if each elution volume i is monodisperse (see Problem 6.40). Particularly since this chromatography separates molecules by size and not mass, the different elution volumes are not truly monodisperse. In practice, however, Eq. (6.149) is used to calculate M_z.

The universality class of the randomly branched polymers can be determined by constructing a universal molar mass distribution plot, like the ones shown in Fig. 6.26. First, the number density distribution function $n(p, N)$ is determined from the concentration and weight-average molar mass of each elution volume. The concentration detector directly determines the weight fraction of polymer in each elution volume $w_N(V_i)$:

$$w_N(V_i) = \frac{c_i}{\sum_j c_j}. \tag{6.150}$$

This weight fraction must first be converted to a function of molar mass, using the calibration curve[6] for the columns used (Fig. 1.26). The two weight fraction distributions are related to each other as

$$w_N(M_i)\, dM_i = -w_N(V_i)\, dV_i \tag{6.151}$$

where the minus sign comes from the fact that larger molar masses elute at smaller elution volumes:

$$w_N(M_i) = -w_N(V_i) \frac{dV_i}{dM_i} \cong \frac{-c_i}{2.303 M_i \sum_j c_j} \frac{dV_i}{d\log M_i}. \tag{6.152}$$

The final relation uses a derivative with respect to the base ten logarithm of molar mass, as is customary for the calibration curve (see Fig. 1.26). The number density distribution function $n(p, N)$ is related to w_N in a simple way:

$$n(p, N) = \frac{w_N}{N} \cong \frac{-M_0 c_i}{2.303 (M_i)^2 \sum_j c_j} \frac{dV_i}{d\log M_i}. \tag{6.153}$$

The characteristic degree of polymerization N^* is chosen for each sample so that the $n(p, N)$ data for that sample can be superimposed onto the universal curve. For both of the curves in Fig. 6.26, more than 10 samples with different extents of reaction were superimposed in this fashion. Although the Fisher exponent τ is not very different for the two classes of percolation, τ is sufficiently different not to allow the data from one class to be superimposed with the wrong exponent.

The system in Fig. 6.26 corresponding to critical percolation (with $\tau = 2.18$) has an average of $N_0 \cong 2$ monomers between branch points. The system corresponding to mean-field percolation (with $\tau = 5/2$) has long linear chains between branch points, with an average degree of polymerization of $N_0 \cong 900$. Most random branching reactions create chain lengths between branch points that are between these two clean limits. Such systems will exhibit the crossover anticipated by the Ginzburg criterion, between critical percolation close to the gel point and mean-field percolation further away. The critical and mean-field predictions can be simply

---

[6] With branched polymers the simplest procedure is to measure the concentration, intrinsic viscosity, and weight-average molar mass of each elution volume using appropriate detectors, see Section 1.7.4.

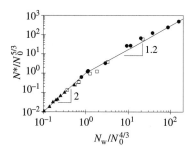

Fig. 6.34

Correlation of the characteristic degree of polymerization and the weight-average degree of polymerization for randomly branched polyesters below the gel point, using the scaling form suggested by Eqs (6.154) and (6.155) (lines with slope $1/(\sigma\gamma)$ for each class). Three different polyester chemistries are utilized, with average degree of polymerization between branch points of $N_0 = 2$ (filled circles), $N_0 = 20$ (open squares), and $N_0 = 900$ (filled triangles). $N_0 = 2$ data from C. P. Lusignan *et al.*, *Phys. Rev. E* **52**, 6271 (1995). $N_0 = 20$ data from E. V. Patton *et al.*, *Macromolecules* **22**, 1946 (1989). $N_0 = 900$ data from C. P. Lusignan *et al.*, *Phys. Rev. E* **60**, 5657 (1999).

matched at the Ginzburg extent of reaction ε_G.

$$N_w \approx N_0^{4/3} \begin{cases} (\varepsilon/\varepsilon_G)^{-\gamma} & \varepsilon < \varepsilon_G, \\ (\varepsilon/\varepsilon_G)^{-1} & \varepsilon > \varepsilon_G, \end{cases} \tag{6.154}$$

$$N^* \approx N_0^{5/3} \begin{cases} (\varepsilon/\varepsilon_G)^{-1/\sigma} & \varepsilon < \varepsilon_G, \\ (\varepsilon/\varepsilon_G)^{-2} & \varepsilon > \varepsilon_G. \end{cases} \tag{6.155}$$

This scaling-level matching expects a plot of $N^*/N_0^{5/3}$ against $N_w/N_0^{4/3}$ to be universal for all polymer gelation reactions. Such a plot is shown in Fig. 6.34 for three randomly branched polyester systems with $N_0 = 2$, 20, and 900. Figure 6.26 already showed that the system with $N_0 = 900$ corresponds to mean-field percolation and the triangles in Fig. 6.34 show that those samples have $N^*/N_0^{5/3} \approx (N_w/N_0^{4/3})^2$. Furthermore, all samples with $N_0 = 900$ have $N_w/N_0^{4/3} < 1$ and $N^*/N_0^{5/3} < 1$, which means that $\varepsilon > \varepsilon_G$, based on Eqs (6.154) and (6.155). Similarly, all samples with $N_0 = 2$ belong to the critical percolation universality class (circles in Fig. 6.26) and demonstrate the scaling expected from Eqs (6.154) and (6.155) in Fig. 6.34. The slope for $\varepsilon < \varepsilon_G$ ($N_w/N_0^{4/3} > 1$ and $N^*/N_0^{5/3} > 1$) in Fig. 6.34 is $1/(\sigma\gamma) \cong 1.2$. The $N_0 = 20$ system falls in the crossover between the two universality classes. For samples with $N_0 = 20$ that are far from the gel point, $\varepsilon > \varepsilon_G$ and the mean-field exponents apply, while for $N_0 = 20$ samples close to the gel point, $\varepsilon < \varepsilon_G$ and critical percolation exponents apply. However, it is important to notice that if the $N_0 = 20$ data in Fig. 6.34 were considered on their own, an apparent exponent intermediate between those of critical and mean-field theories would be observed. Figure 6.34 shows that such apparent exponents must not be used to extrapolate beyond the range of the data!

The extent of reaction p can, in principle, be measured by molecular spectroscopy methods such as FTIR and NMR. However, these methods always have some small relative error (typically of order a few percent). Since the same error has to apply to any determination of the extent of reaction at the gel point p_c, the relative error in the relative extent of reaction $\varepsilon = (p - p_c)/p_c$ *diverges* at the gel point. As a result, plots such as Fig. 6.34, where two quantities with finite relative errors (N_w and N^*) are plotted against each other, are more useful for determining exponent values and hence universality class. It is important to recognize that these plots always give information about combinations of exponents. For example, the slope in Fig. 6.34 is $1/(\sigma\gamma) = 1/(3 - \tau)$, as required by Eq. (6.104).

6.7 Summary of branching and gelation

A wide variety of linking processes can lead to the transformation from a liquid to a solid known as gelation. This transition occurs at a particular extent of reaction, p_c, called the gel point. A rough estimate of the gel point for linking precursor molecules (or monomers) with functionality f is

$p_c = 1/(f - 1)$. As the reaction proceeds, a highly polydisperse mix of molecules is formed with a power law distribution of molar masses. The probability that a randomly chosen polymer at extent of reaction p will have molar mass M is

$$n(p, M) \sim M^{-\tau} f(M/M^*), \qquad (6.156)$$

where τ is the Fisher exponent and the cutoff function $f(M/M^*)$ sharply truncates the distribution at the molar mass of the (largest) characteristic molecule M^*. This characteristic molar mass diverges as the gel point is approached.

$$M^* \sim \left| \frac{p - p_c}{p_c} \right|^{-1/\sigma}. \qquad (6.157)$$

This molar mass distribution applies to all polymers below the gel point and above the gel point it applies to the sol fraction. The weight-average molar mass of this distribution also diverges as the gel point is approached from either side:

$$M_w \sim \left| \frac{p - p_c}{p_c} \right|^{-\gamma} \quad \text{with } \gamma = \frac{3 - \tau}{\sigma}. \qquad (6.158)$$

The number-average molar mass stays finite at the gel point and therefore the polydispersity index of the sol diverges at the gel point proportional to M_w.

At the gel point, the distribution of molar masses is a power law, since M^* diverges. Above the gel point the system consists of a macroscopic gel, permeated by a polydisperse soup of microscopic molecules (the sol) described by the same self-similar distribution of molar masses. The distribution of molar masses in the sol gets progressively narrower as the reaction proceeds further beyond the gel point, since the characteristic molar mass M^* decreases. The fraction of the material that is a part of the gel, called the gel fraction P_{gel}, grows steadily above the gel point:

$$P_{gel} \sim \left(\frac{p - p_c}{p_c} \right)^{\beta} \quad \text{with } \beta = \frac{\tau - 2}{\sigma}. \qquad (6.159)$$

Once the extent of reaction reaches approximately twice the gel point, nearly all monomers are attached to the gel and there is essentially no sol fraction remaining. Such well-developed networks are the subject of Chapter 7.

The numerical values of the critical exponents τ, σ, β, γ, ... and the functional form of the cutoff function $f(M/M^*)$ of gelation depend on the amount of overlap of the linking species. If the linking species have significant overlap, mean-field gelation theory provides a good description of the process with exponents $\tau = 5/2$ and $\sigma = 1/2$, and an exponential cutoff function $f(M/M^*) \sim \exp(-M/M^*)$ over a significant range of the relative extent of reaction. Sufficiently close to the gel point, *all* gelation transitions are believed to be described by the critical percolation model. The exponent values and the cutoff function in critical percolation models depend on the dimension of space in which the linking process takes place. In three-dimensional critical percolation the exponents are $\tau \cong 2.18$ and $\sigma \cong 0.45$.

The precise point where critical percolation starts to apply is determined by the overlap parameter of the linking species. Long linear chains in a melt are strongly overlapping before any crosslinking occurs. Vulcanization of these linear chains creates branched structures with fractal dimension $\mathcal{D} = 4$, which makes the branched structures overlap each other less as they grow. Once the largest branched polymers are just at their overlap concentration, critical percolation begins to apply. In critical percolation, molecules of a given molar mass are just at overlap between themselves and with similar molar mass pieces of larger molecules. This condition is called hyperscaling and leads to a fractal dimension of $\mathcal{D} \cong 2.53$ in three-dimensional percolation.

In the polymerization of a melt of multifunctional monomers A_f with functionality f greater than two, the critical percolation model applies over the entire range of possible extents of reaction. However, such densely branched polymers are often not very useful because they are typically brittle, since the chain sections between branch points are not long enough to entangle. One interesting use of such materials is as electrophotographic toners. Electrophotography is the process by which copiers work. The toner is usually a polymer material that contains the appropriate dye for colour copiers or carbon black for black-and-white copiers. The image resolution in electrophotography is determined in part by how small the toner particles can be made. Hence, the brittle nature of densely branched gels near their gel point make them ideal toner materials, as they are easily broken down to a very small particle size.

Randomly branched polymers are of enormous importance for certain polymer processing operations, such as blow moulding and film blowing. The molar mass distribution of all randomly branched polymers is described by percolation models. For commercial randomly branched polymers, the critical percolation model applies only very close to the gel point. The branching chemistry used in commercial randomly branched polymers is usually stopped far short of the critical region. While critical percolation does not apply to these polymers, the mean-field percolation model does a superb job of describing the molar mass distribution of randomly branched commercial polymers.

Gelation processes, such as crosslinking linear chains or condensation of f-functional monomers A_f (where A reacts with A) with $f > 2$ are quite different from either linear condensation polymers or hyperbranched polymers. Linear condensation polymers (made from AB monomers, where A only reacts with B) and hyperbranched polymers (made from AB_{f-1} monomers, where A only reacts with B) have *both* their number-average and weight-average molar masses diverge as their reaction nears completion and they never make network polymers. The hyperbranching reaction makes the same structure of randomly branched polymers as gelation, but with a very different distribution of molar masses.

Problems

Section 6.1

6.1 Ergodicity refers to a system's ability to access all allowed states. Which of the following gels can access all possible states and are hence ergodic:

(i) weak physical gels;
(ii) strong physical gels;
(iii) condensation polymerized gels;
(iv) addition polymerized gels.

6.2 Prove that for any two-dimensional lattice that is constructed as a slice of a three-dimensional lattice, the percolation threshold is larger in two dimensions than in three dimensions.

6.3 Identify the class and dimension of gelation model that describes the following cases:

(i) addition polymerization in aqueous solution;
(ii) linear condensation polymerization at the air–water interface;
(iii) branched condensation polymerization at the air–water interface;
(iv) branched condensation polymerization in solution.

6.4 Calculate the number-average degree of polymerization N_n and weight-average degree of polymerization N_w for one-dimensional percolation with extent of reaction $p = 0.99$. What is the polydispersity index of the molecules at this extent of reaction?

6.5 In the process of developing a communication network, a start-up company CheepieCom decided to save money on cables and connected each new customer to the nearest existing one by a single cable.

(i) How many connecting cables did they run between N customers? What is the structure of the CheepieCom network?
(ii) What happens to this network if a storm breaks one cable? How can this network be made more robust against catastrophic failure?

6.6 Consider propagation of a forest fire in the presence of a strong wind.

(i) How would you modify the percolation model to take into account strong blowing in one direction.
(ii) Is the percolation threshold (critical probability of ignition of a neighbouring tree) the same, higher, or lower than in the absence of the wind?

Section 6.2

6.7* Prove that the number fraction distribution for hyperbranched polymers is properly normalized [i.e., show that Eq. (6.9) satisfies Eq. (6.12)].

6.8 Derive the following result for the z-average degree of polymerization of hyperbranched polymers assuming no intramolecular reactions:

$$N_z = \frac{[1 - p^2(f-1)]^2 + 2p(f-1)(1-p)^2}{[1 - p(f-1)]^2[1 - p^2(f-1)]}. \tag{6.160}$$

Demonstrate that near complete reaction (for $p \to p_c$) the ratio N_z/N_w asymptotically approaches 3 for all functionalities f.

6.9 Calculate the radius of gyration for an ideal regular dendrimer of generation g with $n = f = 3$, where each generation has linear sections between branch points with one Kuhn monomer of length b. *Hint*: Use Kramers theorem.

6.10* Use Flory theory to determine the end-to-end distance of a linear strand that runs from the core to any end at generation g of the regular dendrimer in Problem 6.9 with excluded volume $v > 0$ (in good solvent).

Section 6.4

6.11 Calculate and plot sol and gel fractions for gelation of tetrafunctional monomers within mean-field theory, as functions of extent of reaction p. *Hint*: Note that one of the solutions of Eq. (6.51) is $P_{sol} = 1$.

6.12 Demonstrate that the mean-field gelation prediction of the polydispersity index below the gel point is

$$\frac{N_w}{N_n} = \frac{(1+p)(1-fp/2)}{1-(f-1)p}. \tag{6.161}$$

How does this polydispersity index depend on functionality f at a given relative extent of reaction $\varepsilon = (p - p_c)/p_c$?

6.13 Draw all different possible structures of a trifunctional randomly branched 4-mer without loops. Count and compare the number of unreacted groups in each 4-mer. Explain the significance of your results.

6.14 Evaluate the mean-field estimate of the gel point p_c for bond percolation on the following lattices:

    (i) Honeycomb ($f = 3$).
    (ii) Square ($f = 4$).
    (iii) Triangular ($f = 6$).
    (iv) Diamond ($f = 4$).
    (v) Simple cubic ($f = 6$).
    (vi) Compare your results with the values reported in Table 6.1. What is the origin of disagreement?
    (vii) Compare your results with the values reported in Table 6.1 for lattices with the same functionalities but with different dimensions. Which dimensions ($d = 2$ or $d = 3$) are closer to your prediction? Explain why?

6.15* Demonstrate that the weight fraction w_{N_B, N_L} of randomly branched polymers of N_B f-functional and N_L bifunctional monomers ($N = N_B + N_L$) is

$$w_{N_B,N_L} = \left[\frac{(1-p)^2}{p} - (f-2)(1-p)\phi_B\right] N \left[p\phi_B(1-p)^{f-2}\right]^{N_B} [p(1-p)]^{N_L}$$

$$\times \frac{[N_L + (f-1)N_B]!}{N_B![(f-2)N_L + 1]!},$$

where ϕ_B is the mole fraction of branched units.

The complexity of a molecule is defined as the number N_B of f-functional units in a molecule. The complexity distribution is therefore $\sum_{N_L} w_{N_B,N_L}$.

Show that the complexity distribution is similar to the simple f-functional case in the absence of bifunctional units (derived in Section 6.4).

6.16* Calculate the distribution function for $\phi_{A_1}, \phi_{A_2}, \ldots, \phi_{A_i}$ moles of reactants with functionalities $f_1, f_2, \ldots, f_i$ of A groups which are allowed to react with $\phi_{B_1}, \phi_{B_2}, \ldots, \phi_{B_j}$ moles of reactants with functionalities $g_1, g_2, \ldots, g_j$ of B groups where condensation occurs only between A and B groups.

6.17 Calculate the weight-average degree of polymerization N_w above the gel point in the mean-field theory for functionality $f=3$ using the recurrence relation approach.

6.18 Calculate the number-average degree of polymerization N_n and the number of finite molecules per site n_{tot} above the gel point in the mean-field theory for functionality $f=3$.

6.19 Compare the asymptotic distribution function for mean-field gelation of tetrafunctional monomers [Eq. (6.77) with $f=4$] with the exact result [Eq. (6.63)] near the gel point. How large does N need to be for the two distributions to agree within 1% for $p=0.330$?

6.20 (i) Derive the general relation between the fraction of monomers that are crosslinked p and the gel fraction P_{gel} for random crosslinking of long linear chains of N monomers:

$$1 - P_{gel} = (1 - pP_{gel})^N. \tag{6.162}$$

Hint: $p(1 - P_{gel})$ is the probability that a given monomer is crosslinked but not part of the gel.

(ii) Show that in the large N limit, this equation becomes the relation known in the literature as the gel curve:

$$1 - P_{gel} \cong \exp(-NpP_{gel}). \tag{6.163}$$

Section 6.5

6.21 Power law distribution of molar masses.
Consider a polymer system (with either linear or branched polymers) with a power law number density of molecules:

$$n(N) \approx N^{-\tau}.$$

(i) For what values of the Fisher exponent τ is the number-average degree of polymerization finite for all extents of reaction?

(ii) For what values of τ is the polydispersity index of these polymers finite for all extents of reaction?

(iii) For what values of τ is the z-average degree of polymerization finite for all extents of reaction?

6.22 Size of randomly branched polymers.
Consider a gelation process stopped at relative extent of reaction

$$\varepsilon = \frac{p - p_c}{p_c} = -0.01.$$

(i) What is the characteristic degree of polymerization N^* at this extent of reaction, assuming that starting molecules were monomers ($N_0 = 1$)?

(ii) What is the ideal size of the characteristic randomly branched molecule, if the monomer size is $b = 3$ Å?

(iii) What is the size of the characteristic molecule in the polymerization reactor?

(iv) What is the size of the characteristic molecule in dilute solution in an athermal solvent?

Hint: Use Flory theory to estimate the swelling of a randomly branched polymer in an athermal solvent.

6.23 Ginzburg criterion

(i) Estimate the size of the critical zone ε_G for vulcanization of precursor chains with degree of polymerization $N_0 = 1000$.

(ii) What is the characteristic degree of polymerization N_G of the branched polymers at the crossover extent of reaction ε_G?

(iii) What is the size ξ of the characteristic branched polymer at ε_G if the monomer size is $b = 3\,\text{Å}$?

(iv) What is the volume fraction of the characteristic branched polymer inside its pervaded volume at ε_G?

(v) What is the gel fraction at ε_G above gel point?

(vi) Compare the answers to (iv) and (v).

6.24 The number density of molecules in the mean-field theory for a general functionality is given by Eq. (6.63). Use Stirling's approximation to estimate this distribution near the gel point in the form

$$n(p, N) = N^{-\tau} f_{\pm}(N/N^*).$$

Show that $\tau = 5/2$, find N^* and evaluate the cutoff function $f_{\pm}(N/N^*)$ for a general functionality f.

6.25 What is the analogue of the scaling form of molar mass distribution

$$n(p, N) = N^{-\tau} f_{-}(N/N^*)$$

for one-dimensional percolation?

How do the number- and weight-average molar masses diverge as the extent of reaction approaches unity (complete reaction)?

6.26 Show that Eq. (6.99) is consistent with Eq. (6.53).

6.27 Consider random crosslinking of monodisperse primary molecules with degree of polymerization N_0. Let $\hat{p}$ be the number of crosslinks per primary molecule. Show that the weight fraction distribution function of molecules containing N/N_0 primary molecules is

$$w(\hat{p}, N) = \frac{(\hat{p}N/N_0)^{N/N_0 - 1}}{(N/N_0)!} \exp(-\hat{p}N/N_0). \qquad (6.164)$$

If the extent of reaction is defined as the number of bonds per monomer $p = \hat{p}/N_0$ show that the gel point corresponds to $p = 1/N_0$ or $(\hat{p} = 1)$.

6.28 Calculate the size of an ideal randomly branched polymer with precursor chains made of N_0 Kuhn monomers with Kuhn length $b = 5\,\text{Å}$ and total number $N = 10^4$ monomers. Estimate this size for

(i) $N_0 = 100$.

(ii) $N_0 = 10$.

(iii) Estimate the volume fraction of Kuhn monomers belonging to this molecule inside its pervaded volume for cases (i) and (ii). Explain your results.

6.29 State whether the following combinations of parameters are constants of order unity, or diverge, or vanish at the gel point for

(i) mean-field percolation,

(ii) critical percolation,

$$\frac{N_w}{N^*} \qquad \frac{N_w^2}{N^*} \qquad \frac{N_w}{P_{gel}N^*} \qquad N_w P_{gel}.$$

Which combinations provide information about universality class and thereby can determine the Ginzburg extent of reaction ε_G?
Which combinations are independent of universality class and hence useful for checking consistency of measurements?

6.30 Consider the total number of monomers per site in a sol (the first moment of the number density distribution function):

$$\int Nn(p, N)\, dN = \int N^{1-\tau}\hat{f}(z)\, dN$$

(i) Show that for negative values of z

$$\int N^{1-\tau}[\hat{f}(z) - \hat{f}(0)]\, dN = 0.$$

(ii) Use the relation between the scaling variable z and N to show that

$$\int |z|^{[(2-\tau)/\sigma]-1}[\hat{f}(z) - \hat{f}(0)]\, dz = 0.$$

(iii) Can $\hat{f}(z)$ be a monotonic function for negative values of z? What is the shape of the simplest cutoff function that satisfies this equation.

6.31 Plot the cutoff function $N^\tau n(p, N)$ as a function of N/N^* for mean-field gelation using the exact expression Eq. (6.63) for $f = 3$ and extents of reaction $p = 0.4, 0.45$, and 0.48. What happens with the maximum of the cut-off function in mean field as $p \to p_c$?
Hint: Plot the cutoff function against N/N_w.

6.32 Calculate the N_0-dependent prefactors for critical percolation predictions of gel fraction, correlation length, and characteristic degree of polymerization for crosslinking short chains with degree of polymerization N_0. *Hint*: These predictions are required to match the mean-field predictions at the Ginzburg point.

6.33 Use hyperscaling to derive the following relation for the correlation length exponent ν in terms of β, γ and the dimension of space d:

$$\nu = \frac{2\beta + \gamma}{d}. \tag{6.164}$$

6.34 Use hyperscaling and the Flory–de Gennes calculation of fractal dimension to derive the following approximate relation for the Fisher exponent τ in any space dimension and compare with the results in Table 6.4 for $d = 2, 3, 4, 5$ and 6:

$$\tau = \frac{3d + 2}{d + 2}. \tag{6.165}$$

6.35 Derive equations for the exponents τ and σ in terms of the exponents β and γ.

6.36 What is the largest possible volume fraction $b^3 g(r)$ of a randomly branched polymer inside its own pervaded volume in the vulcanization of precursor N_0-mers below the gel point?
Hint: Use Eq. (6.145).

6.37 What distinguishes a hyperbranched polymer at complete reaction from a gel formed by random branching?

6.38 Compare random crosslinking of linear precursor N_0-mers (with $f \approx N_0$) and end-linking of linear precursor N_0-mers with tetrafunctional crosslinkers.

(i) What is the qualitative difference between gel points of these two gelation processes?

(ii) How different are the resulting networks at the completion of all possible reactions?

(iii) Do they have Ginzburg zones of similar size $\varepsilon_G \approx N_0^{-1/3}$? Explain your answer.

6.39* How can symmetric amplitude ratios below and above the gel point in the mean-field zone outside the critical gelation regime $|\varepsilon| \gg \varepsilon_G$ match with asymmetric amplitude ratios in the critical region? Here, $\varepsilon = (p - p_c)/p_c$ is the relative extent of the reaction and ε_G is its value at the Ginzburg point (the boundary of the critical regime). Consider, for example, the weight-average number of monomers in the branched molecules:

$$N_w \approx A \left| \frac{\varepsilon}{\varepsilon_G} \right|^{-\gamma}.$$

The amplitude A is the same below and above the gel point in the mean-field regime $|\varepsilon| \gg |\varepsilon_G|$, while the values below and above gel point are quite different in the critical gelation regime. Propose a way of matching these differences at the Ginzburg point ε_G.

Section 6.6

6.40 Equations (6.148) and (6.149) provide a way of obtaining weight- and z-average molar masses from size exclusion chromatography. Are they approximate or exact methods? If the concentration c_i and weight-average molar mass M_{wi} of each elution volume are accurately measured, would the correct M_w and M_z of the whole sample be known?

Bibliography

Burchard, W. Solution properties of branched macromolecules, *Adv. Polym. Sci.* **143**, 113 (1999).

Flory, P. J. *Principles of Polymer Chemistry* (Cornell University Press, Ithaca, New York, 1953).

de Gennes, P. G. *Scaling Concepts in Polymer Physics* (Cornell University, Ithaca, New York, 1979).

Hawker, C. J. Dendritic and hyperbranched macromolecules: precisely controlled macromolecular architectures, *Adv. Polym. Sci.* **147**, 113 (1999).

Stauffer, D. and Aharony, A. *Introduction to Percolation Theory*, 2nd edn (Taylor and Francis, 1992).

Stauffer, D., Coniglio, A., and Adam, M. Gelation and critical phenomena, *Adv. Polym. Sci.* **44**, 103 (1982).

Networks and gels

7

When the crosslinking reactions of Chapter 6 are driven far beyond the gel point, nearly all species are attached to the gel in a single macroscopic network polymer. Such networks, with either chemical or strong physical bonds, are important soft solids. If the glass transition and melting temperatures are below room temperature, the material is a **rubber**. Rubbers are an important class of materials with many practical uses. Common examples are rubber bands, gaskets, adhesives, and automobile tires. Rubber bands can be stretched enormous amounts without breaking or even losing their elasticity. In this chapter, the physics that allow for such marvellous properties will be considered in detail. The entropic nature of elasticity in rubbers is the origin of their remarkable mechanical properties.

7.1 Thermodynamics of rubbers

The first law of thermodynamics states that the change in internal energy of a system, such as a polymeric network, is the sum of all the energy changes: heat added to the system $T\,dS$, work done to change the network volume $-p\,dV$ and work done upon network deformation $f\,dL$:

$$dU = T\,dS - p\,dV + f\,dL. \tag{7.1}$$

The differential dU represents the change in internal energy that arises if there is an entropy change dS, a volume change dV, or sample length change dL. The internal energy U is a thermodynamic state function of variables S, V, and L. The Helmholtz free energy F is defined as internal energy minus the product of temperature and entropy:

$$F = U - TS. \tag{7.2}$$

The change in the Helmholtz free energy is written in differential form:

$$\begin{aligned} dF = dU - d(TS) &= dU - T\,dS - S\,dT \\ &= -S\,dT - p\,dV + f\,dL. \end{aligned} \tag{7.3}$$

The Helmholtz free energy is a thermodynamic state function of variables T, V, and L. The change in the Helmholtz free energy can be written as a complete differential:

$$dF = \left(\frac{\partial F}{\partial T}\right)_{V,L} dT + \left(\frac{\partial F}{\partial V}\right)_{T,L} dV + \left(\frac{\partial F}{\partial L}\right)_{T,V} dL. \qquad (7.4)$$

Comparing Eqs (7.3) and (7.4), we identify the partial derivatives of the Helmholtz free energy:

$$\left(\frac{\partial F}{\partial T}\right)_{V,L} = -S, \qquad (7.5)$$

$$\left(\frac{\partial F}{\partial V}\right)_{T,L} = -p, \qquad (7.6)$$

$$\left(\frac{\partial F}{\partial L}\right)_{T,V} = f. \qquad (7.7)$$

A second derivative of the Helmholtz free energy does not depend on the order of differentiation:

$$\frac{\partial^2 F}{\partial T \partial L} = \frac{\partial^2 F}{\partial L \partial T}. \qquad (7.8)$$

Using Eqs (7.5) and (7.7), Eq. (7.8) can be rewritten as one of the **Maxwell relations**:

$$-\left(\frac{\partial S}{\partial L}\right)_{T,V} = \left(\frac{\partial f}{\partial T}\right)_{V,L}. \qquad (7.9)$$

The force f, applied to deform a network, consists of two contributions:

$$f = \left(\frac{\partial F}{\partial L}\right)_{T,V} = \left[\frac{\partial(U - TS)}{\partial L}\right]_{T,V} = \left(\frac{\partial U}{\partial L}\right)_{T,V} - T\left(\frac{\partial S}{\partial L}\right)_{T,V}. \qquad (7.10)$$

The first term describes how the internal energy changes with the sample length and the second contribution is the product of absolute temperature and the rate of change of entropy with sample length. The second term can be rewritten using the Maxwell relation above [Eq. (7.9)]:

$$f = \left(\frac{\partial U}{\partial L}\right)_{T,V} + T\left(\frac{\partial f}{\partial T}\right)_{V,L} = f_E + f_S. \qquad (7.11)$$

The two contributions to the force are an energetic term that is the change of internal energy with sample length

$$f_E = \left(\frac{\partial U}{\partial L}\right)_{T,V}, \qquad (7.12)$$

and an entropic term that is the product of temperature and the change of entropy with sample length:

$$f_S = T\left(\frac{\partial f}{\partial T}\right)_{V,L} = -T\left(\frac{\partial S}{\partial L}\right)_{T,V}. \tag{7.13}$$

In typical crystalline solids, such as metals, the energetic contribution dominates the force because the internal energy increases when the crystalline lattice spacings are distorted from their equilibrium positions. In rubbers, the entropic contribution to the force is more important than the energetic one. In 'ideal networks' there is no energetic contribution to elasticity, so $f_E = 0$.

The dominance of the entropic part of Eq. (7.11) bestows a peculiar temperature dependence to the force at constant extension. While crystalline solids have the force decrease weakly with increasing temperature, rubbers show the opposite behaviour. The network strands lose conformational entropy when stretched (see Section 2.6) making $\partial S/\partial L < 0$ and the force *increases* with increasing temperature [Eq. (7.13)].

7.1.1   Flory construction

A simple way to separate energetic from entropic contributions to the elastic force was developed by Flory. Consider a typical temperature dependence of a retraction force f for a network of constant volume V at constant elongation L, as shown in Fig. 7.1. The slope of the curve at temperature T is

$$\text{slope} = \left(\frac{\partial f}{\partial T}\right)_{V,L}, \tag{7.14}$$

and the change in the ordinate from the point on the curve to the intercept of the tangent with the f axis is the entropic contribution to the force [Eq. (7.13)]. Therefore, the value at the intercept of the tangent to the curve with the f axis is the energetic contribution to the force [Eq. (7.12)]. Figure 7.1 is schematic, but shows the typical relative importance of the energetic f_E and entropic f_S parts of the force in a stretched polymer network. Note that the entropic component f_S accounts for more than 90% of total force in the rubbery state. **Rubber elasticity** has primarily *entropic* origins. In the rest of this chapter, we ignore the energetic contribution and concentrate exclusively on the entropic one.

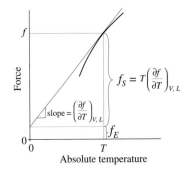

Fig. 7.1
A schematic representation of the Flory construction for a polymer network,

7.2   Unentangled rubber elasticity

7.2.1   Affine network model

Polymer networks are unique in their ability to reversibly deform to several times their size. The enormous deformability of networks arises from the entropic elasticity of the polymer chains that make up the network (the network strands). The simplest model that captures this idea of rubber elasticity is the **affine network model** originally proposed by Kuhn. The main assumption of the affine network model is an **affine deformation**: the relative deformation of each network strand is the same as the macroscopic relative deformation imposed on the whole network.

Consider a rubber network with undeformed dimensions L_{x0}, L_{y0}, and L_{z0} (Fig. 7.2). If the network experiences relative deformations in the x, y, and z directions by the factors λ_x, λ_y, and λ_z, then the dimensions of the deformed network are

$$L_x = \lambda_x L_{x0}, \quad L_y = \lambda_y L_{y0} \quad \text{and} \quad L_z = \lambda_z L_{z0}. \tag{7.15}$$

Assume that each network strand has N monomers. One network strand, shown in Fig. 7.2, has end-to-end vector $\vec{R}_0$ with projections along the x, y, and z directions of R_{x0}, R_{y0}, and R_{z0} in the undeformed state. In the affine network model, the positions of the junction points (the ends of the strands) are always fixed at particular points in space by the deformation and not allowed to fluctuate. For affine deformation, the end-to-end vector of the same chain in the deformed state is $\vec{R}$ (see Fig. 7.2) with projections along the x, y, and z directions of

$$R_x = \lambda_x R_{x0}, \quad R_y = \lambda_y R_{y0} \quad \text{and} \quad R_z = \lambda_z R_{z0}. \tag{7.16}$$

Recall the entropy of a chain of N Kuhn monomers of length b with end-to-end vector $\vec{R}$ [Eq. (2.92)]:

$$S(N, \vec{R}) = -\frac{3}{2} k \frac{\vec{R}^2}{Nb^2} + S(N, 0) = -\frac{3}{2} k \frac{R_x^2 + R_y^2 + R_z^2}{Nb^2} + S(N, 0). \tag{7.17}$$

The entropy change of this chain upon deformation is the difference in entropy of the final and initial states:

$$\begin{aligned} S(N, \vec{R}) - S(N, \vec{R}_0) &= -\frac{3}{2} k \frac{R_x^2 + R_y^2 + R_z^2}{Nb^2} + \frac{3}{2} k \frac{R_{x0}^2 + R_{y0}^2 + R_{z0}^2}{Nb^2} \\ &= -\frac{3}{2} k \frac{(\lambda_x^2 - 1) R_{x0}^2 + (\lambda_y^2 - 1) R_{y0}^2 + (\lambda_z^2 - 1) R_{z0}^2}{Nb^2}. \end{aligned} \tag{7.18}$$

The entropy change of the whole network is the sum of all the entropy changes of the network's n strands:

$$\Delta S_{\text{net}} = -\frac{3}{2}\frac{k}{Nb^2}\left[(\lambda_x^2 - 1)\sum_{i=1}^{n}(R_{x0})_i^2 + (\lambda_y^2 - 1)\sum_{i=1}^{n}(R_{y0})_i^2 \right.$$

$$\left. +(\lambda_z^2 - 1)\sum_{i=1}^{n}(R_{z0})_i^2\right]. \qquad (7.19)$$

If the network is formed by crosslinking chains in their ideal state (in a melt) the components of the mean-square end-to-end distance in the undeformed state are given by Eq. (2.83):

$$\langle R_{x0}^2 \rangle = \frac{1}{n}\sum_{i=1}^{n}(R_{x0})_i^2 = \frac{Nb^2}{3} = \langle R_{y0}^2 \rangle = \langle R_{z0}^2 \rangle. \qquad (7.20)$$

Therefore, the sums of the squares of the components of the end-to-end vectors of all n strands can be written quite simply:

$$\sum_{i=1}^{n}(R_{x0})_i^2 = \sum_{i=1}^{n}(R_{y0})_i^2 = \sum_{i=1}^{n}(R_{z0})_i^2 = \frac{n}{3}Nb^2. \qquad (7.21)$$

The entropy change upon deformation of the network is the central feature of rubber elasticity:

$$\Delta S_{\text{net}} = -\frac{3}{2}k\frac{(\lambda_x^2 - 1)(n/3)Nb^2 + (\lambda_y^2 - 1)(n/3)Nb^2 + (\lambda_z^2 - 1)(n/3)Nb^2}{Nb^2}$$

$$= -\frac{nk}{2}(\lambda_x^2 + \lambda_y^2 + \lambda_z^2 - 3). \qquad (7.22)$$

The main contribution to the free energy of the network comes from the changes in entropy, as discussed in Section 7.1.1. Ignoring any enthalpic contribution, the free energy required to deform a network is minus temperature times the entropy change:

$$\Delta F_{\text{net}} = -T\Delta S_{\text{net}} = \frac{nkT}{2}(\lambda_x^2 + \lambda_y^2 + \lambda_z^2 - 3). \qquad (7.23)$$

Dry networks are typically incompressible, which means that their volume does not change appreciably when they are deformed:

$$V = L_{x0}L_{y0}L_{z0} = L_xL_yL_z = \lambda_x L_{x0}\lambda_y L_{y0}\lambda_z L_{z0} = \lambda_x\lambda_y\lambda_z V. \qquad (7.24)$$

If the volume of the network remains constant, the product of the deformation factors is unity:

$$\lambda_x\lambda_y\lambda_z = 1. \qquad (7.25)$$

In practice, the volume change that occurs when a network is deformed is measurable, but extremely small (see Problem 7.7).

7.2.1.1   Uniaxial deformation

If the network is either stretched or compressed in a single direction (along the x axis), the deformation is termed uniaxial. For uniaxial deformations at constant volume, the other two dimensions of the network adjust to keep the volume constant:

$$\lambda_x = \lambda \quad \lambda_y = \lambda_z = \frac{1}{\sqrt{\lambda}}. \tag{7.26}$$

The free energy change for a uniaxial deformation at constant volume is obtained by simply substituting the deformation factors [Eq. (7.26)] into Eq. (7.23):

$$\Delta F_{\text{net}} = \frac{nkT}{2}\left(\lambda^2 + \frac{2}{\lambda} - 3\right) \quad \text{for uniaxial deformation.} \tag{7.27}$$

The force required to deform a network is the rate of change of its free energy with respect to its size along the axis of deformation. For example, the x component of the force is the derivative of the free energy with respect to length along the x axis [see Eq. (7.7)]:

$$f_x = \frac{\partial \Delta F_{\text{net}}}{\partial L_x} = \frac{\partial \Delta F_{\text{net}}}{\partial (\lambda L_{x0})} = \frac{1}{L_{x0}}\frac{\partial \Delta F_{\text{net}}}{\partial \lambda}$$

$$= \frac{nkT}{L_{x0}}\left(\lambda - \frac{1}{\lambda^2}\right) \quad \text{for uniaxial deformation.} \tag{7.28}$$

If the cross-sectional area $L_y L_z$ of the macroscopic network is doubled, then twice as large a force is required to obtain the same deformation. This leads naturally to a definition of **stress** as the ratio of force and cross-sectional area. Both the force and the cross-sectional area have direction and magnitude (the direction of the cross-sectional area being described by the unit vector normal to its surface), making the stress a tensor. The ij-component of the **stress tensor** is the force applied in the i direction per unit cross-sectional area of a network perpendicular to the j axis. For example, the xx-component σ_{xx} is the force applied in the x direction f_x divided by the area $L_y L_z$ perpendicular to the x axis:

$$\sigma_{xx} = \frac{f_x}{L_y L_z} = \frac{nkT}{L_{x0}L_y L_z}\left(\lambda - \frac{1}{\lambda^2}\right) = \frac{nkT}{L_{x0}L_{y0}L_{z0}}\lambda\left(\lambda - \frac{1}{\lambda^2}\right)$$

$$= \frac{nkT}{V}\left(\lambda^2 - \frac{1}{\lambda}\right) \equiv \sigma_{\text{true}} \quad \text{for uniaxial deformation in } x. \tag{7.29}$$

This is the **true stress** in the network and it is therefore denoted by σ_{true}. Since it is often not easy to measure the cross-sectional area of the deformed network, an engineering stress is also defined. In the **engineering stress** the original cross-sectional area $L_{y0}L_{z0}$ is used instead of the deformed cross-sectional area $L_y L_z$:

$$\sigma_{\text{eng}} = \frac{f_x}{L_{y0} L_{z0}} = \frac{nkT}{L_{x0} L_{y0} L_{z0}} \left(\lambda - \frac{1}{\lambda^2} \right) = \frac{nkT}{V} \left(\lambda - \frac{1}{\lambda^2} \right) = \frac{\sigma_{\text{true}}}{\lambda}. \quad (7.30)$$

The coefficient relating the stress and the deformation is the shear modulus G, as will be shown in Problem 7.43:

$$G = \frac{nkT}{V} = \nu kT = \frac{\rho \mathcal{R} T}{M_s}. \quad (7.31)$$

The number of network strands per unit volume (number density of strands) is $\nu = n/V$. In the last equality, ρ is the network density (mass per unit volume), M_s is the number-average molar mass of a network strand, and $\mathcal{R}$ is the gas constant. The network modulus increases with temperature because its origin is entropic, analogous to the pressure of an ideal gas $p = nkT/V$. The modulus also increases linearly with the number density of network strands $\nu = n/V = \rho N_{\text{Av}}/M_s$. Equation (7.31) states that the modulus of any network polymer is kT per strand.

The affine predictions for both true and engineering stresses in uniaxial deformation at constant network volume can be rewritten using the shear modulus:

$$\sigma_{\text{true}} = G \left(\lambda^2 - \frac{1}{\lambda} \right), \quad (7.32)$$

$$\sigma_{\text{eng}} = G \left(\lambda - \frac{1}{\lambda^2} \right). \quad (7.33)$$

By writing these equations in terms of the shear modulus, the form of the stress–elongation relation becomes quite general. Many other network elasticity models also predict stress–elongation relations of this form, with different predictions for the shear modulus. For this reason, we refer to Eqs (7.32) and (7.33) as the *classical* stress–elongation forms. As demonstrated in Fig. 7.3, this classical form describes the small deformation uniaxial data on polymer networks quite well. The main physics behind such classical models is the *entropic elasticity of polymer chains*.

7.2.2 Phantom network model

The main assumption of the affine network model is that the ends of network strands (the crosslink junctions) are fixed in space and are displaced affinely with the whole network, as if they were permanently attached to some elastic background [see Fig. 7.4(a)]. In real networks, the ends of network strands are attached to other strands at crosslinks [see Fig. 7.4(b)]. These crosslinks are not fixed in space—they can fluctuate around their average positions. These fluctuations lead to a net lowering of the free energy of the system by reducing the cumulative stretching of the network

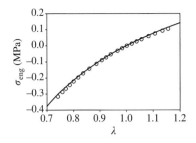

Fig. 7.3
Engineering stress in uniaxial compression ($\lambda < 1$ and $\sigma_{\text{eng}} < 0$) and tension ($\lambda > 1$ and $\sigma_{\text{eng}} > 0$) for a poly(dimethyl siloxane) network prepared by end-linking chains with $M_n = 18\,400$ g mol^{-1}. The curve is the classical prediction of Eq. (7.33), with $G = 0.28$ MPa. Data are from W. Oppermann and N. Rennar, *Prog. Colloid Polym. Sci.* **75**, 49 (1987).

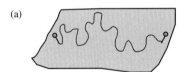

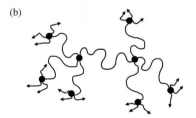

Fig. 7.4
(a) In the affine network model, the ends of each network strand are pinned to an elastic background. (b) In the phantom network model, the ends of network strands are joined at crosslink junctions that can fluctuate. Circles are crosslink junctions and arrows denote attachments to the rest of the macroscopic network.

strands. The simplest model that incorporates these fluctuations is called the **phantom network model**. In a phantom network, the strands are ideal chains with ends joined at crosslinks. The ends of the strands at the surface of the network are attached to the elastic non-fluctuating boundary of the network. This attachment fixes the volume of the phantom network and prevents its collapse that would have been inevitable because such simple models ignore excluded volume interactions between monomers.

Recall from Problem 2.38 that the fluctuations of a single monomer in an ideal chain with fixed ends are identical to the fluctuations of an end monomer of a single **effective chain** of K monomers. For the particular case of the center monomer of an ideal chain with $2N$ monomers, the effective chain has $K = N/2$ monomers. Hence, the constraining effect of the two strands of N monomers is identical to the constraining effect of a single effective chain of $K = N/2$ monomers. More generally, if there are f chains of N monomers connected to a given monomer (such as in the case of the branch point of an f-arm star polymer) the fluctuations of this branch point are the same as the fluctuations of an effective chain of $K = N/f$ monomers.

The fluctuations of junction points in a network are quite similar to those of the branch point of an f-arm star polymer. In order to calculate the amplitude of these fluctuations, start with $f - 1$ strands that are attached at one end to the surface of the network and joined at the other end by a junction point connecting them to a single strand [see the left-most part of Fig. 7.5(a)]. The strands attached to the elastic non-fluctuating network surface are called seniority-zero strands. Each of these $f - 1$ seniority-zero strands are attached to a single seniority-one strand by a f-functional crosslink [see the left-most part of Fig. 7.5(a)]. The seniority of a particular strand is defined by the number of other network strands along the shortest path between it and the network surface. The $f - 1$ seniority-zero strands

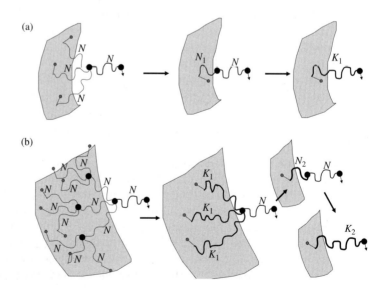

(a)

(b)

Fig. 7.5
Recurrence relation diagrams for the effective chains of phantom networks, shown here for tetrafunctional networks ($f = 4$). In each sketch the leftmost ends of the effective chains are pinned to the macroscopic boundary of the network.

are connected in parallel and can be replaced by a single effective chain containing N_1 monomers with the same constraining effect as the $f-1$ original chains together [see Fig. 7.5(a)]:

$$N_1 = \frac{N}{f-1}. \tag{7.34}$$

This single effective chain containing N_1 monomers is connected in series with a single seniority-one N-mer and together with it can be described by an effective chain of $K_1 = N + N_1$ monomers. In this way, Fig. 7.5(a) sketches how $f-1$ zero-seniority strands together with one seniority-one network strand can be replaced by an effective strand with K_1 monomers:

$$K_1 = N + N_1 = N + \frac{N}{f-1} = N\left(1 + \frac{1}{f-1}\right). \tag{7.35}$$

As can be seen from the left-most part of Fig. 7.5(b), $f-1$ of the seniority-one network strands are connected at a single crosslink junction to one seniority-two strand. Each of these seniority-one strands together with the corresponding $f-1$ seniority-zero strands can be replaced by one effective chain with K_1 monomers [central part of Fig. 7.5(b)]. A parallel combination of $f-1$ of these effective chains can be replaced by one effective chain with N_2 monomers:

$$N_2 = \frac{K_1}{f-1} = \frac{N}{f-1}\left(1 + \frac{1}{f-1}\right). \tag{7.36}$$

Combining this effective chain with the real seniority-two chain connected to it in series, gives an effective chain representing the combined effect of a tree of strands from seniority-zero through seniority-two [Fig. 7.5(b)]:

$$K_2 = N + N_2 = N\left(1 + \frac{1}{f-1} + \frac{1}{(f-1)^2}\right). \tag{7.37}$$

Continuing this procedure gives a geometric series for the number of monomers in an effective chain representing a combined effect of a tree of strands from seniority-zero through an arbitrarily large seniority. This series rapidly converges and each junction point in the bulk of a phantom network can be thought of as connected to the elastic non-fluctuating surface of the network through f effective chains with K monomers in each.

$$K = N\left(1 + \frac{1}{f-1} + \frac{1}{(f-1)^2} + \frac{1}{(f-1)^3} + \cdots\right)$$
$$= \frac{N}{1 - 1/(f-1)} = \frac{f-1}{f-2}N. \tag{7.38}$$

This means that each of the original $f-1$ chains, connecting any network strand to the macroscopic network through a very long tree-like

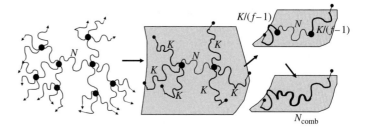

Fig. 7.6
The combined chain of a phantom
network.

sequence of junction points, can be replaced by an effective chain with
K monomers. That effective chain connects one end of each network strand
to the elastic background. There are $f-1$ of these effective chains at each
end of every network strand and they can be replaced by another effective
chain with

$$\frac{K}{f-1} = \frac{N}{f-2} \tag{7.39}$$

monomers (see Fig. 7.6). Every network strand has one of these effective
chains at each of its ends. Effective chains represent the way elasticity is
transmitted from macroscopic scales down to individual chains. The network strand with N monomers together with the two effective chains with
$N/(f-2)$ monomers each, can be considered as one **combined chain** with

$$N_{\text{comb}} = N + 2\frac{N}{f-2} = \frac{f}{f-2}N \tag{7.40}$$

monomers (Fig. 7.6). The ends of this combined chain are not fluctuating
and can be assumed to be attached to an elastic non-fluctuating background just as the ends of the network strand in the affine network model.
Thus, a phantom network model with fluctuating junction points is
equivalent to an affine network model with a combined chain containing
N_{comb} monomers.

The shear modulus of the phantom network is obtained from the
modulus of the affine network [Eq. (7.31)] by replacing N with $Nf/(f-2)$:

$$G = \nu kT\frac{f-2}{f} = \frac{\rho \mathcal{R}T}{M_{\text{s}}}\left(1 - \frac{2}{f}\right). \tag{7.41}$$

For any functionality f, the phantom network modulus is lower than the
affine network modulus [Eq. (7.31)] because allowing the crosslinks to
fluctuate in space makes the network softer. The phantom network has the
same number density of strands as the affine network but only the fraction
$(f-2)/f$ of the combined chain is the real strand [Eq. (7.40)] and only this
fraction supports stress. The phantom network modulus approaches the
affine prediction in the limit of high functionality of crosslinks. Crosslinks
in phantom networks with high functionality f do not fluctuate much and
are almost fixed in space as in the affine network model. Networks typically

have functionalities of 3 or 4. For $f = 3$, the phantom prediction is one third of the affine network modulus and for $f = 4$, the phantom modulus is half of the affine prediction.

These predictions need to be modified because real networks have defects. As shown in Fig. 7.7, some of the network strands are only attached to the network at one end. These dangling ends cannot bear stress and hence do not contribute to the modulus. Similarly, other structures in the network (such as dangling loops) are also not elastically effective. The phantom network prediction can be recast in terms of the number density of elastically effective strands ν and the number density of elastically effective crosslinks μ. For a perfect network without defects, the phantom network modulus is proportional to the difference of the number densities of network strands ν and crosslinks $\mu = 2\nu/f$, since there are $f/2$ network strands per crosslink:

$$G = kT(\nu - \mu). \tag{7.42}$$

This equation applies to networks with defects as well, and hence is more general than Eq. (7.41), but care must be taken to only include elastically effective strands and crosslinks. Elastically effective strands are the ones that deform and store elastic energy upon network deformation. Elastically effective crosslinks are those that connect at least two elastically effective strands.

Experimental estimates of ν and μ usually must rely on a model for the crosslinking chemistry, making quantitative tests of the phantom model difficult. Network defects preclude the use of Eqs (7.31) and (7.41), written in terms of the molar mass of a network strand M_s. Indeed, since M_s is not known for real networks and the affine and phantom models predict the same classical form of the stress–elongation curve [Eqs (7.32) and (7.33)] there is no practical means of determining which (if either) model is correct for small deformations of unentangled networks. For these reasons, we henceforth describe the modulus of all classical models G_x as a network of strands with apparent molar mass M_x:

$$G_x = \frac{\rho \mathcal{R} T}{M_x}. \tag{7.43}$$

For the affine network model, M_x is the actual strand molar mass ($M_x = M_s$) whereas the phantom network model requires a longer combined strand length $M_x = fM_s/(f - 2)$ [Eq. (7.40)].

7.2.3 Finite extensibility

Both the affine and phantom network models predict the same (classical) dependence of stress on deformation [Eqs (7.32) and (7.33)]. Detailed quantitative comparison of the classical form with experiments indicates two major disagreements (see Fig. 7.8). Experiments demonstrate softening at intermediate deformations and hardening at higher deformations. In

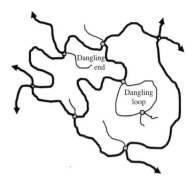

Fig. 7.7
Defects in a randomly crosslinked network are dangling ends and loops, denoted by thin lines. Circles are crosslink junctions and arrows denote attachments to the macroscopic network.

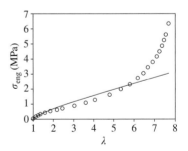

Fig. 7.8
Engineering stress in tension for a crosslinked rubber (data from L. R. G. Treloar, *The Physics of Rubber Elasticity*, 3rd edition, Clarendon Press, Oxford, 1975). The solid curve is the classical form [Eq. (7.33)] fit to the small deformation data.

order to understand the origins of these disagreements, the assumptions made in the classical theories must be re-examined. Here the strain hardening is discussed and in Section 7.3.3 the softening at intermediate deformations will be treated.

Strain hardening at high deformations λ can be explained by the non-Gaussian statistics of strongly deformed chains. Recall that the Gaussian approximation for a freely jointed chain model is valid for end-to-end distances much shorter than that for a fully stretched state $R \ll R_{\mathrm{max}} = bN$. In Section 2.6.2, the Langevin functional dependence of normalized end-to-end distance R/Nb on the normalized force $fb/(kT)$ for a freely jointed chain [Eq. (2.112)] was derived:

$$\frac{R}{Nb} = \mathcal{L}\left(\frac{fb}{kT}\right) = \coth\left(\frac{fb}{kT}\right) - \frac{kT}{fb}. \tag{7.44}$$

The force f required to stretch a single chain to an end-to-end distance R can be expressed through the inverse Langevin function:

$$f = \frac{kT}{b}\mathcal{L}^{-1}\left(\frac{R}{Nb}\right). \tag{7.45}$$

This dependence of the force f on chain elongation R deviates from the Gaussian approximation (Hooke's law) at large elongations, as shown in Fig. 2.15. Owing to the finite extensibility of chains, the force diverges at the maximum end-to-end distance $R_{\mathrm{max}} = bN$ [Eq. (2.116)]. A different relation with an even stronger divergence [Eq. (2.117)] has been used for the worm-like chain model.

Finite chain extensibility is the major reason for strain hardening at high elongations (Fig. 7.8). Another source of hardening in some networks is stress-induced crystallization. For example, vulcanized natural rubber (*cis*-polyisoprene) does not crystallize in the unstretched state at room temperature, but crystallizes rapidly when stretched by a factor of 3 or more. The extent of crystallization increases as the network is stretched more. The amorphous state is fully recovered when the stress is removed. Since the crystals invariably have larger modulus than the surrounding amorphous network, the effective modulus increases with crystallization, which makes the stress increase more rapidly with elongation.

7.3   Entangled rubber elasticity

7.3.1   Chain entanglements and the Edwards tube model

In the 1940s, it was recognized that the classical predictions of network modulus were bounded. A real network could certainly not be expected to have lower modulus than the phantom prediction, since it is based on unrestricted fluctuations of ideal strands that are allowed to pass through each other. At the other extreme, the classical models have no means to attain a higher modulus than the affine prediction, based on junctions that are not allowed to fluctuate at all. However, the modulus of many real networks is considerably larger than the predictions of either classical model!

In both the affine and phantom network models, chains are only aware that they are strands of a network because their ends are constrained by crosslinks. Strand ends are either fixed in space, as in the affine network model, or allowed to fluctuate by a certain amplitude around some fixed position in space, as in the phantom network model. Monomers other than chain ends do not 'feel' any constraining potential in these simple network models.

In real networks made of long linear polymers, network chains impose topological constraints on each other because they cannot cross (see Fig. 7.9). The importance of these topological constraints, called **entanglements**, in polymer networks was discussed by Treloar as early as 1940. Since then many models of *entanglement effects* in polymers have been proposed. Indeed, the focus of Chapter 9 is the consequences of entanglement in polymer liquids. However, a clearer picture of what an entanglement really is remains elusive and the sketch in Fig. 7.9 is a very crude representation of an entanglement[1] between two chains.

Despite this rather vague notion of individual entanglements, Edwards showed that the essence of entanglements can be treated using a tube model. The collective effect of all surrounding chains on a given strand is represented in the Edwards tube model by a quadratic constraining potential acting on every monomer of each network strand. The minima of these constraining potentials lie along the dashed line of Fig. 7.10, called the **primitive path**. Every network strand is effectively confined by constraining potentials to a tube-like region with the primitive path at its centre (see Fig. 7.10).

Each monomer is constrained to stay fairly close to the primitive path, but fluctuations driven by the thermal energy kT are allowed. Strand excursions in the quadratic potential are not likely to have free energies much more than kT above the minimum. Strand excursions that have free energy kT above the minimum at the primitive path define the width of the **confining tube**, called the **tube diameter** a (Fig. 7.10). In the classical affine and phantom network models, the amplitude of the fluctuations of a typical network monomer, that is not adjacent to the crosslinks, is of the order of the unperturbed strand size. In entangled polymer networks, the topological interactions of neighbouring chains restrict the transverse fluctuations of a network strand to the confining tube of diameter a.

This tube diameter can be interpreted as the end-to-end distance of an **entanglement strand** of N_e monomers:

$$a \approx bN_e^{1/2}. \qquad (7.46)$$

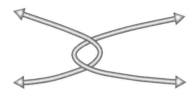

Fig. 7.9
The fact that two chains cannot pass through one another creates topological interactions known as entanglements that raise the network modulus.

---

[1] Entanglement appears to be caused by a collective topological restriction of many neighbouring chains (see Section 9.1).

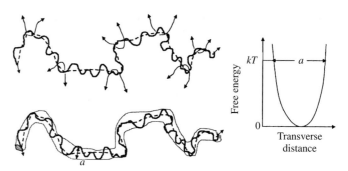

Fig. 7.10

A chain or network strand (thick curve) is topologically constrained to a tube-like region by surrounding chains. The primitive path is shown as the dashed curve. The roughly quadratic potential defining the tube is also sketched.

The entanglement strand has **entanglement molar mass** $M_e = N_e M_0$. The entanglement strand effectively replaces the network strand in the determination of the modulus for networks made from long strands, and also determines the rubbery plateau modulus of high molar mass polymer melts:

$$G_e = \frac{\rho \mathcal{R} T}{M_e}. \tag{7.47}$$

The importance of entanglements in network elasticity is proven beyond any doubt by three experimental observations.

(1) Computer simulations of polymer networks demonstrate that allowing chains to pass through each other (forming a true phantom network with no entanglements possible) lowers the shear modulus greatly, as demonstrated in Fig. 7.11. Computer simulations have the enormous advantage of knowing the number of monomers in a strand N *a priori*. The modulus of three such networks was measured at constant monomer density and plotted against $1/N$ (filled circles in Fig. 7.11). Since $1/N$ is proportional to the number density of strands, both the affine and phantom models predict a straight line going through the origin. However, in contrast to the prediction of either classical model, the straight line has a non-zero intercept! Another advantage of computer simulations, is that the rules of the simulation can be changed to be quite unrealistic. By allowing the chains to artificially pass through each other in the simulation, the open squares in Fig. 7.11 are obtained for the modulus. The straight line describing these phantom networks has an intercept of zero within numerical uncertainties, and agrees with the predictions of the phantom network model. Figure 7.11 proves that *topological interactions between network strands raise the network modulus*. The simplest idea that accounts for interactions of these strands is the notion of entanglements embodied in the Edwards tube model.

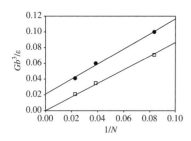

Fig. 7.11

Computer simulations of network modulus for networks with three different strand lengths (filled circles with number of monomers per strand $N = 12$, 26, and 44). The open squares are the same networks but the modulus is measured when the strands are allowed to pass through each other as in a true phantom network. Data from R. Everaers, *New J. Phys.* **1**, 12.1–12.54 (1999), see http://www.njp.org. ε is the energy scale in the Lennard–Jones potential [Eq. (3.96)].

(2) Instantaneously deformed high molar mass polymer melts (long polymer chains in their liquid state) behave at intermediate times as networks with well-defined values of shear modulus, called the plateau modulus G_e, which is independent of molar mass for long-chain polymers. This rubbery plateau is seen for all polymer melts with

molar mass significantly above the molar mass of an entanglement strand M_e, as will be discussed in detail in Chapter 9. Experiments on polymer melts suggest that chains form some sort of temporary entanglement network due to the topological constraints they impose on each other. Only on very long time scales do they 'find out' that they are not permanently crosslinked and the melt begins to flow as a liquid. When such long chains are crosslinked to form a network, there is *no reason to expect these entanglements to disappear*!

(3) The modulus of well-developed networks (having sol fraction near zero) with molar mass of network strands considerably larger than the molar mass of an entanglement strand ($M_s > M_e$) is always significantly larger than either of the classical predictions. Figure 7.12 shows experimental data for networks prepared by end-linking linear poly (dimethyl siloxane) (PDMS) telechelic chains (meaning that there is a reactive group at each end of the chain, but none along the chains). Both of the classical models expect a zero intercept, and hence would allow for fully reacted networks of arbitrarily low modulus, provided that the molar mass of the starting chains is sufficiently large. However, the data suggest instead a non-zero intercept, that is consistent with the plateau modulus of high molar mass linear PDMS melts ($G_e/\mathcal{R}T \cong 80\,\text{mol m}^{-3}$).

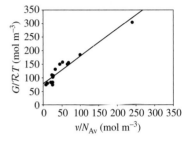

Fig. 7.12
Modulus of end-linked PDMS networks with $f = 4$ at 30 °C from S. K. Patel *et al.*, *Macromolecules* **25**, 5241 (1992). The line has an intercept determined by the plateau modulus of a melt of high molar mass PDMS linear polymers $G_e = 2.0 \times 10^5\,\text{Pa}$.

Therefore, it is well established that topological entanglements dominate and control the modulus of polymer networks with long network strands. The Edwards tube model explains the non-zero intercept in plots of network modulus against number density of strands (see Figs 7.11 and 7.12). The modulus of networks with very long strands between crosslinks approaches the plateau modulus of the linear polymer melt. The modulus of the entangled polymer network can be approximated as a simple sum.

$$G \cong G_x + G_e \approx \rho \mathcal{R} T \left(\frac{1}{M_x} + \frac{1}{M_e} \right) \tag{7.48}$$

The modulus is controlled by crosslinks for low molar mass strands between crosslinks ($G \cong G_x$ for $M_x < M_e$) and by entanglements for high molar mass strands between crosslinks ($G \cong G_e$ for $M_x > M_e$). The modulus becomes nearly independent of the molar mass of the network strands between crosslinks in the limit of very long strands. The straight lines with non-zero intercept in Figs 7.11 and 7.12 are Eq. (7.48).

Equation (7.48) is applicable to well-developed networks with essentially no sol fraction. The effective modulus is of order kT per network strand without entanglements and kT per entanglement strand when entanglements dominate. Equation (7.48) allows no possibility of making a fully developed network with a modulus smaller than the plateau modulus of the corresponding melt of linear chains. However, networks with smaller modulus can of course be made if the crosslinking reaction is kept close to the gel point. The modulus of gels in the gelation regime is discussed in

Section 7.5. Networks with modulus lower than the plateau modulus can also be prepared by crosslinking chains in solution and then removing the solvent, because fewer entanglements are trapped by crosslinking in solution.

The deformation dependence of the stress in the Edwards tube model is the same as in the classical models [Eqs (7.32) and (7.33)] because each entanglement effectively acts as another crosslink junction in the network. Therefore, the Edwards tube model is unable to explain the stress softening at intermediate deformations, demonstrated in Fig. 7.8. The reason for the classical functional form of the stress–strain dependence is that the confining potential is assumed to be independent of deformation.

7.3.2   The Mooney–Rivlin model

As an alternative to the molecular approach of the three models described above, a phenomenological model of elasticity may be used. In such a model, a general expression for the free energy is written without asking any questions about the molecular interpretation of the terms of this free energy.

The model developed by Mooney and Rivlin starts from three **strain invariants**:[2]

$$I_1 = \lambda_x^2 + \lambda_y^2 + \lambda_z^2, \tag{7.49}$$

$$I_2 = \lambda_x^2\lambda_y^2 + \lambda_y^2\lambda_z^2 + \lambda_z^2\lambda_x^2, \tag{7.50}$$

$$I_3 = \lambda_x^2\lambda_y^2\lambda_z^2. \tag{7.51}$$

The free energy density of the network F/V is written as a power series in the difference of these invariants from their values in the undeformed network ($\lambda_x = \lambda_y = \lambda_z = 1$):

$$\frac{F}{V} = C_0 + C_1(I_1 - 3) + C_2(I_2 - 3) + C_3(I_3 - 1) + \cdots \tag{7.52}$$

The second term in this series is analogous to the free energy of the classical models [Eq. (7.23)]

$$C_1(I_1 - 3) = C_1(\lambda_x^2 + \lambda_y^2 + \lambda_z^2 - 3) \tag{7.53}$$

with the identification $C_1 = G_x/2$. The third term in Eq. (7.52) describes the deviations from the classical dependence. For incompressible networks, the third invariant does not change with deformation,

$$I_3 = \lambda_x^2\lambda_y^2\lambda_z^2 = \left(\frac{V}{V_0}\right)^2 = 1, \tag{7.54}$$

making the fourth term of Eq. (7.52) zero.

[2] They are called invariants because they are independent of the choice of coordinate system.

For uniaxial deformation of an incompressible network,

$$\lambda_x = \lambda \quad \lambda_y = \lambda_z = \frac{1}{\sqrt{\lambda}}, \tag{7.55}$$

the Mooney–Rivlin free energy density is written in terms of the stretching factor λ:

$$\frac{F}{V} = C_0 + C_1\left(\lambda^2 + \frac{2}{\lambda} - 3\right) + C_2\left(2\lambda + \frac{1}{\lambda^2} - 3\right) + \cdots \tag{7.56}$$

The true stress in the Mooney–Rivlin model can be obtained from the free energy density:

$$\sigma_{\text{true}} = \frac{1}{L_y L_z}\frac{\partial F}{\partial L_x} = \lambda\frac{\partial (F/V)}{\partial \lambda} = 2C_1\left(\lambda^2 - \frac{1}{\lambda}\right) + 2C_2\left(\lambda - \frac{1}{\lambda^2}\right) + \cdots$$

$$= \left(2C_1 + \frac{2C_2}{\lambda}\right)\left(\lambda^2 - \frac{1}{\lambda}\right) + \cdots \tag{7.57}$$

The engineering stress can be calculated from the true stress [Eq. (7.30)]:

$$\sigma_{\text{eng}} = \frac{\sigma_{\text{true}}}{\lambda} = \left(2C_1 + \frac{2C_2}{\lambda}\right)\left(\lambda - \frac{1}{\lambda^2}\right). \tag{7.58}$$

This leads to the famous **Mooney–Rivlin equation**:

$$\frac{\sigma_{\text{true}}}{\lambda^2 - 1/\lambda} = \frac{\sigma_{\text{eng}}}{\lambda - 1/\lambda^2} = 2C_1 + \frac{2C_2}{\lambda}. \tag{7.59}$$

For classical models, the Mooney–Rivlin coefficients are $2C_1 = G$ and $C_2 = 0$. However, experimental data plotted in Fig. 7.13, in the form suggested by Eq. (7.59), show that $C_2 > 0$. In this Mooney–Rivlin plot, the stress divided by the prediction of the classical models is plotted as a function of the reciprocal deformation $1/\lambda$. The predictions of the affine, phantom, and Edwards tube network models correspond to horizontal lines on the Mooney–Rivlin plot ($C_2 = 0$). Experimental data on uniaxial extension of networks are described well by the Mooney–Rivlin equation with $C_2 > 0$, indicating strain softening as deformation increases (as $1/\lambda$ decreases). Molecular interpretation of this phenomenological result is considered next.

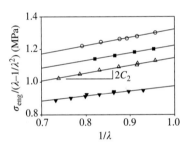

Fig. 7.13
Mooney–Rivlin plots for uniaxial tension data on three networks prepared from radiation-crosslinking a linear polybutadiene melt with $M_w = 344\,000\,\text{g mol}^{-1}$, with four different doses, making four different crosslink densities. The lines are fits of Eq. (7.59) to each data set. Data of L. M. Dossin and W. W. Graessley, *Macromolecules* **12**, 123 (1979).

7.3.3 Constrained fluctuations models

The phantom network model assumes there are no interactions between network strands other than their connectivity at the junction points. It has long been recognized that this is an oversimplification. Chains surrounding a given strand restrict its fluctuations, raising the network modulus. This is a very complicated effect involving interactions of many polymer chains, and hence, is most easily accounted for using a mean-field theory. In the

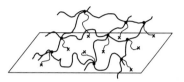

Fig. 7.14
The constrained junction model has virtual chains (thin lines) connecting each network junction (circles) to the elastic background (at the crosses).

mean-field theory, the constraining effect of many surrounding chains is replaced by an effective constraining field acting on a given network strand. The mean-field approximation is reasonable as long as the overlap parameter (the number of chains overlapping with a given chain) is large. Mean-field models of constrained fluctuations are now briefly reviewed.

In the 1970s Ronca, Allegra, Flory, and Erman proposed to take into account the additional interaction between chains by constraining junctions of network strands. The **constrained-junction model** can be represented by a phantom network with an additional harmonic constraining potential acting on network junctions (see Fig. 7.14). This constraining potential can be represented by 'virtual' chains[3] connecting junction points to the elastic non-fluctuating background. The points of attachment of the virtual chains to the elastic background are chosen in such a way to maintain the Gaussian statistics of network strands in their preparation condition.

The constraining potential represented by virtual chains must be set up so that the fluctuations of junction points are restricted, but the virtual chains must not store any stress. If the number of monomers in each virtual chain is independent of network deformation, these virtual chains would act as real chains and would store elastic energy when the network is deformed. A principal assumption of the constrained-junction model is that the constraining potential acting on junction points changes with network deformation. In the virtual chain representation of this constraining potential, this important assumption means that the number of monomers in the virtual chains changes with deformation. In the case of anisotropic deformation, the constraining potential becomes anisotropic and can be represented by virtual chains with different number of monomers n_i constraining fluctuations of junction points in the direction i ($i = x, y$, or z)

$$n_i = \lambda_i^2 n_0, \tag{7.60}$$

where n_0 is the number of monomers in virtual chains constraining junction fluctuations in a given direction in an undeformed network and λ_i is the deformation parameter in the direction i. In Problem 7.21, it is demonstrated that Eq. (7.60) is the only possible assumption leading to no contribution to stress from virtual chains and all elastic energy being stored only in real chains. The physical reason for the deformation dependence of the constraining potential is that strands move further apart from each other in the elongation direction ($\lambda > 1$) and have a weaker constraining effect on strand fluctuations in that direction. In contrast, the strands move closer together in the compressing direction ($\lambda < 1$) where the constraining effect is strengthened. The decrease of constraining potential upon network elongation leads to increased fluctuations and to a non-classical dependence of stress on elongation with strain softening qualitatively similar to that observed in experiments (see Fig. 7.8).

[3] Virtual chains represent topological interactions between strands, as opposed to effective chains that model the effects of strand connections.

The constrained-junction model relies on an additional parameter that determines the strength of the constraining potential, and can be thought of as the ratio of the number of monomers in real network strands and in virtual chains N/n_0. If this ratio is small, the virtual chain is relatively long ($n_0 \gg N$), and surrounding chains have very little effect on the fluctuations of the junction points. If the ratio N/n_0 is large, the virtual chain is relatively short ($n_0 \ll N$) and fluctuations of the junction points are strongly suppressed, practically pinning them to the elastic non-fluctuating background. Hence, the constrained-junction model continuously crosses over from the phantom network model to the affine network model with increasing strength of the constraining potential (as N/n_0 increases from 0 to ∞).

The interactions between long overlapping network strands suppress fluctuations not only of the network junctions, but of all monomers in every network strand. In an attempt to capture this effect, Kloczkowski, Mark, and Erman proposed a **diffused-constraints model**. Instead of the constraining potential acting on network junctions, a constraining potential was applied to a single monomer on each network strand. The location along the strand of this constrained monomer is different in different network strands. The constraining potential of this diffused-constraints model can also be represented by a virtual chain with its number of monomers changing with deformation, as given by Eq. (7.60). This constraint reduces fluctuations of the particular monomer the virtual chain is attached to and its immediate neighbours, but it does not have a significant effect on fluctuations of monomers of the network strand that are far from the virtual chain. Constraining only one monomer per network strand is not enough to represent the extent of topological constraints imposed on a given strand by the surrounding network chains. In the limit of very strong confining potential ($N/n_0 \to \infty$) the diffused-constraints model also reduces to the affine network model.

7.3.3.1 Tube models of constrained fluctuations

Topological entanglements imposed by surrounding chains upon a given network strand reduce fluctuations of all monomers of this strand. Therefore, the constraining potential providing a mean-field representation of these topological interactions should be applied to *all* monomers of the chain. This is the basic assumption of the Edwards tube model, discussed in Section 7.3.1. The parabolic constraining potential acting on all monomers reduces their fluctuations to the confining tube of width a [Eq. (7.46)]. This constraining potential can be described by virtual chains that connect each monomer of the network strand to the elastic non-fluctuating background. The collective effect of these virtual chains is to restrict the fluctuations of each network strand to a confining tube (Fig. 7.10).

The same confinement of monomer fluctuations can be achieved by attaching shorter virtual chains with $n(p)$ 'virtual' monomers to every pth monomer of the network strand as long as the product $pn(p)$ is the same and $p < n(p)$. In Problem 7.22, it is shown that for a comb polymer with side branches containing n monomers attached to every pth monomer of backbone at one end and to the elastic non-fluctuating background at the other end, the mean-square fluctuations of backbone monomers are

$$a^2 \approx b^2 \sqrt{pn(p)}, \qquad (7.61)$$

as long as $p \lesssim n$. Thus, virtual chains with $n \approx N_e^2$ can be attached to every monomer on the network strand (with $p = 1$) or shorter virtual chains with $n \approx N_e$ can be attached to monomers separated by $p \approx N_e$ monomers along the strand (one per entanglement strand). In both cases the fluctuations of network monomers will be constrained to the same confining tube with diameter $a \approx bN_e^{1/2}$. The condition $p \lesssim n$ assures that all monomers of the network strand have similar amplitude of fluctuations independent of how far they are from the points of attachment to virtual chains. This avoids the problem of inhomogeneous fluctuations of monomers seen in the diffused-constraints model.

One of the main assumptions of the Edwards tube model is that the number n of monomers in the virtual chains (the strength of the constraining potential) is independent of network deformation. This assumption implies that the amplitude of monomer fluctuations (the tube diameter a) does not change upon network deformation [Eq. (7.61)]. As mentioned in Section 7.3.1, this assumption of deformation-independent confining potential in the Edwards tube model leads to the classical dependence of stress on deformation.

Non-affine tube model. Rubinstein and Panyukov combined the main ideas of the constrained junction model and the Edwards tube model into a **non-affine tube model.** As in the Edwards tube model, the constraining effect of surrounding network strands on a given strand is represented by a confining potential, modelled by virtual chains. The virtual chain attachments to the elastic non-fluctuating background are not forced to be located along the primitive path line as in the Edwards tube model, but rather are placed randomly in space in such a way to make the primitive path a random walk in the preparation state. This random placement of attachment points represents the randomness of the network crosslinking process and assures that the tube has random walk statistics.

The number of monomers in virtual chains is assumed to change with deformation according to Eq. (7.60), similar to the constrained-junction and diffused-constraints models. If one virtual chain is attached to every entanglement strand of N_e monomers, it contains of order N_e virtual monomers in the undeformed state of the network. The number of monomers in each virtual chain changes as the network is deformed [see Eq. (7.60)].

$$n_i \approx \lambda_i^2 N_e. \qquad (7.62)$$

For anisotropic deformation, it is important to realize that the virtual chains in component directions $i = x, y, z$ have different numbers of monomers because the λ_i are different. As with the constrained-junction model, the constraining potential in the non-affine tube model is chosen so that no stress will be supported by virtual chains when the network is deformed, to be consistent with the microscopic definition of the stress tensor. Combining Eqs (7.61) and (7.62), we discover that the tube diameter changes non-affinely with network deformation:

$$a_i \approx b N_e^{1/2} \lambda_i^{1/2} \approx a \lambda_i^{1/2}, \qquad (7.63)$$

where a is the tube diameter of the network in the undeformed state. This means that fluctuations of monomers increase along the elongation direction and decrease along the compression direction. The physical picture of the confining tube is that entanglements suppress fluctuations of monomers. As the distance between entanglements changes upon network deformation, so should the amplitude of monomer fluctuations. In fact, the amplitude of these fluctuations in the non-affine tube model coincides with the distance between entanglements in the corresponding direction, which is equal to the size of an entanglement strand of N_e monomers (see Problem 7.20).

The deformation dependence of the confining potential [Eq. (7.62)] results in a non-classical stress–strain dependence of the non-affine tube model. The prediction of this model for the stress–elongation relation in tension is qualitatively similar to the Mooney–Rivlin equation [Eq. (7.59)] and is also in excellent agreement with experiments on uniaxial deformation of networks in tension:

$$\frac{\sigma_{true}}{\lambda^2 - 1/\lambda} = \frac{\sigma_{eng}}{\lambda - 1/\lambda^2} = G_x + \frac{G_e}{\lambda - \lambda^{-1/2} + 1} \qquad (7.64)$$

However, this form still overpredicts the stress required to compress a network.

Non-affine slip-tube model. In the non-affine slip-tube model, the stored length of network chains is allowed to redistribute along the contour of the tube (Fig. 7.15). Upon asymmetric (uniaxial or biaxial) deformation, the network is stretched in some directions and compressed in others. Stored length from compressed directions of the tube can redistribute itself into the stretched directions, balancing the tension in all directions and lowering the free energy and the stress in the network. The resulting dependence of stress on the deformation in the non-affine slip-tube model does not have a simple analytical form. However, the model has been solved numerically and its solution in the experimentally relevant range of $0.1 < \lambda < 10$ can be approximated in a form similar to Eq. (7.64):

$$\frac{\sigma_{true}}{\lambda^2 - 1/\lambda} = \frac{\sigma_{eng}}{\lambda - 1/\lambda^2} = G_x + \frac{G_e}{0.74\lambda + 0.61\lambda^{-1/2} - 0.35}. \qquad (7.65)$$

Both Eqs (7.64) and (7.65) reduce to Eq. (7.48) in the small deformation limit ($\lambda \to 1$). This simple additivity separates the crosslink and entanglement contributions to the stress and hence allows them to be determined

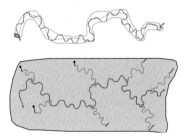

Fig. 7.15
In the non-affine slip-tube model, entanglements are represented by slip-rings that are attached to the elastic background through virtual chains that represent the potential of the confining tube.

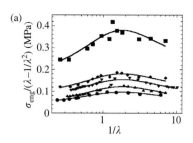

(a)

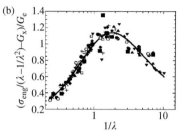

(b)

Fig. 7.16
Comparison of the non-affine slip-tube model [Eq. (7.65) using G_x and G_e as adjustable parameters] with experiments, see M. Rubinstein and S. Panyukov, *Macromolecules* **35**, 6670 (2002). Filled squares and diamonds are data on vulcanized natural rubber from R. S. Rivlin and D. W. Saunders, *Philos. Trans. R. Soc. London A* **243**, 251 (1951). Filled and open circles are PDMS network data of H. Pak and P. J. Flory, *J. Polym. Sci., Polym. Phys.* **17**, 1845 (1979). Triangles and inverted triangles are PDMS network data of P. Xu and J. E. Mark, *Rubber Chem. Technol.* **63**, 276 (1990). Part (b) demonstrates the universal form of Eq. (7.65) and also includes simulation data on end-linked networks with $N_s = 35$ (open squares), $N_s = 100$ (+) and $N_s = 350$ (×) from G. S. Grest *et al.*, *J. Non-Cryst. Solids* **274**, 139 (2000).

from experiment. Experimental data for different networks are plotted in the Mooney–Rivlin form in Fig. 7.16(a). Using G_x and G_e as fitting parameters, Eq. (7.65) provides a reasonable description of the data. Note the maximum on the Mooney–Rivlin plot for uniaxially compressed networks. The simple Mooney–Rivlin expression [Eq. (7.59)] predicts a straight line and does not agree with network compression data. Figure 7.16(b) reduces all data from both experiments and simulations to a common curve in the form suggested by Eq. (7.65).

7.4 Swelling of polymer gels

Another amazing property of polymer networks is their ability to change volume manyfold when exposed to an appropriate solvent. When a network polymer is swollen in a solvent it is called a gel. In this section, the swelling of unentangled gels is treated. At the level of the Edwards tube model, swelling of entangled gels is identical, with the number of monomers in a strand N replaced by the number of monomers in an entanglement strand N_e in the preparation state. However, as the previous section indicates, deformation of entangled networks (including swelling) is more complicated than the affine treatment of the Edwards tube model, with the topological confinement significantly diminished upon swelling.

The volume fraction ϕ of polymer in a swollen (or partly swollen) state can be easily determined experimentally by measuring the volume V of the gel (including the solvent within it) and its volume in the dry state V_{dry}:

$$\phi = \frac{V_{dry}}{V}. \tag{7.66}$$

Let ϕ_0 be the polymer volume fraction in the preparation state where crosslinking was performed, with the gel volume V_0. The total amount of polymer in a well-developed gel (with no sol fraction) does not change upon swelling or deswelling. The change of the volume is due entirely to the change in the amount of solvent within the gel:

$$V_0\phi_0 = V\phi = V_{dry}. \tag{7.67}$$

When an unconstrained macroscopic network polymer is swollen in a solvent, it undergoes uniform swelling by the same amount in all directions. In this case, the linear deformation λ in each direction is simply the 1/3 power of the ratio of final and initial volumes V/V_0, or the 1/3 power of the initial and final volume fractions ϕ_0/ϕ:

$$\lambda = \left(\frac{V}{V_0}\right)^{1/3} = \left(\frac{\phi_0}{\phi}\right)^{1/3}. \tag{7.68}$$

On swelling, each network strand is stretched as the crosslink junctions move further apart. The stretching of an ideal chain was treated in

Section 2.6.1. The free energy required to stretch the ideal chain is quadratic in its end-to-end distance R [see Eqs (2.94) and (2.101)]:

$$F_{el} \approx kT\frac{R^2}{Nb^2}. \qquad (7.69)$$

This equation is the Flory form of the elastic part of the free energy of a network. The mean-square end-to-end distance of network strands in their preparation state is R_0^2. Assuming affine deformation on the length scales of a network strand, the mean-square end-to-end distance in the final state is $R^2 = (\lambda R_0)^2$. Modern treatments of network swelling and elasticity utilize a more general form of Eq. (7.69) for the elastic energy of a swollen or deformed network strand, known as the Panyukov form

$$F_{el} \approx kT\frac{(\lambda R_0)^2}{R_{ref}^2}, \qquad (7.70)$$

where R_{ref}^2 is the mean-square fluctuation of the end-to-end distance of the network strand. In many cases, R_{ref}^2 is equal to the mean-square end-to-end distance of a free chain with the same number of monomers as the strand in the same solution. As the network swells (or the quality of the solvent is changed) the strand elasticity changes because R_{ref}^2 changes. The modulus $G(\phi)$ of the gel in the swollen (or partly swollen) state is proportional to the chain number density $\nu = \phi/(Nb^3)$ times the elastic free energy per chain [Eq. (7.70)] (see Problem 7.29):

$$G(\phi) \approx \nu kT\frac{(\lambda R_0)^2}{R_{ref}^2} \approx \frac{kT}{b^3}\frac{\phi}{N}\frac{(\lambda R_0)^2}{R_{ref}^2}. \qquad (7.71)$$

At swelling equilibrium, the elasticity is balanced by the osmotic pressure Π of a semidilute solution of uncrosslinked chains at the same concentration.[4] Since the modulus is proportional to the elastic free energy per unit volume, any gel swells until the modulus and osmotic pressure are balanced. The **equilibrium swelling ratio** Q is the ratio of the volume in the fully swollen state V_{eq} and the volume in the dry state V_{dry}:

$$Q \equiv \frac{V_{eq}}{V_{dry}} \quad \text{when } G \approx \Pi. \qquad (7.72)$$

It is important to emphasize the fact that the osmotic pressure in Eq. (7.72) is the osmotic pressure of a semidilute solution of linear chains at the same volume fraction as the gel. This is not to be confused with the osmotic pressure of the gel calculated from its definition in Eq. (4.62), which includes effects from the elasticity of the gel.

---

[4] Equilibrium is really attained by minimizing the free energy ($\partial F/\partial V = 0$) but if both dominant terms of the free energy are power laws in concentration, then the scaling method used here is correct up to a numerical prefactor.

7.4.1 Swelling in θ-solvents

The mean-square end-to-end distance of a free chain in a θ-solvent is independent of concentration, with $R_0^2 \approx b^2 N$. The fluctuations of a network strand that control its elasticity are also independent of concentration:

$$R_{\text{ref}}^2 \approx R_0^2 \approx b^2 N. \tag{7.73}$$

Hence, the more general Panyukov form [Eq. (7.70)] reduces to the Flory form [Eq. (7.69)] for swelling in θ-solvents. The gel modulus in a θ-solvent is then proportional to the 1/3 power of concentration:

$$G(\phi) \approx \frac{kT}{b^3} \frac{\phi}{N} \lambda^2 \approx \frac{kT}{b^3} \frac{\phi}{N} \left(\frac{\phi_0}{\phi}\right)^{2/3} \approx \frac{kT}{Nb^3} \phi_0^{2/3} \phi^{1/3}. \tag{7.74}$$

Originally derived by James and Guth, this weak concentration dependence of the network modulus comes from two competing effects. As concentration is lowered, the number density of strands naturally decreases. However, the strands also stretch as concentration is lowered, and this stretching raises the modulus somewhat through the proportionality to λ^2 (see Problem 7.29). The net effect is the weak decrease of the gel modulus upon swelling, given by Eq. (7.74).

The osmotic pressure of a semidilute solution in a θ-solvent was discussed in Section 5.4.2 [Eq. (5.57)]:

$$\Pi \approx \frac{kT}{b^3} \phi^3. \tag{7.75}$$

From the condition of swelling equilibrium [Eq. (7.72)] using Eqs (7.74) and (7.75) for θ-solvents, the equilibrium swelling ratio is obtained:

$$Q \approx \frac{N^{3/8}}{\phi_0^{1/4}}. \tag{7.76}$$

If the network was prepared in the dry state $\phi_0 = 1$, the volume fraction of polymer in the equilibrium swollen state in a θ-solvent provides a direct measure of the average number of monomers N in a network strand:

$$N \approx Q^{8/3}. \tag{7.77}$$

For networks prepared at other preparation concentrations (at the θ-temperature) Eq. (7.76) can be solved for N to estimate the average number of monomers in a network strand from the volume fraction of polymer in the preparation state ϕ_0 and the equilibrium swelling Q. However, extreme caution must be used in estimating N from Eq. (7.77) because of trapped entanglements. For densely crosslinked networks with $N \ll N_e$, entanglements are not important and Eq. (7.77) will provide an excellent estimate of N. On the other hand, for entangled networks with $N \gg N_e$,

Eq. (7.77) would only serve to estimate the entanglement strand in the preparation state assuming the Edwards tube model is correct ($N_e \approx Q^{8/3}$).

The modulus of the network in the dry state is obtained from Eq. (7.74) with $\phi = 1$:

$$G(1) \approx \frac{kT}{Nb^3} \phi_0^{2/3} \approx \frac{kT}{b^3} Q^{-8/3}. \qquad (7.78)$$

The last relation is valid for any preparation concentration, and was obtained using Eq. (7.76). Such *universal* relations that are independent of the details of the preparation state are useful for predicting the equilibrium swelling in a θ-solvent from a measurement of the modulus of the network in the dry state (and vice versa). Equation (7.78) has not been tested in a θ-solvent, but it does describe swelling in concentrated solution, where ideal chain statistics apply (see the small swelling part of Fig. 7.17).

7.4.2 Swelling in athermal solvents

The end-to-end distance of a network strand in the preparation state at initial concentration ϕ_0 in an athermal solvent is given by Eq. (5.26) with $v \approx b^3$.

$$R_0 \approx bN^{1/2}\phi_0^{-(v-1/2)/(3v-1)}. \qquad (7.79)$$

As shown in Section 5.3.1, the size of semidilute polymer chains in an athermal solvent decreases weakly with concentration as $R \sim \phi^{-0.12}$, since the exponent $v \cong 0.588$. The end-to-end distance of a free chain at concentration ϕ in an athermal solvent is also the fluctuation of the strand that determines its elasticity in the swollen (or partly swollen) state:

$$R_{\text{ref}} \approx bN^{1/2}\phi^{-(v-1/2)/(3v-1)}. \qquad (7.80)$$

In contrast, the Flory form [Eq. (7.69)] underestimates the fluctuation size by assuming that R_{ref} is the ideal size $bN^{1/2}$.

The gel modulus in an athermal solvent has a stronger concentration dependence than in a θ-solvent:

$$G(\phi) \approx \frac{kT}{b^3} \frac{\phi}{N} \left(\frac{\lambda R_0}{R_{\text{ref}}}\right)^2 \approx \frac{kT}{b^3} \frac{\phi}{N} \left(\frac{\phi_0}{\phi}\right)^{2/3} \left(\frac{\phi}{\phi_0}\right)^{(2v-1)/(3v-1)}$$

$$\approx \frac{kT}{Nb^3} \phi_0^{1/[3(3v-1)]} \phi^{(9v-4)/[3(3v-1)]}. \qquad (7.81)$$

Since $v \cong 0.588$, the modulus of a gel swollen in an athermal solvent is predicted to decrease as solvent is added as $G(\phi) \sim \phi_0^{0.44}\phi^{0.56}$.

The osmotic pressure in an athermal solvent was considered in Section 5.3.2 [Eq. (5.48) with $v \approx b^3$]:

$$\Pi \approx \frac{kT}{b^3} \phi^{3v/(3v-1)}. \qquad (7.82)$$

The prediction of $\Pi \sim \phi^{2.3}$ [Eq. (7.82) with $\nu \cong 0.588$] was shown to be consistent with experiments in Fig. 5.7. The equilibrium swelling ratio is determined using Eqs (7.81) and (7.82) in Eq. (7.72):

$$Q \approx \frac{N^{3(3\nu-1)/4}}{\phi_0^{1/4}}. \tag{7.83}$$

Comparing Eqs (7.76) and (7.83) shows the expected result that any network swells much more in an athermal solvent than in a θ-solvent.

If the network was prepared in the dry state ($\phi_0 = 1$) the average number of monomers in a network strand can be determined by measurement of the equilibrium swelling in an athermal solvent (with $\nu \cong 0.588$).

$$N \approx Q^{1.75}. \tag{7.84}$$

However, in practice Eq. (7.84) is not as useful as Eq. (7.77). Most good solvents are not in the athermal limit (with $v \approx b^3$) meaning that the exact value of the excluded volume (or χ) must be known to calculate N from equilibrium swelling measurements in a good solvent, as discussed in the next section. As in a θ-solvent, the modulus in the dry state $G(1)$ is universally related to the equilibrium swelling ratio Q in an athermal solvent:

$$G(1) \approx \frac{kT}{Nb^3} \phi_0^{1/[3(3\nu-1)]} \approx \frac{kT}{b^3} Q^{-4/[3(3\nu-1)]}. \tag{7.85}$$

Since $\nu \cong 0.588$, the dry modulus is related to the equilibrium swelling in an athermal solvent as $G(1) \sim Q^{-1.75}$. This dependence is considerably weaker than in a θ-solvent, simply because the same network will not swell nearly as much in a θ-solvent. Equation (7.85) applies to networks that are swollen into the semidilute regime in a good solvent, as shown in the large swelling part of Fig. 7.17.

7.4.3 Swelling in good solvents

In a good solvent that is not in the athermal limit ($0 < v < b^3$), Eq. (5.48) must be used for the osmotic pressure:

$$\Pi \approx \frac{kT}{b^3} \left(\frac{v}{b^3}\right)^{3(2\nu-1)/(3\nu-1)} \phi^{3\nu/(3\nu-1)} \quad \text{for } \phi < \phi^{**} \approx \frac{v}{b^3}. \tag{7.86}$$

Recall from Chapter 5 that the crossover concentration $\phi^{**} \approx v/b^3$ [Eq. (5.36)] denotes the boundary between semidilute and concentrated solutions. For $\phi > \phi^{**}$ chains are nearly ideal in concentrated solutions, whereas for $\phi < \phi^{**}$ chains are swollen on intermediate scales. Network modulus and equilibrium swelling depend on the relative value of preparation and fully swollen concentrations (ϕ_0 and $1/Q$) with respect to the crossover concentration ϕ^{**}. Since the swollen concentration is always lower than the preparation concentration ($1/Q < \phi_0$) there are three possible cases:

(i)   good solvent regime with $1/Q < \phi_0 < \phi^{**}$;
(ii) intermediate regime with $1/Q < \phi^{**} < \phi_0$;
(iii) θ-regime with $\phi^{**} < 1/Q < \phi_0$.

In the good solvent regime (i), the lower excluded volume [see Eq. (5.37)] reduces R_0 and R_{ref} relative to the athermal case [Eqs (7.79) and (7.80)]. However, the modulus in the good solvent regime (with $\phi < \phi_0 < \phi^{**}$) predicted by Eq. (7.81) is independent of the excluded volume, since R_0 and R_{ref} only enter into Eq. (7.81) as the ratio R_0/R_{ref}. Using Eqs (7.81) and (7.86) in equating modulus and osmotic pressure, gives the equilibrium swelling ratio for $\phi_0 < \phi^{**}$:

$$Q \approx \frac{N^{3(3\nu-1)/4}}{\phi_0^{1/4}} \left(\frac{v}{b^3}\right)^{9(2\nu-1)/4} \quad \text{for } 1/Q < \phi_0 < \phi^{**}. \qquad (7.87)$$

In the intermediate regime (ii), the chain is ideal in the preparation state $R_0 \approx bN^{1/2}$, but is swollen in the final state [Eq. (5.37)]. The modulus of the swollen gel is calculated from Eq. (7.71):

$$G(\phi) \approx \frac{kT}{b^3} \frac{\phi}{N} \left(\frac{\lambda R_0}{R_{ref}}\right)^2 \approx \frac{kT}{b^3} \frac{\phi}{N} \left(\frac{\phi_0}{\phi}\right)^{2/3} \left(\frac{\phi}{\phi^{**}}\right)^{(2\nu-1)/(3\nu-1)}$$

$$\approx \frac{kT}{Nb^3} \left(\frac{b^3}{v}\right)^{(2\nu-1)/(3\nu-1)} \phi_0^{2/3} \phi^{(9\nu-4)/[3(3\nu-1)]} \quad \text{for } \phi < \phi^{**} < \phi_0.$$

$$(7.88)$$

Since $\nu \cong 0.588$, the modulus of a gel in the intermediate regime decreases as good solvent is added as $G(\phi) \sim v^{-0.23} \phi_0^{2/3} \phi^{0.56}$. Increasing the excluded volume at constant ϕ, ϕ_0 and T, *lowers* the modulus because the larger excluded volume only increases R_{ref}.

Balancing this network elastic modulus in the intermediate regime [Eq. (7.88)] with the osmotic pressure [Eq. (7.86)] produces the expression for equilibrium swelling in the intermediate regime:

$$Q \approx \frac{N^{3(3\nu-1)/4}}{\phi_0^{(3\nu-1)/2}} \left(\frac{v}{b^3}\right)^{3(2\nu-1)} \quad \text{for } 1/Q < \phi^{**} < \phi_0. \qquad (7.89)$$

The general relation for the equilibrium swelling in any solvent has three branches that correspond to the good solvent case [Eq. (7.87)], the intermediate case [Eq. (7.89)], and the θ-solvent case [Eq. (7.76)].

$$Q \approx \begin{cases} (v/b^3)^{0.40} N^{0.57} \phi_0^{-1/4} & 1/Q < \phi_0 < \phi^{**} \\ (v/b^3)^{0.53} N^{0.57} \phi_0^{-0.38} & 1/Q < \phi^{**} < \phi_0 \\ N^{3/8} \phi_0^{-1/4} & \phi^{**} < 1/Q < \phi_0. \end{cases} \qquad (7.90)$$

The swelling increases steadily as the excluded volume increases. If the network is prepared in the bulk ($\phi_0 = 1$), the general relation between the dry network modulus and the equilibrium swelling depends upon whether

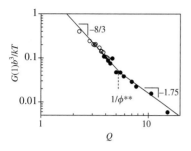

Fig. 7.17
Dry nework modulus $G(1)$ correlated
with equilibrium swelling Q in toluene
for PDMS networks. The open circles
are for model networks made from
end-linking linear chains with two
reactive ends. The filled symbols are
networks with dangling end defects
made by end-linking mixtures of chains
with one and two reactive ends.
Data from S. K. Patel *et al.*,
Macromolecules **25**, 5241 (1992).

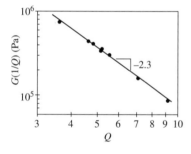

Fig. 7.18
Fully swollen modulus of end-linked
PDMS networks swollen to equilibrium
in toluene at 25 °C. The starting
telechelic chains had $M_n = 4400\,\mathrm{g\,mol}^{-1}$
and they were crosslinked at various
preparation concentrations in the
range $0.3 \leq \phi_0 \leq 1$. The line is the
prediction of Eq. (7.92) with $Q > 1/\phi^{**}$.
Data from K. Urayama *et al.*,
J. Chem. Phys. **105**, 4833 (1996).

the concentration of the swollen state $1/Q$ is above or below the crossover concentration ϕ^{**}:

$$G(1) \approx \frac{kT}{b^3} \begin{cases} Q^{-1.75}(\phi^{**})^{0.69} & Q > 1/\phi^{**} \\ Q^{-8/3} & Q < 1/\phi^{**}. \end{cases} \quad (7.91)$$

In Fig. 7.17, the dry modulus of various PDMS networks is plotted as a function of their equilibrium swelling in toluene. A single curve results for both 'model' networks (open symbols) made by end-linking linear chains with two reactive ends and networks with intentionally introduced defects in the form of dangling ends (filled symbols) made by end-linking mixtures of chains with one and two reactive ends. The data are fit to Eq. (7.91) as the solid lines in Fig. 7.17 and their intersection determines the crossover concentration $\phi^{**} \cong 0.2$, which is typical for good solvents.

A similar 'universal' relation exists (for networks prepared in either melt or concentrated solution with $\phi_0 > \phi^{**}$) between the modulus of the equilibrium fully swollen state $G(1/Q)$ and the equilibrium swelling:

$$G(1/Q) \approx \frac{kT}{b^3} \begin{cases} Q^{-2.3}(\phi^{**})^{0.69} & Q > 1/\phi^{**} \\ Q^{-3} & Q < 1/\phi^{**}. \end{cases} \quad (7.92)$$

This relation is of particular importance for estimating the modulus of a fully swollen gel (which can be challenging to measure) from the equilibrium swelling. Figure 7.18 demonstrates that Eq. (7.92) describes experimental data quite well.

The results in this section were all derived for unentangled networks. The Edwards tube model for entangled networks gives identical results with N replaced by N_e, the number of Kuhn monomers in an entanglement strand in the preparation state, because both entanglement strands and network strands are assumed to deform affinely in the Edwards tube model. If the Edwards tube model were correct, the universal relations [Eqs (7.91) and (7.92)] would still apply for entangled networks, since they are independent of N. However, the non-affine tube models predict that entangled networks will swell considerably more than the Edwards tube model predicts.

7.5   Networks in the gelation regime

For the gels in the gelation regime of Chapter 6, percolation theory predicts the modulus of unentangled gels in their preparation state using the same physics as presented in Section 7.2 (either the phantom or affine network models predict that modulus is proportional to kT per strand). The number density of network strands is determined from the correlation volume ξ^3 (the pervaded volume of a network strand in the gelation regime) and the overlap parameter P [Eq. (6.140), the number of overlapping strands per correlation volume]. The number density of strands inside the correlation volume (P/ξ^3) is the same as the overall number density of strands:

$$G \approx \nu kT \approx \frac{P}{\xi^3} kT \approx \frac{kT}{b^3} \frac{P_{\mathrm{gel}}}{N^*}. \quad (7.93)$$

The final relation was obtained using Eq. (6.140), where P_{gel} is the gel fraction and N^* is the number of monomers in each highly branched network strand in the gelation regime.

The Soxhlet extraction method discussed in Section 6.6 can be used to separate the sol and gel fractions of a gel in the gelation regime, allowing direct determination of the gel fraction P_{gel}. Percolation theory expects the molar mass of a network strand M^* to be the same as the characteristic molar mass in the sol fraction. Hence, M^* can be determined by the size exclusion chromatography methods of Section 6.6, applied to the sol fraction. Equation (7.93) is tested in Fig. 7.19, where the shear modulus is shown to be proportional to P_{gel}/M^*.

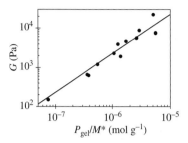

Fig. 7.19
Shear modulus of polyester gels in the gelation regime as a function of the ratio of gel fraction and the characteristic molar mass of the sol. The line has the slope of unity expected by Eq. (7.93). Data of R. H. Colby *et al.*, *Phys. Rev. E* **48**, 3712 (1993).

Recall from Chapter 6, the relative extent of reaction ε, which describes the proximity to the gel point for gels in the gelation regime. At the gel point $\varepsilon = 0$ and the gelation regime ends at $\varepsilon \approx 1$, where the gel fraction is approximately unity. The gel fraction grows above the gel point with exponent β [$P_{gel} \sim \varepsilon^{\beta}$ see Eq. (6.99)]. The number of monomers in the characteristic branched polymer decreases beyond the gel point with exponent $-1/\sigma$ [$N^* \sim \varepsilon^{-1/\sigma}$ see Eq. (6.95)]. Hence, Eq. (7.93) expects the modulus in their preparation state to grow beyond the gel point with exponent $\beta + 1/\sigma$:

$$G \sim \varepsilon^{\beta+1/\sigma}. \qquad (7.94)$$

For critical percolation, $\beta = 0.41$ and $\sigma = 0.45$ (see Table 6.4), so Eq. (7.94) predicts $G \sim \varepsilon^{2.6}$ close to the gel point. Mean-field percolation has $\beta = 1$ and $\sigma = 1/2$, and predicts $G \sim \varepsilon^3$ may apply further from the gel point. These predictions are in reasonable agreement with data on gels in the gelation regime that do not have entanglement effects.

A simple way to think about the effect of entanglements is written in the spirit of Eq. (7.48):

$$G \approx \frac{\rho RT}{M_x} + T_e G_e. \qquad (7.95)$$

The **entanglement trapping factor** T_e changes from zero at the gel point to unity for fully developed networks that have very few defects (formed by end-linking telechelic chains). If long entangled chains are randomly crosslinked in the melt, a fraction T_e of the entanglements will be permanently trapped in the network, and hence, have a similar effect on the modulus as an actual network junction would have. This trapping of entanglements starts in the gelation regime, but since there are many dangling ends and loops in randomly crosslinked networks, they often have many entanglements that are not trapped. Entanglements that involve dangling ends are only temporary and do not contribute to the equilibrium modulus of the network. The details of how T_e grows beyond the gel point are not yet fully established, although some simple ideas exist (see Problem 7.33). Equation (7.95) clearly indicates how to make networks with lower modulus than the plateau modulus of the polymer melt G_e. Either crosslinking chains just barely beyond the gel point or crosslinking in solution and removing the solvent can keep T_e small and result in low modulus networks.

7.6 Linear viscoelasticity

Linear mechanical properties of the networks and gels discussed in this chapter are measured with the same methods of linear viscoelasticity as the polymer liquids (melts and solutions) discussed in Chapters 8 and 9. The various methods are described here, with examples pertaining to each class of materials.

Consider the deformation geometry of simple shear, sketched in Fig. 7.20. The material being sheared is between two flat rigid surfaces. The adhesion between the material and the surfaces is assumed to be strong enough that there is no slippage at either surface. The bottom surface is held so that it does not move, and the upper surface is free to move, apart from the fact that the material between the surfaces may resist that motion. If a force f is applied to the top surface in the x direction, the force will be transmitted through the material to the bottom surface (even if the material is a liquid!). Since the bottom surface is held so that it does not move, this 'holding' must be done so as to apply an equal-and-opposite force $-f$ to the bottom surface. Otherwise, the entire assembly would need to accelerate in the x direction in response to force f. The **shear stress** σ_{xy} (called here σ for short) in this simple shear is defined as the ratio of the applied force and the cross-sectional area of the surfaces A, which is also the area of any plane perpendicular to the y direction within the material being sheared:

$$\sigma \equiv \frac{f}{A}. \tag{7.96}$$

The **shear strain** is defined as the displacement of the top plate Δx relative to the thickness of the sample h (see Fig. 7.20):

$$\gamma \equiv \frac{\Delta x}{h}. \tag{7.97}$$

By defining the stress and strain in this fashion, each part of the entire sample being sheared has identical shear stress σ and shear strain γ in simple shear, *as long as the material shears uniformly*.

If the material between the surfaces is a perfectly elastic solid, the shear stress σ and shear strain γ are proportional, with the constant of proportionality defining the **shear modulus** G:

$$G \equiv \frac{\sigma}{\gamma}. \tag{7.98}$$

Since the stress has units of force/area and the strain is dimensionless, the modulus has units of force/area. Equation (7.98) is **Hooke's law of elasticity** and it is valid for all solids at sufficiently small strains.

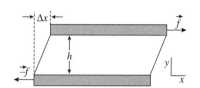

Fig. 7.20
Deformation in simple shear requires application of equal and opposite forces to the two plates. The shear strain is $\gamma = \Delta x/h$. The figure is a two-dimensional representation.

On the other hand, if the material between the surfaces is a simple liquid, the stress is identically zero at any constant strain γ. In liquids, the stress is determined by deformation *rate*. The rate of change of shear strain with time is called the **shear rate**:

$$\dot{\gamma} \equiv \frac{d\gamma}{dt} \qquad (7.99)$$

If the top plate moves with a constant velocity v, the shear rate is $\dot{\gamma} = v/h$. For simple liquids, the shear stress σ is linearly proportional to shear rate $\dot{\gamma}$, with the constant of proportionality defining the **shear viscosity** η:

$$\eta \equiv \frac{\sigma}{\dot{\gamma}}. \qquad (7.100)$$

This relation is **Newton's law of viscosity** and liquids that obey it are referred to as Newtonian liquids. Since the stress has units of force/area and the shear rate has units of reciprocal time, the viscosity has units of force · time/area. The SI unit of stress is the Pascal ($Pa \equiv kg\,m^{-1}\,s^{-2}$) and the SI viscosity unit is $Pa\,s = kg\,m^{-1}\,s^{-1}$.

Polymers are viscoelastic, meaning that they have intermediate properties between Newtonian liquids and Hookean solids. The simplest model of viscoelasticity is the **Maxwell model**, which combines a perfectly elastic element with a perfectly viscous element in series, as shown in Fig. 7.21. Since the elements are in series, the total shear strain γ is the sum of the shear strains in each element:

$$\gamma = \gamma_e + \gamma_v. \qquad (7.101)$$

The shear strain in the elastic element is γ_e and the shear strain in the viscous element is γ_v (see Fig. 7.21). Since the elements are in series, they must each bear the same stress:

$$\sigma = G_M \gamma_e = \eta_M \frac{d\gamma_v}{dt} \qquad (7.102)$$

The ratio of the viscosity η_M of the viscous element and the modulus G_M of the elastic element defines a time scale with special significance, called the **relaxation time**:

$$\tau_M \equiv \frac{\eta_M}{G_M}. \qquad (7.103)$$

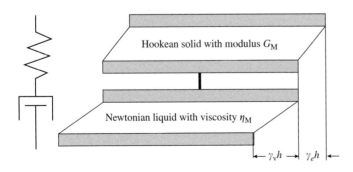

Fig. 7.21
The Maxwell model is a Hookean solid and a Newtonian liquid in series. The shear stress is identical at all points in both the solid and the liquid. The total shear strain is the sum of the strains in the solid and liquid elements. Each element must have the same vertical gap h between the rigid plates that the real sample has.

Networks and gels

In the next few sections, we will see that the Maxwell model responds like a solid on time scales that are short compared with the relaxation time. In contrast, on time scales that are longer than the relaxation time, the Maxwell model flows like a liquid.

7.6.1 Stress relaxation after a step strain

Consider imposing a step strain of magnitude γ at time $t = 0$ (see Fig. 7.20). If the material between the plates is a perfectly elastic solid, the stress will jump up to its equilibrium value $G\gamma$ given by Hooke's law [Eq. (7.98)] and stay there as long as the strain is applied. On the other hand, if the material is a Newtonian liquid, the transient stress response from the jump in strain will be a spike that instantaneously decays to zero. For viscoelastic materials, the stress after such a step strain can have some general time dependence $\sigma(t)$. The **stress relaxation modulus** $G(t)$ is defined as the ratio of the stress remaining at time t (after a step strain was applied at time $t = 0$) and the magnitude of this step strain γ:

$$G(t) \equiv \frac{\sigma(t)}{\gamma}. \tag{7.104}$$

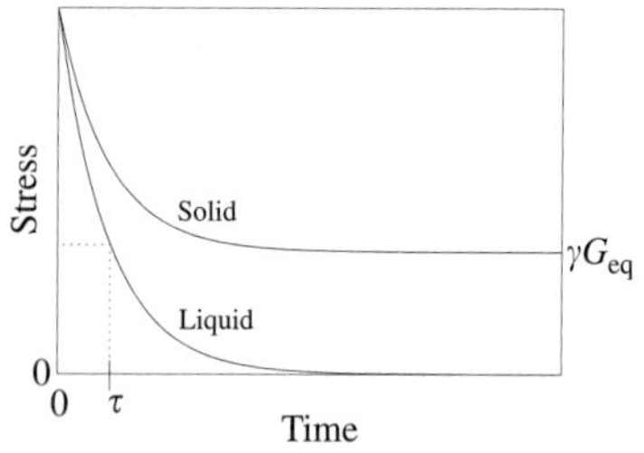

Notice that the above equation is simply a time-dependent generalization of Hooke's law [Eq. (7.98)]. For viscoelastic solids, $G(t)$ relaxes to a finite value, called the **equilibrium shear modulus** G_{eq} (see Fig. 7.22, top curve):

$$G_{eq} = \lim_{t \to \infty} G(t). \tag{7.105}$$

For viscoelastic liquids, the Maxwell model can be used to qualitatively understand the stress relaxation modulus. In the step strain experiment, the total strain γ is constant and Eqs (7.101)–(7.103) can be combined to give a first order differential equation for the time-dependent strain in the viscous element:

$$\tau_M \frac{d\gamma_v(t)}{dt} = \gamma - \gamma_v(t). \tag{7.106}$$

Combined with the initial condition of no strain in the viscous element when the strain is first applied [$\gamma_v(0) = 0$] allows integration of this differential equation:

$$\frac{d\gamma_v(t)}{\gamma - \gamma_v(t)} = \frac{dt}{\tau_M}, \tag{7.107}$$

$$\ln[\gamma - \gamma_v(t)] = \frac{-t}{\tau_M} + C. \tag{7.108}$$

The constant of integration in evaluated from the initial condition, giving $C = \ln\gamma$:

$$\gamma_e(t) = \gamma - \gamma_v(t) = \gamma \exp(-t/\tau_M). \tag{7.109}$$

Fig. 7.22
Stress relaxation in step strain experiments on a viscoelastic solid (upper curve) and a viscoelastic liquid (lower curve). The dashed lines show the value of the stress at the relaxation time τ of the liquid. The solid has the same relaxation time.

The stress relaxes exponentially towards zero on the time scale τ_M, evaluated using Eq. (7.102):

$$\sigma(t) = G_M \gamma_e(t) = G_M \gamma \exp(-t/\tau_M). \qquad (7.110)$$

In the Maxwell model, the stress relaxation modulus has a simple exponential decay.

$$G(t) \equiv \frac{\sigma(t)}{\gamma} = G_M \exp(-t/\tau_M). \qquad (7.111)$$

Beyond their relaxation time, the stress or the relaxation modulus of most viscoelastic liquids has a nearly exponential time decay to zero, sketched as the bottom curve in Fig. 7.22:

$$G(t) \equiv \frac{\sigma(t)}{\gamma} \approx G(\tau) \exp(-t/\tau) \quad \text{for } t > \tau. \qquad (7.112)$$

The relaxation time τ is a fundamental dynamic property of all viscoelastic liquids. Polymer liquids have multiple relaxation modes, each with its own relaxation time. Any stress relaxation modulus can be described by a series combination of Maxwell elements.

All materials have a region of **linear response** at sufficiently small values of applied strain, where the relaxation modulus is independent of strain. Thus, doubling the strain in the linear response regime merely doubles the value of the stress at all times, making the stress relaxation modulus $G(t)$ independent of strain at small values of the applied strain.

7.6.2 The Boltzmann superposition principle

Another manifestation of linear response is the **Boltzmann superposition principle**. The stress from any combination of small step strains is simply the linear combination of the stresses resulting from each individual step $\delta\gamma_i$ applied at time t_i:

$$\sigma(t) = \sum_i G(t - t_i)\delta\gamma_i. \qquad (7.113)$$

This equation simply states that, for linear response, the stress resulting from each step is independent of all the other steps. The system remembers the deformations that were imposed on it earlier, and continues to relax from each earlier deformation as new ones are applied. The stress relaxation modulus tells how much stress remains at time t from each past deformation $\delta\gamma_i$ through the elapsed time $t - t_i$ that has passed since that deformation was applied at time t_i.

Using the definition of the shear rate [Eq. (7.99)] the summation increment can be transformed into time, since $\delta\gamma_i = \dot{\gamma}_i \delta t_i$.

$$\sigma(t) = \sum_i G(t - t_i)\dot{\gamma}_i \delta t_i. \qquad (7.114)$$

The stress from any smooth strain history can be written as an integral over the strain history, by replacing the above summation with an integration:

$$\sigma(t) = \int_{-\infty}^{t} G(t - t')\dot{\gamma}(t')\,dt'. \tag{7.115}$$

The lower integration limit is $t = -\infty$ because we must integrate over *all* past times (not just those starting at some arbitrarily defined zero point) to ensure that all past deformations are accounted for. Equation (7.115) can be used to relate many different linear response experiments.

The stress in any material is the result of all past deformations. The memory of each past deformation only decays as the relaxation modulus decays over the *elapsed* time $t - t'$ from the application of that deformation.

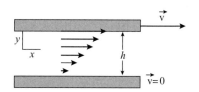

Fig. 7.23
Steady simple shear of a liquid is accomplished by confining the liquid between a moving top plate and a stationary bottom plate. The velocity in the liquid is in the x direction and changes linearly in the y direction, making the shear rate the same everywhere.

7.6.3 Steady shear

In steady simple shear, the top plate in Fig. 7.23 is moved at a constant velocity $\vec{v}$. The shear rate $\dot{\gamma} = |\vec{v}|/h$ is a time-independent constant that can be pulled out of the Boltzmann superposition integral:

$$\sigma(t) = \dot{\gamma} \int_{-\infty}^{t} G(t - t')\,dt' = \dot{\gamma} \int_{0}^{\infty} G(s)\,ds. \tag{7.116}$$

The last relation was obtained using the variable transformation $s \equiv t - t'$, which implies $ds = -dt'$. The integration limits change with this transformation because when $t' = -\infty$, $s = \infty$, and at $t' = t$, $s = 0$. For any liquid, the relaxation modulus $G(t)$ eventually decays to zero fast enough that the integral in the above equation is simply a number with units of stress · time. Thus, the stress at long times in the steady simple shear experiment is constant, and proportional to the shear rate $\dot{\gamma}$. Newton's law of viscosity [Eq. (7.100)] already defined the viscosity in steady shear as the ratio of shear stress and shear rate. Therefore, the viscosity of any liquid is the time integral of its stress relaxation modulus:

$$\eta = \int_{0}^{\infty} G(t)\,dt. \tag{7.117}$$

The Maxwell model [Eq. (7.111)] has a particularly simple viscosity:

$$\eta = G_M \int_{0}^{\infty} \exp(-t/\tau_M)\,dt = G_M \tau_M = \eta_M. \tag{7.118}$$

For viscoelastic solids (Fig. 7.22, top curve) the modulus does not decay to zero, meaning that the viscosity of any solid is infinite. For most viscoelastic liquids (Fig. 7.22, bottom curve) the stress decays to zero in a nearly exponential fashion on time scales longer than their longest relaxation time [Eq. (7.112)]:

$$\eta \approx G(\tau) \int_0^\infty \exp(-t/\tau)\,dt = G(\tau)\tau \int_0^\infty \exp(-s)\,ds = G(\tau)\tau. \quad (7.119)$$

The second integral is simply a number (in this case unity) because it is written in terms of the *dimensionless* integration variable $s \equiv t/\tau$, so $ds = dt/\tau$. The viscosity is proportional to the product of the relaxation time and the value of the modulus at that relaxation time:

$$\eta \approx G(\tau)\tau. \quad (7.120)$$

This relation will be used many times in the remainder of this book to estimate the viscosity from scaling models. It is essentially stating that the area under the bottom curve in Fig. 7.22 is proportional to the area of the rectangle defined by the dashed lines.

In practice, it is more precise to evaluate the viscosity from a step strain experiment by transforming the integration of Eq. (7.117) to a logarithmic time scale using the identity $t\,d\ln t = dt$, and the lower limit of integration changes because when $t = 0$, $\ln t = -\infty$:

$$\eta = \int_{-\infty}^\infty tG(t)\,d\ln t. \quad (7.121)$$

The improved precision of Eq. (7.121) as compared to Eq. (7.117), for determining the viscosity from the step strain experiment, arises from the fact that $tG(t)$ is a function with a well-defined peak, and the relaxation modulus can decay over many decades of time for viscoelastic materials, such as polymers.

If the applied shear rate is too large for linear response, Boltzmann superposition no longer holds in steady shear. An apparent viscosity is still operationally defined as the ratio of shear stress and shear rate, but that apparent viscosity should not be confused with the zero shear rate viscosity η of the liquid. Most polymeric liquids exhibit **shear thinning** of the apparent viscosity at large shear rates, meaning that the viscosity progressively decreases as shear rate is raised. The apparent viscosity has also been observed to increase with shear rate for some materials and such response is called **shear thickening**. The zero shear rate viscosity of the liquid is only measured at low shear rates where $\dot\gamma \ll 1/\tau$. In this book we will only consider the linear response of viscoelastic liquids, and hence our use of the term 'viscosity' always signifies the zero shear rate viscosity.

On some time scale, all liquids display viscoelasticity. Newtonian liquids like water have viscosity independent of shear rate over ordinary ranges of measurement $(10^{-5}\,\mathrm{s}^{-1} < \dot\gamma < 10^5\,\mathrm{s}^{-1})$. Dielectric spectroscopy reveals that water molecules respond to an oscillating electric field at a frequency of 17 GHz at room temperature. Hence, at shear rates of order $10^{10}\,\mathrm{s}^{-1}$, water would be expected to be viscoelastic, and have a shear thinning apparent viscosity.

7.6.4　Creep and creep recovery

Thus far we have imposed a constant strain (the step strain experiment) and constant shear rate (the steady shear experiment). Another simple viscoelastic experiment is accomplished by applying a constant shear stress to the sample. In a creep (step stress) experiment, a constant stress σ is applied to an initially relaxed sample, and the strain is monitored as a function of time $\gamma(t)$. The shear **creep compliance** $J(t)$ is defined as the ratio of the time dependent strain and the applied stress:

$$J(t) \equiv \frac{\gamma(t)}{\sigma}. \tag{7.122}$$

The Maxwell model has a particularly simple response in creep, because for constant stress, Eq. (7.102) requires the strain in the elastic element to be constant

$$\gamma_{e} = \frac{\sigma}{G_{M}}, \tag{7.123}$$

and the strain in the viscous element is simply linear in time:

$$\frac{d\gamma_{v}(t)}{dt} = \frac{\sigma}{\eta_{M}} \quad \Rightarrow \quad \gamma_{v}(t) = \frac{\sigma}{\eta_{M}}t. \tag{7.124}$$

The creep compliance of the Maxwell model is linear in time,

$$J(t) = \frac{\gamma_{e} + \gamma_{v}(t)}{\sigma} = \frac{1}{G_{M}} + \frac{t}{\eta_{M}}, \tag{7.125}$$

with the value at $t=0$ determining the elastic response (G_{M}) and the slope determining the viscosity (η_{M}).

The strain of a viscoelastic liquid in creep is shown as the top curve in Fig. 7.24. The slope in Fig. 7.24 at long times is the shear rate $\dot{\gamma}$ and the viscosity is therefore determined using Newton's law of viscosity [Eq. (7.100)]. For liquids, the long-time creep compliance is linear in time and its form is reminiscent of the Maxwell model [Eq. (7.125)]:

$$J(t) = J_{eq} + \frac{t}{\eta} \quad \text{for } t \gg \tau. \tag{7.126}$$

The $t=0$ intercept of the long-time creep compliance is a measure of the stored elastic energy in flow, and is called the **steady state compliance** J_{eq}.

The time-dependent strain of a viscoelastic solid in creep is sketched as the bottom curve in Fig. 7.24. The long-time creep compliance of any solid is simply a time-independent compliance J_{eq} that is the reciprocal of its equilibrium modulus G_{eq}.

$$J_{eq} = \lim_{t\to\infty} J(t) = \frac{1}{\sigma}\lim_{t\to\infty}\gamma(t) = \frac{1}{G_{eq}}. \tag{7.127}$$

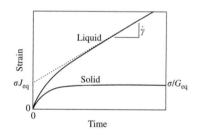

Fig. 7.24
Strain from a creep experiment with constant applied stress σ for a viscoelastic solid (lower curve) and a viscoelastic liquid (upper curve). The slope at long times is the steady shear rate $\dot{\gamma}$, from which the viscosity is calculated as $\eta = \sigma/\dot{\gamma}$ (the viscosity of any solid is infinite, corresponding to zero slope). The extrapolation of this straight line to zero time (dotted line) gives the elastic part of the strain, from which the recoverable compliance J_{eq} is determined.

The final result was obtained using Hooke's law [Eq. (7.98)], which does not discriminate between application of stress with measurement of strain (creep) or application of strain and measurement of stress (step strain).

The Boltzmann superposition principle can be used to relate the steady state compliance to the stress relaxation modulus (see Problem 7.44):

$$J_{eq} = \frac{1}{\eta^2} \int_0^\infty tG(t)\,\mathrm{d}t. \tag{7.128}$$

This integral is dominated by the long-time behaviour of $tG(t)$, given by Eq. (7.112):

$$J_{eq} \approx \frac{G(\tau)}{\eta^2} \int_0^\infty t\exp(-t/\tau)\,\mathrm{d}t. \tag{7.129}$$

This integral is evaluated via integration by parts: $u = t$, making $\mathrm{d}u = \mathrm{d}t$, and $\mathrm{d}v = \exp(-t/\tau)\,\mathrm{d}t$, making $v = -\tau\exp(-t/\tau)$:

$$J_{eq} \approx \frac{G(\tau)}{\eta^2} \left([-t\tau\exp(-t/\tau)]_0^\infty + \tau\int_0^\infty \exp(-t/\tau)\,\mathrm{d}t \right). \tag{7.130}$$

The term in square brackets is zero and the integral is evaluated by making the variable transformation $s \equiv t/\tau$, so $\mathrm{d}s = \mathrm{d}t/\tau$:

$$J_{eq} \approx \frac{G(\tau)\tau^2}{\eta^2} \int_0^\infty \exp(-s)\,\mathrm{d}s = \frac{G(\tau)\tau^2}{\eta^2}. \tag{7.131}$$

Combining with Eq. (7.120) provides a very important relation for the relaxation time:

$$\tau \approx \eta J_{eq}. \tag{7.132}$$

Furthermore, substitution of this relaxation time into Eq. (7.131) shows the significance of the steady state compliance:

$$J_{eq} \approx \frac{1}{G(\tau)}. \tag{7.133}$$

Using Eq. (7.132) for the relaxation time, the long-time behaviour of a liquid in creep can be rewritten:

$$J(t) = J_{eq} + \frac{t}{\eta} \approx \frac{\tau + t}{\eta} \quad \text{for } t \gg \tau. \tag{7.134}$$

Creep has special intuitive appeal for understanding viscoelasticity because the elastic part $J_{eq} = \tau/\eta$ and the viscous part t/η are simply *additive* in creep.

To evaluate the steady state compliance from a step strain experiment, it is useful to transform the integration of Eq. (7.128) to a logarithmic time scale using the identity $t\,\mathrm{d}\ln t = \mathrm{d}t$, and the lower limit of integration changes because when $t = 0$, $\ln t = -\infty$:

$$J_{\mathrm{eq}} = \frac{1}{\eta^2} \int_{-\infty}^{\infty} t^2 G(t)\,\mathrm{d}\ln t. \tag{7.135}$$

Combining Eqs (7.117), (7.128), and (7.132) allows us to recognize that the relaxation time is a number average on a linear time scale and a weight average on a log time scale, with the distribution in both cases being the stress relaxation modulus $G(t)$:

$$\tau = \frac{\int_0^{\infty} tG(t)\,\mathrm{d}t}{\int_0^{\infty} G(t)\,\mathrm{d}t} = \frac{\int_{-\infty}^{\infty} t^2 G(t)\,\mathrm{d}\ln t}{\int_{-\infty}^{\infty} tG(t)\,\mathrm{d}\ln t}. \tag{7.136}$$

Two steady states are recognized for the long-time creep compliance of materials. Either the sample is a solid and the compliance becomes time independent or the sample is a liquid and the compliance becomes linear in time. Once steady state has been achieved in creep, the stress can be removed ($\sigma = 0$) and the elastic recoil, called creep recovery, can be measured. Recovery strain is defined as $\gamma_{\mathrm{R}}(t) \equiv \gamma(0) - \gamma(t)$ for $t > 0$, where t is defined to be zero at the start of recovery. The **recoverable compliance** is defined as the ratio of the time-dependent recovery strain $\gamma_{\mathrm{R}}(t)$ and the initially applied stress σ, where both γ_{R} and t are now defined to be zero at the start of recovery:

$$J_{\mathrm{R}}(t) \equiv \frac{\gamma_{\mathrm{R}}(t)}{\sigma}. \tag{7.137}$$

Boltzmann superposition relates the recoverable compliance after steady state has been achieved in creep to the creep compliance:[5]

$$J_{\mathrm{R}}(t) = J(t) - \frac{t}{\eta}. \tag{7.138}$$

For a solid, the viscosity is infinite, and $J_{\mathrm{R}}(t) = J(t)$, so all deformation in creep is subsequently recovered in creep recovery, with precisely the same time dependence, as shown in the lower curves in Fig. 7.25. In contrast, only the elastic part of the compliance of a liquid is recovered, as shown in the upper curves of Fig. 7.25:

$$\lim_{t\to\infty} J_{\mathrm{R}}(t) = \lim_{t\to\infty}\left[J(t) - \frac{t}{\eta}\right] = J_{\mathrm{eq}}. \tag{7.139}$$

7.6.5 Oscillatory shear

A simple linear viscoelastic measurement that has become very easy to implement with the advent of modern electronics is oscillatory shear. A sinusoidal strain with angular frequency ω is applied to a sample in simple shear:

$$\gamma(t) = \gamma_0 \sin(\omega t). \tag{7.140}$$

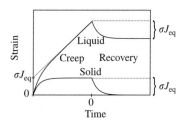

Fig. 7.25
Strain from a creep experiment with constant applied stress σ followed by a creep recovery experiment (starting at $t = 0$) with zero applied stress, for a viscoelastic solid (lower curves) and a viscoelastic liquid (upper curves). The recoverable compliance can be determined from either creep or recovery. All deformation is recovered for solids but only the elastic part of the deformation is recovered for liquids.

---

[5] Note that $t = 0$ in the left hand side of Eq. (7.138) corresponds to the start of recovery, while $t = 0$ on the right hand side corresponds to the start of creep.

The principal advantage of this technique is that the viscoelastic response of any material can be probed directly on different time scales ($1/\omega$) of interest by simply varying the angular frequency ω. If the material studied is a perfectly elastic solid, then the stress in the sample will be related to the strain through Hooke's law [Eq. (7.98)]:

$$\sigma(t) = G\gamma(t) = G\gamma_0 \sin(\omega t). \tag{7.141}$$

The stress is perfectly in-phase with the strain for a Hookean solid, as shown in Fig. 7.26. At times $t = \pi/(2\omega), 5\pi/(2\omega), 9\pi/(2\omega), \ldots$ the strain has a maximum and the stress also has a maximum. Similarly, at times $t = 0, \pi/\omega, 2\pi/\omega, \ldots$ both the strain and the stress are simultaneously zero.

On the other hand, if the material being studied is a Newtonian liquid, the stress in the liquid will be related to the *shear rate* through Newton's law [Eq. (7.100)]:

$$\sigma(t) = \eta \frac{d\gamma(t)}{dt} = \eta\gamma_0\omega\cos(\omega t) = \eta\gamma_0\omega\sin\left(\omega t + \frac{\pi}{2}\right). \tag{7.142}$$

The stress in a Newtonian liquid still oscillates with the same angular frequency ω, but is out-of-phase with the strain by $\pi/2$, as shown in Fig. 7.27. At times $t = \pi/(2\omega), 5\pi/(2\omega), 9\pi/(2\omega), \ldots$ the strain has a maximum, but both the shear rate and the stress are zero at these points. Conversely, at times $t = 0, \pi/\omega, 2\pi/\omega, \ldots$ the strain is zero, but both the shear rate and the stress are either at their maximum or minimum values.

More generally, the linear response of a viscoelastic material always has the stress oscillate at the same frequency as the applied strain, but the stress leads the strain by a **phase angle** δ.

$$\sigma(t) = \sigma_0 \sin(\omega t + \delta). \tag{7.143}$$

In general, δ can be frequency dependent, with any value in the range $0 \le \delta \le \pi/2$. The two simple cases already treated correspond to the limits allowed for the phase angle. Solids that obey Hooke's law have $\delta = 0$ at all frequencies, while liquids that obey Newton's law have $\delta = \pi/2$ at all frequencies. Since the stress is always a sinusoidal function with the same frequency as the strain, we can separate the stress into two orthogonal functions that oscillate with the same frequency, one in-phase with the strain and the other out-of-phase with the strain by $\pi/2$:

$$\sigma(t) = \gamma_0[G'(\omega)\sin(\omega t) + G''(\omega)\cos(\omega t)]. \tag{7.144}$$

Equation (7.144) defines $G'(\omega)$ as the **storage modulus** and $G''(\omega)$ as the **loss modulus**. Equation (7.144) can be related to the previous equation for the stress in oscillatory shear using the trigonometric identity for the sine of a sum:

$$\sin(\omega t + \delta) = \cos\delta\sin(\omega t) + \sin\delta\cos(\omega t). \tag{7.145}$$

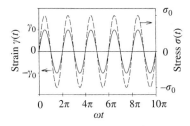

Fig. 7.26
Oscillatory strain (solid curve and left axis) and oscillatory stress (dashed curve and right axis) are in-phase for a Hookean solid.

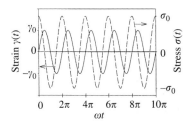

Fig. 7.27
Oscillatory strain (solid curve and left axis) and oscillatory stress (dashed curve and right axis) are out-of-phase for a Newtonian liquid. The stress leads the strain by phase angle $\delta = \pi/2$.

This suggests that the storage and loss moduli can be related to the phase angle and the modulus amplitude σ_0/γ_0 at each frequency ω:

$$G' = \frac{\sigma_0}{\gamma_0} \cos \delta, \tag{7.146}$$

$$G'' = \frac{\sigma_0}{\gamma_0} \sin \delta. \tag{7.147}$$

The ratio of loss and storage moduli is the tangent of the phase angle, called the **loss tangent**:

$$\tan \delta = \frac{G''}{G'}.$$

The storage and loss moduli are the real and imaginary parts of the **complex modulus** $G^*(\omega)$:

$$G^*(\omega) = G'(\omega) + iG''(\omega). \tag{7.148}$$

The Boltzmann superposition principle can be used to show that the storage modulus is related to the sine transform of $G(t)$,

$$G'(\omega) = G_{eq} + \omega \int_0^\infty [G(t) - G_{eq}] \sin(\omega t)\, dt, \tag{7.149}$$

and the loss modulus is obtained from the cosine transform (see Problem 7.41):

$$G''(\omega) = \omega \int_0^\infty [G(t) - G_{eq}] \cos(\omega t)\, dt. \tag{7.150}$$

In these equations, G_{eq} is the equilibrium modulus of the solid at long times (or low frequencies). For a liquid, $G_{eq} \equiv 0$.

Figure 7.22 showed that the relaxation modulus distinguishes viscoelastic solids from viscoelastic liquids via the long-time behaviour of $G(t)$. Since low frequencies correspond to long times, the oscillatory shear experiment makes this distinction in the low frequency response of the material. The terminal (low frequency) response of any solid is dominated by the storage modulus because the stress is very nearly in-phase with the strain. The viscoelastic solid has $G' \gg G''$ at low frequencies, and G' becomes independent of frequency in the low frequency limit, with a value equal to the equilibrium modulus G_{eq}:

$$G_{eq} = \lim_{\omega \to 0} G'(\omega). \tag{7.151}$$

The viscoelastic response of a liquid in oscillatory shear is markedly different. The terminal response (at low frequency) of any liquid is dominated by the loss modulus because the stress is very nearly in-phase with the shear rate. The viscoelastic liquid has $G'' \gg G'$ at low frequencies. G'' is proportional to frequency in the low-frequency limit, and the proportionality constant is the viscosity [compare Eqs (7.142) and (7.144)]:

$$\eta = \lim_{\omega \to 0} \frac{G''(\omega)}{\omega}. \tag{7.152}$$

The storage modulus of a viscoelastic liquid is not zero, but instead is proportional to the square of frequency, and provides a measure of the stored elastic energy. Traditionally, this is written in terms of the steady state compliance J_{eq}:

$$J_{eq} = \frac{1}{\eta^2} \lim_{\omega \to 0} \frac{G'(\omega)}{\omega^2} = \lim_{\omega \to 0} \frac{G'(\omega)}{[G''(\omega)]^2}. \tag{7.153}$$

An example of the linear viscoelastic response in oscillatory shear for a nearly monodisperse linear polybutadiene melt is shown in Fig. 7.28. Extrapolation of the limiting power laws of $G' \sim \omega^2$ and $G'' \sim \omega$ (the dashed lines in Fig. 7.28) to the point where they cross has special significance. The intersection of the power laws $G' = J_{eq}\eta^2\omega^2$ and $G'' = \eta\omega$ using the above two equations allows us to solve for the frequency where they cross $\omega = 1/(\eta J_{eq})$, which is the reciprocal of the relaxation time[5] τ [Eq. (7.132)]. The modulus level where the two extrapolations cross, obtained by setting $\omega = 1/\tau = 1/(\eta J_{eq})$ in either equation, is simply the reciprocal of the steady state recoverable compliance $1/J_{eq}$.

The viscoelastic responses of two polymer networks are shown in Fig. 7.29. One is a nearly perfect network (circles) made by end-linking linear chains with two reactive ends. The storage modulus for this network (filled circles) is independent of frequency and much larger than the loss modulus (open circles). For comparison, an imperfect network made by linking a mixture of chains with one and two reactive ends is also shown. While $G' > G''$ for the imperfect network, the storage modulus (filled squares) has a weak frequency dependence and the loss modulus (open squares) is significantly larger than for the perfect network. Both of these observations are caused by the gradual relaxation of dangling structures in the imperfect network (see Problem 9.43). The loss tangent $\tan \delta = G''/G'$ is much larger for the imperfect network than for the nearly perfect one.

The various experimental methods of linear viscoelasticity are summarized in Table 7.1. All information for linear viscoelastic response can, in principle, be obtained from each method. The oscillatory methods are particularly useful because they directly probe the response of the system on the time scale of the imposed frequency of oscillation $1/\omega$. Commercial rheometers can accomplish this with either applied stress or applied strain, and the two methods are both listed in Table 7.1.

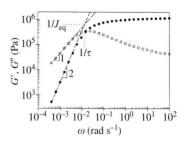

Fig. 7.28
Oscillatory shear response of a linear polybutadiene at 25 °C, with $M_w = 925\,000\,\text{g mol}^{-1}$ and $M_w/M_n < 1.1$. Filled symbols are the storage modulus G' and open symbols are the loss modulus G''. The crossing of the terminal slopes of 1 and 2 (dashed lines) determines the relaxation time and the steady state recoverable compliance.

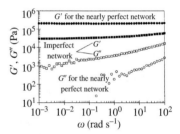

Fig. 7.29
Oscillatory shear response of two PDMS networks at 30 °C. Filled symbols are the storage modulus G' and open symbols are the loss modulus G''. The circles are nearly perfect networks made by end-linking linear chains of $M_n = 20\,000\,\text{g mol}^{-1}$ with two reactive ends. The squares are imperfect networks made by reacting 48 mol% linear chains of $M_n = 71\,000\,\text{g mol}^{-1}$ with one reactive end and 52 mol% linear chains of $M_n = 58\,000\,\text{g mol}^{-1}$ with two reactive ends. Notice that the storage modulus is lower and the loss modulus is higher for the imperfect networks. The increased noise in G'' at low frequencies is caused by the difficulty in precise determination of a phase angle very close to zero.

---

[5] This relaxation time for the Maxwell model is an average relaxation time [see Eq. 7.136] whenever a material has multiple relaxation modes.

Table 7.1 Comparison of linear viscoelastic experimental methods

Technique	Apply	Measure	Determine
Step strain	Constant strain γ	Stress $\sigma(t)$	Relaxation modulus $G(t) = \sigma(t)/\gamma$
Steady shear	Constant shear rate $\dot{\gamma}$	Steady stress σ	Viscosity $\eta = \sigma/\dot{\gamma}$
Step stress (creep)	Constant stress σ	Strain $\gamma(t)$	Compliance $J(t) = \gamma(t)/\sigma$
Oscillatory strain	$\gamma(t) = \gamma_0 \sin \omega t$	$\sigma(t) = \sigma_0 \sin(\omega t + \delta)$	$G'(\omega) = (\sigma_0/\gamma_0)\cos \delta$; $G''(\omega) = (\sigma_0/\gamma_0)\sin \delta$
Oscillatory stress	$\sigma(t) = \sigma_0 \sin \omega t$	$\gamma(t) = \gamma_0 \sin(\omega t - \delta)$	$G'(\omega) = (\sigma_0/\gamma_0)\cos \delta$; $G''(\omega) = (\sigma_0/\gamma_0)\sin \delta$

7.7   Summary of networks and gels

Polymer networks are very important soft solids. As such, they find applications ranging from adhesives to automobile tires. The mechanical properties of network polymers arise primarily from the changes in entropy of the network strands when the network is macroscopically deformed. One direct consequence of entropic elasticity is that the modulus increases as temperature is raised (whereas conventional solids get softer). Network polymers are thus quite different from other solids, such as crystalline metals, where mechanical properties have energetic origins, coming from the change in internal energy when lattice spacings are distorted from their equilibrium values. The modulus G of network polymers in their preparation state is proportional to kT per elastically effective network strand, plus any contribution from trapped entanglements:

$$G \cong G_x + G_e \tag{7.154}$$

The number density of network strands determines G_x and the plateau modulus caused by inter-chain entanglements is G_e. This plateau modulus is understood on a molecular level by imagining surrounding chains confining each network strand to an effective tube.

Many models lead to a classical form for the variation of the true stress with uniaxial elongation:

$$\sigma_{\text{true}} = G\left(\lambda^2 - \frac{1}{\lambda}\right). \tag{7.155}$$

For large stretching ratios, finite extensibility of the network strands becomes important and this strong stretching is better described by finite extensibility models such as the inverse Langevin function for freely jointed chains. Classical theory [Eq. (7.155)] is not applicable for entangled networks because entanglements are qualitatively different than crosslinks, when the network is deformed. This qualitative difference leads to non-affine deformation of the confining tube and a non-classical stress–strain relation [Eq. (7.65)].

Polymer gels are diluted polymer networks. The diluent can be other polymer chains or solvent. The modulus of gels decreases on dilution, in ways that depend on the details of the polymer–solvent interaction. When immersed in an excess of an appropriate solvent, polymer networks can often swell considerably. The swelling is driven by the favorable free energy of mixing with solvent (osmotic pressure) and is resisted by the energy required to stretch network strands (modulus). The swelling reaches an equilibrium state when the osmotic and elastic parts of the free energy balance. The equilibrium swelling of an unentangled network in a θ-solvent can be used to estimate the average number of monomers in a network strand:

$$N \approx Q^{8/3}. \tag{7.156}$$

However, such relations only apply to networks that do not have entanglements. Real networks usually have important effects from entanglements and their swelling is not yet quantitatively understood. However, some qualitative features apply to both entangled and unentangled swollen networks. At a constant extent of crosslinking, the modulus always increases with both the preparation state concentration and the concentration of use. Furthermore, fewer entanglements exist in networks prepared at lower concentrations and deswelling such networks to the melt state can make networks with a modulus *below* the plateau modulus of the melt.

Dynamics in polymer networks and also in polymer liquids can be studied by a variety of viscoelasticity experiments. For networks and gels, these methods determine the modulus from the long time (or low frequency) viscoelastic response and directly measure the relaxation of dangling structures and sol fraction at higher frequencies. For polymer liquids in solution or melt states, the viscosity and relaxation time are determined from the long time (or low frequency) viscoelastic response, with higher frequencies providing information about other relaxation modes. The linear viscoelastic response of polymer liquids will be discussed in Chapters 8 and 9.

Problems

Section 7.1

7.1 Equation (7.9) is an example of a Maxwell relation. Derive the other two Maxwell relations associated with the Helmholtz free energy of a rubber.

7.2 Estimate the fraction of the tensile force at 300 K that has energetic origins by using a Flory construction on the following data for the temperature dependence of tensile force for a crosslinked rubber with 1 cm^2 cross-sectional area held at constant elongation.

T (K)	219	253	293	335
Force (N)	1.13×10^2	1.28×10^2	1.45×10^2	1.64×10^2

7.3 If a crosslinked rubber is rapidly stretched at constant volume, does it get warmer or cooler? Explain your answer.

Section 7.2

7.4 Consider an affine model of an incompressible network with polydisperse strands between crosslinks. Prove that the stress σ in this network due to uniaxial extension still follows the classical dependence on deformation λ [Eqs (7.32) and (7.33)]. Which average molar mass of network strands enters into Eq. (7.31) for the network modulus?

7.5 What would be the size of a phantom network, made from strands with N monomers of size b and with functionality f of crosslinks, if it were not attached to the macroscopic boundaries of the network?

7.6 For any isotropic solid stretched in the x direction, the longitudinal tensile strain ε_{xx} is the change in length divided by the initial length. Similarly, the transverse tensile strains ε_{yy} and ε_{zz} are the changes in width and height divided by the initial values.

(i) Write the longitudinal and transverse strains in terms of the deformation factor λ.

Poisson's ratio ν is defined from the ratio of transverse and longitudinal strains in tension:

$$\nu \equiv \frac{-\varepsilon_{yy}}{\varepsilon_{xx}}. \tag{7.157}$$

(ii) Write Poisson's ratio in terms of λ, and show that for small deformations ($\lambda \to 1$) at constant volume, Poisson's ratio is $1/2$. What happens to Poisson's ratio at larger deformations?

7.7 The **bulk modulus** K is defined as the reciprocal of the isothermal compressibility, and Young's modulus E is defined as the ratio of longitudinal tensile stress and longitudinal tensile strain:

$$K \equiv -V\frac{\partial P}{\partial V}\bigg|_T \qquad E \equiv \frac{\sigma_{xx}}{\varepsilon_{xx}} \tag{7.158}$$

Poisson's ratio relates Young's modulus to either the shear modulus G or the bulk modulus K:

$$E = 2(1+\nu)G = 3(1-2\nu)K. \tag{7.159}$$

A vulcanized natural rubber (*cis*-polyisoprene) has bulk modulus $K = 2 \times 10^9$ Pa and Young's modulus $E = 1.3 \times 10^6$ Pa.

(i) What is Poisson's ratio for this vulcanized natural rubber?
(ii) What is the shear modulus of this vulcanized natural rubber?
(iii) Show that in the limit where the bulk modulus is much larger than the shear modulus, Poisson's ratio is $1/2$. This limit is also achieved for most liquids (including polymer melts and solutions) which is why they are termed incompressible.
(iv) Relate the deviation of Poisson's ratio from its incompressible limit ($\nu = 1/2$) to the ratio of moduli E/K.

7.8 Heating up a strip of natural rubber from $T_1 = 300$ K to $T_2 = 350$ K increases the stress at constant elongation $\lambda = 2$ from $\sigma_1 = 4.3 \times 10^5$ Pa to $\sigma_2 = 5 \times 10^5$ Pa. Assuming a linear dependence of stress on temperature at constant elongation, estimate the energetic σ_E and entropic σ_S contributions to the stress at $T_1 = 300$ K.

7.9 A crosslinked polybutadiene ball of mass 100 g is dropped at 300 K from a height of 1 m and bounces back 92 cm.

(i) Assuming that the floor does not deform and all energy lost in this collision goes into heating up the ball, how much did the ball's temperature increase? The heat capacity of polybutadiene at 300 K is $C_p = 2.2$ kJ kg^{-1} K^{-1}.
(ii) By how much does this temperature change alter the elastic modulus of the ball?

7.10 The rubber in an inflated balloon is stretched biaxially, with initial diameter d_0 and initial thickness of its walls t_0. This incompressible rubber contains ν elastic strands per unit volume.

(i) Derive the tensile stress σ in either of the two stretching directions in the rubber in terms of the relative change of balloon size $\lambda = d/d_0$ during subsequent inflation or deflation. Assume that the rubber obeys the affine network model.

(ii) Relate the internal pressure (p) and the degree of expansion $(\lambda = d/d_0)$. Assume that the ideal gas law $(pV = n\mathcal{R}T)$ is valid for the helium inside the balloon.

(iii) At what value of λ is the helium pressure maximum?

This problem was adapted from U. W. Gedde, *Polymer Physics*, Chapman and Hall, London (1995).

7.11 A steel ball is attached to a rubber band of size $0.2\,\text{cm} \times 0.2\,\text{cm} \times 10\,\text{cm}$ and stretches it in the z direction to a 40 cm length, developing an engineering stress $2 \times 10^7\,\text{dyn}\,\text{cm}^{-2}$ in the rubber at the temperature $20\,^\circ\text{C}$. Assume that the rubber band obeys the affine model.

(i) What is the relative elongation λ of the rubber band?

(ii) How many moles of network chains ν are there per cubic centimeter?

(iii) What is the number-average molar mass M_s between crosslinks, if the density of rubber is $\rho = 1\,\text{g}\,\text{cm}^{-3}$?

(iv) What is the mass m of the steel ball?

(v) What engineering stress σ_1 is required to stretch the rubber to a length 30 cm at $20\,^\circ\text{C}$?

(vi) What mass m_1 do we need to attach to the rubber band to develop this stress σ_1 at $50\,^\circ\text{C}$?

7.12 A strip of elastomer $1\,\text{mm} \times 1\,\text{mm} \times 10\,\text{cm}$ that obeys the affine network model is stretched to 40 cm length at $T = 300\,\text{K}$ at the engineering stress $1 \times 10^7\,\text{Pa}$. The volume occupied by each monomer is $b^3 = 80\,\text{Å}^3$.

(i) What is the relative elongation λ?

(ii) Find the shear modulus G.

(iii) How many moles of network strands are in this rubber strip?

(iv) Estimate the number of monomers between crosslinks.

7.13 Two identical 10 cm long rubber bands, A and B, are tied together at their ends, stretched to a total length of 80 cm and held in this stretched state. Rubber bands A and B are held at different temperatures T_A and T_B.

(i) Is the true stress σ_{true} the same for the two bands? Explain.

(ii) Is the engineering stress σ_{eng} the same for the two bands? Explain.

(iii) What is the temperature T_B of band B if the length of rubber band B is $L_B = 38\,\text{cm}$ and the temperature of band A is $10\,^\circ\text{C}$. Assume the classical form for the stress-elongation relation of the rubber band.

7.14 To stretch a rubber band by the factor of $\lambda = 2$ requires true stress $\sigma_1 = 1.5 \times 10^7\,\text{dyn}\,\text{cm}^{-2}$ at $0\,^\circ\text{C}$ and true stress $\sigma_2 = 1.65 \times 10^7\,\text{dyn}\,\text{cm}^{-2}$ at $30\,^\circ\text{C}$. Assume that the rubber band obeys the affine model.

(i) Estimate the entropic and energetic components of the stress at $30\,^\circ\text{C}$.

(ii) What is the entropic part of the modulus of the rubber band at $30\,^\circ\text{C}$?

(iii) What is the number-average molar mass of chains between crosslinks, if the elastomer density is $\rho = 1.1\,\text{g}\,\text{cm}^{-3}$?

7.15 A rubber band with molar mass between crosslinks $M_s = 3000\,\text{g}\,\text{mol}^{-1}$ is uniaxially stretched to three times its original length. After achieving equilibrium at $21\,^\circ\text{C}$ it is allowed to contract adiabatically back to the unstretched state.

(i) Is the final temperature higher or lower than 21 °C? Why?

(ii) Estimate the temperature change of this rubber band if its heat capacity is $C_p = 1700 \, \mathrm{J \, kg^{-1} \, K^{-1}}$. Assume the classical form for the stress-elongation relation of the rubber band.

7.16   What is the molecular origin of strain hardening at large elongations of polymeric networks that do not crystallize on stretching?

7.17   Demonstrate that the true stress in uniaxially deformed incompressible networks is the derivative of the free energy per unit volume F/V with respect of the logarithm of deformation λ:

$$\sigma = \left. \frac{\partial (F/V)}{\partial (\ln \lambda)} \right|_V. \tag{7.160}$$

Section 7.3

7.18   Demonstrate that I_1, I_2, and I_3 (defined in Eqs. 7.49, 7.50, and 7.51) are invariant with respect to any change in coordinate system.

7.19   Consider the Pincus blob picture of stretching an ideal chain, discussed in Section 2.6.1. For each Pincus blob there is of order kT of entropic free energy stored in the stretched chain. The chain is a sequence of *trans*, *gauche$_+$*, and *gauche$_-$* states (see Fig. 2.1) with the energy difference between *trans* and *gauche* states of order kT. Can the stretching free energy per Pincus blob be thought of as forcing the chain to make one extra trans state (going forward between neighbouring Pincus blobs)? Indeed, at maximum extension, the chain is in the all-*trans* conformation (Fig. 2.2). Why is there no significant energetic contribution to stretching an ideal chain comparable to the entropic contribution (of order kT per Pincus blob)?

7.20*   Consider a non-affine tube model, with the number of monomers in the virtual chains changing with network deformation as $n_i = \lambda_i^2 n_{i0}$, where n_{i0} is the number of monomers in virtual chains constraining fluctuations in the i direction in the undeformed state.

(i) Start with virtual chains with $n_{i0} \approx N_e$ virtual monomers in the undeformed state, with the points of attachments to the network chains separated by N_e real monomers. Demonstrate that after elongation by a factor $\lambda_x > 1$ in the x direction, the affine strand in the x direction contains $\lambda_x N_e$ real monomers. The affine strand is defined as the smallest section of a network strand that deforms affinely. The entire network deforms affinely on length scales larger than the affine strand. Sections of network chain that are smaller than the affine strand are deformed non-affinely.

Hint: Use the invariance of confinement upon appropriate simultaneous change of the length and the number of virtual chains [Eq. (7.61)].

(ii) Derive the following result for the size of the affine strand in the elongation direction:

$$R_{\mathrm{af}}(\lambda_x) \approx b \sqrt{N_e} \lambda_x^{3/2} \quad \text{for } \lambda_x > 1.$$

(iii) Prove that the distance between entanglements that are separated by N_e monomers along the deformed affine strand is

$$a_x \approx b \sqrt{N_e} \lambda_x^{1/2}.$$

Compare this distance between entanglements with the root-mean-square fluctuations of the affine strand. Explain this result.

Hint: Assume that the affine strand is uniformly stretched.

(iv) Show that the contribution to the free energy density from the elongation direction is linear in deformation:

$$\frac{F_x}{V} \approx \frac{\rho \mathcal{R} T}{M_e} \lambda_x$$

Hint: Consider the contribution from an affine strand.

(v) Derive the following result for the size of the affine strand in the compression (y) direction ($\lambda_y < 1$):

$$R_{af}(\lambda_y) \approx b \sqrt{N_e} \lambda_y^{1/2} \quad \text{for } \lambda_y < 1.$$

Show that the affine strand has N_e / λ_y monomers.
Hint: Use the invariance of confinement upon appropriate simultaneous change of the length and the number of virtual chains [Eq. (7.61)].

(vi) Show that the free energy density contribution from the direction of confinement is

$$\frac{F_y}{V} \approx \frac{\rho \mathcal{R} T}{M_e} \frac{1}{\lambda_y}.$$

(vii) Assume that the free energy density cost for each component can be approximated by the expression

$$\frac{F_i}{V} \approx \frac{\rho \mathcal{R} T}{M_e} \left(\lambda_i + \frac{1}{\lambda_i} \right), \tag{7.161}$$

and the total free energy density of the network can be approximated by the sum of the phantom and entanglement contributions

$$\frac{F}{V} \approx \frac{\rho \mathcal{R} T}{M_x} \sum_i \frac{\lambda_i^2}{2} + \frac{\rho \mathcal{R} T}{M_e} \sum_i \left(\lambda_i + \frac{1}{\lambda_i} \right). \tag{7.162}$$

Demonstrate that for uniaxial deformation of an incompressible network, the stress is given by Eq. (7.64) where the crosslink and entanglement moduli G_x and G_e, respectively, are defined in Eqs (7.43) and (7.47).

7.21* Consider a combined linear chain consisting of a 'real' N-mer and a 'virtual' m-mer. The ends of this combined chain are attached to an elastic non-fluctuating background. The x coordinate of the attachment points in the undeformed state is X_1 at the virtual chain end and X_2 at the real chain end of the combined chain. The ends of the combined chain deform affinely with the elastic background $X_1' = \lambda X_1$ and $X_2' = \lambda X_2$.

(i) Show that if the stress were supported exclusively by the real chain it would be equal to

$$\sigma = \frac{3kT}{b^2 N} \langle (R' - X_2')^2 \rangle - kT.$$

(ii) Derive the expression of mean-square end-to-end distance of the N-mer part of the combined chain.

$$\langle (R' - X_2')^2 \rangle = \frac{N^2}{(N+m)^2} (X_2' - X_1')^2 + \frac{b^2 N m}{N+m}.$$

(iii) Show that the stress supported exclusively by the real chain is

$$\sigma = kT\frac{N}{N+m}\left(\frac{(X_2' - X_1')^2}{b^2(N+m)} - 1\right).$$

(iv) Assume that the number of monomers in the virtual chain changes as a power law of the deformation:

$$m = m_0\lambda^\alpha.$$

Show that the contribution of this combined chain to the stress is

$$\sigma = kT\left(\frac{3(X_2' - X_1')^2}{b^2(N + m_0\lambda^\alpha)} - 1\right)\frac{N}{N + m_0\lambda^\alpha} + kT\left(1 - \frac{\alpha}{2}\right) \tag{7.163}$$

Determine the exponent α for which the virtual chain does not contribute to the stress and all of the stress is supported by the real chain.

7.22 Consider a comb polymer with p Kuhn monomers of length b between neighbouring points of attachment of n-mer side branches. Assume the number of side branches is large and that the end of each side branch is randomly attached to the elastic non-fluctuating background. Show that the mean-square fluctuation of the junction points is $b^2\sqrt{pn}$ in the limit $n \gg p$. What is the mean-square fluctuation of the junction point in the opposite limit $(p \gg n)$?

Section 7.4

7.23 Consider a network formed by end-linking chains with degree of polymerization $N = 100$ and monomer volume $b^3 = 30\,\text{Å}^3$ in a dry melt state $(\phi_0 = 1)$.

(i) What is the modulus $G(1)$ of the network at preparation (dry) conditions at room temperature?

(ii) What is the equilibrium swelling ratio of this network at the θ-temperature $(\theta = 35\,°\text{C})$?

(iii) What is the equilibrium swelling ratio of this network in an athermal (good) solvent?

(iv) What would be the modulus $G(1/Q)$ of the network in a fully swollen state in the athermal solvent?

7.24 (i) Calculate the concentration dependence of the osmotic pressure of a gel in an athermal solvent, being sure to include both mixing and strand elasticity parts of the free energy.

(ii) Plot the osmotic pressure of a gel prepared in the melt state $(\phi_0 = 1)$ with $N = 100$ monomers between crosslinks and the osmotic pressure of a solution of very long linear chains as functions of volume fraction of polymer. Why is the osmotic pressure of the gel smaller than that of the solution?

(iii) At what concentration is the osmotic pressure of the gel zero? What is the physical meaning of this concentration?

7.25 Swelling a thin film network.

The photoactive elements in photographic film are dispersed in a crosslinked gelatin network. During processing, this network is swollen to allow the developer solution to enter the gel. Since the network is strongly bonded to a solid substrate, it is constrained to only swell in one dimension (perpendicular to the substrate) with the other two dimensions unchanged by swelling.

(i) Derive the expression below for the concentration dependence of the modulus of the swollen gel, assuming the developer solution is an athermal solvent and that the network is also prepared in an athermal solvent:

$$G(\phi) \approx \frac{kT}{b^3 N} \frac{\phi_0^{7/4}}{\phi^{3/4}}. \qquad (7.164)$$

(ii) Why does this modulus *increase* as the network is diluted?

(iii) Derive an expression for the equilibrium swelling.

(iv) For a photographic film with $N = 100$ Kuhn monomers between crosslinks, prepared at volume fraction $\phi_0 = 0.1$, what is the equilibrium swelling?

7.26 What is the true osmotic pressure of a polymer gel swollen to equilibrium in a good solvent?

7.27 Estimate the average number of monomers in a network strand for gels prepared in the dry state and swollen to equilibrium in a θ-solvent with

(i) $Q = 2$,

(ii) $Q = 4$

7.28 At room temperature, compare the equilibrium swelling, dry modulus and modulus at swelling equilibrium for a network with $N = 10$ Kuhn monomers of length $b = 4\,\text{Å}$ per strand prepared in the dry state, swollen in

(i) athermal solvent,

(ii) θ-solvent

7.29 The modulus of unentangled networks is kT per combined strand [Eq. (7.43)]. Explain why the stretching caused by isotropic swelling increases the contribution to the modulus from each combined strand. Why can the elastic modulus be approximated by an elastic energy density?

7.30 A network with strands of number-average degree of polymerization $N = 100$ has equilibrium swelling ratio $Q = 25$ in an athermal solvent. Assuming the network was prepared in an athermal solvent, estimate the preparation concentration.

7.31 Consider a PDMS network prepared in the dry state that swells to an equilibrium swelling ratio of $Q = 4$ in the concentrated regime (θ-like swelling). Estimate the dry modulus of the network at room temperature. Compare your result with Fig. 7.17 and estimate the numerical prefactor in Eq. (7.91).

7.32 Consider a PDMS network prepared in the dry state that swells in toluene at 25 °C to an equilibrium swelling ratio of $Q = 7$. Estimate the dry and fully swollen moduli if the boundary of the concentrated regime is $\phi^{**} \approx 0.2$ and $b = 13\,\text{Å}$ for PDMS.

Section 7.5

7.33 Langley used the mean-field percolation model to derive the entanglement trapping factor, assuming that the probability of entanglement between two network strands is proportional to the square of their concentration. The probability $\mathcal{P}_2$ that a randomly chosen monomer in a network strand is connected to the macroscopic gel along both linear chain paths emanating from it is determined by integrating the gel curve [Eq. (6.163)] for each direction over the entire strand:

$$\mathcal{P}_2 = \frac{1}{N} \int_0^N [1 - \exp(-spP_{\text{gel}})][1 - \exp(-[N - s]pP_{\text{gel}})]\, ds. \qquad (7.165)$$

The integration variable s represents the contour coordinate that counts monomers along the strand (running from 0 at one end to N at the other end) and p is the probability that a monomer has been randomly crosslinked. This integration assures that each monomer in the strand of N monomers is connected in both directions to the gel.

(i) Obtain the probability $\mathcal{P}_2$ by integrating Eq. (7.165) and use Eq. (6.163) to write your result as a function solely of gel fraction.
(ii) For an entanglement to be permanently trapped by the crosslinking, all four strand ends in Fig. 7.9 must be connected to the macroscopic gel, making the entanglement trapping factor of Eq. (7.95) $T_e \approx (\mathcal{P}_2)^2$. If any of the four paths are simply a network defect (such as a dangling end or loop, see Fig. 7.7) the entanglement will eventually be abandoned and hence not contribute to the modulus of the network. Use the result from part (i) to derive the Langley prediction for the entanglement trapping factor:

$$T_e = \left[2 - P_{\text{gel}} + \frac{2P_{\text{gel}}}{\ln(1 - P_{\text{gel}})}\right]^2. \qquad (7.166)$$

7.34 (i) Calculate the probability $\mathcal{P}_1$ that a randomly chosen monomer in a strand is connected to the macroscopic gel along only one of the linear chain paths emanating from it.
(ii) Calculate the probability $\mathcal{P}_0$ that a randomly chosen monomer in a strand is not connected to the macroscopic gel along either of the two linear paths emanating from it.
(iii) Show that the probabilities $\mathcal{P}_0$, $\mathcal{P}_1$, and $\mathcal{P}_2$ (from Problem 7.33) are properly normalized (i.e., that $\mathcal{P}_0 + \mathcal{P}_1 + \mathcal{P}_2 = 1$).

7.35 When a lightly crosslinked gel first swells in a solvent, it jumps around. Why does it do it?

Section 7.6

7.36 Stress relaxation.

(i) Use the following table of data for the shear stress relaxation modulus of a polymer melt to determine the viscosity η, the steady state compliance J_{eq}, and the longest relaxation time ηJ_{eq}. Be sure to include any graphs that may be relevant for these calculations.

t(s)	$G(t)$ (Pa)	t(s)	$G(t)$ (Pa)	t(s)	$G(t)$ (Pa)
0.0133	1.0×10^6	20.6	4.8×10^5	106	1.2×10^5
0.365	9.4×10^5	28.6	4.1×10^5	122	9.6×10^4
1.00	8.8×10^5	41.9	3.2×10^5	139	7.6×10^4
1.90	8.3×10^5	50.0	2.8×10^5	155	6.2×10^4
2.79	7.9×10^5	55.3	2.6×10^5	189	4.0×10^4
3.69	7.5×10^5	60.6	2.4×10^5	238	2.2×10^4
5.50	7.0×10^5	68.6	2.1×10^5	305	1.0×10^4
7.29	6.6×10^5	73.9	1.9×10^5	356	5.0×10^3
10.0	6.1×10^5	81.9	1.7×10^5	422	2.2×10^3
15.3	5.4×10^5	90.0	1.5×10^5	472	7.1×10^2

(ii) How does the longest relaxation time compare with the maxima in the functions $tG(t)$ and $t^2G(t)$?

(iii) Sketch the time dependences of the creep compliance and recoverable compliance that you would expect to see for the polymer melt in part (i), assuming creep and recovery were measured at the same temperature. Indicate values of the rubbery compliance and the steady state compliance on the plot.

7.37 Creep.

(i) Determine the viscosity η, the steady state compliance J_{eq}, and the longest relaxation time $\eta_0 J_{eq}$ from the following creep data on a polymer solution at a shear stress of 20 Pa:

Time (s)	100	200	300	400	500	600	700	800	900	1000
Strain	0.31	0.52	0.74	0.92	1.10	1.28	1.46	1.64	1.82	2.00

(ii) Small-angle scattering reveals that the polymer's radius of gyration in this solution is 400 Å. Individual chains in polymer liquids move randomly in solution by a process known as diffusion. For diffusive motion, the square of the typical distance x moved is proportional to the time t allowed for motion, with the coefficient of proportionality being the diffusion coefficient D:

$$D \approx \frac{x^2}{t}.$$

Use the fact that the polymer roughly diffuses a distance equal to its coil size during a time interval equal to its longest relaxation time, to estimate the diffusion coefficient of the polymer.

7.38 Determine the viscosity η, the steady state compliance J_{eq}, and the longest relaxation time $\tau \approx \eta J_{eq}$ from the following data for monodisperse linear polystyrene with $M = 60\,600\ \mathrm{g\,mol^{-1}}$ at $180\,^\circ\mathrm{C}$.

ω (rad s^{-1})	G' (Pa)	G'' (Pa)
57.54	17 400	49 000
36.31	7590	33 100
22.91	3090	21 900
14.45	1260	14 100
9.12	525	9120
5.75	200	5750
3.63	83.2	3720
2.91	33.1	2900

7.39 At low frequencies G'' of a liquid contains information on viscosity η, while G' of a liquid contains information on recoverable compliance J_{eq}. At low frequencies G' of a solid contains information on its modulus G_{eq}. What does G'' tell us about the network at low frequency?

7.40 If it is necessary to signal average for three full cycles, how much time was required to obtain the lowest frequency data points of Fig. 7.28?

7.41 Boltzmann superposition in oscillatory shear.

Use the Boltzmann superposition integral to derive the storage modulus of a viscoelastic liquid as a sine transform of the stress relaxation modulus $G(t)$ [Eq. (7.149) with $G_{eq} = 0$)]. Also derive the loss modulus as a cosine transform of $G(t)$ [Eq. (7.150) with $G_{eq} = 0$] for a viscoelastic liquid.

7.42 Oscillatory shear.

Show that the dissipated energy per unit volume in each cycle of oscillatory shear is $\pi\gamma_0^2 G''$.

7.43 Prove that, for an affine network in uniaxial tension, the ratio

$$\frac{\sigma_{true}}{\lambda^2 - 1/\lambda}$$

is the shear modulus defined in Eq. (7.98), making use of the fact that Young's modulus is three times the shear modulus.

Hint: Hooke's law in tension is $\sigma_{true} = E\varepsilon$, where $E = 3G$ is Young's modulus and $\varepsilon = (L - L_0)/L_0$ is the extensional strain (L and L_0 are the final and initial lengths of the sample).

7.44* Use the Boltzmann superposition integral to derive Eq. (7.128) for the recoverable compliance of a viscoelastic liquid.

7.45 Diblock copolymers with roughly equal block lengths can microphase separate into a lamellar phase, with alternating layers of mostly A monomer and mostly B monomer. When rapidly cooled into the lamellar phase from the isotropic phase, the layers formed by diblock copolymers are roughly parallel to each other locally. A polydomain texture is created from this quenching, with a typical grain size of order 0.1 μm. The oscillatory shear response of such a quenched sample is observed to have $G' \sim G'' \sim \omega^{1/2}$ at the lowest measurable frequencies. Can this observed response be the real terminal response of the sample? Is this sample a viscoelastic solid or a viscoelastic liquid?

7.46 Consider a liquid with a single relaxation mode and stress relaxation modulus given by the Maxwell model:

$$G(t) = G_0 \exp(-t/\tau).$$

(i) Calculate the viscosity of this liquid.
(ii) Estimate the viscosity of the liquid using $\eta \approx G(\tau)\tau$ [Eq. (7.120)]. What is the relative error of this estimate?

7.47 Find the relation between creep compliance $J(t)$ and recoverable compliance $J_R(t)$ using the Boltzmann superposition principle.

7.48 Dielectric spectroscopy indicates that water molecules respond to an oscillating electric field at a frequency of 17 GHz at room temperature. Is water still a Newtonian liquid at this high a frequency or is it viscoelastic? If it is viscoelastic, at what time scales can viscoelasticity be observed?

7.49 The Voigt model is similar to the Maxwell model, but the elastic and viscous elements are in parallel instead of in series. In the Voigt model each element has the same strain and the stress is the sum of the stresses in the two elements.

(i) Derive the creep compliance of the Voigt model. What type of material does the Voigt model describe? What is the behaviour of the Voigt model during creep recovery?
(ii) Derive the stress relaxation modulus of the Voigt model. What goes wrong with the Voigt model in describing stress relaxation?
(iii) Derive the stress relaxation modulus of a series combination of the Voigt model (with modulus G_V and viscosity η_V) and an elastic element (with modulus G_E).
(iv) Describe in words what happens to the three elements in part (iii) as a function of time after the step strain.

(v) What is the relaxation time for the three-element model with constants described in part (iii) and what is its physical significance?

7.50 Consider a series combination of two Maxwell models, whose stress relaxation modulus is a simple sum:

$$G(t) = G_A \exp(-t/\tau_A) + G_B \exp(-t/\tau_B).$$

(i) Calculate the storage and loss moduli of this system.
(ii) What is the longest relaxation time of the system if $\tau_A > \tau_B$? What is the average relaxation time τ determined from Eq. (7.136)?
(iii) What are the viscosity and the recoverable compliance of the series combination of two Maxwell models?
(iv) Calculate the low frequency behaviour of the storage and loss moduli. Verify that $\omega = 1/\tau$ corresponds to the intersection of the extrapolations of the low frequency power laws of $G'(\omega)$ and $G''(\omega)$. Verify that $\tau = \eta J_{eq}$.

Bibliography

Boyd, R. H. and Phillips, P. J. *The Science of Polymer Molecules* (Cambridge University Press, Cambridge, 1993).

Doi, M. and Edwards, S. F. *The Theory of Polymer Dynamics* (Clarendon Press, Oxford, 1986).

Erman, B. and Mark, J. E. *Structures and Properties of Rubberlike Networks* (Oxford University Press, Oxford, 1997).

Ferry, J. D. *Viscoelastic Properties of Polymers*, 3rd edition (Wiley, New York, 1980).

Flory, P. J. *Principles of Polymer Chemistry* (Cornell University Press, Ithaca, New York, 1953).

Graessley, W. W. The entanglement concept in polymer rheology. *Advances in Polymer Science* **16**, 1 (1974).

Heinrich, G., Straube, E., and Helmis, G. Rubber elasticity of polymer networks: theories. *Advances in Polymer Science* **85**, 33 (1988).

Queslel, J. P. and Mark, J. E. Molecular interpretation of the moduli of elastomeric polymer networks of known structure. *Advances in Polymer Science* **65**, 135 (1984).

Staverman, A. J. Properties of phantom networks and real networks. *Advances in Polymer Science* **44**, 73 (1982).

Treloar, L. R. G. *The Physics of Rubber Elasticity*, 3rd edition (Clarendon Press, Oxford, 1975).

Dynamics

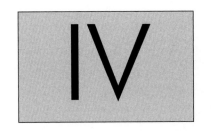

Unentangled polymer dynamics

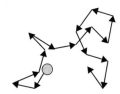

A small colloidal particle in any liquid diffuses due to the fluctuations of the number of molecules hitting it randomly from different directions. Colloidal particles are significantly larger than the molecules in the liquid, but small enough that collisions with molecules noticeably move the particle.[1] The trajectory of the particle, shown in Fig. 8.1, is another example of a random walk. The three-dimensional mean-square displacement of the colloidal particle during time t is proportional to t, with the coefficient of proportionality related to the **diffusion coefficient** D:

$$\langle[\vec{r}(t) - \vec{r}(0)]^2\rangle = 6Dt. \qquad (8.1)$$

The average distance the particle has moved is proportional to the square root of time:

$$\langle[\vec{r}(t) - \vec{r}(0)]^2\rangle^{1/2} = (6Dt)^{1/2}. \qquad (8.2)$$

Whereas the motion of the particle obeys Eq. (8.1) at all times, we shall see that the motion of monomers in a polymer is not always described by Eq. (8.1) [or Eq. (8.2)]. When the motion of a molecule obeys Eq. (8.1), it is called a simple **diffusive motion**. The random motion of small particles in a liquid was observed long ago using a microscope by a biologist named Brown and is often referred to as **Brownian motion**.

If a constant force $\vec{f}$ is applied to a small particle, pulling it through a liquid, the particle will achieve a constant velocity $\vec{v}$ in the same direction as the applied force. For a given particle and a given liquid, the coefficient relating force and velocity is the **friction coefficient** ζ:

$$\vec{f} = \zeta\vec{v}. \qquad (8.3)$$

Since the constant force acting on the particle results in a constant velocity, there must be an equal and opposite viscous drag force of the liquid acting on the particle with magnitude ζv. The diffusion coefficient D and the friction coefficient ζ are related through the **Einstein relation**:

$$D = \frac{kT}{\zeta}. \qquad (8.4)$$

Fig. 8.1
Motion of a particle in a liquid is a random walk that results from random collisions with molecules in the liquid.

---

[1] Colloidal particles have sizes between 1 nm and 10 μm.

The physics behind this relation is the fluctuation–dissipation theorem: the same random kicks of the surrounding molecules cause both Brownian diffusion and the viscous dissipation leading to the frictional force. It is instructive to calculate the time scale τ required for the particle to move a distance of order of its own size R:

$$\tau \approx \frac{R^2}{D} \approx \frac{R^2 \zeta}{kT}. \tag{8.5}$$

The time scale for diffusive motion is proportional to the friction coefficient.

   The mechanical properties of a liquid are fundamentally different from the solids discussed in Chapter 7. Solids have stress proportional to deformation (for small deformations). However, the stress in liquids depends only on the rate of deformation, not the total amount of deformation. If we pour water from one bucket into another bucket, there is only resistance during the flow, but there is no shear stress in the water in either bucket at rest. We describe the deformation rate of a liquid in shear by the shear rate $\dot{\gamma} = d\gamma/dt$ [Eq. (7.99)]. For the steady simple shear flow of Fig. 7.23, the shear rate is the same everywhere, equal to the way in which velocity changes with vertical position. The stress σ in a Newtonian liquid is proportional to this shear rate [Newton's law of viscosity Eq. (7.100), $\sigma = \eta\dot{\gamma}$], with the viscosity η being the coefficient of proportionality.

   If a sphere of radius R moves in a Newtonian liquid of viscosity η, a simple dimensional argument can determine the friction coefficient of the sphere. The friction should depend only on the viscosity of the surrounding liquid and the sphere size:

$$\zeta(\eta, R). \tag{8.6}$$

The friction coefficient is the ratio of force and velocity, with units of $\mathrm{kg\,s^{-1}}$. The viscosity is the ratio of stress and shear rate, with units of $\mathrm{kg\,m^{-1}s^{-1}}$ and the sphere radius has units of length (m). The only functional form that is dimensionally correct gives a very simple relation:

$$\zeta \approx \eta R. \tag{8.7}$$

The full calculation of the slow flow of a Newtonian liquid past a sphere was published by Stokes in 1880, yielding the numerical prefactor of 6π that results in **Stokes law**:

$$\zeta = 6\pi\eta R. \tag{8.8}$$

Combining Stokes law with the Einstein relation [Eq. (8.4)] gives a simple equation for the diffusion coefficient of a spherical particle in a liquid, known as the **Stokes–Einstein relation**:

$$D = \frac{kT}{6\pi\eta R}. \tag{8.9}$$

This important relation is used to determine coil size from measured diffusion coefficient (for example, by dynamic light scattering—see Section 8.9, or by pulsed-field gradient NMR). The size determined from a measurement of diffusion coefficient is the **hydrodynamic radius**:

$$R_h \equiv \frac{kT}{6\pi\eta D}. \tag{8.10}$$

8.1 Rouse model

The first successful molecular model of polymer dynamics was developed by Rouse. The chain in the Rouse model is represented as N beads connected by springs of root-mean-square size b, as shown in Fig. 8.2. The beads in the Rouse model only interact with each other through the connecting springs. Each bead is characterized by its own independent friction with friction coefficient ζ. Solvent is assumed to be freely draining through the chain as it moves.

The total friction coefficient of the whole Rouse chain is the sum of the contributions of each of the N beads:

$$\zeta_R = N\zeta. \tag{8.11}$$

The viscous frictional force the chain experiences if it is pulled with velocity $\vec{v}$ is $\vec{f} = -N\zeta\vec{v}$. The diffusion coefficient of the Rouse chain is obtained from the Einstein relation [Eq. (8.4)].

$$D_R = \frac{kT}{\zeta_R} = \frac{kT}{N\zeta}. \tag{8.12}$$

The polymer diffuses a distance of the order of its size during a characteristic time, called the **Rouse time**, τ_R:

$$\tau_R \approx \frac{R^2}{D_R} \approx \frac{R^2}{kT/(N\zeta)} = \frac{\zeta}{kT}NR^2. \tag{8.13}$$

The Rouse time has special significance. On time scales shorter than the Rouse time, the chain exhibits viscoelastic modes that shall be described in Section 8.4. However, on time scales longer than the Rouse time, the motion of the chain is simply diffusive.

Polymers are fractal objects, with size related to the number of monomers in the chain[2] by a power law:

$$R \approx bN^\nu \tag{8.14}$$

The reciprocal of the fractal dimension of the polymer (see Section 1.4) is ν. For an ideal linear chain $\nu = 1/2$ and the fractal dimension is $1/\nu = 2$. The Rouse time of such a fractal chain can be written as the product of

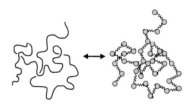

Fig. 8.2
In the Rouse model, a chain of N monomers is mapped onto a bead–spring chain of N beads connected by springs.

---

[2] There are $N-1$ springs in the Rouse model and, for long chains, the number of springs is approximated by N.

the time scale for motion of individual beads, the **Kuhn monomer relaxation time**

$$\tau_0 \approx \frac{\zeta b^2}{kT}, \tag{8.15}$$

and a power law in the number of monomers in the chain:

$$\tau_R \approx \frac{\zeta}{kT} NR^2 = \frac{\zeta b^2}{kT} N^{1+2\nu} \approx \tau_0 N^{1+2\nu}. \tag{8.16}$$

For an ideal linear chain, $\nu = 1/2$ and the Rouse time is proportional to the square of the number of monomers in the chain:

$$\tau_R \approx \tau_0 N^2. \tag{8.17}$$

The full calculation of the relaxation time of an ideal chain was published by Rouse in 1953, with a coefficient of $1/(6\pi^2)$:

$$\tau_R = \frac{\zeta b^2}{6\pi^2 kT} N^2. \tag{8.18}$$

This Rouse stress relaxation time is half of the end-to-end vector correlation time because stress relaxation is determined from a quadratic function of the amplitudes of normal modes (see Problem 8.36).

The time scale for motion of individual monomers τ_0, is the time scale at which a monomer would diffuse a distance of order of its size b if it were not attached to the chain. In a polymer solution with solvent viscosity η_s, each monomer's friction coefficient is given by Stokes law [Eq. (8.8)]:

$$\zeta \approx \eta_s b. \tag{8.19}$$

The monomer relaxation time τ_0 and the chain relaxation time of the Rouse model τ_R can be rewritten in terms of the solvent viscosity η_s:

$$\tau_0 \approx \frac{\eta_s b^3}{kT}, \tag{8.20}$$

$$\tau_R \approx \frac{\eta_s b^3}{kT} N^2. \tag{8.21}$$

When probed on time scales smaller than τ_0, the polymer essentially does not move and exhibits elastic response. On time scales longer than τ_R, the polymer moves diffusively and exhibits the response of a simple liquid. For intermediate time scales $\tau_0 < t < \tau_R$, the chain exhibits interesting viscoelasticity discussed in Section 8.4.1.

8.2 Zimm model

The viscous resistance imparted by the solvent when a particle moves through it arises from the fact that the particle must drag some of the surrounding solvent with it. The force acting on a solvent molecule at distance r from the particle becomes smaller as r increases, but only slowly (decaying roughly as $1/r$). This long-range force acting on solvent

(and other particles) that arises from motion of one particle is called
hydrodynamic interaction. In the case of the bead–spring model of a
polymer chain, when one bead moves, there are hydrodynamic interaction
forces acting on the other beads of the chain. The Rouse model ignores
hydrodynamic interaction forces, and assumes the beads only interact
through the springs that connect them. We shall see later that this
assumption is reasonable for polymer melts, but is not correct for a
polymer in a dilute solution.

In dilute solutions, hydrodynamic interactions between the monomers
in the polymer chain are strong. These hydrodynamic interactions also are
strong between the monomers and the solvent within the pervaded volume
of the chain. When the polymer moves, it effectively drags the solvent
within its pervaded volume with it. For this reason, the best model of
polymer dynamics in a dilute solution is the Zimm model, which effectively
treats the pervaded volume of the chain as a solid object moving through
the surrounding solvent.

Assume that the chain (and any section of the chain) drags with it the
solvent in its pervaded volume. Thus the chain moves as a solid object of
size $R \approx bN^\nu$. The friction coefficient of the chain of size R being pulled
through a solvent of viscosity η_s is given by Stokes law:

$$\zeta_Z \approx \eta_s R. \tag{8.22}$$

There is a coefficient 6π in Stokes law [Eq. (8.8)] for a spherical
object $\zeta = 6\pi\eta_s R$, but chains are not spheres and we drop all numerical
coefficients.

From the Einstein relation [Eq. (8.4)] the diffusion coefficient of a chain
in the Zimm model is reciprocally proportional to its size R:

$$D_Z = \frac{kT}{\zeta_Z} \approx \frac{kT}{\eta_s R} \approx \frac{kT}{\eta_s b N^\nu}. \tag{8.23}$$

This is simply the Stokes–Einstein relation [Eq. (8.9)] for a polymer in
dilute solution. The Zimm model predicts that the chain diffuses as a
particle with volume proportional to the chain's pervaded volume in
solution. In 1956, Zimm published a full calculation, where he preaveraged
the hydrodynamic interactions to obtain this result with an extra coefficient of $8/(3\sqrt{6\pi^3})$ for an ideal chain:

$$D_Z = \frac{8}{3\sqrt{6\pi^3}} \frac{kT}{\eta_s R} \cong 0.196 \frac{kT}{\eta_s R}. \tag{8.24}$$

In the Zimm model, the chain diffuses a distance of order of its own size
during the **Zimm time** τ_Z:

$$\tau_Z \approx \frac{R^2}{D_Z} \approx \frac{\eta_s}{kT} R^3 \approx \frac{\eta_s b^3}{kT} N^{3\nu} \approx \tau_0 N^{3\nu}. \tag{8.25}$$

The coefficient relating the relaxation time to a power of the number
of monomers in the chain is once again the monomer relaxation time τ_0

[Eq. (8.20)]. Zimm's full calculation of the chain relaxation time provides an extra coefficient of $1/(2\sqrt{3\pi})$ for an ideal chain:

$$\tau_Z = \frac{1}{2\sqrt{3\pi}} \frac{\eta_s}{kT} R^3 \cong 0.163 \frac{\eta_s}{kT} R^3. \tag{8.26}$$

This Zimm stress relaxation time is half of the Zimm end-to-end vector correlation time.

The Zimm time is proportional to the pervaded volume of the chain. Note that the Zimm time τ_Z has a weaker dependence on chain length than the Rouse time τ_R [Eq. (8.16)].

$$3\nu < 2\nu + 1 \quad \text{for } \nu < 1. \tag{8.27}$$

Comparison of Eqs (8.16) and (8.25) reveals that the Zimm time is shorter than the Rouse time in dilute solution. In principle, a chain in dilute solution could move a distance of order of its size by Rouse motion, by Zimm motion, or some combination of the two. The chain could simply move its monomers by Rouse motion through the solvent without dragging any of the solvent molecules with it, or it could drag all of the solvent in its pervaded volume with it, thereby moving by Zimm motion. In dilute solution, Zimm motion has less frictional resistance than Rouse motion, and therefore, the faster process is Zimm motion. The chain effectively moves as though it were a solid particle with volume of order of its pervaded volume (with linear size R). The solvent within the pervaded volume of the chain is hydrodynamically coupled to the chain.[3] When the chain moves in response to its monomers being randomly hit by solvent from different directions, it effectively drags the surrounding solvent with it.

Using Eq. (3.77) for the size of the chain in a good solvent with intermediate excluded volume v in Eq. (8.25), and combining with the θ-solvent result of Eq. (8.25) with $\nu = 1/2$, yields a general expression for the Zimm time in dilute polymer solutions:

$$\tau_Z \approx \frac{\eta_s}{kT} R^3 \approx \begin{cases} \tau_0 N^{3/2} & N < b^6/v^2 \\ \tau_0 (v/b^3)^{6\nu-3} N^{3\nu} & N > b^6/v^2 \end{cases} \tag{8.28}$$

Using $\nu = 0.588$, the Zimm relaxation time for long chains is $\tau_0 (v/b^3)^{0.53} N^{1.76}$.

8.3 Intrinsic viscosity

In solution, a confusing plethora of viscosities have been defined over the years. The ratio of solution viscosity η to solvent viscosity η_s is the **relative viscosity**:

$$\eta_r \equiv \frac{\eta}{\eta_s}. \tag{8.29}$$

---

[3] While some solvent does move with the chain, solvent molecules diffuse into and out of the pervaded volume on a faster time scale than the diffusion of the polymer (see Problem 8.5).

The relative viscosity is the simplest dimensionless measure of solution viscosity. The difference of the relative viscosity from unity is the **specific viscosity**:

$$\eta_{sp} \equiv \eta_r - 1 = \frac{\eta - \eta_s}{\eta_s}. \tag{8.30}$$

The numerator $(\eta - \eta_s)$ is the polymer contribution to the solution viscosity, so the specific viscosity is a dimensionless measure of the polymer contribution to the solution viscosity.

The ratio of specific viscosity to polymer concentration is the **reduced viscosity**, η_{sp}/c, which has units of reciprocal concentration. In the limit of very low concentrations (far below the overlap concentration) the reduced viscosity becomes a very important material property called the intrinsic viscosity (see Section 1.7.3, and in particular Fig. 1.24):

$$[\eta] = \lim_{c \to 0} \frac{\eta_{sp}}{c}. \tag{8.31}$$

The intrinsic viscosity is the initial slope of specific viscosity as a function of concentration, and has units of reciprocal concentration [see Eq. (1.97)].

The value of the stress relaxation modulus at the relaxation time $G(\tau)$ is of the order of kT per chain in either the Rouse or Zimm models, just as the strands of a network in Chapter 7 stored of order kT of elastic energy:

$$G(\tau) \approx kT \frac{\phi}{Nb^3}. \tag{8.32}$$

The polymer contribution to the viscosity in either the Rouse or the Zimm model is proportional to $G(\tau)\tau$ [Eq. (7.120)]:

$$\eta - \eta_s \approx kT \frac{\phi}{Nb^3} \tau. \tag{8.33}$$

The typical experimental concentration used in defining intrinsic viscosity is the polymer mass per unit volume of solution, $c = \phi M_0/(b^3 \mathcal{N}_{Av})$ where M_0 is the molar mass of a Kuhn monomer [see Eq. (1.18)]. The intrinsic viscosity then follows:

$$[\eta] \approx \frac{kT \mathcal{N}_{Av}}{\eta_s M_0 N} \tau. \tag{8.34}$$

The expression for the relaxation time in the Rouse model of an ideal chain $\tau_R \approx \eta_s b^3 N^2/(kT)$ [Eq. (8.21)] leads to the Rouse prediction for the intrinsic viscosity:

$$[\eta] \approx \frac{b^3 \mathcal{N}_{Av}}{M_0} N \quad \text{Rouse model.} \tag{8.35}$$

The Rouse model predicts that the intrinsic viscosity in a θ-solvent is proportional to molar mass. However, the Rouse model assumes no hydrodynamic interactions and is not expected to be valid in dilute solutions where intrinsic viscosity is defined.

Substituting the prediction for the relaxation time of the Zimm model $\tau_Z \approx \eta_s R^3/(kT)$ [Eq. (8.25)] into the expression for intrinsic viscosity [Eq. (8.34)] leads to the Zimm prediction for intrinsic viscosity:

$$[\eta] \approx \frac{R^3 \mathcal{N}_{Av}}{M_0 N} \approx \frac{b^3 \mathcal{N}_{Av}}{M_0} N^{3\nu-1} \quad \text{Zimm model.} \tag{8.36}$$

The Zimm model assumes that as the polymer moves it drags the solvent inside its pervaded volume with it. The Zimm model has the correct physics for the intrinsic viscosity. Equation (8.36) is more commonly written in terms of the molar mass $M = M_0 N$,

$$[\eta] = \Phi \frac{R^3}{M}, \tag{8.37}$$

where $\Phi = 0.425 \mathcal{N}_{Av} = 2.5 \times 10^{23}\,\mathrm{mol}^{-1}$ is a universal constant for all polymer–solvent systems. This famous relation between intrinsic viscosity, coil size and molar mass is known as the Fox–Flory equation.

Equation (8.36) predicts that the intrinsic viscosity obeys a power law in molar mass. This power law was empirically recognized long ago, and is known as the Mark–Houwink equation [Eq. (1.100)]:

$$[\eta] = KM^a. \tag{8.38}$$

From the derivation of the Fox–Flory equation, based on the Zimm model, the Mark–Houwink exponent a is related to the exponent describing the molar mass dependence of coil size in solution ν:

$$a = 3\nu - 1. \tag{8.39}$$

The Mark–Houwink equation provides an indirect estimate of molar mass from a measurement of intrinsic viscosity $[\eta]$, if the two Mark–Houwink constants K and a, are known. The predictions of Mark–Houwink constants are summarized in Table 8.1. Comparison with Table 1.4 shows that the Zimm model agrees reasonably well with experimental results, as $a = 0.50$ is observed in θ-solvent and $0.7 < a < 0.8$ is usually observed in good solvents.

Using the Zimm time [Eq. (8.28)] in Eq. (8.34), yields a general expression for the intrinsic viscosity, valid for any solvent with $T \geq \theta$:

$$[\eta] \approx \frac{kT \mathcal{N}_{Av}}{\eta_s M_0 N} \tau_Z \approx \frac{R^3 \mathcal{N}_{Av}}{M_0 N}$$

$$\approx \frac{b^3 \mathcal{N}_{Av}}{M_0} \begin{cases} N^{1/2} & N < b^6/v^2 \\ (v/b^3)^{6\nu-3} N^{3\nu-1} & N > b^6/v^2 \end{cases} \tag{8.40}$$

Table 8.1 Predictions of Mark–Houwink constants

	K	a
Rouse model in θ-solvent	$b^3 \mathcal{N}_{Av}/M_0^2$	1
Zimm model in θ-solvent	$b^3 \mathcal{N}_{Av}/M_0^{3/2}$	$3\nu - 1 = 1/2$
Zimm model in good solvent	$b^3 \mathcal{N}_{Av}/M_0^{1.764}$	$3\nu - 1 \cong 0.76$

For long chains in good solvent, $\nu = 0.588$ and the intrinsic viscosity *universally* scales as $v^{0.53} N^{0.76}$.

This relation is tested with experimental data in Fig. 8.3.[4] It is important to point out the fact that the intrinsic viscosity of polystyrene in toluene (filled squares in Fig. 8.3) crosses over to the θ-solvent result at $M \approx 30\,000$ g mol^{-1}. This provides a direct measure of the number of Kuhn monomers in a thermal blob $g_T \approx (30\,000 \text{ g mol}^{-1})/(720 \text{ g mol}^{-1}) \approx 40$ for polystyrene in toluene. The crossover between the θ-solvent and good solvent cases of Eq. (8.40) is at $N = g_T \approx (b^3/v)^2$ [Eq. (3.75)], so $g_T \approx 40$ means that the excluded volume is estimated to be $v \approx 0.16b^3$ for polystyrene in toluene. Hence, although toluene is a quite good solvent for polystyrene, it is nowhere near the athermal solvent limit, which would have even higher intrinsic viscosity that would maintain the power law with 0.76 slope to even lower molar masses. Polystyrene in methyl ethyl ketone (open circles in Fig. 8.3) has even smaller excluded volume, as $g_T \approx (100\,000 \text{ g mol}^{-1})/(720 \text{ g mol}^{-1}) \approx 140$ and $v \approx 0.08b^3$.

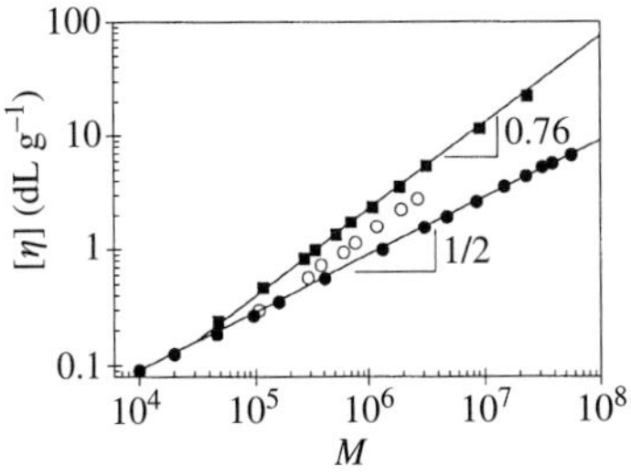

Figure 8.3 also shows clearly that caution is needed when using Mark–Houwink equations from the literature that have intermediate exponents in the range $0.5 < a < 0.76$. Such intermediate exponents correspond to the crossover between regimes and are only valid for the range of molar masses they were measured in.

The fact that the intrinsic viscosity measurement is simultaneously simple and precise makes it an extremely popular molecular characterization tool. Intrinsic viscosity can easily be measured to $\pm 0.1\%$ precision, which is far superior to osmotic pressure and light scattering, which have precisions of $\pm 5\%$ under the best of circumstances. Furthermore, if intrinsic viscosity and absolute molar mass are measured over a sufficiently wide range, the thermodynamic nature of the polymer solvent interaction, reflected in the excluded volume v, can be estimated using Eq. (8.40).

The temperature dependence of intrinsic viscosity enters Eq. (8.40) through the excluded volume $v \approx b^3 (T - \theta)/T$. For chains that are smaller than the thermal blob, the short chain branch of Eq. (8.40) (with $N < b^6/v^2$) applies. For such short chains, the intrinsic viscosity is independent of temperature and $[\eta]/N^{1/2}$ reduces data for different lengths of short chains to a common temperature-independent line, demonstrated in Fig. 8.4(a) for polyisobutylene in toluene with $M < 11\,000$ g mol^{-1}. On the other hand, chains with size far exceeding the thermal blob size have important excluded volume effects. The long chain branch of Eq. (8.40) (with $N > b^6/v^2$) applies to long chains and $[\eta]/N^{0.764}$ reduces data for different lengths of long chains to a common curve, as shown in Fig. 8.4(b) for polyisobutylene in toluene with $M > 400\,000$ g mol^{-1}. The curve in Fig. 8.4(b) is determined by the temperature dependence of excluded volume $(v/b^3)^{0.53} \approx (1 - \theta/T)^{0.53}$ with $\theta = 245\,K \cong -28\,°C$ determined from the fit. Intermediate molar masses (not shown) with $M = 48\,000$ g mol^{-1} and

Fig. 8.3
Intrinsic viscosities of polystyrenes in three solvents. Cyclohexane is a θ-solvent ($v = 0$, filled circles, from Y. Einaga *et al.*, *J. Polym. Sci., Polym. Phys.* **17**, 2103, 1979), with Mark–Houwink exponent $a = 1/2$. Methyl ethyl ketone is a better solvent ($v \approx 0.08b^3$, open circles, from R. Okada *et al.*, *Makromol. Chem.* **59**, 137, 1963) and toluene is a good solvent ($v \approx 0.16b^3$, filled squares, from R. Kniewske and W.-M. Kulicke, *Makromol. Chem.* **184**, 2173, 1983) with $a = 0.76$.

---

[4] The customary units for intrinsic viscosity are dL g^{-1}, where 1 dL $= 0.1$ L.

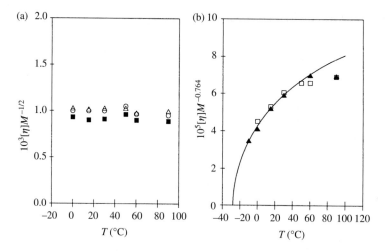

Fig. 8.4

Temperature dependence of intrinsic viscosity for polyisobutylene fractions in toluene.
(a) data for the three lowest molar masses (open triangles are $M = 7080 \, \mathrm{g \, mol^{-1}}$, filled squares
are $M = 9550 \, \mathrm{g \, mol^{-1}}$ and open circles are $M = 10\,200 \, \mathrm{g \, mol^{-1}}$) that all are smaller than the
thermal blob and hence have unperturbed size. The short chain branch of Eq. (8.40) reduces
these data to a common temperature-independent line. (b) data for the two highest molar
masses (open squares are $M = 463\,000 \, \mathrm{g \, mol^{-1}}$ and filled triangles are $M = 1\,260\,000 \, \mathrm{g \, mol^{-1}}$
that obey the long chain branch of Eq. (8.40). The curve is fitted to the data using Eq. (8.40)
and the temperature dependence of excluded volume with $\theta = -28\,^\circ\mathrm{C}$ determined from the
fit. The data are from T. G. Fox and P. J. Flory, *J. Phys. Chem.* **53**, 197 (1949).

$M = 110\,000 \, \mathrm{g \, mol^{-1}}$ fall in the crossover between the two limiting cases of
Eq. (8.40) and do not obey the scaling of either clean limit. Unfortunately,
the molar mass range of $20\,000 < M < 200\,000 \, \mathrm{g \, mol^{-1}}$ is the important
range for commercial polymers, and it corresponds to the crossover for
most good solvent/polymer solutions.

The R^3 in Eq. (8.37) comes from the relaxation time in Eq. (8.34). This
Zimm time really has two size scales within it. The hydrodynamic radius
R_h enters through the diffusion coefficient [Eq. (8.10)] and the radius of
gyration R_g enters through the length scale that the molecule moves in its
relaxation time:

$$[\eta] \approx \frac{\mathcal{R}T}{\eta_\mathrm{s} M} \tau_\mathrm{Z} \approx \frac{\mathcal{R}T}{\eta_\mathrm{s} M} \frac{R_\mathrm{g}^2}{D_\mathrm{Z}} \approx \frac{\mathcal{N}_\mathrm{Av} R_\mathrm{g}^2 R_\mathrm{h}}{M}. \qquad (8.41)$$

Using data for polystyrene in two good solvents[5] (ethylbenzene and
tetrahydrofuran) Eq. (8.41) is found to apply reasonably with

$$\frac{[\eta] M_\mathrm{w}}{\mathcal{N}_\mathrm{Av} R_\mathrm{g}^2 R_\mathrm{h}} \cong 7, \qquad (8.42)$$

in the range $93\,000 \, \mathrm{g \, mol^{-1}} \leq M_\mathrm{w} \leq 4\,800\,000 \, \mathrm{g \, mol^{-1}}$.

---

[5] K. Venkataswamy *et al.*, *Macromolecules* **19**, 124 (1986).

8.4 Relaxation modes

In Sections 8.1 and 8.2, we calculated the longest relaxation time of unentangled polymers using molecular models. The linear viscoelastic response of polymeric liquids, discussed in Section 7.6, measures the full spectrum of relaxation times. Since polymer chains are self-similar objects, they also exhibit dynamic self-similarity. *Smaller sections of a polymer chain with g monomers relax just like a whole polymer chain that has g monomers.* In all unentangled molecular models for polymer dynamics (both Rouse and Zimm and combinations thereof) the relaxations are described by N different **relaxation modes**. The modes are numbered by **mode index** $p = 1, 2, 3, \ldots, N$. These modes are analogous to the modes of a vibrating guitar string. Mode p involves coherent motion of sections of the whole chain with N/p monomers, and the corresponding relaxation time of this mode τ_p is similar to the longest relaxation time of a chain with N/p monomers. For all unentangled molecular models of flexible polymer dynamics, the shortest mode has mode index $p = N$ with relaxation time τ_0, the relaxation time of a monomer [Eq. (8.20)]. The longer modes depend on whether hydrodynamic interactions are important or not, as discussed below.

Consider a polymer liquid subjected to a unit step strain at time $t = 0$. The **equipartition principle** states that $kT/2$ of free energy is associated with each degree of freedom at equilibrium.[6] Immediately following the unit step strain, the entire chain stores of order NkT of elastic energy, since there are N independent modes that each store of order kT. To determine the time dependent viscoelastic response, we simply need to determine the relaxation time of each mode.

8.4.1 Rouse modes

In the Rouse model, the (longest) relaxation time of the ideal chain is given by Eq. (8.17):

$$\tau_R \approx \tau_0 N^2. \tag{8.43}$$

Since the pth mode involves relaxation on the scale of chain sections with N/p monomers, the relaxation time of the pth mode has a similar form to the longest mode:

$$\tau_p \approx \tau_0 \left(\frac{N}{p}\right)^2 \quad \text{for } p = 1, 2, \ldots, N. \tag{8.44}$$

The relaxation time of a monomer, τ_0 [Eq. (8.15)] is the shortest relaxation time of the Rouse model, with mode index $p = N$, making $\tau_N = \tau_0$. The mode with index $p = 1$ is the longest relaxation mode of the chain with relaxation time equal to the Rouse time $\tau_1 = \tau_R$, and corresponds to relaxation on the scale of the entire chain. The mode with index $p = 2$ corresponds to the two halves of the chain with $N/2$ monomers, each

---

[6] In three-dimensional space, each mode has three degrees of freedom.

relaxing independently. The mode with index p breaks the chain into p sections of N/p monomers, and each of these sections relax as independent chains of N/p monomers on the time scale τ_p.

As expected, higher index modes, involving fewer monomers, relax faster than lower index modes. Therefore, at time τ_p after a step strain, all modes with index higher than p have mostly relaxed, but modes with index lower than p have not yet relaxed.

The number of unrelaxed modes per chain at time $t = \tau_p$ is equal to the mode index p. Each unrelaxed mode contributes energy of order kT to the stress relaxation modulus. The stress relaxation modulus at time $t = \tau_p$ is proportional to the thermal energy kT and the number density of sections with N/p monomers, $\phi/(b^3 N/p)$:

$$G(\tau_p) \approx \frac{kT}{b^3} \frac{\phi}{N} p. \tag{8.45}$$

The time dependence of the mode index p for the mode that relaxes at time $t = \tau_p$ can be found from Eq. (8.44).

$$p \approx \left(\frac{\tau_p}{\tau_0}\right)^{-1/2} N. \tag{8.46}$$

Combining Eqs (8.45) and (8.46) approximates the stress relaxation modulus for the Rouse model at intermediate time scales:

$$G(t) \approx \frac{kT}{b^3} \phi \left(\frac{t}{\tau_0}\right)^{-1/2} \quad \text{for } \tau_0 < t < \tau_R. \tag{8.47}$$

This expression effectively interpolates between a modulus level of order kT per monomer at the shortest Rouse mode ($t \approx \tau_0$) to a modulus level of order kT per chain at the longest Rouse mode ($t = \tau_R \approx \tau_0 N^2$) using a power law. We already know that the stress relaxation modulus has an exponential decay beyond its longest relaxation time [Eq. (7.112)]. Therefore, an approximate description of the stress relaxation modulus of the Rouse model is the product of [Eq. (8.47)] and an exponential cutoff:

$$G(t) \approx \frac{kT}{b^3} \phi \left(\frac{t}{\tau_0}\right)^{-1/2} \exp(-t/\tau_R) \quad \text{for } t > \tau_0. \tag{8.48}$$

The Rouse time τ_R is the longest stress relaxation time [Eq. (8.18)].

For oscillatory shear, Eqs (7.149) and (7.150) allow calculation of the storage and loss moduli of a solution of linear Rouse chains (see Problem 8.14):

$$G'(\omega) \approx \frac{\phi kT}{b^3 N} \frac{(\omega\tau_R)^2}{\sqrt{[1 + (\omega\tau_R)^2]}\left[\sqrt{1 + (\omega\tau_R)^2} + 1\right]} \quad \text{for } \omega < 1/\tau_0, \tag{8.49}$$

$$G''(\omega) \approx \frac{\phi kT}{b^3 N} \omega \tau_R \sqrt{\frac{\sqrt{1 + (\omega \tau_R)^2} + 1}{1 + (\omega \tau_R)^2}} \quad \text{for } \omega < 1/\tau_0. \quad (8.50)$$

In the frequency range, $1/\tau_R \ll \omega \ll 1/\tau_0$, the storage and loss moduli of the Rouse model are equal to each other and scale as the square root of frequency:

$$G'(\omega) \cong G''(\omega) \sim \omega^{1/2} \quad \text{for } 1/\tau_R \ll \omega \ll 1/\tau_0. \quad (8.51)$$

For high frequencies $\omega > 1/\tau_0$, there are no relaxation modes in the Rouse model. The storage modulus becomes independent of frequency, and equal to the short time stress relaxation modulus, which is kT per monomer $G'(\omega) \approx \phi kT/b^3$. This high-frequency saturation is not included in Eqs (8.49) and (8.50). At low frequencies $\omega < 1/\tau_R$, the storage modulus is proportional to the square of frequency and the loss modulus is proportional to frequency, as is the case for the terminal response of any viscoelastic liquid.

Figure 8.5 shows that experimental data on unentangled polyelectrolyte solutions are described quite well by the Rouse model. Polyelectrolytes are charged polymers that have a wide range of concentrations where dynamics obey the Rouse model.

The viscosity of the Rouse model is obtained by integrating $G(t)$ [Eq. (7.117)]:

$$\eta = \int_0^\infty G(t)\,dt \approx \frac{kT}{b^3}\phi \int_0^\infty \left(\frac{t}{\tau_0}\right)^{-1/2} \exp(-t/\tau_R)\,dt$$

$$\approx \frac{kT}{b^3}\phi\sqrt{\tau_0\tau_R}\int_0^\infty x^{-1/2}\exp(-x)\,dx \approx \frac{kT}{b^3}\phi\sqrt{\tau_0\tau_R} \approx \frac{kT}{b^3}\tau_0 N\phi \approx \frac{\zeta}{b}N\phi. \quad (8.52)$$

Equation (8.52) made use of the variable transformation $x \equiv t/\tau_R$, and the integral involving x is simply a numerical coefficient. Notice that the final relation is identical to that expected by Eq. (7.120), the product of $G(\tau_R)$ [Eq. (8.32)] and τ_R [Eq. (8.17)]. The Rouse model applies to melts of short unentangled chains (for which hydrodynamic interactions are screened). The Rouse viscosity has a very simple form for an unentangled polymer melt:

$$\eta \approx \frac{\zeta}{b}N. \quad (8.53)$$

The viscosity of the Rouse model is proportional to the number of monomers in the chain. The Rouse model has been solved exactly (by Rouse), and the full calculation gives an extra coefficient of $1/36$:

$$\eta = \frac{\zeta}{36b}N. \quad (8.54)$$

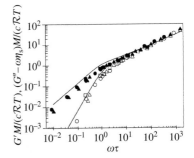

Fig. 8.5
Oscillatory shear data for solutions of poly(2-vinyl pyridine) in 0.0023 M HCl in water. Open symbols are the storage modulus G' and filled symbols are the loss modulus G''. Squares have $c = 0.5\,\mathrm{g\,L^{-1}}$, triangles have $c = 1.0\,\mathrm{g\,L^{-1}}$, and circles have $c = 2.0\,\mathrm{g\,L^{-1}}$. The curves are the predictions of the Rouse model [Eqs (8.49) and (8.50)]. Data from D. F. Hodgson and E. J. Amis, *J. Chem. Phys.* **94**, 4581 (1991).

Rouse also derived an exact relation for the stress relaxation modulus,

$$G(t) = kT\frac{\phi}{Nb^3}\sum_{p=1}^{N}\exp(-t/\tau_p),\tag{8.55}$$

with

$$\tau_p = \frac{\zeta b^2 N^2}{6\pi^2 k T p^2}.\tag{8.56}$$

The stress relaxation times τ_p of the Rouse model are half of the correlation times of normal modes (see Problem 8.36).

This exact form demonstrates that each mode ($p = 1, 2, \ldots, N$) relaxes as a Maxwell element [Eq. (7.111)]. The exact [Eq. (8.55)] and approximate [Eq. (8.48)] Rouse predictions of the stress relaxation modulus of an unentangled polymer melt are compared in Fig. 8.6. This figure clearly shows that Eq. (8.48) is an excellent approximation of the exact Rouse result for long chains ($N \gg 1$).

Chain sections containing N/p monomers move a distance of order of their size $b(N/p)^{1/2}$ during the mode relaxation time τ_p. The position vector of monomer j at time t is $\vec{r}_j(t)$. The mean-square displacement of monomer j during time τ_p is of the order of the mean-square size of the sections involved in coherent motion on this time scale:

$$\langle[\vec{r}_j(\tau_p) - \vec{r}_j(0)]^2\rangle \approx b^2\frac{N}{p} \approx b^2\left(\frac{\tau_p}{\tau_0}\right)^{1/2}.\tag{8.57}$$

In the final relation we used the time dependence of the mode index p [Eq. (8.46)]. The mean-square displacement of a monomer on intermediate time scales thus increases as the square root of time:

$$\langle[\vec{r}_j(t) - \vec{r}_j(0)]^2\rangle \approx b^2\left(\frac{t}{\tau_0}\right)^{1/2} \quad \text{for } \tau_0 < t < \tau_R.\tag{8.58}$$

For the motion to be diffusive, the mean-square displacement must be linear in time [see Eq. (8.1)]. Since the mean-square displacement on intermediate time scales is a weaker-than-linear power of time, the motion is referred to as **subdiffusive motion**. Individual monomers are not 'aware' that they belong to an N-mer on times shorter than the Rouse time of the chain. At each moment of time $t < \tau_R$, sections of a chain containing $g(t)$ monomers move coherently. Thus monomers only 'realize' that their chain contains at least $g(t)$ monomers at time scale $t < \tau_R$. The diffusion coefficient of these coherent sections is $D(t) \approx kT/(\zeta g)$. The number of monomers in sections that coherently participate in Rouse motion increases proportional to the square root of time $g(t) \approx (t/\tau_0)^{1/2}$ [Eq. (8.46) with $g = N/p$] and their effective diffusion coefficient decreases with time:

$$D(t) \approx \frac{kT}{\zeta g(t)} \approx \frac{kT}{\zeta}\left(\frac{t}{\tau_0}\right)^{-1/2} \quad \text{for } \tau_0 < t < \tau_R.\tag{8.59}$$

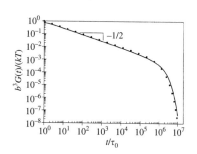

Fig. 8.6

Stress relaxation modulus predicted by the Rouse model for a melt of unentangled chains with $N = 10^3$. The solid curve is the exact Rouse result [Eq. (8.55)] and the dotted curve is the approximate Rouse result [Eq. (8.48)].

At longer times, monomers participate in collective motion of larger sections with smaller effective diffusion coefficient $D(t)$. Therefore the mean-square displacement of monomers is not a linear function of time, but instead subdiffusive:

$$\langle [\vec{r}_j(t) - \vec{r}_j(0)]^2 \rangle \approx D(t)t \sim t^{1/2} \quad \text{for } \tau_0 < t < \tau_R. \quad (8.60)$$

Only on time scales longer than the Rouse time of the chain, is the motion of the chain diffusive, with mean-square displacement proportional to time [Eq. (8.1)].

8.4.2 Zimm modes

Similar scaling analysis of the mode structure can be applied to the Zimm model. The relaxation time of the pth mode is of the order of the Zimm relaxation time of the chain containing N/p monomers [Eq. (8.25)]:

$$\tau_p \approx \tau_0 \left(\frac{N}{p}\right)^{3\nu}. \quad (8.61)$$

The index p of the mode relaxing at time $t = \tau_p$ after a step strain imposed at time $t = 0$ is obtained by solving the above equation for p:

$$p \approx N \left(\frac{\tau_p}{\tau_0}\right)^{-1/(3\nu)} = N \left(\frac{t}{\tau_0}\right)^{-1/(3\nu)}. \quad (8.62)$$

The number of unrelaxed modes per chain at time $t = \tau_p$ is p. The stress relaxation modulus is proportional to the number density of chain sections with N/p monomers:

$$G(t) \approx \frac{kT}{b^3} \frac{\phi}{N} p \approx \frac{kT}{b^3} \phi \left(\frac{t}{\tau_0}\right)^{-1/(3\nu)} \quad \text{for } \tau_0 < t < \tau_Z. \quad (8.63)$$

In θ–solvents ($\nu = 1/2$), the stress relaxation modulus decays as the $-2/3$ power of time, while in good solvents ($\nu \cong 0.588$) $G(t)$ decays approximately as the -0.57 power of time. Like the stress relaxation modulus of the Rouse model [Eq. (8.47)], Eq. (8.63) crosses over from kT per monomer at the monomer relaxation time τ_0 to kT per chain at the relaxation time of the chain $\tau_Z \approx \tau_0 N^{3\nu}$ [Eq. (8.25)]. Once again, an excellent approximation to the stress relaxation modulus predicted by the Zimm model is the product of the power law of Eq. (8.63) and an exponential cutoff:

$$G(t) \approx \frac{kT}{b^3} \phi \left(\frac{t}{\tau_0}\right)^{-1/(3\nu)} \exp(-t/\tau_Z) \quad \text{for } t > \tau_0. \quad (8.64)$$

The polymer contribution to the solution viscosity is obtained by integrating $G(t)$ [Eq. (7.117)]:

$$\eta - \eta_s = \int_0^\infty G(t)\,dt \approx \frac{kT}{b^3}\phi \int_0^\infty \left(\frac{t}{\tau_0}\right)^{-1/3\nu} \exp(-t/\tau_Z)\,dt$$

$$\approx \frac{kT}{b^3}\phi\tau_Z \left(\frac{\tau_Z}{\tau_0}\right)^{-1/3\nu} \int_0^\infty x^{-1/3\nu} \exp(-x)\,dx \approx \frac{kT}{b^3}\phi\tau_0 N^{3\nu-1}$$

$$\approx \eta_s \phi N^{3\nu-1}. \tag{8.65}$$

The variable transformation $x \equiv t/\tau_Z$ was used, and the integral involving x is simply a numerical coefficient. The second-to-last relation was obtained using $\tau_Z \approx \tau_0 N^{3\nu}$ [Eq. (8.25)] and the final relation used Eq. (8.20). The final relation is identical to that expected by Eq. (7.120), the product of $G(\tau_Z)$ [Eq. (8.32)] and τ_Z [Eq. (8.25)]. The Zimm model applies to the relaxation of the entire chain in dilute solution (where hydrodynamic interactions dominate). The intrinsic viscosity is calculated from the polymer contribution to the solution viscosity using Eq. (8.31) and the relation between mass concentration and volume fraction $c = \phi M_0/(b^3 \mathcal{N}_{Av})$ [(see Eq. (1.18)]:

$$[\eta] = \lim_{c \to 0} \frac{\eta - \eta_s}{\eta_s c} \approx \frac{b^3 \mathcal{N}_{Av}}{M_0} N^{3\nu-1} \approx \frac{R^3 \mathcal{N}_{Av}}{M}. \tag{8.66}$$

This result is identical to Eqs (8.36) and (8.37), derived previously.

Using Eqs (7.149) and (7.150) with the approximate Zimm model prediction for the stress relaxation modulus [Eq. (8.64)] provides predictions of the storage and loss moduli that are valid for dilute solutions of linear chains (see Problem 8.16):

$$G'(\omega) \approx \frac{\phi kT \omega\tau_Z}{b^3 N} \frac{\sin\left[(1 - 1/(3\nu))\arctan(\omega\tau_Z)\right]}{[1 + (\omega\tau_Z)^2]^{(1-1/(3\nu))/2}}, \tag{8.67}$$

$$G''(\omega) \approx \frac{\phi kT \omega\tau_Z}{b^3 N} \frac{\cos\left[(1 - 1/(3\nu))\arctan(\omega\tau_Z)\right]}{[1 + (\omega\tau_Z)^2]^{(1-1/(3\nu))/2}}. \tag{8.68}$$

These predictions of the Zimm model are compared with experimental data on dilute polystyrene solutions in two θ-solvents in Fig. 8.7. The Zimm model gives an excellent description of the viscoelasticity of dilute solutions of linear polymers.

As in the Rouse model, the mean-square displacement of monomer j during time τ_p is of the order of the mean-square size of the section containing N/p monomers involved in a coherent motion at this time:

$$\langle [\vec{r}_j(\tau_p) - \vec{r}_j(0)]^2 \rangle \approx b^2 \left(\frac{N}{p}\right)^{2\nu} \approx b^2 \left(\frac{\tau_p}{\tau_0}\right)^{2/3}. \tag{8.69}$$

The time dependence of mode index p [Eq. (8.62)] was used to get the final result. Notice that this final result has an exponent that does not depend on

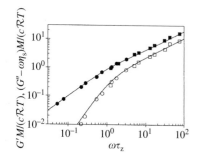

Fig. 8.7
Oscillatory shear data on dilute solutions of polystyrene with $M = 860\,000$ g mol^{-1} in two θ-solvents (circles are in decalin at 16 °C and squares are in di-2-ethylhexyl phthalate at 22 °C). Open symbols are the dimensionless storage modulus and filled symbols are the dimensionless loss modulus, both extrapolated to zero concentration. The curves are the predictions of the Zimm model [Eqs (8.67) and (8.68)]. Data from R. M. Johnson *et al.*, *Polym. J.* **1**, 742 (1970).

solvent quality. The mean-square displacement of a monomer in the Zimm model is subdiffusive on intermediate time scales:

$$\langle[\vec{r}_j(t) - \vec{r}_j(0)]^2\rangle \approx b^2\left(\frac{t}{\tau_0}\right)^{2/3} \quad \text{for } \tau_0 < t < \tau_Z. \quad (8.70)$$

Consistent with the fact that the longest relaxation time of the Zimm model is shorter than the Rouse model, the subdiffusive monomer motion of the Zimm model [(Eq. (8.70)] is always faster than in the Rouse model [Eq. (8.58)] with the same monomer relaxation time τ_0. This is demonstrated in Fig. 8.8, where the mean-square monomer displacements predicted by the Rouse and Zimm models are compared. Each model exhibits subdiffusive motion on length scales smaller than the size of the chain, but motion becomes diffusive on larger scales, corresponding to times longer than the longest relaxation time.

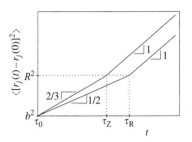

Fig. 8.8
Time dependence of the mean-square monomer displacements predicted by the Rouse and Zimm models on logarithmic scales.

8.5 Semidilute unentangled solutions

There are two limits for unentangled polymer dynamics:

(1) The Zimm limit applies to dilute solutions, where the solvent within the pervaded volume of the polymer is hydrodynamically coupled to the polymer. Polymer dynamics are described by the Zimm model in dilute solutions.

(2) The Rouse limit applies to unentangled polymer melts because hydrodynamic interactions are screened in melts (just as excluded volume interactions are screened in melts). Polymer dynamics in the melt state (with no solvent) are described by the Rouse model, for short chains that are not entangled.

In semidilute solutions there is a length scale, called the **hydrodynamic screening length** ξ_h, separating these two types of dynamics. On length scales shorter than the hydrodynamic screening length (for $r < \xi_h$), the hydrodynamic interactions dominate and dynamics are described by the Zimm model. On length scales larger than the screening length (for $r > \xi_h$) the hydrodynamic interactions are screened by surrounding chains and the dynamics are described by the Rouse model.

In Section 5.3, the static correlation length ξ was defined for semidilute solutions. This correlation length separates single-chain (dilute-like) conformations at shorter length scales ($r < \xi$) from many-chain (melt-like) statistics at longer length scales (for $r > \xi$). The concentration correlation blob of size ξ contains g monomers of a chain, with conformation similar to dilute solutions:

$$\xi \approx bg^\nu. \quad (8.71)$$

The exponent $\nu = 1/2$ in θ-solvents and $\nu \approx 0.588$ in good solvents. The correlation volumes are densely packed, so the volume fraction within each correlation volume (gb^3/ξ^3) must be the same as the overall volume fraction of the solution ϕ:

$$\phi \approx \frac{gb^3}{\xi^3}. \tag{8.72}$$

The correlation length decreases with increasing concentration [Eq. (5.23)]:

$$\xi \approx b\phi^{-\nu/(3\nu-1)}. \tag{8.73}$$

The scaling exponent $\nu/(3\nu - 1) = 1$ in θ-solvents $(\nu = 1/2)$ and $\nu/(3\nu - 1) \cong 0.76$ in good solvents $(\nu \cong 0.588)$.

The hydrodynamic screening length ξ_h in semidilute solutions is expected to be proportional to the static correlation length[7] ξ:

$$\xi_h \approx \xi. \tag{8.74}$$

This proportionality makes sense in both limits. In the melt $(\phi = 1)$, both excluded volume and hydrodynamic interactions are fully screened to the level of individual monomers, so $\xi_h \approx \xi \approx b$. At the overlap concentration $(\phi = \phi^*)$, both excluded volume and hydrodynamic interactions apply over length scales comparable to the size of the entire chain, with $\xi_h \approx \xi \approx R$.

The hydrodynamic screening length can neither be much larger nor much smaller that the static correlation length. Each of the N modes of a chain can, in principle, relax by either Rouse or Zimm motion. On small length scales, Zimm modes are faster than Rouse modes (see Fig. 8.8) because only solvent and other monomers on the same chain are hydrodynamically coupled. However, this situation changes beyond the correlation length, because Zimm motion would couple the motion of monomers from different chains. This extra coupling makes Rouse motion faster than Zimm motion for sections of chain that are larger than the static correlation length, so Rouse dynamics apply on larger length scales.

In semidilute solutions, both statics and dynamics are similar to dilute solutions on length scales shorter than the screening length. For short distances from a given monomer $(r < \xi)$, essentially all other monomers are from the same chain (see Fig. 5.4). The chain conformation is similar to dilute solution and the dynamics are controlled by strong hydrodynamic interactions. Therefore, the relaxation time τ_ξ of a chain section of size ξ is described by the Zimm model and proportional to the correlation volume ξ^3:

$$\tau_\xi \approx \frac{\eta_s}{kT}\xi^3 \approx \frac{\eta_s b^3}{kT}\phi^{-3\nu/(3\nu-1)}. \tag{8.75}$$

On length scales larger than the screening length ξ the dynamics are many-chain-like, with both excluded volume and hydrodynamic interactions

---

[7] Experimental results appear to be consistent with the expectation that hydrodynamic interactions and excluded volume interactions are screened on similar length scales.

screened. The Rouse model applies to the random walk chain of N/g correlation blobs. The relaxation time of the whole chain τ_{chain} is given by Eq. (8.17), with τ_ξ the effective 'monomer' relaxation time, N/g the effective number of 'monomers':

$$\tau_{\text{chain}} \approx \tau_\xi \left(\frac{N}{g}\right)^2 \approx \frac{\eta_s}{kT} \xi^3 \left(\frac{N}{g}\right)^2. \tag{8.76}$$

The number of monomers in a correlation blob is determined by combining Eqs (8.72) and (8.73) [as was done previously in deriving Eq. (5.24)]:

$$g \approx \phi \left(\frac{\xi}{b}\right)^3 \approx \phi^{-1/(3\nu-1)}. \tag{8.77}$$

From Eqs (8.76) and (8.77), the concentration dependence of the relaxation time of the chain in semidilute solution is obtained:

$$\tau_{\text{chain}} \approx \frac{\eta_s b^3}{kT} N^2 \phi^{(2-3\nu)/(3\nu-1)}. \tag{8.78}$$

The concentration dependence of the polymer's relaxation time is a power law with exponent

$$\frac{2-3\nu}{3\nu-1} = 1 \quad \text{in } \theta-\text{solvents } (\nu=1/2), \tag{8.79}$$

and

$$\frac{2-3\nu}{3\nu-1} \cong 0.31 \quad \text{in good solvents } (\nu \cong 0.588). \tag{8.80}$$

Note that if the polymer in dilute solution were highly extended with exponent $\nu > 2/3$, the relaxation time in unentangled semidilute solutions would be predicted to *decrease* with increasing concentration. This is actually observed for semidilute unentangled solutions of charged polymers, called polyelectrolytes, which have $\nu = 1$ in dilute solutions because of charge repulsion. However, for the neutral flexible polymers discussed here, the relaxation time of the chain always increases with concentration.

Polymers diffuse a distance of the order of their size R during their relaxation time τ_{chain}. Recall the size of a linear polymer chain in a semidilute solution [Eq. (5.26) with $v = b^3$]:

$$R \approx \xi \left(\frac{N}{g}\right)^{1/2} \approx bN^{1/2} \phi^{-(2\nu-1)/(6\nu-2)}. \tag{8.81}$$

The exponent

$$\frac{2\nu-1}{6\nu-2} = 0 \quad \text{in } \theta-\text{solvents } (\nu=1/2), \tag{8.82}$$

because the chain maintains a nearly ideal conformation at all concentrations and

$$\frac{2\nu - 1}{6\nu - 2} \cong 0.12 \quad \text{in good solvents } (\nu \cong 0.588). \tag{8.83}$$

The diffusion coefficient D in semidilute solutions decreases as a power law in concentration:

$$D \approx \frac{R^2}{\tau_{\text{chain}}} \approx \frac{kT}{\eta_s b}\frac{\phi^{-(1-\nu)/(3\nu-1)}}{N}. \tag{8.84}$$

The semidilute diffusion coefficient can be written in terms of the Zimm diffusion coefficient of the chain D_Z [Eq. (8.23) valid for diffusion in dilute solutions] and the overlap concentration $\phi^* \approx N^{-(3\nu-1)}$ [Eq. (5.19)]:

$$D \approx D_Z \left(\frac{\phi}{\phi^*}\right)^{-(1-\nu)/(3\nu-1)}. \tag{8.85}$$

The scaling exponent

$$\frac{1-\nu}{3\nu-1} = 1 \quad \text{in } \theta - \text{solvents } (\nu = 1/2), \tag{8.86}$$

and

$$\frac{1-\nu}{3\nu-1} \cong 0.54 \quad \text{in good solvents } (\nu \cong 0.588).$$

The concentration dependence of the diffusion coefficient is plotted in Fig. 8.9 in the scaling form suggested by Eq. (8.85) for polymer solutions in good solvents. The expected exponent is observed over a limited range of approximately one decade above the overlap concentration ϕ^* and a stronger concentration dependence is seen at higher concentrations, where entanglements become important.

In semidilute solutions, hydrodynamic interactions are not screened on length scales smaller than the correlation length ξ. Each mode involves coherent motion of N/p monomers. If N/p is smaller than the g monomers in a correlation blob, motion associated with that mode is described by the Zimm model. On larger length scales, hydrodynamic interactions are screened and modes with index $p < N/g$ are described by the Rouse model. The number of monomers in a correlation blob is given by Eq. (8.77). The crossover mode index for hydrodynamic interaction is

$$p_\xi = \frac{N}{g} \approx N\phi^{1/(3\nu-1)}. \tag{8.87}$$

There are three time scales important for the stress relaxation modulus in semidilute unentangled solutions. The shortest time scale is the relaxation time of a monomer [Eq. (8.20)]. The intermediate time scale is the

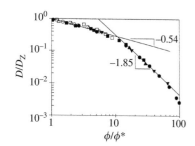

Fig. 8.9
Concentration dependence of diffusion coefficient in good solvent. Filled symbols are four molar masses of polystyrene in benzene spanning the range 78 000–750 000 g mol^{-1}, from L. Leger and J. L. Viovy, *Contemp. Phys.* **29**, 579 (1988). Open symbols are three molar masses of poly(ethylene oxide) in water spanning the range 73 000–660 000 g mol^{-1}, from W. Brown, *Polymer* **25**, 680 (1984). To facilitate comparison, ϕ^* was taken as the volume fraction at which $D = D_Z$ for each data set. The low concentration line is Eq. (8.85) and the high concentration line has the slope expected for entangled solutions in good solvent [Eq. (9.43)].

Zimm relaxation time corresponding to the correlation blob [Eq. (8.75)]. The longest time scale is the Rouse relaxation time of the chain of correlation blobs [Eq. (8.78)].

The stress relaxation modulus follows the Zimm dependence on time scales shorter than τ_ξ, corresponding to motion of chain sections smaller than the correlation length:

$$G(t) \approx \frac{kT}{b^3} \phi \left(\frac{t}{\tau_0}\right)^{-1/(3\nu)} \quad \text{for } \tau_0 < t < \tau_\xi. \tag{8.88}$$

At the crossover time $t = \tau_\xi$ [Eq. (8.75)] the stress relaxation modulus is of the order of the osmotic pressure:

$$G(\tau_\xi) \approx \frac{kT}{b^3} \phi^{3\nu/(3\nu-1)} \approx \frac{kT}{\xi^3} \approx \Pi. \tag{8.89}$$

At longer times, the stress relaxation modulus follows the Rouse dependence:

$$G(t) \approx \frac{kT}{b^3} \phi^{3\nu/(3\nu-1)} \left(\frac{t}{\tau_\xi}\right)^{-1/2} \quad \text{for } \tau_\xi < t < \tau_{\text{chain}}. \tag{8.90}$$

The value of the stress relaxation modulus at the relaxation time of the chain can be determined from Eq. (8.90):

$$G(\tau_{\text{chain}}) \approx \frac{kT}{b^3} \phi^{3\nu/(3\nu-1)} \left(\frac{\tau_{\text{chain}}}{\tau_\xi}\right)^{-1/2} \approx \frac{kT}{b^3} \phi^{3\nu/(3\nu-1)} \frac{g}{N} \approx \frac{kT}{b^3 N} \phi. \tag{8.91}$$

Equations (8.76) and (8.77) were used to simplify this expression for $G(\tau_{\text{chain}})$. The terminal modulus is of order kT per chain, as it must be for any unentangled flexible chain [see Eq. (8.32)]. The stress relaxation modulus at long times is approximated well by the product of the power law and an exponential cutoff:

$$G(t) \approx \frac{kT}{b^3 N} \phi \left(\frac{t}{\tau_{\text{chain}}}\right)^{-1/2} \exp(-t/\tau_{\text{chain}}) \quad \text{for } t > \tau_\xi. \tag{8.92}$$

The time dependence of the stress relaxation modulus in semidilute unentangled solution is sketched in Fig. 8.10. Experimental verification of Rouse dynamics for frequencies smaller than $1/\tau_\xi$ was shown in Fig. 8.5, for a semidilute unentangled polyelectrolyte solution.

The polymer contribution to viscosity in semidilute unentangled solutions is obtained by integrating the stress relaxation modulus over time [Eq. (7.117)].

$$\eta - \eta_s = \int_0^\infty G(t)\, dt \approx \frac{kT}{b^3} \frac{\phi}{N} \tau_{\text{chain}} \approx \eta_s N \phi^{1/(3\nu-1)}. \tag{8.93}$$

In Problem 8.21, the integration is shown to be controlled by the longest relaxation time τ_{chain}.

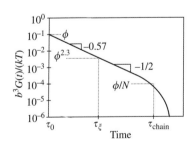

Fig. 8.10
Stress relaxation modulus of an unentangled semidilute solution of chains with $N = 10^3$ monomers at volume fraction $\phi = 0.1$ in an athermal solvent (logarithmic scales).

This result can alternatively be obtained from a de Gennes scaling argument. At the overlap concentration $\phi^* \approx N^{1-3\nu}$, the polymer contribution to viscosity is of the order of the solvent viscosity, and grows as a power law in concentration in semidilute solution:

$$\eta - \eta_s \approx \eta_s \left(\frac{\phi}{\phi^*}\right)^x. \tag{8.94}$$

The exponent x can be determined from the condition that the long-time modes are Rouse-like, and therefore the polymer contribution to solution viscosity should be linearly proportional to polymer molar mass:

$$\eta - \eta_s \approx \eta_s N^{(3\nu-1)x}\phi^x, \tag{8.95}$$

$$(3\nu - 1)x = 1 \quad \Rightarrow \quad x = \frac{1}{3\nu - 1}. \tag{8.96}$$

In θ-solvents ($\nu = 1/2$), the exponent $1/(3\nu - 1) = 2$, and the viscosity is predicted to grow as the square of polymer concentration in unentangled semidilute θ-solutions:

$$\eta_{sp} \approx N\phi^2 \approx \left(\frac{\phi}{\phi^*}\right)^2. \tag{8.97}$$

This concentration dependence is demonstrated in Fig. 8.11. In good solvents ($\nu \cong 0.588$), the exponent $1/(3\nu - 1) \cong 1.3$, and the viscosity is predicted to grow as a weaker power of concentration:

$$\eta_{sp} \approx N\phi^{1.3} \approx \left(\frac{\phi}{\phi^*}\right)^{1.3}. \tag{8.98}$$

8.6 Modes of a semiflexible chain

Polymer dynamics discussed in the previous sections of this chapter correspond to completely flexible chains and are related to modes on length scales larger than the Kuhn length. The relaxation mode structure on length scales shorter than the Kuhn length is significantly different. Many chains, in particular biopolymers, are locally quite stiff. A large part of the relaxation spectrum of such semiflexible chains corresponds to modes with wavelengths shorter than their Kuhn length. In this section, the mode spectrum of semiflexible chains without any intrinsic curvature or twist is described.

8.6.1 Bending energy and dynamics

Consider an elastic beam of length L, thickness L_y and width L_z with Young's modulus E. It is instructive to calculate the elastic energy of bending this beam by a small angle θ (see Fig. 8.12):

$$\theta \approx \sin\left(\frac{h_y}{L}\right) \approx \frac{h_y}{L}. \tag{8.99}$$

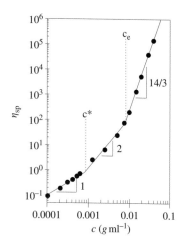

Fig. 8.11

Concentration dependence of specific viscosity for linear poly(ethylene oxide) with $M_w = 5 \times 10^6 \,\mathrm{g\,mol^{-1}}$ in water at $25.0\,^{\circ}$C. Data courtesy of S. Singh.

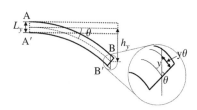

Fig. 8.12

Bending of a rod by angle θ. Insert: the elongation along a surface that is a distance y above the undeformed middle surface is $y\theta$ (and $y < 0$ below the middle surface, in compression).

The central part of the beam (dashed line in Fig. 8.12) is undeformed, the upper half of the beam (AB) is under tension, while the lower half (A'B') is under compression. The deformation along the plane of the bent beam a distance y away from the undeformed central surface is $\Delta L = y\theta$ (see insert in Fig. 8.12). The corresponding extensional strain is $\varepsilon(y) \equiv \Delta L/L = y\theta/L$. The elastic energy density is the work done by deformation per unit volume. Stress is force per unit cross-sectional area and strain is the deformation per unit length, so elastic energy density is proportional to the product of stress and strain $\sigma\varepsilon \approx E\varepsilon^2$, where E is Young's modulus. The elastic energy of a thin slice of the beam of thickness dy and cross-sectional area LL_z is $E(y\theta/L)^2 LL_z dy$. The total elastic energy of a bent beam is obtained by integrating the contribution from each slice over the thickness of the beam:

$$U_L(\theta) \approx \int_{-L_y/2}^{L_y/2} E\left(\frac{y\theta}{L}\right)^2 LL_z \, dy \approx EL_z \frac{\theta^2}{L} \int_{-L_y/2}^{L_y/2} y^2 \, dy$$

$$\approx EL_z \frac{\theta^2}{L} L_y^3. \tag{8.100}$$

The Kuhn length b determines the crossover between stiff and flexible length scales. For rods or beams with length L of the order of the Kuhn length b, the angle of thermally induced fluctuations is of the order of unity $\theta \approx 1$:

$$U_b(1) \approx E\frac{L_y^3 L_z}{b} \approx kT. \tag{8.101}$$

This equation can be solved for the Kuhn length:

$$b \approx E\frac{L_y^3 L_z}{kT}. \tag{8.102}$$

The bending energy of a bent beam [Eq. (8.100)] then can be rewritten in terms of the Kuhn length:

$$U_L(\theta) \approx kT\frac{b}{L}\theta^2 \approx kT\frac{b}{L}\left(\frac{h_y}{L}\right)^2. \tag{8.103}$$

The last relation was obtained using Eq. (8.99) for the deformation angle θ. By writing Eq. (8.103) in terms of the Kuhn length, it becomes much more general and applies to beams with cross-sections that are not rectangular (such as the bending of a cylindrical rod).

Since the beam or rod is a solid, it has natural modes of bending with wavelengths that allow the ends of the beam to be stationary. The first (longest wavelength) mode has wavelength $\lambda = 2L$, the second mode has $\lambda = L$, the third mode has $\lambda = 2L/3$, etc. The fourth mode (with $\lambda = L/2$) is illustrated in Fig. 8.13. Spontaneous thermally induced vibration modes of the beam will form at these wavelengths, and the amplitude of each mode

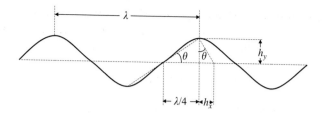

Fig. 8.13
Schematic of the fourth vibration mode (with wavelength $\lambda = L/2$) of a rigid rod of length L.
The transverse oscillation with amplitude h_y reduces the projected rod length along the x-axis.
The amount that the rod length is reduced, per wavelength λ, oscillates with longitudinal
amplitude h_x.

is determined by setting the bending energy from Eq. (8.103) at length
scale λ equal to the thermal energy kT:

$$U_\lambda \approx kTb\frac{h_y^2}{\lambda^3} \approx kT. \tag{8.104}$$

This equation can be solved for the mean-square amplitude of these modes
in the transverse direction:

$$h_y^2 \approx \frac{\lambda^3}{b}. \tag{8.105}$$

To understand the dynamics of the bending fluctuations associated with
these natural modes, a force balance per unit length is required. The force
per unit length associated with the bending mode of wavelength λ is calculated by differentiating the energy U_λ and it is resisted by the frictional
dissipation:

$$\frac{1}{\lambda}\frac{dU_\lambda}{dh_y} \approx kTb\frac{h_y}{\lambda^4} \approx -\frac{\zeta}{b}\frac{dh_y}{dt}. \tag{8.106}$$

To understand the frictional dissipation term, recall that the friction
coefficient ζ of a Kuhn segment of length b is the ratio of force and velocity
in a liquid. Hence, ζ/b is the ratio of force per unit length and the
velocity dh_y/dt. This equation can be solved by separation of variables
and integration:

$$\int \frac{dh_y}{h_y} \approx -\frac{kTb^2}{\zeta\lambda^4}\int dt. \tag{8.107}$$

The solution is exponentially decaying in time $h_y \sim \exp\left(-t/\tau\right)$ with
relaxation time τ proportional to the fourth power of the wavelength λ of
the mode:

$$\tau \approx \frac{\zeta}{kTb^2}\lambda^4. \tag{8.108}$$

Alternatively, the wavelength of a bending mode is proportional to the 1/4
power of its relaxation time:

$$\lambda \approx \left(\frac{kTb^2}{\zeta}t\right)^{1/4} \approx b\left(\frac{kT}{\zeta b^2}t\right)^{1/4} \approx b\left(\frac{t}{\tau_0}\right)^{1/4}. \tag{8.109}$$

The final relation uses the relaxation time of a Kuhn monomer of length b [Eq. (8.108) with $\lambda \approx b$]:

$$\tau_0 \approx \frac{\zeta b^2}{kT}. \tag{8.110}$$

This longest bending mode matches the shortest relaxation time in the Rouse and Zimm models [the relaxation time of a monomer τ_0, Eq. (8.15)].

8.6.2 Tensile modulus and stress relaxation

The two similar triangles in Fig. 8.13 can be used to relate the transverse amplitude h_y and the longitudinal amplitude h_x:

$$\frac{h_x}{h_y} = \frac{h_y}{\lambda/4}. \tag{8.111}$$

The longitudinal amplitude of each mode can then be written in terms of the mode wavelength and the Kuhn length:

$$h_x \approx \frac{h_y^2}{\lambda} \approx \frac{\lambda^2}{b}. \tag{8.112}$$

The final relation made use of Eq. (8.105). The **spring constant** κ_λ (units of force/length) arising from the mode with wavelength λ can be estimated from the fact that the energy of each mode $\kappa_\lambda h_x^2$ is of the order of kT:

$$\kappa_\lambda \approx \frac{kT}{h_x^2} \approx kT\frac{b^2}{\lambda^4}. \tag{8.113}$$

The tensile force f_λ from the mode with wavelength λ arising from application of a small stretch δL is the product of the spring constant and the stretch:

$$f_\lambda \approx \kappa_\lambda \delta L \approx kT\frac{b^2}{\lambda^4}\delta L \approx kT\frac{b^2}{\lambda^3}\frac{\delta L}{\lambda} \approx kT\frac{b^2}{\lambda^3}\varepsilon_\lambda. \tag{8.114}$$

The last relation introduces the extensional strain $\varepsilon_\lambda \equiv \delta L/\lambda$.

The stress borne by a liquid of semiflexible chains from the mode with wavelength λ is the product of the tensile force f_λ in a chain and the line density of the chains in the liquid. The line density can be estimated from the number density of chains c_n/N, where c_n is the monomer number density (the number density of Kuhn segments). The contour length of each chain is Nb and therefore the line density of chains in the liquid is $c_n b$:

$$\sigma_\lambda \approx f_\lambda c_n b \approx kTc_n\left(\frac{b}{\lambda}\right)^3 \varepsilon_\lambda. \tag{8.115}$$

Young's modulus due to the mode with wavelength λ is the ratio of stress and strain:

$$E_\lambda = \frac{\sigma_\lambda}{\varepsilon_\lambda} \approx kTc_n \left(\frac{b}{\lambda}\right)^3. \tag{8.116}$$

Substituting the relation between relaxation time and mode wavelength [Eq. (8.109)] into the expression for modulus [Eq. (8.116)] leads to the time-dependent stress relaxation modulus that decays as the $-3/4$ power of time:

$$E(t) \approx kTc_n \left(\frac{b}{\lambda}\right)^3 \approx kTc_n \left(\frac{t}{\tau_0}\right)^{-3/4} \quad \text{for } \tau_g < t < \tau_0. \tag{8.117}$$

The short time limit τ_g corresponds to the fastest stiff mode, with shortest wavelength determined by the smallest physical length scale using Eq. (8.108). For polymer chains this smallest length is of the order of the bond length l:

$$\tau_g \approx \frac{\zeta l^4}{kTb^2} \approx \tau_0 \left(\frac{l}{b}\right)^4. \tag{8.118}$$

Note that the stress decay is faster than that of the Rouse and Zimm models. Figure 8.14 compares the prediction of Eq. (8.117) with oscillatory shear data on the stiff biopolymer myosin in dilute solution. Myosin clearly exhibits a stronger frequency dependence than the flexible chain models predict, and is reasonably consistent with the 3/4 slope expected from Eq. (8.117).

8.7   Temperature dependence of dynamics

8.7.1   Time–temperature superposition

Both the Rouse and Zimm models, as well as other molecular models to be discussed in Chapter 9, tacitly assume that the relaxation time associated with each mode has the same temperature dependence. Each mode's relaxation time is the product of temperature-independent factors[8] and the monomer relaxation time τ_0 [see Eqs (8.44) and (8.61)]. This has two important consequences for polymer melts and solutions.

First of all, the temperature dependence of all relaxation times is controlled by the ratio of friction coefficient and absolute temperature [see Eq. (8.15)]:

$$\tau \sim \frac{\zeta}{T}. \tag{8.119}$$

The temperature dependence of the modulus at any relaxation time τ is proportional to the product of the polymer mass density ρ and absolute

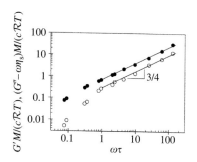

Fig. 8.14
Oscillatory shear data on dilute solutions of the stiff biopolymer myosin, from R. W. Rosser *et al.*, *Macromolecules* **11**, 1239 (1978). Open circles are the dimensionless storage modulus and filled circles are the dimensionless loss modulus, both extrapolated to zero concentration.

---

[8] These factors include the coil size, which can have a weak temperature dependence that is ignored in this discussion.

temperature T [see Eq. (8.45)]:

$$G(\tau) \sim \rho T. \qquad (8.120)$$

Viscosity is the product of relaxation time and the modulus at the relaxation time [Eq. (7.120)]. The temperature dependence of viscosity is proportional to the product of liquid density and friction coefficient:

$$\eta \approx \tau G(\tau) \sim \rho \zeta. \qquad (8.121)$$

The ratio of viscosity and density is the **kinematic viscosity**, which is directly measured in gravity-driven flows. The kinematic viscosity has the same temperature dependence as the friction coefficient. The density of polymer melts weakly decreases as temperature is raised,[9] imparting a weak temperature dependence to the modulus at any relaxation time τ. The temperature dependence of the viscosity of polymer melts is dominated by the strong temperature dependence of the friction coefficient. Near the glass transition temperature T_g, the friction coefficient changes by roughly a factor of 10 when temperature is changed by 3 K, while far above T_g (at $T > T_g + 100\,K$) approximately $25\,K$ temperature change is needed to change the friction coefficient by a factor of 10.

The second important consequence of the relaxation times of all modes having the same temperature dependence is the expectation that it should be possible to superimpose linear viscoelastic data taken at different temperatures. This is commonly known as the **time–temperature superposition** principle. Stress relaxation modulus data at any given temperature T can be superimposed on data at a reference temperature T_0 using a time scale multiplicative shift factor a_T and a much smaller modulus scale multiplicative shift factor b_T:

$$G(t, T) = b_T G\left(\frac{t}{a_T}, T_0\right). \qquad (8.122)$$

The reference temperature T_0 can be chosen to be any convenient temperature, such as a temperature where the liquid is used or the glass transition temperature of the liquid. Equation (8.119) determines the time scale shift factor,

$$a_T = \frac{\zeta T_0}{\zeta_0 T}, \qquad (8.123)$$

where ζ_0 is the friction coefficient at the reference temperature T_0. This time scale shift factor describes the (model independent) temperature dependence of diffusion coefficient:

$$\frac{D(T_0)}{D(T)} = \frac{\zeta T_0}{\zeta_0 T} = a_T. \qquad (8.124)$$

---

[9] The density of a liquid changes by about 10% when temperature is changed by 100 K.

The modulus scale shift factor is determined from Eq. (8.120),

$$b_T = \frac{\rho T}{\rho_0 T_0},$$
(8.125)

where ρ_0 is the density at the reference temperature. Often, Eqs (8.123) and (8.125) do not work perfectly, and the shift factors are treated as adjustable parameters. The temperature dependence of the various dynamic properties can be calculated in terms of the shift factors a_T and b_T. For example, the temperature dependence of viscosity is given by their product (η_0 is the viscosity at the reference temperature):

$$\frac{\eta}{\eta_0} = \frac{\rho \zeta}{\rho_0 \zeta_0} = a_T b_T.$$
(8.126)

Time–temperature superposition also applies to other linear viscoelastic data, with the same shift factors. Two examples are the complex modulus in oscillatory shear,

$$G^*(\omega, T) = b_T G^*(\omega a_T, T_0),$$
(8.127)

and the creep compliance:

$$J(t, T) = \frac{J((t/a_T), T_0)}{b_T}.$$
(8.128)

Using data on logarithmic scales, Eqs (8.122), (8.127), and (8.128) allow simple shifting of data at different temperatures to obtain superposition. The strong temperature dependence of the time (or frequency) scale shift factor a_T makes time–temperature superposition a valuable tool for extending the time (or frequency) range of data at a given temperature. The resulting superimposed curve is a **master curve** for that particular polymer at the chosen reference temperature.

An example of time–temperature superposition is shown in Fig. 8.15 for a poly(vinyl methyl ether) (PVME) melt at a reference temperature of $T_0 = -24\,°C$. Data for G' and G'' were measured as a function of frequency roughly in the range $10^{-2}\,\mathrm{rad\,s^{-1}} < \omega < 10^2\,\mathrm{rad\,s^{-1}}$ at six temperatures (shown in the right-hand side of Fig. 8.14). $T_0 = T_g = -24\,°C$ was selected as the reference temperature and the data at the five other temperatures were superimposed to make the master curve on the left-hand side of Fig. 8.15. In practice, this shifting is first applied to the temperatures closest to T_0, and data sets at temperatures further from T_0 are subsequently added to build up the master curve. The master curve on the left-hand side of Fig. 8.15 has the rheological response at the reference temperature for over 11 decades in frequency! These data extend from the **glassy modulus** G_g at high frequencies all the way to the terminal response (with $G' \sim \omega^2$ and $G'' \sim \omega$) at low frequencies, thereby characterizing the full linear viscoelastic response of this polymer melt. The longest relaxation time is estimated as the reciprocal of the lowest frequency where $G' = G''$. Figure 8.15 shows that this relaxation time is $\tau \cong 2 \times 10^7\,\mathrm{s}$. To directly

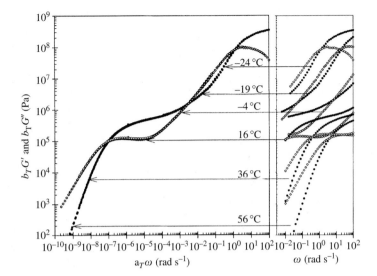

Fig. 8.15
Demonstration of the time–temperature superposition principle, using oscillatory shear data (G', filled circles and G'', open diamonds) on a PVME melt with $M_w = 124\,000\ \mathrm{g\,mol}^{-1}$. The right-hand plot shows the data that were acquired at the six temperatures indicated, with $T_g = -24\,°C$ chosen as the reference temperature. All data were shifted empirically on the modulus and frequency scales to superimpose, constructing master curves for G' and G'' in the left-hand plot. Data and plot courtesy of J. A. Pathak.

probe this frequency at the reference temperature of $T_0 = T_g = -24\,°C$, one complete oscillation period would take $2\pi\tau \cong 10^8\ \mathrm{s} \cong 4$ years! Even graduate students are not so patient. Experiments in the frequency range $10^{-2}\ \mathrm{rad\,s}^{-1} < \omega < 10^2\ \mathrm{rad\,s}^{-1}$ can be done in 2 h (allowing time for thermal equilibration) so the six temperatures studied in Fig. 8.15 can easily be measured in one day by a properly motivated graduate student.

The temperature dependence of the friction coefficient is not very well understood. The simplest model is a thermally activated **Arrhenius equation**:

$$\eta \sim \exp\left(\frac{E_a}{kT}\right). \tag{8.129}$$

The activation energy for flow E_a is a constant at sufficiently high temperatures (typically more than 100 K above the glass transition). At these high temperatures where Eq. (8.129) applies, the activation energy is usually in the range $2kT < E_a < 20kT$. At lower temperatures, the density of the liquid becomes high enough that monomers get in each others way when they try to move.

One simple way to account for this crowding is based on the concept of **free volume**. The molecules in the liquid 'occupy' the vast majority of the liquid's volume V, partly as the atoms that make up the molecules and partly as inaccessible volume that is blocked from access by steric factors. The remaining small fraction of the volume fV is 'free' to be used for molecular motion. As the liquid is cooled, the density increases and the free volume decreases. This slows molecular motion and increases the effective activation energy for flow. The **Doolittle equation** relates the viscosity to the fraction f of the liquid's volume that is free,

$$\eta \sim \exp\left(\frac{B}{f}\right), \tag{8.130}$$

where B is an empirical constant of order unity. The Doolittle equation effectively assumes that the activation energy for flow is reciprocally related to the fractional free volume f. The simplest assumption is that the free volume should have a linear temperature dependence,

$$f = \alpha_f (T - T_\infty).\qquad(8.131)$$

where α_f is the thermal expansion coefficient of the free volume (a constant with dimensions of reciprocal temperature) and T_∞ is the **Vogel temperature** where the free volume is zero. The Vogel temperature is typically about 50 K below the glass transition (see Table 8.3). At the glass transition, most polymers have roughly the same free volume:

$$f_g \cong 0.025 \quad \text{at } T_g.\qquad(8.132)$$

The Doolittle equation [Eq. (8.130)] can be combined with the assumed linear temperature dependence of free volume [Eq. (8.131)] to get the **WLF equation**, so-named for Williams, Landel, and Ferry, who first applied it to polymer melts in 1955:

$$\frac{\eta}{\eta_0} = \exp\left(B\left[\frac{1}{f} - \frac{1}{f_0}\right]\right)\qquad(8.133)$$

$$= \exp\left(\frac{B}{\alpha_f}\left[\frac{1}{T - T_\infty} - \frac{1}{T_0 - T_\infty}\right]\right)$$

$$= \exp\left(\frac{B}{\alpha_f}\left[\frac{T_0 - T}{(T - T_\infty)(T_0 - T_\infty)}\right]\right)$$

$$= \exp\left(\frac{B}{f_0}\frac{(T_0 - T)}{(T - T_\infty)}\right).\qquad(8.134)$$

The final result used Eq. (8.131) for the free volume f_0 at the reference temperature T_0. This 'derivation' relies on several unsubstantiated assumptions [most notably Eqs (8.130) and (8.131)] and hence the WLF equation is a phenomenological description of the temperature dependence of viscosity.

In practice, the small modulus scale shift is often ignored and the time (or frequency) scale shift factor a_T is directly fit to Eq. (8.134) instead of the viscosity ratio [effectively assuming $b_T = 1$ in Eq. (8.126)]. A plot of $\log a_T$ (or $\log [\eta/\eta_0]$) against $(T - T_0)/(T - T_\infty)$ is prepared using the Vogel temperature T_∞ to linearize the plot and the constant B/f_0 is determined from the slope. An example of a linearized Vogel plot is shown in Fig. 8.16 for high molar mass linear chain melts of poly(vinyl methyl ether). The WLF equation provides a good description of the temperature dependence of dynamics for non-crystalline polymers in the temperature range $T_g < T < T_g + 100$ K. The choice of reference temperature is completely arbitrary, but often it is chosen to be T_g and then the free volume in Eq. (8.134) is f_g.

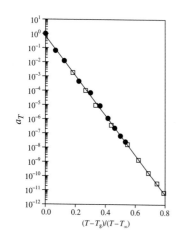

Fig. 8.16
Vogel plot used to determine the coefficients of the WLF equation, using data for high molar mass linear PVME melts. Open squares are data of R. M. Kannan and T. P. Lodge, *Macromolecules* **30**, 3694 (1997) and filled circles are data of J. A. Pathak *et al.*, *Macromolecules* **32**, 2553 (1999). The Vogel temperature T_∞ is adjusted to make this plot linear and B/f_g is determined from the slope.

8.7.2 Transition zone of polymer melts

The transition zone is the range of frequency (or time) over which the storage modulus (or stress relaxation modulus) changes from the glassy modulus to the modulus at the terminal relaxation time for unentangled polymer liquids. For entangled polymer liquids, such as the PVME melt in Fig. 8.15, the transition zone is the range over which the modulus changes from the rubbery plateau modulus to the glassy modulus ($0.003 \, \text{rad s}^{-1} < a_T \omega < 100 \, \text{rad s}^{-1}$ in Fig. 8.15).

The low-frequency end of the transition zone is qualitatively described by the Rouse model of Section 8.4.1 with $G(t) \sim t^{-1/2}$ and $G'(\omega) \sim \omega^{1/2}$, and relaxation modes corresponding to coherent motion of many Kuhn monomers. The higher frequency part of the transition zone has a stronger frequency dependence of the storage modulus $G'(\omega) \sim \omega^{\alpha}$ with $\alpha > 1/2$. It is tempting to associate the part of the transition zone chracterized by the 3/4 power law frequency dependence of the storage modulus $G'(\omega) \sim \omega^{3/4}$ with the modes described by the semiflexible chain model of Section 8.6. Indeed, the $G'(\omega)$ data in the frequency range $0.003 \, \text{rad s}^{-1} < a_T \omega < 0.03 \, \text{rad s}^{-1}$ approximately obey the Rouse scaling and the $G'(\omega)$ data in the frequency range $0.03 \, \text{rad s}^{-1} < a_T \omega < 1 \, \text{rad s}^{-1}$ roughly show the semiflexible chain scaling.

However, the crossover between these two limiting regimes is not well understood, so there is no quantitative description of viscoelasticity for the entire transition zone. Quite generally, the crossover between various scaling regimes is fairly broad and caution must be used when trying to apply power laws in a crossover between regimes. The transition zone is even worse than most crossovers because dynamics of flexible polymers at small length scales (shorter than Kuhn monomer size b) start to be influenced by polymer-specific chemical details. Hence, the crossover from flexible to stiff modes in the transition zone is not universal—each monomer type has a distinct viscoelastic response. For a given monomer, the transition zone does not depend on either chain length or large-scale molecular architecture. The observed viscoelastic response typically has a frequency dependence that is intermediate between the Rouse and semiflexible chain limits $G(t) \sim t^{-\alpha}$ with $0.5 \leq \alpha \leq 0.75$.

Our discussion above focuses on intramolecular effects, but leaves out any intermolecular effects. Current ideas about relaxation in glass-forming liquids speculate that monomer motion is cooperative, involving multiple monomers collectively rearranging in a cooperative fashion. While there is some evidence for this cooperative motion, it is an area of active research that has not yet yielded a generally accepted model. Intermolecular constraints such as the requirement of cooperative relaxations may change the dynamic modulus at very high frequencies.

At still higher frequencies than the highest frequency in Fig. 8.14, polymer liquids exhibit a solid response, with storage modulus G' independent of frequency and equal to the glassy modulus G_g. A typical value of the glassy modulus is of order 10^9Pa (see Table 8.2). The glassy modulus

Table 8.2 Glassy modulus of amorphous polymers

Polymer	G_g (GPa)	T_g (K)
Polyisobutylene	2.4	203
1,4-Polyisoprene	1.5	210
Polypropylene	0.86	259
Polystyrene	1.1	373

is considerably larger than the starting value of the Rouse model $G(\tau_0) \approx kT/b^3$ of kT/v_0 which is of order 10^7 Pa.

This same glassy modulus describes the linear elastic response of the polymer at temperatures below its glass transition temperature T_g, [see Hooke's law, Eq. (7.98)]. The physical reason that the liquid's response becomes similar to that of the glass is that at such high frequencies (or short time scales) monomers (and even small parts of monomers) do not have time to move and relax stress.

8.7.3 Short linear polymer melts

The Rouse model describes the terminal viscoelastic response of polymer melts consisting of linear chains that are too short to form entanglements. For example, Eq. (8.53) describes the viscosity of such short linear polymer melts. However, the viscosity is proportional to the product of friction coefficient of a monomer and the number of monomers per chain. Isothermal viscosity data are *not* proportional to the number of monomers in the chain because the friction coefficient depends on chain length for short chains. Indeed, the glass transition temperature depends on chain length for melts of short chains. The physical reason is that the monomers near the chain ends have more free volume than monomers in the middle of the chain. Assuming Eq. (8.132) applies, there should be a lowering of the glass transition temperature that is proportional to the number density of chain ends $(2\rho N_{Av}/M_n)$, see Problem 8.31:

$$T_g = T_{g\infty} - \frac{C}{M_n}. \tag{8.135}$$

The constant C is typically in the range 10^4–10^5 K g mol^{-1}, which means that for long chains with $M_n > 10^4$–10^5 g mol^{-1}, the glass transition is practically independent of chain length, adopting its long chain limiting value $T_{g\infty}$. The physical significance of the constant C is that it is the molar mass at which the glass transition temperature is 1 K lower than the high molar mass limit $T_{g\infty}$. Representative values of C and $T_{g\infty}$ are listed in Table 8.3.

To test the Rouse prediction that viscosity is proportional to chain length, viscosity data at constant friction coefficient must be used instead of viscosity data at constant temperature. If the coefficient of thermal expansion of the free volume α_f in Eq. (8.131) were independent of chain

Table 8.3 Molar mass dependence of the glass transition temperature [see Eq. (8.136)] and WLF coefficients of high molar mass polymers [see Eq. (8.134) with $T_0 = T_g$]

Polymer	C (K g mol^{-1})	$T_{g\infty}$ (K)	B/f_g	T_∞ (K)
Polybutadiene	1.2×10^4	174	25.6	112
Poly(methyl methacrylate)	6.9×10^4	388	76.9	308
Polystyrene	1.7×10^5	373	30.3	325

length, the simplest procedure to compare viscosities at the same free volume and hence constant friction coefficient would be to measure viscosity at a certain temperature increment above the glass transition (constant $T - T_g$). However, α_f increases as the chains get shorter, precluding the use of a constant $T - T_g$ to attain a constant friction coefficient.[10]

The simplest way to correct viscosity data to constant friction coefficient is to first fit the temperature dependence of viscosity of each individual sample to the WLF equation [Eq. (8.134)], which determines B/f_0. At a given reference temperature, sufficiently long chains have the same B/f_0 and progressively lower values of B/f_0 are obtained for shorter chains, since they have more free volume at a given temperature. The viscosity data at the reference temperature can then be corrected to the friction coefficient of the long chains at the reference temperature using Eq. (8.133). Viscosity data subjected to such a correction are shown in Fig. 8.17 for polybutadiene, polyisobutylene and polystyrene, roughly 120 K above their glass transitions. All linear polymer melts have viscosity proportional to molar mass ($\eta \sim M$) for sufficiently short chains, when the data are determined at a constant friction coefficient as opposed to isothermal data. Longer chains have entanglement effects (discussed in Chapter 9) and have $\eta \sim M^{3.4}$. The full chain length dependence of the viscosity (at constant friction coefficient) of all three polymers are quantitatively described by a simple crossover function:

$$\eta \sim \zeta M \left[1 + \left(\frac{M}{M_c} \right)^{2.4} \right]. \qquad (8.136)$$

Equation (8.136) is tested in Fig. 8.17 (solid curves) and found to describe the molar mass dependence of constant friction coefficient viscosity data for all three of these linear polymers. The **critical molar mass** M_c for entanglement effects in viscosity is always a factor of 2–4 larger than the entanglement molar mass M_e that was defined in Eq. (7.47).

8.8   Randomly branched polymers

As an illustration of the Rouse model, consider the polydisperse mixture of polymers produced by random branching with short chains between branch points. The molar mass distribution and size of the branched polymers in this critical percolation limit were discussed in Section 6.5. Close to the gel point, some very large branched polymers (with $M > 10^6$ g/mol) are formed and the intuitive expectation is that such large branched polymers would be entangled. However, recall that hyperscaling requires polymers of a given size in the critical percolation class only overlap with shorter molecules. Since these shorter polymers relax much more rapidly, each polymer relaxes with no effective topological constraints.

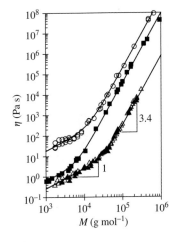

Fig. 8.17
Viscosity data for three linear polymers corrected to the friction coefficient of high molar mass polymer at roughly $T_g + 120$ K, fit to Eq. (8.137) (curves). Open circles are polyisobutylene ($T = 50\,°C$) with $M_c = 14\,000$ g mol^{-1}, filled squares are polybutadiene ($T = 25\,°C$) with $M_c = 6700$ g mol^{-1}, open triangles are free radically prepared polystyrene ($T = 217\,°C$), and filled triangles are anionically prepared polystyrene ($T = 217\,°C$) with $M_c = 35\,000$ g mol^{-1}. Polybutadiene data are from R. H. Colby *et al.*, *Macromolecules* **20**, 2226 (1987). Polyisobutylene and polystyrene data are from T. G. Fox *et al.*, *J. Am. Chem. Soc.* **70**, 2384 (1948), *J. Phys. Chem.* **55**, 221 (1951), and *J. Chem. Phys.* **41**, 344 (1964).

---

[10] The thermal expansion coefficient of the free volume α_f for the shortest polybutadiene chains in Fig. 8.17 is larger than the long chain value by a factor of 1.4.

Unfortunately, there is no accurate theoretical estimate of the strength of hydrodynamic interaction or the extent of hydrodynamic screening of polydisperse branched polymers. Hydrodynamic screening usually correlates well with excluded volume screening. As was demonstrated in Chapter 6, excluded volume interactions are partially screened in the critical percolation limit. Therefore, at least partial screening of hydrodynamic interactions might be expected in these systems. Experiments indicate that hydrodynamic interactions are completely screened in melts of polydisperse branched polymers near the gel point and they can be described by the Rouse model.

Regardless of its complex architecture, any polymer relaxing with no topological constraints and no hydrodynamic interactions is well-represented by the Rouse model, with friction proportional to molar mass. To estimate the terminal response of randomly branched polymers, we apply this reasoning to the characteristic polymers, with size ξ consisting of N^* monomers. The diffusion coefficient of these chains is given by the Einstein relation [Eq. (8.4)]:

$$D_R \approx \frac{kT}{N^* \zeta}. \tag{8.137}$$

Equation (8.137) is perfectly analogous to Eq. (8.12) for a linear polymer (in both cases ζ is the friction coefficient of a single monomer). The Rouse relaxation time of this characteristic polymer, τ^* will be the longest relaxation time in the ensemble of branched polymers. It is determined as the time required for the characteristic polymer to move a distance of order of its own size ξ:

$$\tau^* \approx \frac{\xi^2}{D_R} \approx \frac{\zeta}{kT} N^* \xi^2. \tag{8.138}$$

Equation (8.138) is perfectly analogous to Eq. (8.13) for a linear polymer. This simple argument for the relaxation time will require the relaxation time to be proportional to the product of degree of polymerization and the square of size for the Rouse model of *any* polymer. The scaling of the characteristic polymer's degree of polymerization is $N^* \sim |\varepsilon|^{-1/\sigma}$ with $\sigma \cong 0.45$ [Eq. (6.95)] and the scaling of its size is $\xi \sim |\varepsilon|^{-\nu}$, where $\nu \cong 0.88$ is the critical exponent for the correlation length ξ near the gel point [Eq. (6.125)]. These power laws determine the divergence of the longest relaxation time as the gel point is approached (as the relative extent of reaction $\varepsilon \to 0$):

$$\tau^* \sim |\varepsilon|^{-(2\nu + 1/\sigma)} \sim |\varepsilon|^{-4.0}. \tag{8.139}$$

The value of the relaxation modulus at this terminal relaxation time is of order kT per characteristic polymer. The hyperscaling ideas discussed in Section 6.5.3 require that the characteristic polymers are just at their overlap concentration (they are space-filling but not overlapping). Hence, the Rouse terminal modulus [Eq. (7.93) with $P = 1$] is

$$G(\tau^*) \approx \frac{kT}{\xi^3} \sim |\varepsilon|^{3\nu} \sim |\varepsilon|^{2.6}, \tag{8.140}$$

and the viscosity is the product of this modulus and the relaxation time [Eq. (7.120)].

$$\eta \approx G(\tau^*)\tau^* \sim |\varepsilon|^{\nu-1/\sigma} \sim |\varepsilon|^{-1.3}. \tag{8.141}$$

In practice, the gel point is often difficult to determine with sufficient precision to test these scaling laws. Instead, viscosity and relaxation time can be correlated with weight-average molar mass ($M_{\mathrm{w}} \sim |\varepsilon|^{-\gamma}$ with $\gamma \cong 1.82$) [Eq. (6.103)]:

$$\eta \sim M_{\mathrm{w}}^{(1/\sigma-\nu)/\gamma} \sim M_{\mathrm{w}}^{0.75}, \tag{8.142}$$

$$\tau^* \sim M_{\mathrm{w}}^{(2\nu+1/\sigma)/\gamma} \sim M_{\mathrm{w}}^{2.2}. \tag{8.143}$$

Comparison of these prediction with experimental data on randomly branched polyesters is shown in Fig. 8.18. The line in part (a) has the slope of 0.75 expected by Eq. (8.142) and the line in part (b) has a slope of 2.28 which is slightly larger than the value expected by Eq. (8.143). The agreement is quite good, and indicates that the Rouse model applies to these polymers up to weight-average molar masses exceeding $10^6 \, \mathrm{g \, mol^{-1}}$.

The Rouse relaxation time τ_{R} of a branched polymer of N monomers with size R is a generalization of Eq. (8.138):

$$\tau_{\mathrm{R}} \approx \frac{\zeta}{kT} N R^2 \approx \frac{\zeta b^2}{kT} N^{1+2/\mathcal{D}} \approx \tau_0 N^{1+2/\mathcal{D}}. \tag{8.144}$$

The final relation uses the fractal dimension of the randomly branched polymer, which is $\mathcal{D} \cong 2.53$ for critical percolation in three dimensions. The fact that randomly branched polymers are fractal means that the size r of a polymer section has the same dependence on the number g of

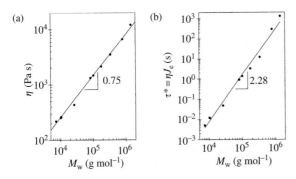

Fig. 8.18
Viscosity (a) and longest relaxation time (b) of randomly branched polyesters with average degree of polymerization between branch points of $N_0 = 2$, at $T = T_{\mathrm{g}} + 64$ K. Data are from C. P. Lusignan *et al.*, *Phys. Rev. E* **52**, 6271 (1995).

monomers in the section $r \approx bg^{1/\mathcal{D}}$ as the size R and degree of polymerization N of the whole polymer $R \approx bN^{1/\mathcal{D}}$ (see Section 1.4). The Rouse model of randomly branched polymers exhibits **fractal dynamics**: the relaxation time $\tau(g)$ of a polymer section of g monomers has the same dependence on the number of monomers g as the whole chain [Eq. (8.144)]:

$$\tau(g) \approx \tau_0 g^{1+2/\mathcal{D}}. \tag{8.145}$$

At time $t = \tau(g)$ smaller randomly branched polymers with degree of polymerization $N < g$ are almost completely relaxed, while larger branched polymers with degree of polymerization $N > g$ consist of N/g unrelaxed sections of size r, each storing elastic energy of order kT. The time dependence of the number of monomers g in these crossover sections is obtained from (Eq. 8.145):

$$g \approx \left(\frac{t}{\tau_0}\right)^{1/(1+2/\mathcal{D})}. \tag{8.146}$$

Equation (8.140) states that the terminal modulus $G(\tau^*)$ is kT per characteristic polymer. Hyperscaling also requires polymers of a given size r and sections of larger polymers of the same size to be just at their overlap concentration, so their pervaded volumes r^3 densely fill all space. This means that Eq. (8.140) can be generalized for the stress relaxation modulus at the time scale where each chain section of size r relaxes:

$$G(t) \approx \frac{kT}{r^3} \approx \frac{kT}{b^3} g^{-3/\mathcal{D}} \approx \frac{kT}{b^3} \left(\frac{t}{\tau_0}\right)^{-3/(\mathcal{D}+2)} \quad \text{for } \tau_0 < t < \tau^*. \tag{8.147}$$

The stress relaxation modulus decays as a power law in time with exponent $3/(\mathcal{D}+2) \cong 0.66$. This power law dependence continues up to the longest relaxation time τ^* of the characteristic branched polymer. At the gel point this power law extends forever because τ^* diverges [see Eq. (8.139)].

For oscillatory shear, Eqs (7.149) and (7.150) allow calculation of the intermediate frequency behaviour of storage and loss moduli of randomly branched polymers below the gel point:

$$G'(\omega) \sim G''(\omega) \sim \omega^{3/(\mathcal{D}+2)} \quad \text{for } 1/\tau^* < \omega < 1/\tau_0. \tag{8.148}$$

This power law character extends over the entire frequency range $\omega < 1/\tau_0$ at the gel point, where τ^* diverges. Critical percolation expects $\mathcal{D} \cong 2.53$, so the storage and loss moduli at the gel point are parallel power laws with exponent $3/(\mathcal{D}+2) \cong 0.66$. The loss tangent at the gel point is

$$\tan \delta \equiv \frac{G''}{G'} = \tan\left(\frac{3\pi}{2(\mathcal{D}+2)}\right) \cong 1.70, \tag{8.149}$$

for all sufficiently small frequencies ($\omega < 1/\tau_0$). The frequency dependence of storage and loss moduli for unentangled randomly branched polymers are compared with the predictions of the Rouse model in Fig. 8.19 (see Problem 8.35 for the full functional form).

8.9 Dynamic scattering

Scattering techniques for measuring various static and thermodynamic properties of polymers, such as molar mass, size, conformations, interaction parameters, etc. were described in experimental sections of Chapters 1–5. In addition to static properties, scattering can provide important information about dynamic properties of polymeric systems. This section focuses on dynamic scattering from dilute solutions, but similar methods are used in semidilute and concentrated solutions.[11]

The instantaneous scattering intensity $I(q, t)$ at wavevectors of magnitude q [see Eq. (2.131)] depends on the spacial arrangement of scattering centres (positions and conformations of molecules) at time t. As molecules move, changing their conformations and locations in space, the scattering intensity $I(q, t)$ fluctuates in time [see Fig. 8.20(a)]. The value of scattering intensity, averaged over a long time interval t, is the static scattering intensity $I(q)$ discussed in Chapters 1–5:

$$I(q) = \langle I(q, 0) \rangle \equiv \lim_{t \to \infty} \frac{1}{t} \int_0^t I(q, t') \, dt'. \tag{8.150}$$

Fluctuations of the instantaneous scattering intensity $I(q, t)$ about its average value $I(q)$ contain information about polymer dynamics on the length scale $1/q$. In order to extract this information it is useful to consider a memory of the instantaneous intensity $I(q, t'')$ at time t'' that still remains after an elapsed time t (at time $t'' + t$). This memory is defined mathematically through a **time autocorrelation function** [see Fig. 8.20(b)]:

$$\langle I(q, 0) I(q, t) \rangle \equiv \lim_{t' \to \infty} \frac{1}{t'} \int_0^{t'} I(q, t'') I(q, t'' + t) \, dt''. \tag{8.151}$$

At $t = 0$, the autocorrelation function $\langle [I(q, 0)]^2 \rangle$ is the mean-square value of the intensity. The limiting value of the autocorrelation function at times much longer than the correlation time τ is the square of the average intensity because the values of $I(q, 0)$ and $I(q, t)$ are independent of each other for $t \gg \tau$:

$$\lim_{t \to \infty} \langle I(q, 0) I(q, t) \rangle = \langle I(q, 0) \rangle \langle I(q, t) \rangle = \langle I(q, 0) \rangle^2. \tag{8.152}$$

[11] Here we only consider dynamic scattering from isotropic solutions.

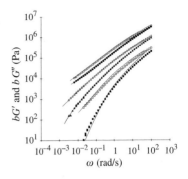

Fig. 8.19
Storage modulus (filled symbols) and loss modulus (open symbols) for three randomly branched polyesters 41 K above their glass transition temperature. Squares have $M_w = 57\,000$ g mol^{-1} ($b = 1$), triangles have $M_w = 380\,000$ g mol^{-1} ($b = 3$), and the circles correspond to a sample extremely close to the gel point ($b = 10$). The curves are predictions from fractal dynamics based on the Rouse model.

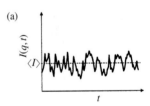

(a)

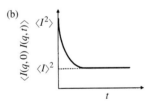

(b)

Fig. 8.20
(a) Intensity of scattered light as a function of time. (b) Time autocorrelation function of the scattered intensity as a function of time.

Hence, the autocorrelation function always decays from the mean of the squared static intensity to the square of the mean intensity [see Fig. 8.20(b)]. In the simplest case of a single relaxation mode, the time autocorrelation function decays like a single exponential with correlation time τ:

$$\langle I(q, 0)I(q, t)\rangle = \langle I(q, 0)\rangle^2 + [\langle [I(q, 0)]^2\rangle - \langle I(q, 0)\rangle^2]\exp\left(-\frac{t}{\tau}\right).$$

$$(8.153)$$

The memory of the initial intensity $I(q, 0)$ decays at the correlation time τ.

The autocorrelation function depends on how molecules move and rearrange on length scales $1/q$ during time t. Therefore, the correlation time τ is expected to depend on q.

Scattering from dilute solutions was discussed in Section 3.5. The intermolecular scattering regime was defined for reciprocal wavevectors $1/q$ larger than the distance between molecules. In this regime the scattering intensity is proportional to the square of the difference in the number of molecules in the neighbouring volumes q^{-3}. The average of the square of this difference determines the static scattering intensity [Eq. (3.126)]. The time variations in the scattering intensity are directly related to the time variations in the number of molecules in volumes q^{-3}. The memory of the number of molecules in a given volume q^{-3} persists as long as most of the molecules that were in this volume initially did not have time to leave it. The correlation time τ is therefore of the order of the time required for molecules to diffuse out of the volume. Since the diffusion distance is q^{-1}, the correlation time is $\tau \approx (q^{-1})^2/D$ [Eq. (8.1)]. A more careful analysis of the problem determines the numerical prefactor:

$$\tau = \frac{1}{2q^2 D}.$$

$$(8.154)$$

Hence, the time autocorrelation function provides a direct means to determine the diffusion coefficient D in dilute monodisperse solutions:

$$\langle I(q, 0)\ I(q, t)\rangle = \langle I(q, 0)\rangle^2 + [\langle [I(q, 0)]^2\rangle - \langle I(q, 0)\rangle^2]\exp(-2q^2 Dt).$$

$$(8.155)$$

In practice, the autocorrelation function is fit to a simple expression with three fitting parameters—the amplitude A, the base line B and the diffusion coefficient D:

$$\langle I(q, 0)I(q, t)\rangle = [A\exp(-q^2 Dt)]^2 + B.$$

$$(8.156)$$

It is crucial to realize that the baseline at long times has to equal the square of the mean static intensity ($B = \langle I(q, 0)\rangle^2$). If this criterion is not realized, then artefacts such as dust are influencing the data! Another criterion for proper measurement of the diffusion coefficient is to make sure that it is independent of q by following the intensity correlations at different scattering angles.

The diffusion coefficients of dilute solutions of polystyrene in toluene are plotted in Fig. 8.21. Data on dilute solutions of flexible polymers obey Eq. (8.23). The data in Fig. 8.21 exhibit the expected crossover from θ-solvent scaling ($\nu = 1/2$) at low molar masses where the coils are smaller than the thermal blob to athermal solvent scaling ($\nu = 0.588$) at high molar masses where there are many thermal blobs per chain.

The hydrodynamic radius of polymers can be obtained from the measured diffusion coefficients in dilute solution and the known solvent viscosity η_s using the Stokes–Einstein relation [Eq. (8.9)]:

$$R_h = \frac{kT}{6\pi\eta_s D} = \frac{kTq^2\tau}{3\pi\eta_s}. \tag{8.157}$$

Toluene at 25 °C has a viscosity $\eta_s \cong 5.6 \times 10^{-4}\,\mathrm{Pa\,s}$ and Fig. 8.21 shows that polystyrene with $M_w = 10^6\,\mathrm{g\,mol^{-1}}$ has diffusion coefficient $D \cong 1.3 \times 10^{-11}\,\mathrm{m^2\,s^{-1}}$. Equation (8.157) determines the hydrodynamic radius of this polymer to be $R_h \cong 30\,\mathrm{nm}$. A typical wavevector in a light scattering experiment is $q \approx 10^7\,\mathrm{m^{-1}}$ the correlation time is $\tau \approx 0.38\,\mathrm{ms}$.

Experimental results for the ratio R_g/R_h are compared with expectations based on the Zimm model for various polymer structures in Table 8.4. In general, increasing the density of monomers decreases the ratio R_g/R_h towards the value for hard spheres of $R_g/R_h = \sqrt{3/5} \cong 0.77$ (see Table 2.4), as seen by increasing the number f of arms in a star polymer, for instance. Experiment shows that the $f = 270$-arm star polymer is practically in the hard sphere limit (see Table 8.4), owing to crowding. Large generation number dendrimers also correspond to the hard sphere limit. Increasing the solvent quality generally increases R_g/R_h. Polydispersity also increases the ratio R_g/R_h, and randomly branched polymers near their

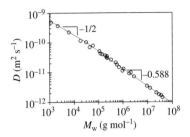

Fig. 8.21
Diffusion coefficient from dynamic light scattering for dilute polystyrene solutions in the good solvent toluene (open circles). The lines show the limiting power laws with slopes expected by Eq. (8.23) ($\nu = 1/2$ inside the thermal blob and $\nu = 0.588$ at higher molar masses). The lines cross at $M_w \approx 30\,000$ g mol^{-1}, consistent with the molar mass inside a thermal blob estimated from intrinsic viscosity of polystyrene in toluene in Fig. 8.3. Data from D. W. Schaefer and C. C. Han, in *Dynamic Light Scattering* (R. Pecora, editor) Plenum, New York (1985).

Table 8.4 Ratio of radius of gyration and hydrodynamic radius for different polymer architectures

Polymer structure	Solvent	R_g/R_h Zimm theory	R_g/R_h experiment
Randomly branched	Good	—	2.0
Linear monodisperse	Good	1.6	1.5
Randomly branched	θ	1.7	—
Linear $M_w/M_n = 2$	θ	1.7	—
Linear monodisperse	θ	1.5	1.3
3-Arm star	θ	1.4	1.2
4-Arm star	θ	1.3	1.05
12-Arm star	θ	1.17	0.93
18-Arm star	θ	1.14	0.82
270-Arm star	θ	1.08	0.77
Hard sphere	—	0.77	0.77

gel point (with very large polydispersity) in good solvent have the largest $R_g/R_h \cong 2.0$.

The experimental values of R_g/R_h are always smaller than the theoretical estimates, indicating that the Zimm model underestimates the hydrodynamic radius. More recent calculations predict R_g/R_h in far better agreement with experiment (for example, linear chains are predicted by Oono[12] to have $R_g/R_h \cong 1.56$ in good solvent have $R_g/R_h \cong 1.24$ in θ-solvent).

In polydisperse solutions, the decay of the autocorrelation function is not a single exponential and it is challenging to extract the distributions of diffusion coefficients and sizes from the non-exponential decay of the intensity correlations. In the case of a bidisperse distribution with diffusion coefficients that differ by at least a factor of 2, it is possible to fit the decay of the intensity correlations by a sum of two exponentials and obtain the corresponding sizes and relative concentrations of the two components:

$$\langle I(q,0)I(q,t)\rangle = [A_1 \exp(-q^2 D_1 t) + A_2 \exp(-q^2 D_2 t)]^2 + B. \quad (8.158)$$

The coefficients A_1 and A_2 are proportional to the relative concentrations of the two components present in the bidisperse solution.[13] The intensity autocorrelation function $\langle I(q,0)\,I(q,t)\rangle$ is fit to the *square* of the sum of modes because it can be related to the square of the electric field autocorrelation function $\langle E^*(q,0)\,E(q,t)\rangle$:

$$\langle I(q,0)\,I(q,t)\rangle \sim \langle |E(q,0)|^2 |E(q,t)|^2 \rangle$$
$$\approx \langle |E(q,0)|^2 \rangle^2 + \langle E^*(q,0)\,E(q,t)\rangle^2. \quad (8.159)$$

Equation (8.159) is strictly valid for a Gaussian distribution of electric fields. The electric field autocorrelation function is related to the **dynamic structure factor** $S(q,t)$ [compare it with the static scattering function $S(q)$ in Eq. (3.121)]:

$$\langle E^*(q,0)\,E(q,t)\rangle \sim S(q,t)$$
$$\equiv \frac{1}{n}\sum_{j=1}^{n}\sum_{k=1}^{n}\langle\exp[-i\vec{q}\cdot(\vec{r}_j(t)-\vec{r}_k(0))]\rangle. \quad (8.160)$$

The number of monomers in the scattering volume (see Section 1.7.2) is n and $\vec{r}_j(t)$ is the position of monomer j at time t. The dynamic structure factor of a bidisperse solution can be represented as a sum of two modes [Eq. (8.158)].

In the case of a more general polydispersity, the time autocorrelation function corresponds to the sum of many modes:

$$\langle I(q,0)\,I(q,t)\rangle = \left[\sum_i A_i \exp(-q^2 D_i t)\right]^2 + B. \quad (8.161)$$

---

[12] Y. Oono, *J. Chem. Phys.* **79**, 520 (1983) and *Adv. Chem. Phys.* **61**, 301 (1985).
[13] $c_1/c_2 = A_1/A_2$, where c_i is the concentration of species i.

In a continuous representation it can be written as a Laplace transform of the distribution $A(\Gamma)$ of decay rates $\Gamma = q^2 D$:

$$\langle I(q,0) I(q,t) \rangle = \left[\int_0^\infty A(\Gamma) \exp(-\Gamma t)\, d\Gamma \right]^2 + B. \qquad (8.162)$$

Determination of the distribution of modes $A(\Gamma)$ and the related distribution of sizes requires inversion of the Laplace transform, which is an ill-defined problem for a limited data set containing any noise. There are some numerical programs (such as CONTIN©) that attempt to perform this inverse transformation. The resulting distributions do sometimes (but not always) correlate (but not coincide) with the actual distribution of hydrodynamic radii in solution.

Dynamic modes of semidilute solutions can also be studied using dynamic scattering. The dynamic structure factors $S(q,t)$ of these more complicated systems still have a simple diffusive behaviour at low values of the wavevector q [Eqs (8.156), (8.159), and (8.160)]:

$$S(q,t) = S(q,0) \exp(-q^2 Dt) \quad \text{for } qR < 1. \qquad (8.163)$$

However, for non-dilute systems, the diffusion coefficient obtained from the low q time dependence of $S(q,t)$ may not be the diffusion coefficient of the polymers. For example, in semidilute solutions the dominant decay in $S(q,t)$ corresponds to correlations disappearing at the scale of the correlation length. In such cases, the diffusion coefficient is called the **cooperative diffusion coefficient**.

The logarithm of the ratio of the dynamic structure factor and the static structure factor is linear in time for diffusive motion described by a single diffusion coefficient:

$$\ln\left[\frac{S(q,t)}{S(q,0)}\right] = -q^2 Dt \approx -q^2 \langle [r(t) - r(0)]^2 \rangle \quad \text{for } qR < 1. \qquad (8.164)$$

The mean-square displacement of the monomers is diffusive [Eq. (8.1)] on large length scales $(qR < 1)$ corresponding to times longer than the relaxation time:

$$\langle [r(t) - r(0)]^2 \rangle \approx Dt. \qquad (8.165)$$

Dynamic scattering can also provide information about relaxation modes of polymers at higher values of the wavevector q $(qR > 1)$. Equation (8.164) can be generalized to higher wavevectors and to semidilute and concentrated solutions by noticing that the decay of the dynamic structure factor is determined by the ratio of mean-square monomer displacement and the square of the reciprocal wavevector $(1/q)^2$:

$$\ln\left[\frac{S(q,t)}{S(q,0)}\right] \approx -q^2 \langle [r(t) - r(0)]^2 \rangle. \qquad (8.166)$$

For the Rouse model the mean-square displacement is proportional to the square root of time [Eq. (8.58)]. The logarithm of the dynamic structure factor of the Rouse model then also scales as the square root of time for $\tau_0 < t < \tau_R$:

$$\ln\left[\frac{S(q,t)}{S(q,0)}\right] \approx -q^2 b^2 \left(\frac{t}{\tau_0}\right)^{1/2} \quad \text{for Rouse model } 1/R < q < 1/b.$$

(8.167)

For the Zimm model the mean-square displacement of monomers is faster [Eq. (8.70)] leading to the logarithm of the Zimm dynamic structure factor scaling as the 2/3 power of time for $\tau_0 < t < \tau_Z$:

$$\ln\left[\frac{S(q,t)}{S(q,0)}\right] \approx -q^2 b^2 \left(\frac{t}{\tau_0}\right)^{2/3} \quad \text{for Zimm model } 1/R < q < 1/b.$$

(8.168)

More generally, dynamic scattering methods are used to study many aspects of polymer dynamics. Full discussion of those methods is beyond the scope of this book.

8.10   Summary of unentangled dynamics

The Rouse model is the simplest molecular model of polymer dynamics. The chain is mapped onto a system of beads connected by springs. There are no hydrodynamic interactions between beads. The surrounding medium only affects the motion of the chain through the friction coefficient of the beads. In polymer melts, hydrodynamic interactions are screened by the presence of other chains. Unentangled chains in a polymer melt relax by Rouse motion, with monomer friction coefficient ζ. The friction coefficient of the whole chain is $N\zeta$, making the diffusion coefficient inversely proportional to chain length:

$$D_R = \frac{kT}{N\zeta}.$$

(8.169)

In contrast, the Zimm model considers the motion of beads (or monomers) to be hydrodynamically coupled with other monomers. Both the polymer and the solvent molecules within the pervaded volume of the chain move together in dilute solutions. The diffusion coefficient of a chain in the Zimm model is of the same form as the Stokes–Einstein relation [Eq. (8.9)] for diffusion of a colloidal particle in a liquid:

$$D_Z \approx \frac{kT}{\eta_s R}.$$

(8.170)

The intrinsic viscosity of polymers in dilute solutions is an extremely important measure of the coil size, owing to its simplicity and precision. The Zimm model leads directly to the Fox–Flory equation for intrinsic viscosity, which in turn leads to the Mark–Houwink equation:

$$[\eta] = \Phi \frac{R^3}{M} = KM^{3\nu-1}. \qquad (8.171)$$

During their relaxation time τ, polymers diffuse a distance of order their own size ($\tau \approx R^2/D$). The relaxation times of the Rouse and Zimm models are then easily obtained from the diffusion coefficients:

$$\tau_R \approx \frac{\zeta N R^2}{kT} \approx \tau_0 N \left(\frac{R}{b}\right)^2 \quad \text{and} \quad \tau_Z \approx \frac{\eta_s R^3}{kT} \approx \tau_0 \left(\frac{R}{b}\right)^3. \qquad (8.172)$$

The final relations are written in terms of the relaxation time of a monomer $\tau_0 \approx \eta_s b^3/kT \approx \zeta b^2/kT$. These relations are very general, and can be applied whenever the topological interactions, called entanglements (discussed in detail in Chapter 9), can be ignored.

In semidilute solutions, the hydrodynamic interactions affect dynamics only up to the scale of the hydrodynamic screening length, which is of the order of the correlation length. On length scales larger than the correlation length, both excluded volume and hydrodynamic interactions are screened by the presence of other chains. This screening becomes increasingly important as concentration is raised, and eventually the melt state is reached where hydrodynamic interactions are fully screened down to the scale of individual monomers. For unentangled chains in semidilute solution, the chain sections within a correlation volume relax by Zimm motion and the sections of chain larger than the correlation length relax by Rouse motion. The relaxation time of the chain in unentangled semidilute solution is our first (and the simplest) example of a hierarchy of relaxation processes. The correlation blob relaxes by Zimm motion on the time scale τ_ξ, while the whole chain relaxes as a Rouse chain of blobs at the time scale τ_{chain}:

$$\tau_\xi \approx \frac{\eta_s \xi^3}{kT} \approx \tau_0 \left(\frac{\xi}{b}\right)^3 \quad \text{and} \quad \tau_{\text{chain}} \approx \tau_\xi \left(\frac{N}{g}\right)^2 \approx \tau_0 \left(\frac{\xi}{b}\right)^3 \left(\frac{N}{g}\right)^2. \qquad (8.173)$$

There are many examples of hierarchies of relaxation processes in Chapter 9.

The time-dependent viscoelastic response of polymers is broken down into individual modes that relax on the scale of subsections of the chain with N/p monomers. The Rouse and Zimm models have different structure of their mode spectra, which translates into different power law exponents for the stress relaxation modulus $G(t)$:

$$G(t) \approx \frac{kT}{b^3} \phi \left(\frac{t}{\tau_0}\right)^{-\kappa} \exp(-t/\tau). \qquad (8.174)$$

The longest mode relaxes at time τ (τ_Z for the Zimm model, with exponent $\kappa = 1/(3\nu)$ and τ_R for the Rouse model, with exponent $\kappa = 1/2$). While the difference between these exponents is small, they can be measured quite precisely, allowing unambiguous identification of Rouse and Zimm motion.

Problems

Section 8.1

8.1 In unentangled polymer melts

    (i) Are hydrodynamic interactions important or screened?
    (ii) Which model describes dynamics?
    (iii) Is the friction coefficient of the polymer proportional to the chain size R or the number of monomers in the chain N?

8.2 Consider a polymer chain represented by $N = 20$ beads connected by springs of root-mean-square size $b = 5\,\text{Å}$ diffusing in a melt of similar chains with bead friction coefficient $\zeta = 3 \times 10^{-10}\,\text{g}\,\text{s}^{-1}$ at 22 °C.

    (i) What is the root-mean-square end-to-end distance R of this chain?
    (ii) What is best model describing the dynamics of the chain in a melt? What is the friction coefficient of the chain?
    (iii) What is the diffusion coefficient D of the chain?
    (iv) Estimate the longest relaxation time τ of the chain.
    (v) Estimate the viscosity of the melt.

Section 8.2

8.3 In dilute polymer solutions

    (i) Are hydrodynamic interactions important or screened?
    (ii) Which model describes dynamics?
    (iii) Is the friction coefficient of the polymer proportional to the chain size R or the number of monomers in the chain N?

8.4 Consider a polymer chain represented by $N = 100$ beads connected by springs of root-mean-square size $b = 5\,\text{Å}$ in a dilute θ-solution at 22.5 °C with solvent viscosity $\eta_s = 0.6\,\text{cP} = 6 \times 10^{-3}\,\text{g}\,(\text{cm s})^{-1}$.

    (i) What is best model describing the dynamics of the chain in a dilute solution? What is the friction coefficient of the chain?
    (ii) What is the diffusion coefficient D of the chain?
    (iii) What is the longest relaxation time τ of the chain?
    (iv) Estimate the specific viscosity $\eta_{sp} = (\eta - \eta_s)/\eta_s$ of a solution at volume fraction $\phi = 0.01$?

8.5 Comparison of polymer and solvent diffusion in dilute solution.

    (i) Calculate the time scale of a solvent molecule, initially located at the centre of a polymer's pervaded volume, to diffuse out of the pervaded volume. Assume the solvent is the same size b as a monomer.
    (ii) Use the Zimm model to show that the time scale for the polymer to diffuse out of the same volume is a factor of N^{ν} longer than the solvent diffusion time.
    (iii) Show that the ratio of diffusion coefficients of the polymer and the solvent is $N^{-\nu}$.
    (iv) Why is the ratio of time scales (part ii) reciprocally related to the ratio of diffusion coefficients (part iii)?

Section 8.3

8.6 (i) Calculate the scaling predictions of the Rouse and Zimm models for the intrinsic viscosity $[\eta]$ and relaxation time τ of dilute solutions of rigid rod polymers of length L.

(ii) Compare the calculated relaxation times in part (i) with the result for the rotational relaxation time of a rod in dilute solution that includes hydrodynamic interactions of monomers along the rod,

$$\tau = \frac{\pi \eta_s L^3}{3kT \ln (L/2b)}, \qquad (8.175)$$

where b is the diameter of the rod.

(iii) Calculate Rouse and Zimm model predictions for $[\eta]$ and τ for a general fractal with fractal dimension $\mathcal{D}$ and show that the rigid rod limit and the ideal chain have the expected results when $\mathcal{D} = 1$ and $\mathcal{D} = 2$.

8.7 Intrinsic viscosity of a dendrimer

(i) Use the Fox–Flory equation to determine the intrinsic viscosity of a dendrimer of generation g, functionality f, and core functionality n, assuming that the degree of polymerization is given by Eq. (6.35) and the size increases linearly with generation number ($R = bg$).

(ii) Using monomer size $b = 7\,\text{Å}$, a monomer mass of $100\,\text{g mol}^{-1}$, functionality $f = 3$, and core functionality $n = 6$, plot the intrinsic viscosity as a function of generation g from $g = 1$ to $g = 10$.

(iii) Qualitatively explain the shape of the plot in part (ii).

8.8 Derive the following general relation between intrinsic viscosity and overlap concentration:

$$c^* \approx \frac{1}{[\eta]}. \qquad (8.176)$$

8.9 Estimate the specific viscosity at the overlap concentration using the following equations:

(i) Huggins equation [Eq. (1.97)] for good solvent with $k_H = 0.3$;
(ii) Huggins equation for θ-solvent with $k_H = 0.8$;
(iii) The Zimm model [Eq. (8.33)] for good solvent;
(iv) The Zimm model for θ-solvent;
(v) The Rouse model for semidilute unentangled solutions in good solvent [Eq. (8.98)];
(vi) The Rouse model for semidilute unentangled solutions in θ-solvent [Eq. (8.97)].

Hint: Remember $[\eta] \approx 1/c^*$.

Section 8.4

8.10 Calculate the stress relaxation modulus $G(t)$, valid for all times longer than the relaxation time of a monomer, for a monodisperse three-dimensional melt of unentangled flexible fractal polymers that have fractal dimension $\mathcal{D} < 3$. Assume complete hydrodynamic screening. *Hint*: Keep the fractal dimension general and make sure your result coincides with the Rouse model for $\mathcal{D} = 2$.

8.11 The Fox-Flory prefactor

The Zimm stress relaxation modulus has the same form as Eq. (8.55) with the Zimm relaxation times [Eqs. (8.26) and (8.61)] replacing the Rouse times.

(i) Show that the intrinsic viscosity can be expressed as a sum over the stress relaxation times

$$[\eta] = \frac{RT}{\eta_s M} \sum_{p=1}^{\infty} \tau_p$$

(ii) Derive the Fox-Flory prefactor Φ of Eq. (8.37) in a θ-solvent from the Zimm stress relaxation times. *Hint*: Use the approximation of a sum $\sum_{p=1}^{\infty} p^{-3/2} \cong 2.6124$.

8.12 Diffusion of a long chain in a melt of shorter chains

Consider dilute long probe chains with N_A Kuhn monomers of length b in a melt of chemically identical unentangled shorter chains with N_B Kuhn monomers.

(i) What is the root-mean-square end-to-end distance R_A of probe chains? *Hint*: Consider two separate cases $N_A < N_B^2$ and $N_A > N_B^2$ and review section 4.5.2.

(ii) What would be the diffusion coefficient of the probe chain D_R, if its dynamics were Rouse-like with monomeric friction coefficient ζ?

(iii) Show that if the dynamics were Zimm-like, the diffusion coefficient of the probe chain for the case $N_A < N_B^2$ would be

$$D_Z \approx \frac{kT}{\zeta N_A} \left(\frac{N_A}{N_B^2} \right)^{1/2} \quad \text{for } N_A < N_B^2. \tag{8.177}$$

Compare this Zimm diffusion coefficient D_Z with the Rouse diffusion coefficient D_R of part (ii). *Hint*: The viscosity of an unentangled melt of shorter N_B-chains is predicted by the Rouse model [Eq. (8.53)].

(iv) Show that in the case $N_A > N_B^2$ the Zimm-like diffusion coefficient of the probe chain would be

$$D_Z \approx \frac{kT}{\zeta N_A} \left(\frac{N_A}{N_B^2} \right)^{2/5} \quad \text{for } N_A > N_B^2. \tag{8.178}$$

Compare this Zimm diffusion coefficient D_Z with the Rouse diffusion coefficient D_R of part (ii).

(v) What is the relation between the degrees of polymerization of the probe chain N_A and melt chains N_B for which $D_Z \approx D_R$? Compare this hydrodynamic crossover to the excluded volume crossover.

(vi) If the faster modes dominate dynamics, what is the mode structure of a very long probe chain in unentangled melt of shorter chains at short and long time scales (and correspondingly short and long length scales)? Compare your answer with the mode structure in semidilute unentangled solutions.

8.13 Consider a steady shear flow (with shear rate $\dot{\gamma}$) of a monodisperse melt of unentangled N-mers with monomeric friction coefficient ζ.

(i) Show that a relative drift velocity of a chain with respect to a typical chain it overlaps with is

$$V \approx \dot{\gamma} R \approx \dot{\gamma} b \sqrt{N}.$$

(ii) Using the energy dissipation rate per monomer ζV^2, where V is the relative velocity of monomers, show that the rate of energy dissipation per chain is

$$\dot{w} = \zeta b^2 N^2 \dot{\gamma}^2$$

for a chain with friction coefficient $N\zeta$.

(iii) Estimate the rate of energy dissipation per unit volume $\dot{W}$ if the volume per monomer in the melt is v_0.

(iv) Using the relation between the rate of energy dissipation per unit volume $\dot{W}$ and viscosity η

$$\dot{W} = \eta \dot{\gamma}^2$$

obtain the expression for viscosity of unentangled melts.

8.14   (i) Derive the approximate Rouse model predictions for G' and G'' [Eqs (8.49) and (8.50)] from the approximate Rouse prediction for the stress relaxation modulus [Eq. (8.48)].

(ii) Show that at intermediate frequencies $(1/\tau_R \ll \omega \ll 1/\tau_0)$ the Rouse model of an ideal linear chain predicts G' and G'' [Eqs (8.49) and (8.50)] scaling as $\sqrt{\omega}$ [Eq. (8.51)].

(iii) What is the value of the loss tangent $\tan \delta$ in this intermediate frequency range of the Rouse model?

8.15   (i) Starting from the exact Rouse stress relaxation modulus [Eq. (8.55)], derive the exact expressions for storage and loss moduli of the Rouse model

$$G'(\omega) = kT \frac{\phi}{Nb^3} \sum_{p=1}^{N} \frac{(\omega\tau_p)^2}{1 + (\omega\tau_p)^2}, \tag{8.179}$$

$$G''(\omega) = kT \frac{\phi}{Nb^3} \sum_{p=1}^{N} \frac{\omega\tau_p}{1 + (\omega\tau_p)^2}, \tag{8.180}$$

with stress relaxation time of the p-th mode $\tau_p = \tau_R/p^2$ related to the Rouse stress relaxation time τ_R of the chain [Eq. (8.18)].

(ii) Demonstrate that the exact solution follows the asymptotic behaviour of Eq. (8.51), $G'(\omega) \sim G''(\omega) \sim \omega^{1/2}$ at intermediate frequency scales $\tau_R^{-1} \ll \omega \ll \tau_0^{-1}$, where the monomer relaxation time is $\tau_0 = \tau_R/N^2$.

(iii) Compare the exact storage and loss moduli with the approximate ones [Eqs. (8.49) and (8.50)].

8.16   (i) Derive the approximate Zimm model predictions for G' and G'' [Eqs (8.67) and (8.68)] from the approximate Zimm stress relaxation modulus [Eq. (8.64)].

(ii) Show that at intermediate frequencies $(1/\tau_Z \ll \omega \ll 1/\tau_0)$ the Zimm model for a linear polymer in a θ-solvent predicts G' and G'' [Eqs (8.67) and (8.68)] scale as $\omega^{2/3}$.

(iii) What is the value of the loss tangent $\tan \delta$ in this intermediate frequency range of the Zimm model?

8.17   What is the physical reason that the 2/3 exponent in Eq. (8.70) for sub-diffusive Zimm motion is independent of solvent quality?

Section 8.5

8.18   In semidilute polymer solutions:

(i) Are hydrodynamic interactions important or screened?

(ii) On which scales do Rouse and Zimm models apply for semidilute unentangled solutions?

8.19   Derive a general expression for the viscosity of a blend of dilute long chains with N_A monomers (with volume fraction ϕ) in a matrix of shorter chains of the same species, but with N_B monomers. Be sure to include both the cases where the long chains swell $(N_A > N_B^2)$ and where they do not $(N_A < N_B^2)$.

8.20   What is the physical significance of the fact that the value of the stress relaxation modulus at the relaxation time of a correlation blob $G(\tau_\xi)$ is proportional to the osmotic pressure in semidilute solutions?

8.21   (i) What is the prediction of the Rouse–Zimm model of semidilute unentangled solutions for the stress relaxation modulus $G(t)$ at very short times $t < \tau_0$?

(ii) Show that the time integral of the stress relaxation modulus in unentangled semidilute solutions over all time regimes (and therefore the solution viscosity) is dominated by the longest relaxation time [Eq. (8.93)].

8.22   Estimate the time dependence of the mean-square displacement of a monomer in an unentangled semidilute solution.

(i) For time scales $t < \tau_\xi$, where τ_ξ is the relaxation time of a chain section inside a correlation volume.

(ii) For time scales $\tau_\xi < t < \tau_{\text{chain}}$, where τ_{chain} is the longest relaxation time of the polymer.

(iii) For time scales $t \geq \tau_{\text{chain}}$.

Section 8.6

8.23   Consider a semi-flexible rod with Young's modulus $E = 10^8$ Pa and cross-section $L_y = L_z = 1$ nm. Estimate the Kuhn length of this rod at room temperature. How does the Kuhn length change if the cross-section changes to $L_y = L_z = 5$ nm?

8.24   Estimate the frequency dependence of the storage and loss modulus of semiflexible chains at intermediate frequencies $\tau_g^{-1} \ll \omega \ll \tau_0^{-1}$ from Eq. (8.117), keeping in mind that Young's modulus is three times the shear modulus ($E(t) = 3G(t)$). What is the value of the loss tangent $\tan \delta$ in this intermediate frequency range?

Section 8.7

8.25   Derive the relation between WLF coefficients [B/f_0 and T_∞ in Eq. (8.134)] for two choices of the reference temperature T_0. *Hint*: The Vogel temperature T_∞ will not change when the reference temperature is changed, but f_0 will change.

8.26   (i) Estimate B/f_g for high molar mass linear PVME melts from Fig. 8.16, ignoring any modulus scale shift b_T.

(ii) Estimate the value of the empirical constant B in the Doolittle equation (8.130).

8.27   The relaxation time of a polybutadiene melt at room temperature (298 K) is 1 s. Estimate the relaxation time of this melt at the glass transition temperature using Table 8.3 and ignoring any modulus scale shift.

8.28   If the temperature dependence of the free volume is assumed to be

$$f = -\frac{B}{9 \ln((T - T_c)/T_c)},$$

derive the following dynamic scaling expression for the temperature dependence of viscosity:

$$\frac{\eta}{\eta_0} = \left(\frac{T_0 - T_c}{T - T_c} \right)^9. \qquad (8.181)$$

8.29 (i) Determine the temperature dependence of the apparent activation energy E_a in the WLF equation [Eq. (8.134)] using the definition

$$E_a = \mathcal{R} \frac{d \ln(\eta/\eta_0)}{d(1/T)}. \tag{8.182}$$

(ii) Show that the high temperature limit of the WLF activation energy is

$$E_\infty = \frac{B\mathcal{R}}{\alpha_f} \tag{8.183}$$

where α_f is the thermal expansion coefficient of the free volume [Eq. (8.131)].

(iii) A typical value of this high-temperature limiting activation energy is $E_\infty \cong 130 \, \text{kJ mol}^{-1}$. By how much does the temperature need to increase from initial temperature $T_1 = 400$ K in the high temperature limit to have the viscosity decrease by a factor of ten?

8.30 Polymer melts typically have their Vogel temperature 50 K below their glass transition temperature.

(i) Assuming that the fractional free volume $f_g \cong 0.025$ at $T_g = T_\infty + 50$ K [Eq. (8.132)], what is the coefficient of thermal expansion for the free volume?

(ii) What is the fractional free volume at $T_g + 10$ K and at $T_g + 100$ K?

(iii) How much do we need to raise the temperature to lower the viscosity by a factor of 10 from its value at T_g? Assume $B = 1$.

(iv) How much do we need to raise the temperature to lower the viscosity by a factor of 10 from its value at $T_g + 100$ K? Assume $B = 1$.

8.31 Assuming that the chain ends have more free volume than monomers in the middle of the chain, derive the molar mass dependence of the glass transition temperature of polymer melts [Eq. (8.135)].

8.32* Demonstrate that the glass transition temperature of ring polymers decreases with increasing molar mass.

Hint: Consider the molar mass dependence of the entropy of ring polymers.

Section 8.8

8.33* Rouse model for polydisperse fractals

The arguments of Section 8.8 can be generalized to describe the relaxation modulus of any polydisperse mixture of fractal polymers with distribution function $n(N)$, the number of N-mers per monomer. The relaxation modulus has contribution kT from each unrelaxed section of g monomers at time $t = \tau(g)$ [see Eq. (8.145)]. The number density of N-mers is $n(N)/b^3$. The number density of unrelaxed sections of g monomers consists of contributions of N/g from all chains with degree of polymerization $N > g$:

$$G(t) \approx kT \int_g^\infty \frac{N}{g} \frac{n(N)}{b^3} \, dN.$$

Use this equation for a general polydisperse fractal polymer and the distribution function of Eq. (6.93) to rederive the Rouse stress relaxation modulus of the melt randomly branched polymers found in the three-dimensional percolation polymerization reactor [Eq. (8.147)]. *Hint:* Use hyperscaling.

8.34 Calculate the stress relaxation modulus $G(t)$ for a polydisperse unentangled melt of linear polymers with a power law distribution of chain lengths

$n(N) = AN^{-\tau}$, where $A = \tau - 1$ is a normalization constant and $\tau > 2$ is the polydispersity exponent. Assume that all excluded volume and hydrodynamic interactions are screened.

8.35* Rouse model of randomly branched polymers

The full-time dependence of the stress relaxation modulus of randomly branched unentangled polymers is best derived from the fractal dynamics of Section 8.8 using the relaxation rate spectrum $P(\varepsilon)$:

$$G(t) = \int_0^\infty P(\varepsilon) \exp(-\varepsilon t)\, d\varepsilon. \qquad (8.184)$$

(i) Show that the relaxation rate spectrum of a single branched polymer between the relaxation rate of the linear section between branch points ε_x and the relaxation rate of the entire chain ε_N is

$$P(\varepsilon)\, d\varepsilon \approx \left(\frac{\varepsilon}{\varepsilon_x}\right)^{1/(1+2/\mathcal{D})} \frac{d\varepsilon}{\varepsilon},$$

where $\mathcal{D}$ is the fractal dimension.

(ii) Show that the contribution to $G(t)$ from one branched polymer with N monomers is

$$G_N(t) = \int_{\varepsilon_N}^{\varepsilon_x} \left(\frac{\varepsilon}{\varepsilon_x}\right)^{1/(1+2/\mathcal{D})} \exp(-\varepsilon t)\, \frac{d\varepsilon}{\varepsilon}.$$

(iii) Derive the stress relaxation modulus of the polydisperse ensemble of randomly branched polymers by summing the response from each molecule in the distribution using

$$G(t) = \int_{N_x}^{N^*} Nn(N)\, G_N(t)\, dN,$$

where N_x is the number of monomers between branch points, N^* is the characteristic degree of polymerization, and $n(N) \sim N^{-\tau}$ is the number density distribution function.

(iv) What is tacitly assumed in the calculation of part (iii)?

(v) Determine the contribution to the complex modulus of one branched polymer from the result of part (ii).

(vi) Using an integration similar to part (iii), show that the complex modulus of the polydisperse ensemble of randomly branched polymers is

$$G^*(\omega) \sim \frac{i\omega}{1 - (N_x/N^*)^{\tau-2}} \times$$

$$\left[\int_{\varepsilon_{N^*}}^{\varepsilon_x} \frac{(\varepsilon/\varepsilon_x)^{(\tau-1)/(1+2/\mathcal{D})}}{i\omega + \varepsilon} \frac{d\varepsilon}{\varepsilon} - \left(\frac{N_x}{N^*}\right)^{\tau-2} \int_{\varepsilon_{N^*}}^{\varepsilon_x} \frac{(\varepsilon/\varepsilon_x)^{1/(1+2/\mathcal{D})}}{i\omega + \varepsilon} \frac{d\varepsilon}{\varepsilon} \right].$$

This form was used for the curves in Fig. 8.19.

8.36* The Rouse model (see the book by Doi and Edwards)

A polymer in the Rouse model is represented by a chain of N beads, each with friction coefficient ζ, located at positions $\{\vec{R}_1, \vec{R}_2, \ldots, \vec{R}_N\}$ and connected by springs with spring constant $3kT/b^2$, where k is the Boltzmann constant, T is the absolute temperature and b^2 is the mean-square unperturbed distance between neighboring beads. The frictional force due to motion of the j-th bead through the solvent $-\zeta(d\vec{R}_j/dt)$ is balanced by the sum of the forces acting on it from the two neighboring beads $(3kT/b^2)(\vec{R}_{j+1} - \vec{R}_j)$ and $(3kT/b^2)(\vec{R}_{j-1} - \vec{R}_j)$ as well as the random Brownian force $\vec{f}_j$. This random force is assumed to be Gaussian with zero average $\langle \vec{f}_j \rangle = 0$ and with no correlations between different moments of time nor between different components of the force. The mean-square value of each component of the random Brownian force is related to the friction coefficient ζ of the beads by the fluctuation-dissipation theorem and is equal to $2\zeta kT$.

(i) Show that the continuum version of the equation of motion for the j-th bead is

$$\zeta \frac{\partial \vec{R}_j}{\partial t} = \frac{3kT}{b^2} \frac{\partial^2 \vec{R}_j}{\partial j^2} + \vec{f}_j \qquad (8.185)$$

with zero average and delta-function correlated random force $\langle f_{j\alpha}(t)f_{i\beta}(t') \rangle = 2\zeta kT\delta(j-i)\delta_{\alpha\beta}\delta(t-t')$, where $f_{j\alpha}$ is the α-component of the random force acting on bead j, $\delta_{\alpha\beta}$ is the Kronecker delta and $\delta(n-m)$ is the Dirac delta function.

(ii) The end beads are attached to the rest of the chain by only one spring and these free ends can be modeled by adding two hypothetical beads at both ends ($j=0$ and $j=N \mid 1$) with their positions coinciding with those of real end beads ($\vec{R}_0 = \vec{R}_1$ and $\vec{R}_{N+1} = \vec{R}_N$). Show that the boundary conditions in the continuum limit become $\partial \vec{R}_j/\partial j = 0$ for $j=0$ and $j=N$.

(iii) The motion of beads is directly coupled to the neighboring beads through the springs, reflected by the presence of the second derivative of bead position vector in the equation of motion for the Rouse model (Eq. 8.185). It is easier to solve this equation in terms of uncoupled variables, called normal modes. Normal modes for the Rouse model are defined by the cosine transform of the real coordinates

$$\vec{X}_p = \frac{1}{N} \int_0^N \cos\left(\frac{\pi jp}{N}\right) \vec{R}_j dj \qquad (8.186)$$

Demonstrate that the Rouse equations for different normal modes are decoupled from each other with the form

$$\zeta_p \frac{\partial \vec{X}_p}{\partial t} = -k_p \vec{X}_p + \vec{f}_p \qquad (8.187)$$

where $k_p = 6\pi^2 kTp^2/(Nb^2)$ and the friction coefficient of zeroth mode (corresponding to the center of mass position) is the friction of the whole chain $\zeta_0 = N\zeta$, while the friction coefficient of all other normal modes is twice as large $\zeta_p = 2N\zeta$ for $p=1,2,3,\ldots$. Show that the average cosine transform of the random force is zero $\langle \vec{f}_p \rangle = 0$, while its correlation function is $\langle f_{p\alpha}(t)f_{q\beta}(t') \rangle = 2\zeta_p kT\delta_{pq}\delta_{\alpha\beta}\delta(t-t')$, where $f_{p\alpha}$ is the α-component of the random force acting on mode p.

(iv) Prove that the time correlation function of each normal mode decays exponentially with time

$$\langle X_{p\alpha}(t)X_{q\beta}(0) \rangle = \delta_{pq}\delta_{\alpha\beta} \frac{kT}{k_p} \exp\left(-\frac{t}{2\tau_p}\right) \qquad (8.188)$$

where the times $\tau_p = \tau_R/p^2$ and $\tau_1 = \tau_R$ is the Rouse stress relaxation time of the chain (Eq. 8.18).

(v) Show that the end-to-end vector of the chain can be expressed as the sum over odd normal modes $\vec{R} \equiv \vec{R}_N - \vec{R}_0 = -4 \sum_{odd\ p} \vec{X}_p$ and therefore its time correlation function is

$$\langle \vec{R}(t) \bullet \vec{R}(0) \rangle = \frac{8}{\pi^2} Nb^2 \sum_{odd\ p} \frac{1}{p^2} \exp\left(-\frac{p^2}{2\tau_R}t\right) \qquad (8.189)$$

(vi) The polymeric contribution to stress can be expressed as the sum over all monomers in the polymer $\sigma_{\alpha\beta} = -\phi/(Nb^3) \sum_j \langle F_{j\alpha} R_{j\beta} \rangle$, where $\phi/(Nb^3)$ is the number density of the chains and $\vec{F}_{j\alpha} = (3kT/b^2)(\vec{R}_{j+1} + \vec{R}_{j-1} - 2\vec{R}_j)$ is the force acting on the j-th monomer from other monomers. Demonstrate that the polymeric contribution to the stress can be written in the continuum limit and expressed in terms of normal modes as

$$\sigma_{\alpha\beta} = \frac{\phi}{Nb^3}\frac{3kT}{b^2}\int_0^N \left\langle \frac{\partial R_{j\alpha}}{\partial j}\frac{\partial R_{j\beta}}{\partial j}\right\rangle dj = \frac{\phi}{Nb^3}\sum_{j=1}^N k_p \langle X_{p\alpha}X_{q\beta}\rangle \qquad (8.190)$$

Calculate the stress relaxation modulus of the Rouse model (Eq. 8.55) by showing that after a small step shear strain γ at time $t=0$ the correlation function of normal modes decays as $\langle X_{px}(t)X_{qy}(t)\rangle = (\gamma kT/k_p)\exp(-t/\tau_p)$.

Section 8.9

8.37 Does the diffusion coefficient measured in dilute solution using Eq. (8.156) rely on proper instrumental calibration to get the absolute intensity?

8.38 The correlation length in semidilute solution can be experimentally determined by measuring the diffusion coefficient of very dilute colloidal spheres of various sizes, provided that the spheres do not interact with the polymers. Consider diffusion of a non-interacting sphere in a semidilute unentangled solution.

   (i) What equation determines the diffusion coefficient D of the sphere, if the sphere radius R is much smaller than the correlation length?

   (ii) What equation determines the diffusion coefficient of the sphere, if it is much larger than the chain size?

   (iii) Sketch the dependence of the product of sphere diffusion coefficient and sphere radius (DR) on sphere radius and explain how such a plot can be used to estimate the correlation length.

8.39* Calculate the Rouse model prediction for the dynamic structure factor for gels in the gelation regime.

Bibliography

Berry, G. C. and Fox, T. G. The viscosity of polymers and their concentrated solutions, *Adv. Polym. Sci.* **5**, 261 (1968).

Chu, B. (ed.) *Selected Papers on Quasielastic Light Scattering by Macromolecular, Supramolecular and Fluid Systems* (SPIE Press, New York, 1990).

Chu, B. *Laser Light Scattering: Basic Principles and Practice*, 2nd edn (Academic Press, New York, 1991).

Doi, M. *Introduction to Polymer Physics* (Clarendon Press, Oxford, 1996).

Doi, M. and Edwards, S. F. *The Theory of Polymer Dynamics* (Clarendon Press, Oxford, 1986).

Ferry, J. D. *Viscoelastic Properties of Polymers*, 3rd edn (Wiley, New York, 1980).

Graessley, W. W. The entanglement concept in polymer rheology, *Adv. Polym. Sci.* **16**, 1 (1974).

Kurata, M. and Stockmayer, W. H. Intrinsic viscosities and unperturbed dimensions of long chain molecules, *Fortschr. Hochpolym.-Forsch* **3**, 196 (1963).

Pecora, R. (ed.) *Dynamic Light Scattering: Applications of Photon Correlation Spectroscopy* (Plenum, New York, 1985).

Schmitz, K. S. *An Introduction to Dynamic Light Scattering from Macromolecules* (Academic Press, New York, 1990).

Entangled polymer dynamics

9.1 Entanglements in polymer melts

The Edwards tube model of polymer entanglements was already discussed in Section 7.3.1. The topological constraints imposed by neighbouring chains on a given chain restrict its motion to a tube-like region (see Fig. 7.10) called the confining tube. The motion of the chain along the contour of the tube (the primitive path) is unhindered by topological interactions. Displacement of monomers in the direction perpendicular to the primitive path is restricted by surrounding chains to an average distance a, called the tube diameter. The number of Kuhn monomers in a strand of size equal to the amplitude of transverse fluctuations (the tube diameter) is N_e, the number of monomers in an entanglement strand. For melts, excluded volume interactions are screened (see Section 4.5.2) and the tube diameter is determined by ideal chain statistics:

$$a \approx b\sqrt{N_e}. \qquad (9.1)$$

The tube can be thought of as being composed of N/N_e sections of size a, with each section containing N_e monomers. The chain can be considered as either a random walk of entanglement strands (N/N_e strands of size a) or a random walk of monomers (N monomers of size b).

$$R \approx a\sqrt{\frac{N}{N_e}} \approx b\sqrt{N}. \qquad (9.2)$$

The average contour length $\langle L \rangle$ of the primitive path (the centre of the confining tube, see Fig. 7.10) is the product of the entanglement strand length a and the average number of entanglement strands per chain N/N_e.

$$\langle L \rangle \approx a\frac{N}{N_e} \approx \frac{b^2 N}{a} \approx \frac{bN}{\sqrt{N_e}}. \qquad (9.3)$$

The average primitive path contour length $\langle L \rangle$ is shorter than the contour length of the chain bN by the factor $a/b \approx \sqrt{N_e}$ because each entanglement strand in a melt is a random walk of N_e Kuhn monomers.

One manifestation of entanglement in long chains ($N \gg N_e$) is the appearance of a wide region in time (or frequency) where the modulus is almost constant in a stress relaxation (or oscillatory shear) experiment. In analogy with crosslinked rubbers, this region is referred to as the **rubbery plateau**, and the nearly constant value of the modulus in this plateau regime is called the **plateau modulus** G_e. In analogy with an affine network, whose modulus is of order kT per network strand [Eq. (7.31)], the plateau modulus is of order kT per entanglement strand [Eq.(7.47)]. The number-average molar mass of an entanglement strand is called the entanglement molar mass M_e. The occupied volume of an entanglement strand with molar mass M_e in a melt with density ρ is the product of the number of Kuhn monomers per strand N_e and the Kuhn monomer volume v_0:

$$\frac{M_e}{\rho \mathcal{N}_{Av}} = v_0 N_e \approx v_0 \frac{a^2}{b^2} \approx \frac{v_0}{b^3} a^2 b. \tag{9.4}$$

Since monomers are space-filling in the melt, the number density of entanglement strands is just the reciprocal of the entanglement strand volume, leading to a simple expression for the plateau modulus of an entangled polymer melt [Eq. (7.47)].

$$G_e \approx \frac{\rho \mathcal{R} T}{M_e} \approx \frac{kT}{v_0 N_e} \approx \frac{b^3}{v_0} \frac{kT}{a^2 b}. \tag{9.5}$$

The number of chains P_e within the **confinement volume** a^3 is determined from the fact that monomers in the melt are space-filling:

$$P_e \approx \frac{a^3}{v_0 N_e} \approx \frac{b^3}{v_0} \sqrt{N_e}. \tag{9.6}$$

Table 9.1 shows N_e and P_e calculated from the measured plateau modulus. All flexible polymers have $P_e \cong 20$ overlapping entanglement strands defining the entanglement volume a^3, which is the **overlap criterion for entanglement** in polymer melts.

Table 9.1 Entanglement parameters for flexible linear polymer melts

Polymer	G_e (MPa)	M_e (g mol^{-1})	N_e	b (Å)	a (Å)	v_0 (Å^3)	P_e
Polyethylene at 140 °C	2.6	1000	7	14	36	320	21
Poly(ethylene oxide) at 80 °C	1.8	1700	13	11	37	210	19
1,4-Polybutadiene at 25 °C	1.15	1900	18	10	41	190	19
Polypropylene at 140 °C	0.47	5800	32	11	62	380	20
1,4-Polyisoprene at 25 °C	0.35	6400	53	8.4	62	220	20
Polyisobutylene at 25 °C	0.34	6700	24	13	62	500	19
Polydimethylsiloxane at 25 °C	0.20	12 000	32	13	74	650	20
Polystyrene at 140 °C	0.20	17 000	23	18	85	1200	22
Polyvinylcyclohexane at 160 °C	0.068	49 000	81	14	130	1100	22

9.2 Reptation in polymer melts

At first glance, understanding the motion of a polymer in the melt is daunting. Since roughly $\sqrt{N}$ other polymers share the pervaded volume of a given chain in the melt, chain motion appears to be a difficult many-body problem. However, by utilizing the Edwards tube concept, de Gennes cleverly reduced this many-body problem to the motion of a single chain confined to a tube of surrounding chains. Models that consider chain motion as being restricted to a tube-like region are referred to as **tube models**. The simplest tube model was proposed by de Gennes in 1971 for the motion of linear entangled polymers, and is called the **reptation model**.

9.2.1 Relaxation times and diffusion

In de Gennes' reptation model, an entangled chain diffuses along its confining tube in a way analogous to the motion of a snake or a worm (see Fig. 9.1). This motion of the chain consists of diffusion of small loops, along the contour of the primitive path. This curvilinear motion of a polymer along its tube satisfies the topological constraints imposed by surrounding chains and is characterized by the Rouse friction coefficient $N\zeta$. The **curvilinear diffusion coefficient** D_c that describes motion of the chain along its tube is simply the Rouse diffusion coefficient of the chain [Eq. (8.12)].

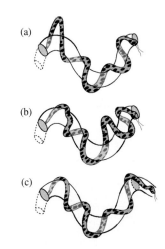

$$D_c = \frac{kT}{N\zeta}. \tag{9.7}$$

The time it takes for the chain to diffuse out of the original tube of average length $\langle L \rangle$ is the **reptation time**:

$$\tau_{\mathrm{rep}} \approx \frac{\langle L \rangle^2}{D_c} \approx \frac{\zeta b^2}{kT} \frac{N^3}{N_e} = \frac{\zeta b^2}{kT} N_e^2 \left(\frac{N}{N_e}\right)^3. \tag{9.8}$$

Here, Eq. (9.3) was used for the average contour length of the tube. The reptation time τ_{rep} is predicted to be proportional to the cube of the molar mass. The experimentally measured scaling exponent is higher than 3:

$$\tau \sim M^{3.4}. \tag{9.9}$$

We will discuss the possible reasons for the disagreement between the simple reptation model and experiments in Section 9.4.5.

The first part of the final relation of Eq. (9.8) is the Rouse time of an entanglement strand containing N_e monomers:

$$\tau_e \approx \frac{\zeta b^2}{kT} N_e^2. \tag{9.10}$$

(a)

(b)

(c)

Fig. 9.1
Reptation steps: (a) formation of a loop at the tail of the snake and elimination of the tail segment of the confining tube; (b) propagation of the loop along the contour of the tube; (c) release of the loop at the head of the snake and formation of a new section of the confining tube.

The ratio of the reptation time τ_{rep} and τ_e is the cube of the number of entanglements along the chain:

$$\frac{\tau_{rep}}{\tau_e} \approx \left(\frac{N}{N_e}\right)^3.$$ (9.11)

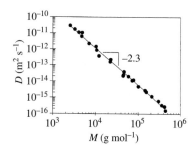

Fig. 9.2
Molar mass dependence of the diffusion coefficient for melts of hydrogenated polybutadiene at 175 °C. Data compiled in T. P. Lodge *Phys. Rev. Lett.* **83**, 3218 (1999).

The chain moves a distance of order of its own size R in its reptation time τ_{rep}, since this is the time scale at which the tube is abandoned:

$$D_{rep} \approx \frac{R^2}{\tau_{rep}} \approx \frac{kT}{\zeta}\frac{N_e}{N^2}.$$ (9.12)

The diffusion coefficient of entangled linear polymers is predicted to be reciprocally proportional to the square of the molar mass, which also disagrees with experiments, as shown in Fig. (9.2):

$$D \approx \frac{R^2}{\tau} \sim M^{-2.3}.$$ (9.13)

9.2.2 Stress relaxation and viscosity

The reptation ideas discussed above will now be combined with the relaxation ideas discussed in Chapter 8 to describe the stress relaxation modulus $G(t)$ for an entangled polymer melt. On length scales smaller than the tube diameter a, topological interactions are unimportant and the dynamics are similar to those in unentangled polymer melts and are described by the Rouse model. The entanglement strand of N_e monomers relaxes by Rouse motion with relaxation time τ_e [Eq. (9.10)]:

$$\tau_e = \tau_0 N_e^2.$$ (9.14)

The Rouse model predicts that the stress relaxation modulus on these short time scales decays inversely proportional to the square root of time [Eq. (8.47)]:

$$G(t) \approx G_0 \left(\frac{t}{\tau_0}\right)^{-1/2} \quad \text{for } \tau_0 < t < \tau_e.$$ (9.15)

The relaxation time of the Kuhn monomer τ_0 is the shortest stress relaxation time in the Rouse model, given by Eq. (8.56) with $p = N$:

$$\tau_0 = \frac{\zeta b^2}{6\pi^2 kT} \approx \frac{\zeta b^2}{kT}.$$ (9.16)

The stress relaxation modulus at τ_0 is the Kuhn modulus (kT per Kuhn monomer):

$$G_0 \approx G(\tau_0) \approx \frac{kT}{v_0}.$$ (9.17)

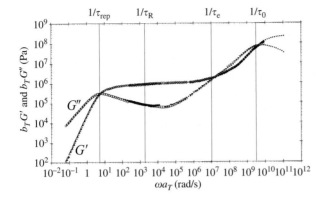

Fig. 9.3
Master curve at 25 °C from oscillatory shear data at six temperatures for a 1,4-polybutadiene sample with $M_w = 130\,000\,\mathrm{g\,mol^{-1}}$. Data from R. H. Colby, L. J. Fetters and W. W. Graessley, *Macromolecules* **20**, 2226 (1987).

Consider for example, a melt of 1,4-polybutadiene linear chains with $M = 130\,000\,\mathrm{g\,mol^{-1}}$. The molar mass of a polybutadiene Kuhn monomer is $M_0 = 105\,\mathrm{g\,mol^{-1}}$ (see Table 2.1) so this chain has $N = M/M_0 = 1240$ Kuhn monomers. At 25 °C, this polymer is 124 K above its glass transition and its oscillatory shear master curve is shown in Fig. 9.3. The time scale for monomer motion is $\tau_0 \cong 0.3\,\mathrm{ns}$, An entanglement strand of 1,4-polybutadiene has molar mass $M_e = 1900\,\mathrm{g\,mol^{-1}}$ (see Table 9.1) and therefore contains $N_e = M_e/M_o = 18$ Kuhn monomers. The whole chain has $N/N_e = M/M_e = 68$ entanglements. The Rouse time of the entanglement strand $\tau_e \cong 0.1\,\mu\mathrm{s}$ [Eq. (9.10)].

At the Rouse time of an entanglement strand τ_e, the chain 'finds out' that its motion is topologically hindered by surrounding chains. Free Rouse motion of the chain is no longer possible on time scales $t > \tau_e$. The value of the stress relaxation modulus at τ_e is the plateau modulus G_e, which is kT per entanglement strand [Eq. (9.5)]:

$$G_e = G(\tau_e) = \frac{G_0}{N_e} = \frac{kT}{v_0 N_e}. \qquad (9.18)$$

The Rouse time of the chain is $\tau_R \cong 0.5\,\mathrm{ms}$:

$$\tau_R = \tau_0 N^2 = \tau_e \left(\frac{N}{N_e}\right)^2. \qquad (9.19)$$

In the simple reptation model, there is a delay in relaxation (the rubbery plateau) between τ_e and the reptation time of the chain τ_{rep} [Eq. (9.11)]. By restricting the chain's Rouse motions to the tube, the time the chain takes to diffuse a distance of order of its size is longer than its Rouse time by a factor of $6\,N/N_e$. This slowing arises because the chain must move along the confining tube. The reptation time of the chain $\tau_{\mathrm{rep}} = 0.2\,\mathrm{s}$ is measured experimentally as the reciprocal of the frequency at which $G' = G''$ in Fig. 9.3 at low frequency (see Problem 9.8). In practice, this time is determined experimentally and τ_0, τ_e and τ_R are determined from τ_{rep}.

The stress relaxation modulus is summarized schematically in Fig. 9.4. For long linear chains, the rubbery plateau can span many decades in time.

The diffusion coefficient of the chain is controlled by the reptation time [Eq. (9.12)]. The linear polybutadiene chain with $M = 130\,000\,\mathrm{g\,mol^{-1}}$ has $N = 1240$ Kuhn monomers, with Kuhn length $b = 10\,\text{Å}$ and coil size $R = b\sqrt{N} \cong 350\,\text{Å}$. Since linear polymers move a distance of order their own size in their reptation time, the reptation time of $\tau_{\mathrm{rep}} \cong 0.2\,\mathrm{s}$ at 25 °C enables estimation of the diffusion coefficient $D \approx R^2/\tau_{\mathrm{rep}} \cong 6 \times 10^{-15}\,\mathrm{m^2\,s^{-1}}$. Physically, this means that at 25 °C this polybutadiene chain moves about 350 Å in a random direction every 0.2 s.

The stress relaxation modulus in the reptation model is proportional to the fraction of original tube remaining at time t (see Fig. 9.1). As time goes on, sections of the original tube are abandoned when the chain end first visits them. Such a problem is called a first-passage time problem. The stress relaxation modulus $G(t)$ for the reptation model was calculated by Doi and Edwards in 1978 by solving the first-passage problem for the diffusion of a chain in a tube (see Problem 9.6):

$$G(t) = \frac{8}{\pi^2} G_e \sum_{\mathrm{odd}\,p} \frac{1}{p^2} \exp\left(-\frac{p^2 t}{\tau_{\mathrm{rep}}}\right). \qquad (9.20)$$

The longest relaxation time in this model is the reptation time required for the chain to escape from its tube

$$\tau_{\mathrm{rep}} = 6\tau_0 \frac{N^3}{N_e} = 6\tau_e \left(\frac{N}{N_e}\right)^3 = 6\tau_R \frac{N}{N_e} \qquad (9.21)$$

where the Rouse time τ_R is the longest relaxation time of the Rouse model [Eq. (8.18)], which is half the end-to-end vector correlation time.

The main contribution[1] comes from the first mode $p = 1$ and the function is almost a single exponential [Eq. (7.111)]:

$$G(t) \approx G_e \exp(-t/\tau_{\mathrm{rep}}). \qquad (9.22)$$

The **Doi–Edwards equation** [Eq. (9.20)] is the first attempt at a molecular model for viscoelasticity of entangled polymers. It ignores tube length fluctuation modes that relax some stress on shorter time scales. These modes significantly modify dynamics of entangled polymers, as described in Section 9.4.5.

The reptation model prediction for the viscosity of an entangled polymer melt is determined by integrating Eq. (9.20):

$$\eta = \int_0^\infty G(t)\,\mathrm{d}t = \frac{8}{\pi^2} G_e \sum_{\mathrm{odd}\,p} \frac{1}{p^2} \int_0^\infty \exp\left(-\frac{p^2 t}{\tau_{\mathrm{rep}}}\right) \mathrm{d}t = \frac{\pi^2}{12} G_e \tau_{\mathrm{rep}}. \qquad (9.23)$$

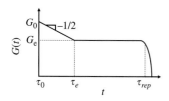

Fig. 9.4
Schematic representation of the stress relaxation modulus of entangled linear polymers on logarithmic scales.

---

[1] Fraction $8/\pi^2$ of the terminal relaxation is associated with the first mode with relaxation time τ_{rep} (see Problem 9.7).

The final result was obtained from the fact that $\sum\limits_{\text{odd } p} 1/p^4 = \pi^4/96$. Since the stress relaxation is nearly a single exponential, the scaling prediction of the viscosity as the product of the plateau modulus [Eq. (9.5)] and the reptation time [Eq. (9.8)] is nearly quantitative:

$$\eta \approx G_e \tau_{\text{rep}} \approx G_e \tau_e \left(\frac{N}{N_e}\right)^3 \approx \frac{kT}{v_0 N_e} \frac{\zeta b^2 N_e^2}{kT} \left(\frac{N}{N_e}\right)^3 \approx \frac{\zeta b^2}{v_0} \frac{N^3}{N_e^2}. \qquad (9.24)$$

The viscosity of a polymer melt is predicted to be proportional to molar mass for unentangled melts (the Rouse model) and proportional to the cube of molar mass for entangled melts (the reptation model).

$$\eta \sim \begin{cases} M & \text{for } M < M_c, \\ M^3 & \text{for } M > M_c. \end{cases} \qquad (9.25)$$

As Fig. 8.17 shows, the critical molar mass M_c for entanglement effects in viscosity [defined in Eq. (8.136)] is typically a factor of 2–4 larger than the entanglement molar mass M_e [defined in Eq. (9.5)]. As shown in Fig. 9.5, the exponent in the entangled regime is $\cong 3.4$ for all linear entangled polymers. This exponent is significantly larger than the prediction of 3 by the simple reptation model [Eq. (9.24)]:

$$\eta \approx G_e \tau \sim M^{3.4}. \qquad (9.26)$$

The deviations from the 3.4 power law at low molar masses ($M < M_c$) are because those chains are too short to be entangled (see Section 8.7.3). The deviations at very high molar mass are consistent with a crossover to pure reptation (see Section 9.4.5).

The simple reptation model does not properly account for all the relaxation modes of a chain confined in a tube. This manifests itself in all measures of terminal dynamics, as the longest relaxation time, diffusion coefficient and viscosity all have stronger molar mass dependences than the reptation model predicts. In Sections 9.4.5 and 9.6.2, more accurate analytical and numerical treatments of this problem are given with results that are in reasonable agreement with the experimental dependence of terminal dynamics on the molar mass of the chain [Eqs (9.9), (9.13), and (9.26)].

9.3   Reptation in semidilute solutions

9.3.1   Length scales

Consider a semidilute solution with polymer volume fraction ϕ. The concentration dependence of the correlation length was discussed in Chapter 5:

$$\xi \approx b\phi^{-\nu/(3\nu-1)}. \qquad (9.27)$$

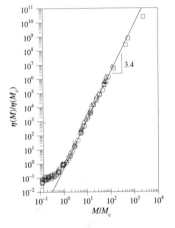

Fig. 9.5
Molar mass dependence of viscosity for polymer melts reduced by their critical molar mass. Open circles are polyisobutylene with $M_c = 14\,000\,\text{g mol}^{-1}$, from T. G. Fox and P. J. Flory, *J. Am. Chem. Soc.* **70**, 2384 (1948) and *J. Phys. Chem.* **55**, 221 (1951). Open squares are polybutadiene with $M_c = 6700\,\text{g mol}^{-1}$, from R. H. Colby *et al.*, *Macromolecules* **20**, 2226 (1987). Open triangles are hydrogenated polybutadiene with $M_c = 8100\,\text{g mol}^{-1}$, from D. S. Pearson *et al.*, *Macromolecules* **27**, 711 (1994).

In an athermal solvent the exponent $\nu \cong 0.588$ and the correlation length decreases with concentration as $\xi \approx b\phi^{-0.76}$, while in a θ-solvent the exponent $\nu = 1/2$ and the correlation length has a stronger concentration dependence $\xi \approx b\phi^{-1}$ [Eq. (5.52)]. The number of monomers g in a correlation volume ξ^3 was also determined in Chapter 5 [Eq. (5.24)]:

$$g \approx \frac{\phi\xi^3}{b^3} \approx \phi^{-1/(3\nu-1)}. \qquad (9.28)$$

In an athermal solvent, the number of monomers in a correlation volume decreases with concentration as $g \approx \phi^{-1.3}$, while in a θ-solvent a stronger concentration dependence is expected with $g \approx \phi^{-2}$. The chain is always a random walk of correlation blobs, with end-to-end distance R [Eq. (5.26)]:

$$R \approx \xi\left(\frac{N}{g}\right)^{1/2} \approx bN^{1/2}\phi^{-(2\nu-1)/(6\nu-2)}. \qquad (9.29)$$

In a good solvent, the chain size decreases with concentration as $R \approx bN^{1/2}\phi^{-0.12}$. In a θ-solvent ($\nu = 1/2$) there is no concentration dependence of chain size, as the chain is nearly ideal at all concentrations $R \approx bN^{1/2}$.

To understand the dynamics of entangled solutions, another length scale, the tube diameter a, must be specified. Just as in the melt, the confinement volume a^3 must contain multiple chains. Entanglements between chains are controlled by binary intermolecular contacts. In the athermal solvent limit, the number density of binary intermolecular contacts is proportional to the reciprocal of the correlation volume $\xi^{-3} \sim \phi^{3\nu/(3\nu-1)}$, and the distance between binary contacts is the reciprocal cube root of this number density $\xi \sim \phi^{-\nu/(3\nu-1)}$. Hence, the correlation length describes the distance between binary intermolecular contacts. *The tube diameter a in an athermal solvent is proportional to, but larger than, the correlation length ξ*:

$$a(\phi) \approx a(1)\phi^{-\nu/(3\nu-1)} \approx a(1)\phi^{-0.76} \quad \text{for an athermal solvent.} \qquad (9.30)$$

The tube diameter in the melt, $a(1) \approx b\sqrt{N_e(1)}$, is given by Eq. (9.1) in terms of the number of Kuhn monomers in an entanglement strand in the melt $N_e(1)$. Notice that $a(1) > b$, which makes $a > \xi$ at all concentrations. Since the chain is a random walk of correlation blobs on scales larger than ξ, the entanglement strand is a random walk of correlation blobs, as depicted in Fig. 9.6.

In a θ-solvent, the correlation length is determined by ternary contacts between chains (see Section 5.4). This is because the effects of binary contacts on the free energy (or osmotic pressure) exactly cancel at the θ-temperature. The solvent-mediated energetic interaction between monomers exactly compensates for the hard core repulsion at the θ-temperature. Binary contacts between chains still occur, they simply have

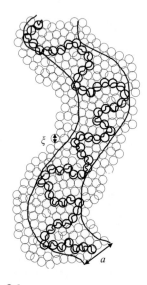

Fig. 9.6
The confining tube in a semidilute solution. Thick circles are the correlation blobs of the chain. Thin empty circles are the correlation blobs of surrounding chains.

no effect on the free energy, and this directly leads to nearly ideal chain statistics at all concentrations in a θ-solvent, and the applicability of mean-field theory. However, binary contacts still control entanglements between chains. The number density of space-filling correlation volumes in a θ-solvent is given by the mean-field result $\xi^{-3} \sim \phi^3$. The same mean-field ideas determine the number density of binary intermolecular contacts to be proportional to ϕ^2. Just as in the good solvent, the average distance between binary contacts is given by the reciprocal cube root of this number density, and in a θ-solvent the distance is proportional to $\phi^{-2/3}$. Once again, we expect the tube diameter to be proportional to, but larger than, the distance between binary contacts:

$$a(\phi) \approx a(1)\phi^{-2/3} \quad \text{for a } \theta\text{-solvent.} \tag{9.31}$$

The length scales ξ, a, and R are plotted as a function of concentration for a typical good solvent in Fig. 9.7. All three length scales change their concentration dependences from athermal to ideal at the concentration $\phi^{**} \approx v/b^3$ separating semidilute and concentrated solutions.

9.3.2 Entanglement concentration

The concentration at which the correlation length ξ is of the order of the coil size $R \approx bN^\nu$ is the overlap concentration ϕ^*, given by Eq. (5.19):

$$\phi^* \approx \frac{Nb^3}{R^3} \approx N^{1-3\nu}. \tag{9.32}$$

In an athermal solvent $\phi^* \approx N^{-0.76}$, while in a θ-solvent $\phi^* \approx N^{-1/2}$.

The concentration at which the tube diameter a [from Eqs (9.30) or (9.31)] equals the coil size R [Eq. (9.29)] is the **entanglement concentration** ϕ_e:

$$\phi_e \approx \begin{cases} [N_e(1)/N]^{(3\nu-1)} \approx [N_e(1)/N]^{0.76} & \text{for an athermal solvent,} \\ [a(1)/b]^{3/2}N^{-3/4} \approx [N_e(1)/N]^{3/4} & \text{for a } \theta\text{-solvent,} \end{cases} \tag{9.33}$$

where $N_e(1)$ is the number of Kuhn monomers in an entanglement strand in the melt. Note that the predictions for both solvents are very similar.

For $\phi > \phi_e$, entanglement effects control chain dynamics and the reptation model must be used as described below. Between the overlap concentration and the entanglement concentration ($\phi^* < \phi < \phi_e$), the solution is semidilute but not entangled, and the unentangled solution model of Section 8.5 describes dynamics. The width of this semidilute unentangled regime is given by the ratio of Eqs (9.33) and (9.32):

$$\frac{\phi_e}{\phi^*} \approx \begin{cases} [N_e(1)]^{3\nu-1} \approx [N_e(1)]^{0.76} & \text{for an athermal solvent,} \\ ([N_e(1)]^{3/4})/N^{1/4} & \text{for a } \theta\text{-solvent.} \end{cases} \tag{9.34}$$

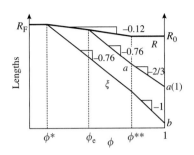

Fig. 9.7
Chain size R, tube diameter a, and correlation length ξ in a good solvent. The semidilute unentangled regime is $\phi^* < \phi < \phi_e$; the semidilute entangled regime is $\phi_e < \phi < \phi^{**}$ and; the concentrated regime is $\phi^{**} < \phi < 1$.

Table 9.1 shows that the number of Kuhn monomers in an entanglement strand in the melt state varies over a wide range ($7 < N_e(1) < 80$) making $4 < \phi_e/\phi^* < 30$ for solutions in an athermal solvent. Since the entanglement concentration ϕ_e cannot be lower than the overlap concentration ϕ^*, the expressions for a θ-solvent [Eqs (9.31), (9.33), and (9.34)] are valid for $N < [N_e(1)]^3$. This condition is not very restrictive and it is satisfied for all experimental studies to date.

9.3.3   Plateau modulus

Owing to the fact that the tube diameter is always larger than the correlation length ($a > \xi$), the entanglement strand is a random walk of correlation volumes in any solvent:

$$a \approx \xi \sqrt{\frac{N_e}{g}}, \tag{9.35}$$

where N_e/g is the number of correlation volumes per entanglement strand. The above relation can be solved for the concentration dependence of the number of monomers in an entanglement strand:

$$N_e(\phi) \approx g\left(\frac{a}{\xi}\right)^2 \approx N_e(1)\begin{cases} \phi^{-1/(3\nu-1)} & \text{for an athermal solvent,} \\ \phi^{-4/3} & \text{for a } \theta\text{-solvent.} \end{cases} \tag{9.36}$$

The two predictions are nearly identical, since $1/(3\nu - 1) \cong 1.3$.

The occupied volume of an entanglement strand is $\xi^3 N_e/g \approx a^2\xi$. Since the correlation volumes are space-filling in solution, the number density of entanglement strands is simply the reciprocal of this volume. Analogous to Eq. (9.5), the plateau modulus of an entangled polymer solution is once again of the order of kT per entanglement strand,

$$G_e(\phi) \approx \frac{kT}{a^2\xi} \approx \frac{kT\phi}{b^3 N_e(\phi)} \approx G_e(1)\begin{cases} \phi^{3\nu/(3\nu-1)} & \text{for an athermal solvent,} \\ \phi^{7/3} & \text{for a } \theta\text{-solvent,} \end{cases}$$

$$\tag{9.37}$$

where $G_e(1)$ is the plateau modulus of the melt, given by Eq. (9.5). The predictions for athermal and θ-solvents are essentially the same (the concentration dependence exponents are $\cong 2.3$ in both cases). This interesting result is experimentally confirmed, as shown in Fig. 9.8.

9.3.4   Relaxation times and diffusion

Topological constraints do not influence polymer motion on length scales smaller than the size of an entanglement strand. In entangled polymer solutions, chain sections with end-to-end distance shorter than the tube

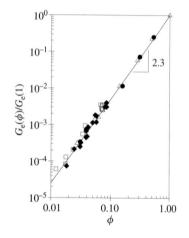

Fig. 9.8
Dilution effect on the plateau modulus of linear polymers. Filled diamonds are polystyrene in cyclohexane at 34.5 °C (θ-solvent), open squares are polystyrene in benzene at 25 °C (good solvent), filled circles are polybutadiene in dioctylphthalate at 25 °C (near θ-solvent) and open triangles are polybutadiene in phenyloctane (good solvent). PS data from M. Adam and M. Delsanti, *J. Phys. France* **44**, 1185 (1983); **45**, 1513 (1984). PB data from R. H. Colby *et al.*, *Macromolecules* **24**, 3873 (1991).

diameter a move as they would in an unentangled solution. On length scales smaller than the correlation length ξ, hydrodynamic interactions are not screened. As with unentangled chains, the relaxation time τ_ξ of the strand within each correlation volume is determined by the Zimm result [Eq. (8.75)]:

$$\tau_\xi \approx \frac{\eta_s}{kT}\xi^3 \approx \frac{\eta_s b^3}{kT}\left(\frac{\xi}{b}\right)^3 \approx \tau_0\left(\frac{\xi}{b}\right)^3 \approx \tau_0 \phi^{-3\nu/(3\nu-1)}. \qquad (9.38)$$

On length scales larger than the correlation length ξ but smaller than the tube diameter a, hydrodynamic interactions are screened, and topological interactions are unimportant. Polymer motion on these length scales is described by the Rouse model. The relaxation time τ_e of an entanglement strand of N_e monomers is that of a Rouse chain of N_e/g correlation volumes [Eq. (8.76)]:

$$\tau_e \approx \tau_\xi\left(\frac{N_e}{g}\right)^2 \approx \tau_0[N_e(1)]^2\begin{cases} \phi^{-3\nu/(3\nu-1)} & \text{for an athermal solvent,} \\ \phi^{-5/3} & \text{for a }\theta\text{-solvent.} \end{cases}$$
$$(9.39)$$

On length scales larger than the tube diameter, topological interactions are important and the motion is described by the reptation model with the chain relaxation time given by the reptation time:

$$\tau_{rep} \approx \tau_e\left(\frac{N}{N_e}\right)^3 \approx \tau_0\left(\frac{\xi}{b}\right)^3\left(\frac{N_e}{g}\right)^2\left(\frac{N}{N_e}\right)^3. \qquad (9.40)$$

Using Eqs (9.27), (9.28), and (9.36) transforms this into a simple relation for the concentration dependence of the reptation time:

$$\tau_{rep} \approx \tau_0\frac{N^3}{N_e(1)}\begin{cases} \phi^{3(1-\nu)/(3\nu-1)} & \text{for an athermal solvent,} \\ \phi^{7/3} & \text{for a }\theta\text{-solvent.} \end{cases} \qquad (9.41)$$

The reptation time has a considerably weaker concentration dependence in athermal solvent than in θ-solvent, since $3(1-\nu)/(3\nu-1) \cong 1.6$. Note that Eq. (9.41) reduces to Eq. (9.20) when $\phi = 1$.

The diffusion coefficient in semidilute polymer solutions is determined from the fact that the chain diffuses a distance of order of its own size in its reptation time:

$$D \approx \frac{R^2}{\tau_{rep}} \approx \frac{b^2}{\tau_0}\frac{N_e(1)}{N^2}\begin{cases} \phi^{-(2-\nu)/(3\nu-1)} & \text{for an athermal solvent,} \\ \phi^{-7/3} & \text{for a }\theta\text{-solvent.} \end{cases} \qquad (9.42)$$

The reptation prediction of the concentration dependence of diffusion coefficient in athermal solvent is slightly weaker than in θ-solvent, since

$(2-\nu)/(3\nu-1)\cong 1.85$. Figure 8.9 already showed that there is a low concentration regime that is semidilute but unentangled that is described by Eq. (8.85). That regime persists for roughly one decade in good solvent, as expected by Eq. (9.34). Above the entanglement concentration ϕ_e, the athermal solvent prediction of Eq. (9.42) applies for a range of concentration (see Fig. 8.9). At still higher concentrations, an even stronger concentration dependence is noted for the two highest concentrations in Fig. 8.9, consistent with the θ-solvent scaling prediction of Eq. (9.42) in concentrated solution (for $\phi > \phi^{**}$):

$$D \sim \begin{cases} \phi^{-0.54} & \text{for } \phi^* < \phi < \phi_e, \\ \phi^{-1.85} & \text{for } \phi_e < \phi < \phi^{**}, \\ \phi^{-7/3} & \text{for } \phi^{**} < \phi < 1. \end{cases} \qquad (9.43)$$

9.3.5 Stress relaxation and viscosity

There are three different regimes of polymer dynamics on three different length and time scales for an entangled polymer solution in an athermal solvent. The stress relaxation modulus of such a solution is shown in Fig. 9.9. Two of the regimes are identical to those discussed in Section 9.2.2 and the other regime was discussed in Section 8.5.

Between τ_0 and the relaxation time of a correlation blob τ_ξ, both static and dynamic properties are similar to those in a dilute solution. Hydrodynamic interactions are important and dynamics of these small sections of chains are described by the Zimm model. The stress relaxation modulus on time scales between τ_0 and τ_ξ is similar to the Zimm result for unentangled solutions discussed in Section 8.5 [Eq. (8.88)]. The stress relaxation modulus decays with time as a power law with exponent $-1/(3\nu)$. This time dependence is $G(t) \sim t^{-0.57}$ in a good solvent with Flory exponent $\nu \cong 0.588$ and is $G(t) \sim t^{-2/3}$ in a θ-solvent. The stress relaxation modulus in this regime decays from the Kuhn modulus G_0 (kT per Kuhn monomer) to the osmotic pressure Π [kT per correlation blob, see Eq. (8.89)].

On intermediate length scales between the correlation length ξ and the tube diameter a, hydrodynamic interactions are screened and topological interactions are not important. The dynamics on these intermediate scales (for $\tau_\xi < t < \tau_e$) are described by the Rouse model with stress relaxation modulus similar to the Rouse result for unentangled solutions [Eq. (8.90) with the long time limit the Rouse time of an entanglement strand τ_e]. At τ_e, the stress relaxation modulus has decayed to the plateau modulus G_e [kT per entanglement strand, Eq. [(9.37), see Fig. 9.9)]. The ratio of osmotic pressure and plateau modulus at any concentration in semidilute solution in athermal solvents is proportional to the number of Kuhn monomers in an entanglement strand in the melt. In θ-solvents this ratio is considerably smaller and concentration dependent:

$$\frac{\Pi}{G_e} \approx \left(\frac{a}{\xi}\right)^2 \approx \begin{cases} N_e(1) & \text{for an athermal solvent,} \\ N_e(1)\phi^{2/3} & \text{for a } \theta\text{-solvent.} \end{cases} \qquad (9.44)$$

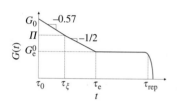

Fig. 9.9
Schematic representation of the stress relaxation modulus of an entangled polymer solution in an athermal solvent on logarithmic scales.

At the Rouse time of an entanglement strand τ_e, the chain in semidilute solution 'finds out' that it is trapped in the confining tube. The stress relaxation modulus between τ_e and the reptation time τ_{rep} is almost constant and equal to the plateau modulus (see Fig. 9.9). At the reptation time [Eq. (9.41)], the stress relaxation modulus decays to zero exponentially [Eq. (9.22)].

The polymer contribution to the viscosity of an entangled polymer solution is estimated as the product of the plateau modulus [Eq. (9.37)] and the reptation time [Eq. (9.41)]:

$$\eta - \eta_s \approx G_e \tau_{rep} \approx \eta_s \frac{N^3}{[N_e(1)]^2} \begin{cases} \phi^{3/(3\nu-1)} & \text{for an athermal solvent,} \\ \phi^{14/3} & \text{for a } \theta\text{-solvent.} \end{cases}$$
(9.45)

The concentration dependence of viscosity is $\eta \sim \phi^{3.9}$ in an athermal solvent with Flory exponent $\nu \cong 0.588$ and $\eta \sim \phi^{4.7}$ in a θ-solvent. The $14/3 \cong 4.7$ exponent is demonstrated for poly(ethylene oxide) in water at $25.0\,°C$ in Fig. 8.11. There are two different scaling regimes for the specific viscosity in an athermal solvent, corresponding to unentangled and entangled semidilute solutions:

$$\eta_{sp} \approx \begin{cases} (\phi/\phi^*)^{1/(3\nu-1)} & \text{for } \phi^* < \phi < \phi_e, \\ (\phi/\phi^*)^{3/(3\nu-1)}/[N_e(1)]^2 & \text{for } \phi_e < \phi < 1. \end{cases}$$
(9.46)

Data for different molar masses of the same polymer species combine into a single plot in good solvent [Fig. 9.10(a)] if specific viscosity $\eta_{sp} = (\eta - \eta_s)/\eta_s$ is plotted as a function of reduced concentration ϕ/ϕ^*. This simple data collapse works in a good solvent because the correlation length and the tube diameter are proportional to each other, with the same concentration exponents. The line in Fig. 9.10(a) has the slope of 3.9 expected by Eq. (9.46) for semidilute entangled solutions.

In a θ-solvent, the correlation length ξ and the tube diameter a have different concentration dependences $[\xi \approx b\phi^{-1}$, Eq. (9.27), with $\nu = 1/2$ and $a \approx a\,(1)\,\phi^{-2/3}$, Eq. (9.31)]. The simple plot of relative viscosity η/η_s vs. ϕ/ϕ^* will only collapse data for different molar masses in unentangled solutions, but *not* in entangled solutions in a θ-solvent.

$$\eta_{sp} \approx \begin{cases} (\phi/\phi^*)^2 & \text{for } \phi^* < \phi < \phi_e, \\ (\phi/\phi^*)^{14/3} N^{2/3}/[N_e(1)]^2 & \text{for } \phi_e < \phi < 1. \end{cases}$$
(9.47)

Construction of a reduced data plot for the viscosity of entangled solutions of a given type of polymer in θ-solvents requires plotting $\eta_{sp}/N^{2/3}$ as a function of reduced concentration, as demonstrated in Fig. 9.10(b). This complicated form of data reduction is a direct consequence of the two length scales a and ξ having different concentration dependences in θ-solvent. The line in Fig. 9.10(b) has the slope predicted by Eq. (9.47).

(a)
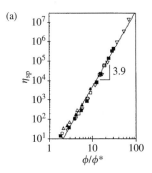

(b)
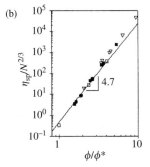

Fig. 9.10
Concentration dependence of viscosity in semidilute solutions of polystyrene at $35\,°C$. (a) Solutions in the good solvent toluene have ϕ/ϕ^* reduce data for different molar masses to a universal curve, using data from M. Adam and M. Delsanti, *J. Phys. France* **44**, 1185 (1983). (b) Solutions in the θ-solvent cyclohexane must have specific viscosity divided by $N^{2/3}$ for ϕ/ϕ^* to reduce data to a universal curve, using data from M. Adam and M. Delsanti, *J. Phys. France* **45**, 1513 (1984). Open triangles are $M = 171\,000\,\text{g mol}^{-1}$, filled triangles are $M = 422\,000\,\text{g mol}^{-1}$, open circles are $M = 1\,260\,000\,\text{g mol}^{-1}$, filled circles are $M = 2\,890\,000\,\text{g mol}^{-1}$, open squares are $M = 3\,840\,000\,\text{g mol}^{-1}$, filled squares are $M = 6\,770\,000\,\text{g mol}^{-1}$, and open inverted triangles are $M = 20\,600\,000\,\text{g mol}^{-1}$.

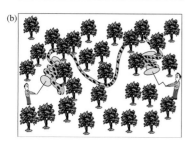

The entangled viscosity data in both good solvent and θ-solvent show stronger concentration dependences than predicted by the simple reptation model. The steeper experimental slopes are consistent with the additional relaxation modes discussed in Section 9.4.5 (see Problem 9.14).

In order to construct a universal plot for the viscosity of all entangled polymer solutions in a given class of solvent, it is necessary to also multiply the ordinates of Fig. 9.10 by $[N_e(1)]^2$ because different polymers have different numbers of Kuhn monomers in their entanglement strands in the melt (see Table 9.1). Such universal plots have indeed been constructed successfully in the literature.

9.4   Dynamics of a single entangled chain

9.4.1   Chain in an array of fixed obstacles

Chains in polymer melts and entangled polymer solutions form an effective entanglement network. Since chains in melts and solutions are free to diffuse, the entanglements they form with their neighbours are temporary and have finite lifetime. Any given chain can disentangle from its neighbours by its own motion (reptate away) or by the motion of its neighbours. Effects of the motion of surrounding chains on the dynamics of a given chain will be discussed in Section 9.5.

A simple case to consider first is a single chain diffusing through a network, where the network only imposes permanent topological obstacles[2] that retard the motion of the chain. Consider an ideal chain in an array of fixed topological obstacles. A two-dimensional schematic representation of this problem, a giant snake in a forest, is presented in Fig. 9.11a. The snake randomly meanders through the forest and each of its conformations are assumed to be as likely as any other (an ideal snake). If the snake gets tired of being in a certain conformation, it is difficult for it to get into a completely different one because of the trees in the forest. These trees constrain this poor reptile to move primarily along its contour. Sideways excursions, although possible, put the snake into uncomfortable conformations with loops. The topological constraints imposed by the trees determine that the preferred path for the motion of the snake is along the confining tube.

Fig 9.11
Frame (a) shows a two-dimensional model of a chain in a permanent entanglement network: a giant snake in a forest. Frame (b) shows two students reeling-in the ends of the snake to construct the primitive path.

The primitive path is the centre line of the confining tube. It can be visualized by hiring one smart student and one brave student to reel in the snake at its ends (Fig. 9.11b). The final contour of the snake, when pulled taut, is the primitive path—the shortest path with the same 'topology' as the original conformation of the snake. A long-exposure photograph of the wiggling snake, taken by a curious student, depicts the whole confining tube in Fig. 9.12.

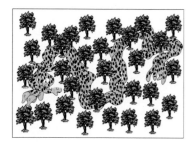

Fig. 9.12
A long-exposure photograph of the giant snake in the forest clearly defines its confining tube.

The reptation model assumes the contour length of the primitive path is fixed at its average value $\langle L \rangle$. In reality, the primitive path length

---

[2] Real networks can lead to correlation and excluded volume effects on chain conformation that we ignore here.

L fluctuates in time as the chain (or snake) moves. A full description of chain dynamics requires knowledge of the probability distribution of the primitive path lengths. This problem has been solved exactly by Helfand and Pearson in 1983 for a lattice model of a chain in a regular array of topological obstacles, but here we present a simple estimate of the probability distribution of primitive path lengths.

If an ideal linear chain is confined to a cylindrical pore of diameter a, it occupies a section of the pore of length $R_\parallel \approx b\sqrt{N}$ [Eq. (3.47)]. Entangled chains occupy a much longer length of confining tube $\langle L \rangle \approx bN/\sqrt{N_e} \gg b\sqrt{N}$, making them strongly stretched. The source of this stretching is the entropy gain at each tube end because each end segment of the primitive path is free to choose from multiple possible directions. This entropy gain leads to an approximately linear contribution to the free energy of a chain in a confining tube[3] of order kT per primitive path step,

$$F_{ent}(L) \approx -\Upsilon kT \frac{L}{a}, \qquad (9.48)$$

where Υ is a numerical constant of order unity. This approximately linear potential can be thought of as arising from nearly constant entropic forces of order kT/a acting on the chain at the tube ends. The chain in its confining tube is effectively under a tension kT/a and can be represented as an array of Pincus blobs of size a (see Section 3.2.1). Stretching an ideal chain along the contour of its tube to length L raises its free energy by $\gamma kTL^2/(2Nb^2)$, where γ is an effective dimensionless spring constant of order unity. The total free energy of a chain in a tube is the sum of these two effects:

$$
\begin{aligned}
F(L) &\approx \frac{\gamma kTL^2}{2Nb^2} - \Upsilon kT \frac{L}{a} \\
&\approx \frac{\gamma kT}{2Nb^2}\left[L^2 - \frac{2\Upsilon Nb^2}{\gamma a}L + \left(\frac{\Upsilon Nb^2}{\gamma a}\right)^2 \right] - \frac{kT\Upsilon^2 Nb^2}{2\gamma a^2} \\
&\approx \frac{\gamma kT}{2Nb^2}(L - \langle L \rangle)^2 - \frac{kT\Upsilon^2}{2\gamma}\frac{N}{N_e}. \qquad (9.49)
\end{aligned}
$$

In the second line of Eq. (9.49), the term $kT\Upsilon^2 Nb^2/(2\gamma a^2)$ was added and subtracted so as to complete the square inside the square brackets, in order to recover the expression for the equilibrium tube length $\langle L \rangle \approx \Upsilon Nb^2/(\gamma a)$ [Eq. (9.3)]. This quadratic approximation for the free energy of **tube length fluctuations** around the average value $\langle L \rangle$ was first proposed by Doi and Kuzuu in 1980:

$$F(L) = \frac{\gamma kT}{2Nb^2}(L - \langle L \rangle)^2. \qquad (9.50)$$

The constant term $kT\Upsilon^2 N/(2\gamma N_e)$ in Eq. (9.49) does not affect the dependence of the free energy $F(L)$ on the contour length L of the primitive

---

[3] The contribution to the free energy is not strictly linear because for each primitive path length L, the entropy of not only the ends, but of the rest of the chain in an entanglement network needs to be considered.

path. The quadratic approximation of the free energy leads to a Gaussian probability distribution of the tube length L for a chain with N monomers:

$$p(N, L) \sim \exp\left[-\frac{F(L)}{kT}\right] \sim \exp\left[-\frac{\gamma}{2Nb^2}(L - \langle L\rangle)^2\right]. \tag{9.51}$$

The average length of a tube with diameter a is $\langle L\rangle \approx aN/N_e$ [Eq. (9.3)]. A typical fluctuation in the tube length corresponds to a free-energy change of order $F(L) - F(\langle L\rangle) \approx kT$:

$$\sqrt{\langle (L - \langle L\rangle)^2\rangle} \approx b\sqrt{N} = R \approx a\sqrt{N/N_e}. \tag{9.52}$$

Thus, a typical tube length fluctuation is of the order of the root-mean-square end-to-end distance R of the chain and the confining tube has a wide range of typical lengths:

$$L \approx \langle L\rangle \pm R \approx a\left(\frac{N}{N_e} \pm \sqrt{\frac{N}{N_e}}\right). \tag{9.53}$$

These thermal fluctuations of the tube length are the basis of the **Doi fluctuation model**, leading to significant modifications of reptation dynamics for entangled linear chains. Linear chains in a permanent network relax stress by abandoning tube sections via tube length fluctuations and reptation. Since the branch point of a branched polymer prohibits its reptation, branches relax only by fluctuations in tube length. For this reason, we next consider relaxation of simple branched polymers: star polymers (next section), H-polymers and comb polymers (Section 9.4.3). Star polymers in particular relax primarily by fluctuations in tube length. The ideas of tube length fluctuations and reptation will be combined in Sections 9.4.4 and 9.4.5 to treat linear polymers relaxing in a permanent network.

9.4.2   Entangled star polymers

All the discussions of entangled polymer dynamics above were limited to linear chains. The molecular architecture of the chain (star vs. linear vs. ring) significantly modifies polymer dynamics. Snake-like reptation is impossible for f-arm star polymers because they would have to drag $f - 1$ arms along the tube of a single arm, significantly reducing the entropy of the star polymer. Therefore, the branch point of a star is usually localized in one cell of an entanglement net. Stars relax stress and diffuse by arm retractions, which are large (exponentially unlikely) fluctuations of the tube lengths of their arms. This is analogous to an octopus entangled in an array of topological constraints (a fishing net), sketched in Fig. 9.13.

The easiest way for the octopus to change the conformation of any of its arms without crossing the obstacles, represented by gray circles in Fig. 9.13, is by retracting that arm. Such arm retraction reduces the length L_a of its primitive path by forming loops. In Section 9.4.1, we demonstrated that such conformations with primitive path reduced by

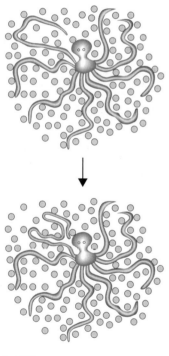

Fig. 9.13

Arm retraction of entangled star polymers demonstrated by an octopus in a fishing net. The circles are permanent topological constraints.

more than the root-mean-square fluctuation R from its equilibrium length $\langle L_a \rangle$ are exponentially unlikely [Eq. (9.51)]. Arm retraction by distance $s = \langle L_a \rangle - L_a$ along the contour of the tube can be analysed as a thermally activated process in an effective potential $U(s) \equiv F(L_a)$ (see Fig. 9.14). This potential is typically approximated by a parabola [Eq. (9.50)]:

$$U(s) \approx \frac{\gamma kT}{2} \frac{(L_a - \langle L_a \rangle)^2}{N_a b^2} \approx \frac{\gamma kT}{2} \frac{s^2}{N_a b^2}. \tag{9.54}$$

The number of Kuhn monomers in each arm of the star is N_a and the effective spring constant of this harmonic potential is γ. Most of the time, the length L_a of the confining tube of an arm is close to its equilibrium value $\langle L_a \rangle$ with deviations from it $|s| \lesssim R = b\sqrt{N_a}$ [Eq. (9.53)] corresponding to an effective potential change of order kT.

Occasionally, there are large atypical fluctuations of the tube length (with $|s| = |L - \langle L \rangle| \gg R$) that are exponentially unlikely [Eq. (9.51)] because of the restricted number of conformations that allow such a state. The probability of the tube length to be reduced by s can be estimated by the Boltzmann weight in the effective potential $U(s)$ [Eq. (9.51)]:

$$p(s) \sim \exp\left(-\frac{U(s)}{kT}\right) \sim \exp\left(-\frac{\gamma}{2} \frac{s^2}{N_a b^2}\right). \tag{9.55}$$

The average time between these large fluctuations $\tau(s)$ is inversely proportional to their probability $p(s)$:

$$\tau(s) \sim \exp\left(\frac{\gamma}{2} \frac{s^2}{N_a b^2}\right). \tag{9.56}$$

The coefficient in front of the exponential depends on the degree of polymerization N of the arm as well as on the magnitude of arm retraction s, but the average retraction time is dominated by the exponential [Eq. (9.56)]. For these large tube length fluctuations, it is important to remember that the quadratic potential [Eq. (9.50)] and the related Gaussian distribution [Eq. (9.51)] are approximations valid for small tube length fluctuations $|L - \langle L \rangle| \ll L$. The probability of large tube length fluctuations deviates from the simple Gaussian form [Eq. (9.51)]. For example, Eq. (9.51) predicts that the probability for the primitive path to be reduced to $L = 0$ and for the chain to form a single loop is exponentially low in the number of entanglements per chain ($\exp[-\gamma N/(2N_e)]$). The actual probability indeed has an exponential dependence on the average number of entanglements per chain,

$$p(N, 0) \sim \exp\left(-\frac{\gamma' N_a}{2 N_e}\right), \tag{9.57}$$

but with a different coefficient in the exponential $\gamma' \neq \gamma$.

The relaxation time of a star in an array of fixed topological obstacles is equal to the time it takes to completely retract its arms, written here

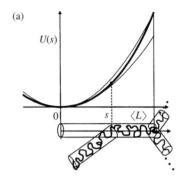

(a)

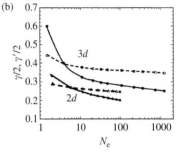

(b)

Fig. 9.14
(a) Effective potential for arm retraction for an entangled star polymer. The thin curves are harmonic approximations for small and large arm retractions. (b) Numerical results for the dependence of the effective spring constants γ (solid curves) and γ' (dashed curves) on the number of Kuhn monomers per entanglement strand on square (two-dimensional) and cubic (three-dimensional) lattices.

by including the power law 'prefactor' in the number of entanglements per arm:[4]

$$\tau_{\text{arm}} = \tau(\langle L \rangle) \sim \left(\frac{N_a}{N_e}\right)^{5/2} \exp\left(\frac{\gamma' \langle L \rangle^2}{2 N_a b^2}\right) \sim \left(\frac{N_a}{N_e}\right)^{5/2} \exp\left(\frac{\gamma' N_a}{2 N_e}\right). \quad (9.58)$$

The relaxation time of a star grows exponentially with the number of entanglements N_a/N_e per arm and is independent of the number of arms f in the star. The coefficient in the exponential is weakly dependent on the relative amount of arm retraction $s/\langle L \rangle$, changing from γ at small retractions to γ' at full retraction, because the harmonic potential is only an approximation of the actual potential. For polystyrene (with $N_e = 23$), the cubic lattice model predicts the spring constant of the harmonic potential to increase from $\gamma = 0.63$ for abandoning the first few tube sections to $\gamma' = 0.75$ for complete retraction of the arm $s = \langle L_a \rangle$ [see Fig. 9.14(b)]. However, this small change of γ to γ' changes the relaxation time of strongly entangled star polymers enormously. For example, a star with $N_a/N_e = 100$ entanglements per arm changes its relaxation time by a factor of $\exp(6) = 400$.

The stress relaxation modulus is proportional to the average fraction of entanglements per arm that have not yet relaxed by having the free end of the arm visiting that tube section. If s is the length of the tube that has been retracted and relaxed during time $t = \tau(s)$ then the stress relaxation modulus at time t is

$$G(t) \approx G_e \frac{\langle L_a \rangle - s}{\langle L_a \rangle} \quad \text{for } \tau_e < t < \tau_{\text{arm}}, \quad (9.59)$$

where G_e is the plateau modulus [Eq. (9.18)]. The stress relaxation modulus of a star polymer has a time dependence similar to that of a linear polymer with molar mass $2 M_a$ (the span molar mass of the star polymer) for times shorter than the Rouse time of the span, as shown in the frequency dependence of the complex modulus in Fig. 9.15. At the terminal time τ_{arm},

Fig. 9.15
Storage modulus (filled symbols) and loss modulus (open symbols) for linear 1,4-polybutadiene with $M_w = 160\,000\,\text{g mol}^{-1}$ (squares) and a 6-arm star 1,4-polybutadiene with $M_a = 77\,000\,\text{g mol}^{-1}$ (circles), both at a reference temperature of 28 °C. The linear polymer was chosen because it's molar mass is approximately the span molar mass of the star polymer. Data courtesy of L. Archer.

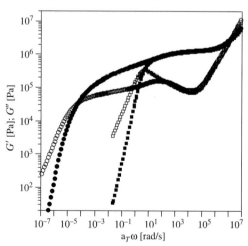

[4] See the Appendix of L. J. Fetters, *et al.*, *Macromolecules* **26**, 647 (1993).

there is of order one unrelaxed entanglement left per arm and the stress relaxation modulus is lower than the plateau modulus by the number of entanglements per arm:

$$G(\tau_{\text{arm}}) \approx \frac{N_e}{N_a} G_e. \qquad (9.60)$$

The viscosity of entangled stars can be estimated as the product of the relaxation time and the terminal modulus:

$$\eta \approx G(\tau_{\text{arm}})\tau_{\text{arm}} \sim \left(\frac{N_a}{N_e}\right)^{3/2} \exp\left(\frac{\gamma'}{2} \frac{N_a}{N_e}\right). \qquad (9.61)$$

The main feature is the exponential growth of the viscosity with the number of entanglements per arm N_a/N_e. Another interesting feature of the viscosity of entangled stars is that it is independent of the number of arms f. An experimental verification of this prediction is presented in Fig. 9.16. Viscosity of three-arm stars is $\sim 30\%$ lower than for stars with the same arm molar mass, but larger number of arms $f \geq 4$. This effect might be due to additional diving modes of a branch point down the tube of a three-arm star (see Problem 9.28).

Naively, one may think that for a branch point to hop between neighbouring entanglement cells, $f - 2$ of the arms must simultaneously retract, forming essentially linear tube and $f - 2$ large loops. This simultaneous retraction is an extremely unlikely event and its probability is the product of the already very low retraction probabilities for each of the $f - 2$ arms. The problem with this naive approach is that it is indeed hard for an octopus to put on a sweater by pulling in all arms and then pushing them all out at the same time. It would be much easier for the octopus to retract one arm at a time. This way, in several steps of arm retraction it could form a favourable arrangement of tubes near the branch point for a successful hop of this branch point between neighbouring cells of an entanglement net.

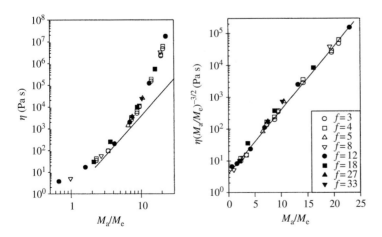

Fig. 9.16
Viscosity of polyisoprene star polymers with various numbers of arms at 60 °C. The left plot shows that viscosity is only a function of the number of entanglements per arm and that the viscosity of entangled linear polyisoprene (line with $M_a \equiv M/2$) is always lower. The right plot shows that Eq. (9.61) describes all star polymer viscosity data with the effective spring constant of the quadratic potential $\gamma' = 0.96$. Data from L. J. Fetters *et al.*, *Macromolecules* **26**, 647 (1993).

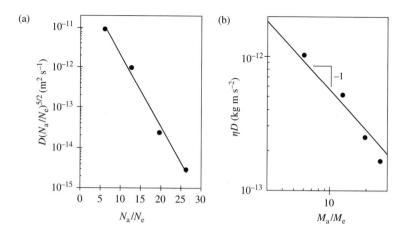

Fig. 9.17
(a) Diffusion coefficients of three-arm
star hydrogenated polybutadienes at
$165\,^\circ\mathrm{C}$. The slope determines $\gamma' = 0.82$.
(b) The product of viscosity and
diffusion coefficient is inversely
proportional to the number of
entanglements on each arm. Data are
from C. R. Bartels *et al.*,
Macromolecules **19**, 785 (1986).

Linear polymers move a distance of order of their own size during their
relaxation time, leading to a diffusion coefficient $D \approx R^2/\tau$ [Eq. (9.12)].
However, the diffusion of entangled stars is different because at the time
scale of successful arm retraction, the branch point can only randomly hop
between neighbouring entanglement cells by a distance of order one tube
diameter a. For this reason, diffusion of an entangled star is *much slower*
than diffusion of a linear polymer with the same number of monomers:

$$D \approx \frac{a^2}{\tau_{\mathrm{arm}}} \sim \left(\frac{N_a}{N_e}\right)^{-5/2} \exp\left(-\frac{\gamma' N_a}{2 N_e}\right). \tag{9.62}$$

The main feature of the diffusion coefficient of stars [Eq. (9.62)] is its
exponential dependence on the number of entanglements per arm N_a/N_e
related to the arm retraction time τ_{arm}. This prediction is in good agreement with experiments, as illustrated in Fig. 9.17(a) for diffusion of three-
arm star hydrogenated polybutadienes. The product of viscosity [Eq.
(9.61)] and diffusion coefficient [Eq. (9.62)] decreases with the number of
entanglements per arm:

$$\eta D \approx G(\tau_{\mathrm{arm}})a^2 \sim \frac{N_e}{N_a} \tag{9.63}$$

as shown in Fig. 9.17(b).

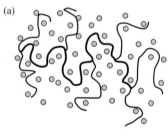

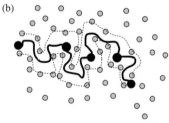

Fig. 9.18
(a) Entangled comb polymer with $q = 7$
branches and N_a monomers per branch
and a backbone (thick line) with
N_{bb} monomers. (b) Reptating backbone
of a comb with N_{bb} monomers (thick
line) and $q - 2 = 5$ high friction points
(black circles) in its confining tube
(dashed lines).

9.4.3   H-polymers and combs

The arm retraction mechanism of star dynamics can be applied to other
entangled branched polymers, such as H-polymers and comb polymers
(see Fig. 1.5) in an array of fixed topological obstacles. In the simplest case,
all side branches of an H-polymer or a comb polymer are the same and
contain N_a monomers (Fig. 9.18). The delineation of the comb into the
backbone (thick line) and branches is done so that the ends of the backbone
coincide with the branch points at the two ends of the polymer.

The retraction time of an arm τ_{arm} in an array of fixed obstacles is the same as the relaxation time of a star polymer with N_a monomers per arm [Eq. (9.58)]. On time scales shorter than τ_{arm}, the branch points are localized and cannot move between neighbour cells of the entanglement net. The branch points begin to hop between neighbouring cells of the entanglement net on the time scale of arm retraction τ_{arm}. Similar to star polymers, the length scale of these hops is of the order of the tube diameter a, allowing the effective friction coefficient for motion of the branch points to be determined by the retraction time of an arm:

$$\zeta_{br} \approx kT\frac{\tau_{arm}}{a^2}. \qquad (9.64)$$

The backbone of the polymer moves by reptation along the contour of its tube, with curvilinear diffusion dominated by the branch point friction ζ_{br}. An H-polymer is the simplest comb polymer with $q=4$ branches per molecule. For any trifunctional comb polymer ($q \geq 4$) the number of branch points is $q-2$ since each end of the backbone has two branches. The total number of monomers in the reptating backbone is N_{bb}. We assume that branches are well-entangled, so that the branch points dominate the friction:

$$(q-2)\zeta_{br} \gg \zeta(N_{bb}+qN_a), \qquad (9.65)$$

where ζ is the monomeric friction coefficient and $N_{bb}+qN_a$ is the total number of monomers in the whole chain. The curvilinear diffusion coefficient of the backbone along its confining tube is given by the Rouse model [Eq. (9.7)] with friction from the $q-2$ branch points:

$$D_c \approx \frac{kT}{(q-2)\zeta_{br}} \approx \frac{a^2}{(q-2)\tau_{arm}}. \qquad (9.66)$$

The length of the confining tube of the backbone is $L_{bb} \approx aN_{bb}/N_e$ leading to the reptation time of the backbone:

$$\tau_{rep} \approx \frac{L_{bb}^2}{D_c} \approx \tau_{arm}(q-2)\left(\frac{N_{bb}}{N_e}\right)^2. \qquad (9.67)$$

The diffusion coefficient of entangled H-polymers and combs is the mean-square size of the backbone divided by its reptation time:

$$D \approx \frac{N_{bb}b^2}{\tau_{rep}} \approx \frac{a^2}{\tau_{arm}(q-2)}\left(\frac{N_e}{N_{bb}}\right). \qquad (9.68)$$

The stress relaxation modulus of combs and H-polymers consists of an arm-retraction part at shorter times ($t < \tau_{arm}$) and a reptation part at longer times ($\tau_{arm} < t < \tau_{rep}$).

9.4.4 Monomer displacement in entangled linear melts

On time scales shorter than the relaxation time of an entanglement strand τ_e, the sections of a linear chain involved in coherent motion are shorter than the entanglement strand and are not aware of the topological constraints. Since hydrodynamic interactions are screened in polymer melts, the motion on very short time scales $t < \tau_e$ is Rouse-like with mean-square monomer displacement given by the subdiffusive motion of the Rouse model [Eq. (8.58)].

$$\langle[\vec{r}(t) - \vec{r}(0)]^2\rangle \approx b^2 \left(\frac{t}{\tau_0}\right)^{1/2} \quad \text{for } t < \tau_e. \tag{9.69}$$

On longer time scales $t > \tau_e$, topological constraints restrict polymer motion to the confining tube. Displacements of monomers tangential to the axis of the tube (primitive path) on length scales larger than the tube diameter a are suppressed by surrounding chains. Monomer displacement along the contour of the tube is unconstrained and follows the subdiffusive motion of the Rouse model [Eq. (8.58)] *along the primitive path*.

For times shorter than the Rouse time of the chain ($t < \tau_R$), each monomer participates in coherent motion of a chain segment consisting of $\sqrt{t/\tau_0}$ neighbouring monomers. The time-dependent curvilinear coordinate of a monomer along the contour of the tube is $s(t)$ (Fig. 9.19). The mean-square monomer displacement *along the tube* is of the order of the mean-square size of this section in three-dimensional space [Eq. (8.58)]:

$$\langle[s(t) - s(0)]^2\rangle \approx b^2 \left(\frac{t}{\tau_0}\right)^{1/2} \approx a^2 \left(\frac{t}{\tau_e}\right)^{1/2} \quad \text{for } \tau_e < t < \tau_R. \tag{9.70}$$

Since the tube itself is a random walk with step length a, the mean-square displacement of a monomer in three-dimensional space $\langle\Delta r^2\rangle$ is the product of the step length a and the contour length displacement $\sqrt{\langle\Delta s^2\rangle}$:

$$\langle\Delta r^2\rangle \approx a\sqrt{\langle\Delta s^2\rangle}. \tag{9.71}$$

Thus, the mean-square monomer displacement in space exhibits a weak one-fourth power law in time when the chain is confined to a tube:

$$\langle[\vec{r}(t) - \vec{r}(0)]^2\rangle \approx a\sqrt{\langle[s(t) - s(0)]^2\rangle} \approx a^2 \left(\frac{t}{\tau_e}\right)^{1/4} \quad \text{for } \tau_e < t < \tau_R. \tag{9.72}$$

This time dependence is slower than for unrestricted Rouse motion [Eq. (8.58)] because displacement along the contour of the tube leads to a smaller displacement in space [Eq. (9.71)]. At the Rouse time of the chain,

$$\tau_R \approx \tau_0 N^2 \approx \tau_e \left(\frac{N}{N_e}\right)^2, \tag{9.73}$$

each monomer participates in coherent Rouse motion of the whole chain along the tube. The mean-square displacement of a monomer along the

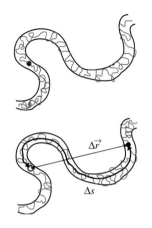

Fig. 9.19
Curvilinear displacement of a monomer (labelled by a dark circle) along the contour of the tube between two conformations is Δs. Only a short section of the tube is shown.

tube at the Rouse time of the chain is of the order of the mean-square size of the *whole chain*:

$$\langle [s(\tau_R) - s(0)]^2 \rangle \approx b^2 N \approx R^2. \tag{9.74}$$

Note that the root-mean-square magnitude of these fluctuations is in perfect agreement with the value of tube length fluctuations derived above [Eq. (9.52)]. Even though this magnitude seems large, it is a small fraction of the contour length of the tube [Eq. (9.3)]:

$$\frac{R}{L} \approx \frac{bN^{1/2}}{bNN_e^{-1/2}} \approx \left(\frac{N_e}{N}\right)^{1/2}. \tag{9.75}$$

At times longer than the Rouse time τ_R, all monomers move coherently with the chain. The chain diffuses along the tube, with a curvilinear diffusion coefficient given by the Rouse model $D_c \approx R^2/\tau_R$:

$$\langle [s(t) - s(0)]^2 \rangle \approx D_c t \approx b^2 N \frac{t}{\tau_R} \approx a^2 \frac{N}{N_e} \frac{t}{\tau_R} \quad \text{for } t > \tau_R. \tag{9.76}$$

In entangled polymer melts this diffusion occurs along the contour of the tube, with the mean-square monomer displacement in space determined using Eq. (9.71):

$$\left\langle [\vec{r}(t) - \vec{r}(0)]^2 \right\rangle \approx a\sqrt{\langle [s(t) - s(0)]^2 \rangle}$$
$$\approx a^2 \left(\frac{N}{N_e}\right)^{1/2} \left(\frac{t}{\tau_R}\right)^{1/2} \quad \text{for } \tau_R < t < \tau_{rep}. \tag{9.77}$$

This curvilinear motion continues up to the reptation time τ_{rep} where the chain has curvilinearly diffused the complete length of the tube, of order aN/N_e. At times longer than the reptation time ($t > \tau_{rep}$) the mean-square displacement of a monomer is approximately the same as the centre of mass of the chain and is a simple diffusion with diffusion coefficient D [Eq. (9.12)].

There are four different regimes of monomer displacement in entangled linear polymer melts, shown in Fig. 9.20. The $t^{1/4}$ subdiffusive regime for the mean-square monomer displacement is a unique characteristic of Rouse motion of a chain confined to a tube, which has been found in both NMR experiments and computer simulations.

9.4.5 Tube length fluctuations

Displacements of monomers at the two ends of the tube are unrelated to each other on time scales shorter than the Rouse time of the chain ($t < \tau_R$). These incoherent curvilinear displacements lead to tube length fluctuations [Eq. (9.70)]:

$$\langle [L(t) - L(0)]^2 \rangle \approx b^2 \left(\frac{t}{\tau_0}\right)^{1/2} \approx a^2 \left(\frac{t}{\tau_e}\right)^{1/2} \quad \text{for } \tau_e < t < \tau_R. \tag{9.78}$$

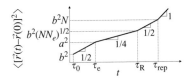

Fig. 9.20
Time dependence of the mean-square monomer displacement predicted by the reptation model for a melt of long entangled linear chains, on logarithmic scales.

Doi was the first to point out that the decrease of tube length due to these fluctuations leads to partial relaxation of stress. The stress relaxation modulus $G(t)$ is not quite constant in the rubbery plateau, but decreases slightly with time. The weak time dependence of the stress relaxation modulus corresponds to the rate at which sections of the tube are vacated by the fluctuating chain. Subdiffusive Rouse dynamics along the contour of the tube [Eq. (8.58)] implies a $t^{1/4}$ time dependence of vacated sections of the tube [see Eq. (9.70)]:

$$G(t) \approx G_e \frac{L(t)}{\langle L \rangle} \approx G_e \frac{\langle L \rangle - \sqrt{\langle [L(t) - L(0)]^2 \rangle}}{\langle L \rangle}$$

$$\approx G_e \left[1 - \frac{b}{\langle L \rangle} \left(\frac{t}{\tau_0} \right)^{1/4} \right]$$

$$\approx G_e \left[1 - \frac{N_e}{N} \left(\frac{t}{\tau_e} \right)^{1/4} \right] \quad \text{for } \tau_e < t < \tau_R. \tag{9.79}$$

The last relation made use of the fact that $\langle L \rangle \approx aN/N_e \approx bN/\sqrt{N_e}$ and $\tau_e \approx \tau_0 N_e^2$. The tube length fluctuations grow and the stress relaxation modulus decreases up to the Rouse time of the whole chain [Eq. (8.17)]. Consequently, the stress relaxation modulus at the Rouse time of the chain is lower than G_e:

$$G(\tau_R) \approx G_e \left[1 - \frac{N_e}{N} \left(\frac{\tau_R}{\tau_e} \right)^{1/4} \right] \approx G_e \left[1 - \mu \sqrt{\frac{N_e}{N}} \right]. \tag{9.80}$$

The final result was obtained using Eq. (9.19) $(\tau_R/\tau_e \approx (N/N_e)^2)$ and μ is a coefficient of order unity. The fraction $\sqrt{N_e/N}$ of the tube is vacated, and therefore relaxed, at the Rouse time of the chain by tube length fluctuations. The modulus at the relaxation time of the chain is also lower by the same factor:

$$G(\tau_{\text{rep}}) \approx G_e \left[1 - \mu \sqrt{\frac{N_e}{N}} \right]. \tag{9.81}$$

Since the distance that the chain must diffuse along the tube has been shortened by tube length fluctuations, the relaxation time is shorter than in the Doi–Edwards reptation model [Eq. (9.8)]:

$$\tau_{\text{rep}} \approx \frac{\langle L \rangle^2 [1 - \mu \sqrt{N_e/N}]^2}{D_c} \approx \tau_0 \frac{N^3}{N_e} \left[1 - \mu \sqrt{\frac{N_e}{N}} \right]^2. \tag{9.82}$$

The stress relaxation modulus then decays exponentially at the reptation time [Eq. (9.22)]. The terminal relaxation time can be measured quite precisely in linear viscoelastic experiments.[5] Hence, Eq. (9.82) provides the simplest direct means of testing the Doi fluctuation model and evaluating

---

[5] The modulus scale typically has a ±5% uncertainty owing to imperfect sample geometry which also affects viscosity but not relaxation times.

the parameter μ, as shown in Fig. 9.21. Requiring each data set to have an intercept of unity in Fig. 9.21 provides a correction to Eq. (9.21) for estimating τ_0. We conclude from Fig. 9.21 that $\mu \cong 1.0$, based primarily on computer simulations of the repton model (Section 9.6.2.6) because the experimental data are noisy owing to the usual $\pm 5\%$ uncertainties in determination of molar mass.

Recall that Fig. 9.3 showed the linear viscoelastic response of a polybutadiene melt with $M/M_e \cong 68$. The squared term in brackets in Eq. (9.82) is the tube length fluctuation correction to the reptation time. With $\mu = 1.0$ and $N/N_e = 68$, this correction is 0.77. Hence, the Doi fluctuation model makes a very subtle correction to the terminal relaxation time of a typical linear polymer melt. However, this subtle correction imparts stronger molar mass dependences for relaxation time, diffusion coefficient, and viscosity.

Tube length fluctuation modes significantly modify the rheological response of entangled polymers. The effect of these modes is most clearly observed in the shape of the loss modulus $G''(\omega)$. The Doi–Edwards equation ignores tube length fluctuations and predicts an almost single exponential stress relaxation modulus with small contribution from higher order modes [Eq. (9.21)]. The corresponding loss modulus is obtained from the Doi–Edwards equation by integration using Eq. (7.150) (see Problem 9.8):

$$G''(\omega) = \frac{8G_e}{\pi^2} \sum_{p:odd} \frac{\omega\tau_{rep}}{(\omega\tau_{rep})^2 + p^4}. \qquad (9.83)$$

In the rubbery plateau (for $3 \lesssim \omega\tau_{rep} \lesssim 300$ in Fig. 9.22), this Doi–Edwards reptation prediction gives $G'' \sim \omega^{-1/2}$. In contrast, a single exponential $G(t)$ leads to $G'' \sim \omega^{-1}$ at high frequencies. The shorter time modes, corresponding to $p = 3, 5, 7, \ldots$, make the reptation prediction of the loss modulus larger than that of a single exponential relaxation at high frequencies. The Doi fluctuation model has even more relaxation in the rubbery plateau, with $G'' \sim \omega^{-1/4}$ for frequencies larger than the reciprocal of the Rouse time of the chain (see Problem 9.36). Experimental data appear to obey a power law that is independent of polymer species (see Fig. 9.22) but with an intermediate exponent[6] ($G'' \sim \omega^{-0.3}$). At higher frequencies, differences between the two polymers are noted (particularly in G'') that are consistent with their 20% difference in M/M_e that creates a factor of 1.8 difference in τ_{rep}/τ_e. These differences show up at high frequency because the normalization of the axes in Fig. 9.22 requires overlap at low frequencies.

In Fig. 9.23(a), the loss moduli of two nearly monodisperse polybutadiene samples are simultaneously fitted by the predictions of the Doi–Edwards reptation model [Eq. (9.83)]. Experimental peaks are much

---

[6] Longitudinal Rouse modes of the chain along the tube may affect the value of this exponent.

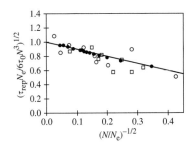

Fig. 9.21
Experimental verification of the Doi fluctuation model using data for polystyrene as open squares, from S. Onogi *et al.*, *Macromolecules* **3**, 109 (1970) and A. Schausberger *et al.*, *Rheol. Acta* **24**, 220 (1985), polybutadiene as open circles from R. H. Colby *et al.*, *Macromolecules* **20**, 2226 (1987) and filled circles from the repton model described in Section 9.6.2.6, courtesy of D. Shirvanyants. The line is Eq. (9.82) with $\mu = 1.0$.

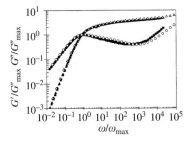

Fig. 9.22
Oscillatory shear data for two nearly monodisperse linear polymers with $M/M_e = 40$, reduced by their terminal loss modulus maximum. Triangles are the storage modulus G' and circles are the loss modulus G''. Filled symbols are for polybutadiene with $M_w = 70\,900\ \mathrm{g\,mol^{-1}}$ ($M/M_e = 37$) from M. Rubinstein and R. H. Colby, *J. Chem. Phys.* **89**, 5291 (1988). Open symbols are for polystyrene with $M_w = 750\,000\ \mathrm{g\,mol^{-1}}$ ($M/M_e = 44$) from A. Schausberger *et al.*, *Rheol. Acta* **24**, 220 (1985).

Fig. 9.23
Simultaneous fit of loss modulus
data for the two monodisperse
polybutadiene samples at 30 °C by
(a) the Doi–Edwards equation and
(b) the Doi tube length fluctuation
model. Lines are the fitting results.
Open circles are data for
$M = 355\,000$ g mol^{-1}. Filled squares
are data for $M = 70\,900$ g mol^{-1}.
Data from M. Rubinstein and
R. H. Colby, *J. Chem. Phys.* **89**,
5291 (1988).

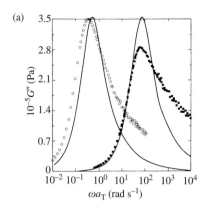

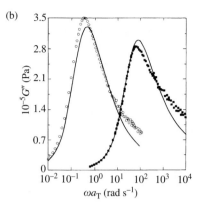

broader, especially at high frequencies (short times). The breadth of the experimental peaks of the loss modulus increases with decreasing molar mass. This comparison suggests that the Doi–Edwards equation underestimates stress relaxation by ignoring tube length fluctuation modes [Eq. (9.79)]. The complete stress relaxation modulus due to motion of a chain in its tube consists of two parts:

(1) Rouse modes of the chain, including tube length fluctuations and longitudinal Rouse modes, are active at times shorter than the Rouse time of the chain.

(2) Reptation modes are active at times longer than the Rouse time of the chain.

The results of models that include tube length fluctuation modes [Fig. 9.23(b)] are in much better agreement with the experimentally measured loss modulus $G''(\omega)$ of monodisperse melts than the prediction of the Doi–Edwards reptation model [Eq. (9.83)]. Tube length fluctuation corrections predict that the loss peak broadens with decreasing molar mass because the fraction of the stress released by fluctuations is larger for shorter chains.

The viscosity can again be estimated as the product of the terminal modulus and the reptation time:

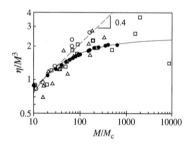

Fig. 9.24
Dependence of viscosity, reduced by the
cube of molar mass, on the number of
entanglements per chain. Filled circles
are data from the 'Repton model'
of Section 9.6.2, courtesy of
D. Shirvanyants. Open symbols are
experimental data for the three
polymers in Fig. 9.5, shifted parallel to
the η_0/M^3 axis to coincide with the
'Repton model' data. The curve is the
Doi fluctuation model [Eq. (9.84)] with
$\mu = 1.0$.

$$\eta \approx \tau_{\text{rep}} G(\tau_{\text{rep}}) \approx \frac{\tau_0 k T}{v_0} \frac{N^3}{N_e^2} \left[1 - \mu \sqrt{\frac{N_e}{N}} \right]^3. \qquad (9.84)$$

Doi's estimate of the effect of tube length fluctuations [Eq. (9.84)] predicts a molar mass dependence that approximates $\eta \sim N^{3.4}$ over a reasonable range of molar masses. Viscosity data from experiments and Repton model simulations are compared with the predictions of the Doi fluctuation model in Fig. 9.24. The Doi fluctuation model with $\mu = 1.0$ (solid curve) is in good agreement with both experimental and simulation data for $M > 10 M_e$. The data exhibit departures from the 3.4 power law (dashed line in Fig. 9.24) for long chains ($M > 300 M_e$) that are well described by the Doi fluctuation model.

Other computer simulations, such as the Evans–Edwards model of a chain in an array of fixed obstacles (described in detail in Section 9.6.2) exhibit fluctuations of the tube length and also find stronger molar mass dependences of relaxation time $\tau \sim M^{3.3\pm0.2}$ and diffusion coefficient $D \sim M^{-2.4\pm0.1}$ than the simple reptation model without tube length fluctuations [Eqs (9.8) and (9.12)]. These results of computer simulations of a single chain in an array of fixed obstacles are in good agreement with experiments on entangled polymer solutions and melts over the entire range of molar masses covered by simulations ($M < 600 M_e$). *Tube length fluctuations are responsible for the stronger molar mass dependences of diffusion coefficient* (Fig. 9.2), *relaxation time* [Eq. (9.9)], *and viscosity* (Fig. 9.5) *than predicted by the simple reptation model.*

9.5 Many-chain effects: constraint release

In Section 9.4, the motion of a single chain in an array of fixed topological constraints was discussed. Such models apply to dynamics of a chain in a network or in a melt of extremely long chains. In a melt of shorter chains, the topological constraints that define the confining tube are formed by neighbouring chains, which also move along their respective tubes. As chain B moves away, the topological constraint it once imposed on chain A disappears (Fig. 9.25). A new set of conformations is now available for chain A. A third chain moves in and imposes a new topological constraint on chain A. The constraints hence fluctuate in time, keeping the time-average total number of topological constraints imposed on a given chain by its neighbours constant. As some neighbours move away and remove their constraints from a given chain, others move in and place new constraints on it.

The exchange of neighbours and their topological constraints imposed on a given chain leads to a modification of the tube that a given polymer is confined to and is called **constraint release**. When a neighbouring chain B moves away, chain A can explore an additional volume of the order of an entanglement mesh size a^3. If a new chain C moves in, it can locally confine chain A to this new volume, changing the conformation of the tube of A. This process can be modelled by a local jump of the tube, analogous to an elementary move of the Rouse model. The rate of these local jumps of the primitive path is reciprocally proportional to the lifetime τ of the topological constraints. Thus, *constraint release leads to Rouse-like motion of the confining tube and its primitive path.*

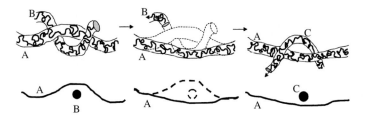

Fig. 9.25
Constraint release mechanism: when chain B reptates away, it releases the constraint on chain A. Later, this constraint is replaced by chain C, which confines chain A in a displaced tube.

9.5.1 Relaxation times and diffusion

Consider a single linear chain with P monomers in a melt of shorter N-mers. The P-mer has two relaxation mechanisms occurring simultaneously:

(1) Single-chain motion of the P-mer within its confining tube by reptation and tube length fluctuations.
(2) Constraint release as a Rouse motion of the tube confining the P-mer.

Whichever process relaxes the chain faster is the one that controls terminal dynamics.

The constraint release process for the P-mer can be modelled by Rouse motion of its tube, consisting of P/N_e segments, where N_e is the average number of monomers in an entanglement strand. The average lifetime of a topological constraint imposed on a probe P-mer by surrounding N-mers is the reptation time of the N-mers $\tau_{rep}(N)$. The relaxation time of the tube confining the probe chain by constraint release is the Rouse time of P/N_e tube segments [Eq. (8.17)] with segment relaxation time $\tau_{rep}(N)$ dictated by the reptation time of the surrounding N-mers:

$$\tau_{tube} \approx \tau_{rep}(N)\left(\frac{P}{N_e}\right)^2. \tag{9.85}$$

The diffusion coefficient of a P-mer in a melt of N-mers can be written as a sum of contributions from each of these two types of motion, assuming that each contributes independently to diffusion:

$$D \approx \frac{R^2}{\tau_{rep}(P)} + \frac{R^2}{\tau_{tube}}. \tag{9.86}$$

The reptation time of the P-mer is $\tau_{rep}(P)$ and the constraint release time τ_{tube} given in Eq. (9.85). The faster of the two types of motion controls the diffusion of the P-mer. For constraint release to significantly affect terminal dynamics, the Rouse relaxation time of the confining tube τ_{tube} must be shorter than the reptation time of the P-mer $\tau_{rep}(P)$:

$$\tau_{rep}(P) > \tau_{tube} \approx \tau_{rep}(N)\left(\frac{P}{N_e}\right)^2. \tag{9.87}$$

Very long P-mers have the constraint release time [Eq. (9.85)] shorter than their reptation time.[7] Such very long P-mers relax and diffuse by constraint release (Rouse motion of their tubes) before they get a chance to reptate out of their confining tubes. For shorter P-mers, the reptation time $\tau_{rep}(P)$ is shorter than the constraint release time τ_{tube} and reptation dominates the diffusion of these chains. Reptation certainly dominates diffusion in monodisperse solutions and melts (for $P = N$).

---

[7] Constraint release controls the terminal relaxation in the reptation model if $P/N_e > (N/N_e)^3$ and in the Doi fluctuation model if $(P/N_e)^{1.4} > (N/N_e)^{3.4}$.

Experiments on diffusion of deuterated polystyrene into a melt of hydrogenated polystyrene (Fig. 9.26) confirm the crossover assumed in Eq. (9.86). For very long matrix chains (large N), the terminal dynamics of the P-mer are controlled by reptation and consequently the diffusion coefficient of the P-mer only depends on the molar mass of the P-chains and is independent of N:

$$D \approx \frac{R^2}{\tau_{\text{rep}}(P)} \quad \text{for large } N. \tag{9.88}$$

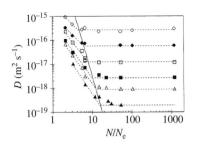

Fig. 9.26
Diffusion coefficient of trace amounts of deuterium-labelled polystyrene P-mers into polystyrene N-mer melts at $174\,^{\circ}$C for six P-mers: open circles $M = 55\,000\,\text{g mol}^{-1}$; filled circles $M = 110\,000\,\text{g mol}^{-1}$; open squares $M = 255\,000\,\text{g mol}^{-1}$; filled squares $M = 520\,000\,\text{g mol}^{-1}$; open triangles $M = 915\,000\,\text{g mol}^{-1}$; and filled triangles have $M = 2\,000\,000\,\text{g mol}^{-1}$. Data from P. F. Green and E. J. Kramer, *Macromolecules* **19**, 1108 (1986).

On the other hand, if the matrix chains are short enough (small N) constraint release controls the terminal dynamics of the P-chains [Eq. (9.85)] and the diffusion coefficient of the P-mers depends strongly on N:

$$D \approx \frac{R^2}{\tau_{\text{tube}}} \quad \text{for small } N. \tag{9.89}$$

The solid line in Fig. 9.26 is the crossover between Eqs (9.88) and (9.89), and divides the data nicely into a regime of constraint release control, where D is strongly dependent on N/N_e for short-chain matrices [described by Eq. (9.89)] and a regime of reptation control, where D is independent of N for diffusion into long-chain matrices [described by Eq. (9.88)].

9.5.2 Stress relaxation

Constraint release has a limited effect on the diffusion coefficient: it is important only for the diffusion of very long chains in a matrix of much shorter chains and can be neglected in monodisperse solutions and melts. The effect of constraint release on stress relaxation is much more important than on the diffusion and cannot be neglected even for monodisperse systems. Constraint release can be described by Rouse motion of the tube. The stress relaxation modulus for the Rouse model decays as the reciprocal square root of time [Eq. (8.47)]:

$$G(t) \sim (t/\tau)^{-1/2}. \tag{9.90}$$

Thus, a finite fraction of the stress relaxes by constraint release at time scales of the order of the constraint lifetime in the Rouse model of constraint release. This is also the time scale at which the stress relaxes by reptation in monodisperse entangled solutions and melts. Both processes simultaneously contribute to the relaxation of stress. Therefore, constraint release has to be taken into account for a quantitative description of stress relaxation even in monodisperse systems. The contribution of constraint release to stress relaxation in polydisperse solutions and melts is even more important as will be discussed below.

9.5.2.1 Stress relaxation in binary blends

Single-chain models, such as the Doi–Edwards reptation model [Eq. (9.21)] or the Doi tube length fluctuation model, assume a linear contribution to the stress relaxation modulus from each component of a polydisperse system:

$$G(t) = \sum_N \phi_N G_N(t), \qquad (9.91)$$

where ϕ_N is the volume fraction of N-mers and $G_N(t)$ is the single-chain stress relaxation modulus of N-mers. For a binary blend of long (L) and short (S) chains, these models predict a simple linear addition of the stress relaxation moduli of the two components weighted by their volume fractions:

$$G(t) = \phi_L G_L(t) + \phi_S G_S(t). \qquad (9.92)$$

However, many experiments observe that the amount of stress relaxed at the time scale of the reptation time τ_S of shorter chains is much larger than the volume fraction of short chains. This is shown in Fig. 9.27(a), where the loss moduli of binary blends are compared with the predictions of Eq. (9.92) using the Doi–Edwards reptation model predictions for $G(t)$ [Eq. (9.21)] for the $G_L(t)$ and $G_S(t)$ relaxation functions. Recall from Section 7.6.5 that the magnitude of $G''(\omega)$ directly reflects the amount of relaxation occurring at each frequency ω. Hence, Eq. (9.92) strongly underestimates the amount of relaxation occurring when the short chains relax [the high-frequency peak in $G''(\omega)$].

Some of the stress relaxed at time scale τ_S occurs by release of constraints imposed on long chains by short ones, which makes a significant contribution to the stress relaxation at the reptation time of the short chains τ_S.

Topological constraints are often assumed to be *pairwise* entanglements between chains. There are three types of these pairwise entanglements in a binary blend: between two long chains (L–L); between two short chains (S–S), and between a short and a long chain (S–L). If the dynamics of each chain along its tube is approximated by the Doi–Edwards reptation model, there are two time scales in the problem—reptation times of long (τ_L) and short (τ_S) polymers. The constraint on a given chain, caused by a long neighbour, has lifetime τ_L, while the constraint imposed by a short neighbour has lifetime τ_S. The constraint release process in a binary blend can be represented by a Rouse model with two mobilities of the effective beads:

(1) slow, corresponding to entanglements with long chains;
(2) fast, corresponding to entanglements with short chains.

These two mobilities can be assumed to be randomly distributed along the tube with relative concentrations corresponding to the probabilities of entanglement with a chain of each type. The simplest assumption is that these relative concentrations are proportional to the volume fractions of each type of chain (for pairwise entanglements).

The combined stress relaxation modulus for both reptation and constraint release of a binary blend is

$$G(t) = \phi_L G_L(t)\Lambda_L(t) + \phi_S G_S(t)\Lambda_S(t), \qquad (9.93)$$

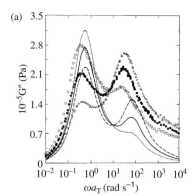

(a)

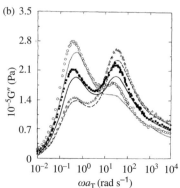

(b)

Fig. 9.27
Comparison of the loss modulus data for three blend compositions of the two polybutadiene samples in Fig. 9.23 at 30 °C with the predictions of (a) Doi–Edwards reptation model and (b) self-consistent constraint release model. Dotted lines are the predictions and open circles are the data for $\phi_L = 0.882$. Solid lines and filled squares are for $\phi_L = 0.768$. Dashed lines and open triangles are for $\phi_L = 0.638$. Data from M. Rubinstein and R. H. Colby, *J. Chem. Phys.* **89**, 5291 (1988).

where $\Lambda_L(t)$ and $\Lambda_S(t)$ are the stress relaxation functions due to constraint release (Rouse motion of tubes) of long and short chains (with two bead friction coefficients $\tau_L kT/a^2$ and $\tau_S kT/a^2$). This model is in reasonable agreement with experiments. Adding constraint release modes to the Doi–Edwards reptation model improves the dependence of the heights of the loss modulus peaks on the volume fractions ϕ_L and ϕ_S of components. But this model lacks the higher frequency modes because it does not include tube length fluctuations.

These tube length fluctuation modes (see Section 9.4.5) of the neighbouring chains affect the constraint release modes of a given chain. If entanglements between chains are assumed to be binary, there should be a **duality** between constraint release events and 'chain in a tube' relaxation events. A release of an entanglement by reptation or tube length fluctuation of one chain in its tube leads to a release of the constraint on the second chain. If this duality is accepted, the distribution of constraint release rates can be determined self-consistently from the stress relaxation modulus of the tube model.

The constraint release process in this self-consistent model is described by a Rouse model with random bead mobilities. The distribution of these mobilities is given by the constraint release rate distribution. The predictions of this self-consistent model are in good agreement with experiments on binary blends [see Fig. 9.27(b)].

The constraint release model represents the effects of surrounding polymers on the motion of a given chain by allowing transverse motion of the tube, controlled by the rate that entanglements are abandoned. The width of the confining tube stays, on average, constant in the constraint release process. While the more general problem of polydispersity effects on constraint release is well posed, it has not been solved thus far for anything beyond the simplest case of a binary blend. Consequently, other models that are easier to solve but are also less accurate have been proposed. For instance, Marrucci and Viovy have suggested that the tube diameter might be considered to increase as entanglements are abandoned. After a step strain the relaxed sections of the chains are assumed to be unable to confine the polymer any more, effectively becoming solvent-like at long time scales. The polymer is confined to a wider dilated tube and this process is called **tube dilation**. Problems 9.45–9.48 show that tube dilation has rather limited utility for linear polymers. Tube dilation has most effectively been applied to the dynamics of branched polymers with an exponentially broad distribution of relaxation rates (see Problem 9.49).

9.6 Computer simulations in polymer physics

Rapid advances in computer technology are making computer simulations powerful tools to study polymer properties. Computer simulations occupy an important intermediate position between theory and experiments. They can provide valuable tests of assumptions and predictions of theoretical models as well as attempt to mimic experimental systems, such as polymer solutions, melts, and networks.

There are two main approaches used to simulate polymer materials: molecular dynamics and Monte Carlo methods. The molecular dynamics approach is based on numerical integration of Newton's equations of motion for a system of particles (or monomers). Particles follow deterministic trajectories in space for a well-defined set of interaction potentials between them. In a qualitatively different simulation technique, called Monte Carlo, phase space is sampled randomly. Molecular dynamics and Monte Carlo simulation approaches are analogous to time and ensemble methods of averaging in statistical mechanics. Some modern computer simulation methods use a combination of the two approaches.

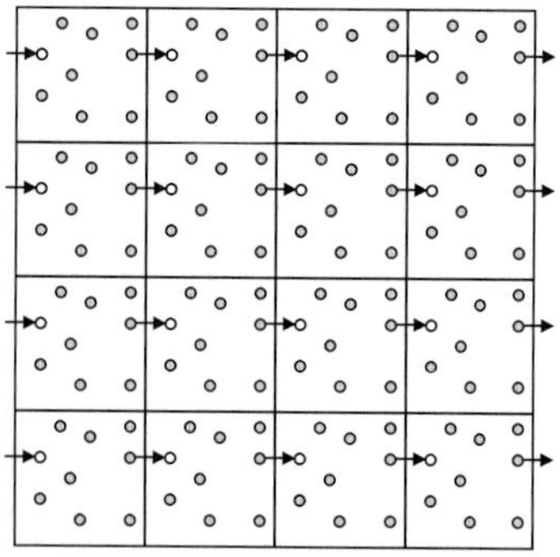

9.6.1 Molecular dynamics

Consider a molecular dynamics simulation of a system consisting of K particles in a cubic box of volume $V = L^3$. Periodic boundary conditions are typically used to minimize surface effects. Periodic boundary conditions correspond to densely filling space with identical copies of the simulation box (see Fig. 9.28 for a two-dimensional sketch of periodic boundary conditions). As particles leave the simulation box from one side, they automatically reenter it from the opposite side.

In one of the simplest models, all particles are identical with mass m and size σ interacting with each other via a Lennard-Jones potential [Eq. (3.96)]. A cutoff in the potential is introduced at r_{cut} in order to speed-up the simulations:

$$U(r) = \begin{cases} 4\epsilon\left[(\sigma/r)^{12} - (\sigma/r)^6\right] - 4\epsilon\left[(\sigma/r_{\text{cut}})^{12} - (\sigma/r_{\text{cut}})^6\right] & \text{for } r < r_{\text{cut}}, \\ 0 & \text{for } r > r_{\text{cut}}. \end{cases}$$

$$(9.94)$$

A typical cutoff used for attractive interactions is $r_{\text{cut}} = 2.5\sigma$ and for purely repulsive interactions the cutoff is at the minimum of the potential $r_{\text{cut}} = 2^{1/6}\sigma$. The potential in Eq. (9.94) is shifted in order to make it continuous at the cutoff $U(r_{\text{cut}}) = 0$.

A simulation starts with an initial set of positions $\{\vec{r}_i\}$ and velocities $\{d\vec{r}_i/dt\}$ of all particles (with $i = 1, 2, \ldots, n$ for a system of n particles). Temperature T is determined from the average kinetic energy of the particles $\sum_i m_i (d\vec{r}_i/dt)^2/(2n)$. This average kinetic energy is conserved during the simulation and is equal to $3kT/2$ in three dimensions:

$$T = \frac{1}{3nk}\sum_{i=1}^{n} m_i \left(\frac{d\vec{r}_i}{dt}\right)^2.$$

$$(9.95)$$

The total force acting on particle i at position $\vec{r}_i$ is the sum of the forces from all other particles in the simulation box and is equal to its mass m_i times its acceleration:

$$m_i \frac{d^2\vec{r}_i}{dt^2} = -\sum_{j \neq i} \frac{\partial U(\vec{r}_i - \vec{r}_j)}{\partial(\vec{r}_i - \vec{r}_j)}.$$

$$(9.96)$$

Fig. 9.28
Periodic boundary conditions for a two-dimensional cell.

Since this is basically a statement of Newton's second law applied to each particle, the n equations of the form of Eq. (9.96) are called **Newton's equations of motion**. These equations of motion are integrated over a small time step δt and new positions and velocities of all the particles are computed. The position of particle i at time $t + \delta t$ can be obtained by the Taylor series expansion in powers of the time step δt:

$$\vec{r}_i(t + \delta t) = \vec{r}_i(t) + \delta t \frac{d\vec{r}_i}{dt}\bigg|_t + \frac{(\delta t)^2}{2}\frac{d^2\vec{r}_i}{dt^2}\bigg|_t + \frac{(\delta t)^3}{6}\frac{d^3\vec{r}_i}{dt^3}\bigg|_t + \cdots \quad (9.97)$$

Similarly, the position of particle i at earlier time $t - \delta t$ can also be written as a series expansion:

$$\vec{r}_i(t - \delta t) = \vec{r}_i(t) - \delta t \frac{d\vec{r}_i}{dt}\bigg|_t + \frac{(\delta t)^2}{2}\frac{d^2\vec{r}_i}{dt^2}\bigg|_t - \frac{(\delta t)^3}{6}\frac{d^3\vec{r}_i}{dt^3}\bigg|_t + \cdots \quad (9.98)$$

Adding Eqs (9.97) and (9.98) provides the **Verlet algorithm** for calculation of the position of particle i at time $t + \delta t$:

$$\vec{r}_i(t + \delta t) = 2\vec{r}_i(t) - \vec{r}_i(t - \delta t) + (\delta t)^2 \frac{d^2\vec{r}_i}{dt^2}\bigg|_t + O((\delta t)^4). \quad (9.99)$$

Acceleration $d^2\vec{r}_i/dt^2|_t$ of particle i is determined by Newton's equation of motion [Eq. (9.96)]. The position of the particle in the Verlet algorithm is calculated with the accuracy of $(\delta t)^4$, as denoted by $O((\delta t)^4)$. The Verlet algorithm for the velocity of particle i at time t is obtained from its positions at times $t + \delta t$ and $t - \delta t$ with accuracy of $(\delta t)^3$ [by subtracting Eq. (9.98) from Eq. (9.97)]:

$$\frac{d\vec{r}_i}{dt}\bigg|_t = \frac{\vec{r}_i(t + \delta t) - \vec{r}_i(t - \delta t)}{2\delta t} + O((\delta t)^3). \quad (9.100)$$

Positions and velocities of all particles are calculated at every step of the molecular dynamics simulation, producing a complete time evolution of the system. In order for this time evolution to be accurate, the integration time step δt has to be much smaller than the shortest characteristic time of the system (the reciprocal Einstein frequency of the Lennard-Jones crystal). The simple Verlet algorithm is used for systems with constant number of particles, volume, and total energy. There are more sophisticated integration algorithms for simulations of systems at constant temperature that allow temperature rescaling to be done concurrently with the calculation of new positions and velocities of particles.

A physical quantity $\langle A \rangle$ is evaluated by the average of its instantaneous value $A(t)$ at time t over a long period of time (large number X of molecular dynamics steps) after initial equilibration during a sufficiently long run (with equilibration time t_0):

$$\langle A \rangle = \frac{1}{X}\sum_{j=1}^{X} A(t_0 + j\delta t). \quad (9.101)$$

If the simulation is long enough for the system to equilibrate (if it is much longer than all relaxation times), this time average is equivalent to the ensemble average.

In a more complicated simulation, particles are connected by bonds into small molecules or even into polymers. The atomic details included in a simulation depend on the specific problem being investigated. It is tempting to include the chemical details of monomers with accurate atomic potentials. Such simulations are carried out to answer detailed questions, such as the temperature dependence of density, or polarizability and solubility in a specific solvent. Including accurate potentials between atoms (e.g., bond-bending and bond-stretching C–C potentials) requires very small integration time steps (much shorter than vibration periods of these potentials) and makes complete relaxation of long chains with hundreds of monomers practically impossible. For example, a common time step of a molecular dynamics simulation is $\delta t = 10^{-15}$ s, while relaxation times of long polymers could be 10^{-3} s or even longer than 1 s. Computer technology is still a long way from simulations covering 12 or 15 orders of magnitude of time.[8] Therefore, multi-scale methods are being currently developed to bridge the gap between atomistic and course-grained models.

A simple generic bead–spring model of chains can be used to study universal polymer properties that do not depend on specific chemical details. Bonds between neighbouring Lennard-Jones particles in a chain can be represented by the finite extension non-linear elastic (FENE) potential,

$$U^{\text{FENE}}(r) = \begin{cases} -\frac{\kappa}{2} r_{\text{bond}}^2 \ln\left[1 - (r/r_{\text{bond}})^2\right] & \text{for } r < r_{\text{bond}}, \\ \infty & \text{for } r > r_{\text{bond}}, \end{cases} \qquad (9.102)$$

with typical values of bond length r_{bond} between 1.5σ and 2σ and typical values of bond strength κ between $5\epsilon/\sigma^2$ and $30\epsilon/\sigma^2$, where ϵ and σ are parameters of the Lennard-Jones potential [Eq. (9.94)]. Polymer solutions can be simulated by connecting some of the particles into polymers, while leaving the rest of the particles to represent solvent molecules. At low polymer concentrations, most of the particles in a simulation box are solvent molecules and most of the simulation time is occupied by calculation of their trajectories. It is necessary to keep explicit solvent molecules if the objective of the molecular dynamics is to simulate hydrodynamic flow or to investigate the details of the polymer–solvent interaction.

Explicit solvent is usually excluded from simulations directed at the study of solutions of longer chains for long times due to computational constraints (simulations with explicit solvent would be prohibitively long). Interactions between monomers are replaced by an effective potential mediated by implicit solvent of a given quality (see Chapter 3). The effects of collisions of explicit solvent molecules with monomers of the chains can be replaced by random forces acting on monomers in implicit solvent. These random forces are usually assumed to be non-correlated and therefore hydrodynamic interaction between monomers in explicit solvent

---

[8] Currently, 7–8 orders of magnitude in time are accessible on modern computers.

is lost as soon as explicit solvent molecules are replaced by random forces acting on monomers. Molecular dynamics simulation with random forces and corresponding viscous friction (called Brownian dynamics) is an example of a hybrid method.

A good example of results from a molecular dynamics simulation of entangled polymers is shown in Fig. 9.29. The 40 configurations of the chain shown are equally spaced in time up to the Rouse time of the chain. The chain is clearly confined to a tube-like region, with only the ends of the chain beginning to explore the rest of the volume.

9.6.2   Monte Carlo

Random sampling of different possible states of the system is called a Monte Carlo simulation technique. Starting from an arbitrary initial state of the system, a transition into another state is attempted following a certain set of transition rules. In the simplest cases, this transition corresponds to a random jump of a particle or several particles.[9] If different states of a system have different energies, the probability P_i of the system to be found in a state i with energy E_i is proportional to its statistical weight. For systems with constant volume, temperature, and number of particles, the statistical weight of a state is given by the Boltzmann factor:

$$P_i \sim \exp\left(-\frac{E_i}{kT}\right). \tag{9.103}$$

In equilibrium, a detailed balance must be satisfied, meaning that the number of transitions per unit time from any state i to any state j is on average equal to the number of transitions per unit time from j to i. Then, the number of transitions from any state i to any state j is proportional to the product of the probability P_i of being in state i, the probability $g_{i \to j}$ of making an attempt to move from state i to state j and the probability $p_{i \to j}$ of accepting this attempted transition. Therefore, detailed balance can be written as a simple equation:

$$P_i g_{i \to j} p_{i \to j} = P_j g_{j \to i} p_{j \to i}. \tag{9.104}$$

The condition of detailed balance can be solved for the ratio of acceptance probabilities of forward and backward transitions between states i and j.

$$\frac{p_{i \to j}}{p_{j \to i}} = \frac{P_j g_{j \to i}}{P_i g_{i \to j}}. \tag{9.105}$$

9.6.2.1   Metropolis algorithm

In the simplest Monte Carlo methods, such as the **Metropolis algorithm**, the probability of attempting to move to state j from state i is the same as the probability of attempting to move to state i from state j:

$$g_{j \to i} = g_{i \to j}. \tag{9.106}$$

[9] In practicce, the transitions often do not correspond to realistic moves, in an attempt to sample all of phase space as rapidly as possible.

Fig. 9.29
Molecular dynamics simulation of a chain with $N = 400$ monomers in an entangled polymer melt. Forty configurations of the chain are shown at equally spaced time intervals up to the Rouse time of the chain. Picture courtesy of G. S. Grest based on data from K. Kremer and G. S. Grest, *J. Chem. Phys.* **92**, 5057 (1990).

In this case, the ratio of acceptance probabilities for attempted moves between states i and j depends on the energy difference $E_j - E_i$ between these states:

$$\frac{p_{i \to j}}{p_{j \to i}} = \frac{P_j}{P_i} = \exp\left(-\frac{E_j - E_i}{kT}\right). \tag{9.107}$$

In the Metropolis algorithm an attempted transition into a state with lower energy is always accepted:

$$p_{i \to j} = 1 \quad \text{if } E_j \leq E_i. \tag{9.108}$$

If the energy of the final state E_j is higher than the energy of the initial state E_i, the attempted transition is accepted with probability

$$p_{i \to j} = \exp\left(-\frac{E_j - E_i}{kT}\right) \quad \text{if } E_j > E_i, \tag{9.109}$$

thereby automatically satisfying detailed balance at equilibrium. If the transition is not accepted, the old state counts one more time for any quantity to be calculated in the Monte Carlo simulation. For example, the average of any quantity over a large number K of Monte Carlo steps (after a large number J of equilibration steps) includes multiple terms with the same value whenever attempted moves were not accepted:

$$\langle A \rangle = \frac{1}{K} \sum_{i=J+1}^{J+K} A_i. \tag{9.110}$$

The Metropolis set of transition rates [Eqs (9.108) and (9.109)] satisfies the detailed balance [Eq. (9.107)] at equilibrium and the average over a long Monte Carlo run approximates the thermodynamic average. The detailed balance is also satisfied and thermodynamic equilibrium is approached in more sophisticated biased Monte Carlo algorithms with the probability of attempting to move to state j from state i not equal to the probability of attempting to move to state i from state $j(g_{j \to i} \neq g_{i \to j})$. These algorithms are used to enhance sampling of highly improbable states. Below, we outline several examples of lattice Monte Carlo models in which the states are either allowed and have equal energies and equal probabilities or forbidden with infinite energies and zero probabilities.

9.6.2.2   Random walk

For a chain modeled by an ideal N-step random walk on a cubic lattice there are 6^N different states with a fixed position of one end (see Fig. 9.30). It is impossible to sample all of these states for large N. Therefore, the simulation is restricted to a smaller, but still representative, subset of all allowed states for long polymers. There are many different methods of generating this subset of states. For example, the first monomer A_0 can be placed at the origin and the bonds of the chain are placed sequentially with their bond directions determined by a random number generator.

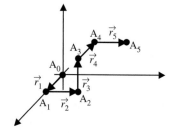

Fig. 9.30
Random walk on a cubic lattice.

The random number generator produces random numbers in the interval between 0 and 1. For a random walk on a cubic lattice, there are six possible directions. If the first random number is less than 1/6, the first bond is chosen to be directed to the right. If it is between 1/6 and 1/3, the bond $\vec{r}_1$ is directed up. If the random number is between 1/3 and 1/2, the bond is directed to the left. If the random number is between 1/2 and 2/3, the bond is directed down. If the random number is between 2/3 and 5/6, the bond is directed forward (out of the page). If the random number is greater than 5/6, the bond $\vec{r}_1$ between A_0 and A_1 is directed backward (into the page). The same procedure (with different random numbers) is repeated for the remaining $N-1$ bonds of the N-step random walk generating one possible conformation of the chain. Repeating the same procedure many times produces a large number of conformations of ideal random walks. The resulting random walks satisfy the ideal chain statistics of Chapter 2.

9.6.2.3 Self-avoiding walk

Monte Carlo simulations of a polymer in an athermal solvent are more difficult. In order to satisfy the excluded volume requirements in a lattice simulation, no lattice site can be occupied by more than one monomer. The simplest sampling technique is based on the algorithm described above for an ideal random walk. The only modification is that whenever an attempted new bond runs into an already occupied site, the whole chain is thrown away and a new chain is grown from the very beginning. The success rate for growing a self-avoiding N-mer by this simple algorithm rapidly decreases with N. In order to simulate long self-avoiding walks, alternative Monte Carlo algorithms have been developed. These alternative approaches include biased sampling (checking one step ahead to avoid self-intersection), dimerization (attempting to connect two shorter self-avoiding chains of $N/2$ monomers each), and the pivot algorithm (rotating sections of a self-avoiding chain by a lattice angle). Similar methods are also used for off-lattice simulations.

9.6.2.4 Verdier–Stockmayer model of unentangled chain dynamics

Verdier and Stockmayer used Monte Carlo simulations to study polymer dynamics using the bond moves shown in Fig. 9.31. The Monte Carlo simulation proceeds by randomly choosing one of the $N+1$ monomers of the chain by a procedure similar to choosing a random direction for a bond (multiplying a random number from an interval between 0 and 1 by $N+1$ and taking an integer part of the product). Next, an attempt is made to move the chosen monomer by selecting the type of move from the set shown in Fig. 9.31 and randomly picking the new potential location for the monomer. Moves 1, 2, and 3 in Fig. 9.31 are used for random walks, while moves 1, 2 and 4 are used for self-avoiding walks.

For example, if the end monomer is chosen to be moved by an end flip (move 1 in Fig. 9.31), its new possible position is selected in a way identical to choosing a random bond vector for a random walk (by choosing one of

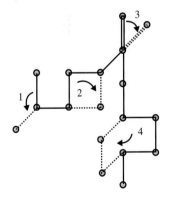

Fig. 9.31
Typical moves in a dynamic Monte Carlo simulation: (1) end flip, (2) corner move, (3) kink jump, and (4) crankshaft move. The solid lines are bonds in one of many possible configurations and the dotted lines are potential new bond positions.

six possible directions for the bond). The final step is to determine whether the move is accepted or not by checking whether it satisfies the conditions of the problem. For example, for self-avoiding walks the move is not accepted if there is an overlap between monomers. Whether the move is accepted or not, the clock of the simulation is moved forward by $1/(N+1)$ time units. A unit time in a Monte Carlo simulation corresponds to $N+1$ such attempted moves (an average of one attempt per monomer). All of the Monte Carlo moves (Fig. 9.31) are local and not directly correlated with each other. They can be thought of as representing uncorrelated monomer displacements of the Rouse model. On larger length and time scales the chain follows Rouse dynamics. The Monte Carlo time unit is shorter than, but proportional to, the monomer relaxation time τ_0.

In the random walk and self-avoiding walk models, described above, all energies are either zero (no interactions) or infinite (complete exclusion for overlapping monomers). In a more general case, finite, but non-zero interaction energies could be considered. In this case, different states (different polymer conformations) would have different energies and therefore different statistical weights. Monte Carlo moves are accepted or rejected according to an algorithm satisfying detailed balance [such as the Metropolis algorithm of Eqs (9.108) and (9.109)]. Some results of off-lattice Monte Carlo simulations of isolated chains in implicit solvents of different quality were presented in Fig. 3.16.

9.6.2.5 Evans–Edwards model of entangled chain dynamics

The dynamics of an entangled chain in an array of fixed obstacles can also be studied by Monte Carlo simulations. An initial unrestricted random walk conformation of a chain on a lattice (representing a chain in a melt) could be obtained using the method of section 9.6.2.2. The topological entanglement net of surrounding chains is represented by obstacles, sketched as solid circles in the middle of each elementary cell in Fig. 9.32.

Motion of the chain, represented by a random walk on a lattice, is defined by the set of allowed elementary moves of monomers between neighbouring sites that preserve chain connectivity and do not allow the chain to cross obstacles. Moves 1, 2, and 3 for an unrestricted two-dimensional random walk on a square lattice are sketched in Fig. 9.32. The corner flip (move 2) crosses obstacles and is therefore forbidden. The two remaining moves (end flips and kink jumps) satisfy topological constraints and therefore are allowed. The Monte Carlo model of chain motion by end flips and kink jumps is called the Evans–Edwards model. Forbidding corner flips leads to a dramatic change of polymer dynamics from free Rouse motion to Rouse motion of a chain confined in a tube (reptation with tube length fluctuations). The only allowed moves for a monomer in the middle of an entangled Evans–Edwards chain are the kink jumps that represent the diffusion of loops in the reptation model (Fig. 9.1). The time in a Monte Carlo simulation is defined by the average number of attempted moves per monomer.

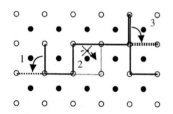

Fig. 9.32

Evans–Edwards model on a two-dimensional square lattice. Corner flips (move 2) cross obstacles and are forbidden. End flips (move 1) and kink flips (move 3) satisfy topological constraints and are allowed.

The diffusion coefficient D of the chain [Eq. (8.1)] is calculated by averaging the square of the displacement of the centre of mass of the chain $\langle(\vec{r}_{cm}(t) - \vec{r}_{cm}(0))^2\rangle$ during some long time t over many independent runs and taking its ratio to $6t$ in a three-dimensional simulation ($4t$ in a two-dimensional simulation):

$$D = \lim_{t \gg \tau} \frac{\langle(\vec{r}_{cm}(t) - \vec{r}_{cm}(0))^2\rangle}{6t}. \tag{9.111}$$

The limit in front of the ratio means that the time t has to be much longer than the longest relaxation time of the chain. The resulting diffusion coefficients obtained by Monte Carlo simulation of the Evans–Edwards model of entangled polymers are presented in Fig. 9.33(a). The diffusion coefficient decreases with the number of monomers in the chain. Another quantity that can be extracted from the Monte Carlo simulations of the Evans–Edwards model is the relaxation time of the chain. It can be defined as the characteristic decay time of the time correlation function of the end-to-end vector $\langle R(t)R(0)\rangle \sim \exp(-t/\tau_{rep})$. Figure 9.33(b) presents the results of such simulations.

Both diffusion coefficient and relaxation time obey stronger power laws in chain length than predicted by the simple reptation model [Eqs (9.8) and (9.12)].

$$D \sim N^{-2.5 \pm 0.1} \qquad \tau_{rep} \sim N^{3.5 \pm 0.1}. \tag{9.112}$$

These stronger molar mass dependences are in excellent agreement with experiments on entangled polymer liquids and the Doi fluctuation model. The important modes missing in the simple reptation model (that uses only the centre of mass diffusion along the contour of the tube) are already included in the Evans–Edwards model. It is important to emphasize that no assumptions about the existence of a confining tube or a primitive path was made in the Evans–Edwards model. The chain knows nothing about such definitions and moves the best way it can within the established rules. In order to verify that the missing modes are indeed related to tube length fluctuations, an even simpler model is discussed next.

9.6.2.6 Repton model of chain motion in a confining tube

The original reptation model of de Gennes assumes that loops diffuse randomly along the tube. The cumulative effect of the diffusion of loops is the reptation motion. A discretized version of the reptation model is the **repton model** that follows the diffusion of loops along the contour of the tube. The confining tube is represented by a set of neighbouring sites on a one-dimensional lattice. The chain in its tube is coarse-grained to a cluster of reptons on these lattice sites. Each tube section must have at least one repton. Longer loops are represented by additional reptons assigned to the section of the tube from which the loop emanates.

Figure 9.34 displays the mapping of a chain in an array of fixed obstacles to the repton model. Topological obstacles form a lattice with cell size equal to the tube diameter. Roughly N_r monomers are in cell I, between the end of the chain A and the point B where the chain finally leaves cell I for

(a)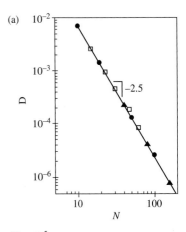

(b)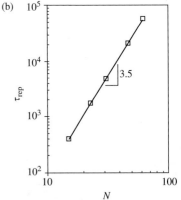

Fig. 9.33
Chain length dependence of diffusion coefficient and relaxation time from Monte Carlo simulations of the Evans–Edwards model. Filled circles are from J. M. Deutsch and T. L. Madden, *J. Chem Phys.* **91**, 3252 (1989). Open squares are from J. Reiter, *J. Chem. Phys.* **94**, 3222 (1991). Filled triangles are from A. Baumgartner *et al.*, *J. Stat. Phys.* **90**, 1375 (1998).

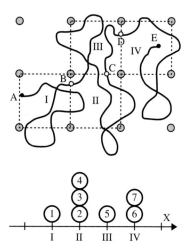

Fig. 9.34

Mapping of a chain in an array of topological entanglements (grey circles) onto a repton model. Cells of the entanglement net, numbered I–IV, are outlined by dotted lines and are mapped onto a one-dimensional lattice. Unentangled loops corresponding to these cells are mapped onto reptons, numbered 1–7, on the lattice sites.

cell II. This is mapped into a single repton at site I of the one-dimensional lattice. The number of monomers assigned to one repton N_r is proportional to, but smaller than, the number of monomers per entanglement strand N_e [see Eq. (9.114)]. Cell II has roughly $3N_r$ monomers in the chain section between points B and C, thereby placing three reptons at site II of the repton model. Similarly, cell III has roughly N_r monomers and cell IV has roughly $2N_r$ monomers, making the number of reptons 1 and 2 at sites III and IV of the one-dimensional lattice, respectively. In this manner, the chain of N monomers in Fig. 9.34 is mapped onto a sequence of $n = 7$ reptons that are placed on $K = 4$ sites of a one-dimensional lattice. This mapping effectively coarsens the chain to an integer number of connected reptons that each represents a section of the chain with N_r monomers.

The dynamics of a single chain in a tube is mapped onto the dynamics of the cluster of reptons along the one-dimensional lattice. To properly describe chain motion along the tube, the motion of reptons must obey a specific set of rules. The reptons can never vacate a site in the middle of the cluster because the real chain always stays connected. This means that repton 5 cannot move at the particular time step shown in Fig. 9.34 and that repton 1 cannot move to a site to the left of site I (but could move to site II). The order of the reptons must always be preserved, since they represent sequential sections of the chain along the confining tube. Hence, repton 2 could move to site I but not to site III, repton 3 cannot move[10] and repton 4 can move to site III, but not to site I.

At each time step, a repton and the direction of its motion ($+$ or $-$) are randomly selected. If the move obeys the simple rules described above (preserving connectivity and order) the move is actually made with probability p. For a chain segment of N_r monomers in Fig. 9.34, corresponding to one repton, there are z possible directions to move, where z is the number of faces in each cell ($z = 4$ for a square lattice and $z = 6$ for a cubic lattice). For a move between neighbouring occupied sites, there is one and only one direction of motion that will move the repton to a new site, making $p = 1/z$. For example, there is one face between cells II and III in Fig. 9.34 for the chain segment, corresponding to repton 4, to go through. All of the other $z - 1$ directions effectively do not move the repton because these moves keep the chain in loops associated with the original cell. However, the reptons at the ends of the chain are different. The end sections of the chain have $z - 1$ ways to move to new cells, making the probability $p - (z - 1)/z$ for a repton to move to an empty site. For example, the last section of the chain, corresponding to repton 7, has $z - 1 = 3$ new cells to choose from on the square lattice ($z = 4$) in Fig. 9.34. The relative probability $p = (z - 1)/z$ for a repton to move to an empty site controls the average number of occupied sites $\langle K \rangle$ for a cluster of N/N_r reptons:

$$\langle K \rangle = \left(\frac{z-1}{z} \right) \frac{N}{N_r} = \frac{N}{N_e}. \tag{9.113}$$

---

[10] The chain segments corresponding to repton 3 could be in cell II or in any cell adjacent to cell II, but the mapping to reptons always places repton 3 in cell II.

Thus, each repton contains slightly fewer than N_e monomers:

$$N_r = \frac{z-1}{z} N_e. \tag{9.114}$$

These simple rules for connectivity, order, and motion allow the repton model to be analysed analytically and easily solved numerically. The time dependence of such motion is shown in Fig. 9.35(a) for the extremities of the repton chain. The repton model allows direct visualization of tube length fluctuations.

The stress relaxation modulus is determined by the set of sites on the one-dimensional lattice (cells of the confining tube) that have not been vacated between time 0 (the moment of the step strain) and time t. The number of these 'still occupied' sites is the difference between the furthest propagation of the right end of the cluster to the left $x_R(t)$ (the upper dashed curve in Fig. 9.35(a)) and the furthest propagation of the left end of the cluster to the right $x_L(t)$ (the lower dashed curve in Fig. 9.35(a)). The stress relaxation modulus is proportional to the number of unrelaxed modes of the chain, equal to the number of cells of the original confining tube that have never been vacated. The stress relaxation modulus is proportional to the average fraction of still occupied sites in the repton model:

$$G(t) = G_e \frac{\langle x_R(t) - x_L(t) \rangle}{\langle K \rangle}. \tag{9.115}$$

The ensemble average utilizes only positive values of $x_R(t) - x_L(t)$, and averages over many different chains. All negative values of $x_R(t) - x_L(t)$ are replaced by zeroes. The stress relaxation modulus for a chain of 80 reptons is shown in Fig. 9.35(b).

Viscosity is determined by integration of the stress relaxation modulus [Eq. (7.117)] $\eta = \int G(t) \, dt$. The numerically calculated dependence of viscosity on the number of reptons (for the coordination number $z = 6$) is

$$\eta \sim N^{3.3} \tag{9.116}$$

for the entire range of repton numbers N studied (more than two decades), as shown in Fig. 9.36. Analogous results were obtained in Monte Carlo simulations of the Evans–Edwards model. The fact that the results of the Evans–Edwards model that does not assume the existence of the tube agree with the results of the repton model validates the concept of the confining tube.

The simple reptation model that takes into account only the centre of mass motion along the contour of the tube, but ignores tube length fluctuations, predicts different results than what are obtained from the Evans–Edwards and repton models, as well as from experiments. This implies that tube length fluctuations lead to important corrections to the simple reptation dynamics of the chain in its confining tube. The apparent molar mass exponent in the Evans–Edwards and repton models (as well as in experiments) is larger than the 'pure' reptation value of 3 because real chains abandon their tube *faster* through fluctuations in tube length and

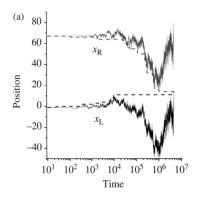

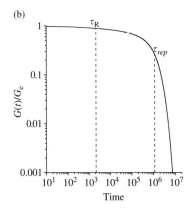

Fig. 9.35
Repton model results for a chain of 80 reptons, courtesy of D. Shirvanyants. (a) Tube length fluctuations represented by the coordinates of the first (lower curve) and the last (upper curve) reptons. The interval $x_R(t) - x_L(t)$ between the dashed lines has always been occupied between times 0 and t. (b) The stress relaxation modulus calculated from ensemble averaging information from part (a) for many runs.

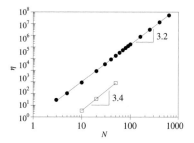

Fig. 9.36

Viscosity of linear polymer calculated from Monte Carlo simulations. Filled circles are repton model data from M. Rubinstein, *Phys. Rev. Lett.* **59**, 1946 (1987), with data range extended by D. Shirvanyants and open squares are Evans–Edwards model data from J. M. Deutsch and T. L. Madden, *J. Chem Phys.* **91**, 3252 (1989).

the effect gets stronger as the chains get shorter. The tube length fluctuations that are evident in Fig. 9.35(a) directly result in the stronger molar mass dependence of viscosity observed in Fig. 9.36.

9.7   Summary of entangled dynamics

Topological constraints, called entanglements, are manifestations of the fact that chains cannot cross one another. There is no first-principles microscopic model of chain entanglements. The most successful existing model of entangled polymers is the Edwards tube model. This model postulates that topological constraints of surrounding chains confine the motion of any long chain to a tube-like region. This postulate separates the complicated many-chain problem of entangled polymer solutions and melts into two simpler problems: the motion of a single chain in its tube and the motion of the tube due to motion of surrounding chains.

The main parameter of the tube model is the tube diameter a, determined by the amplitude of fluctuations that are restricted by the surrounding chains. The tube diameter is related to the number of monomers in an entanglement strand N_e:

$$a \approx b\sqrt{N_e}. \tag{9.117}$$

The tube diameter in an athermal solvent is proportional to, but larger than, the correlation length ξ

$$a(\phi) \approx a(1)\phi^{-0.76} \quad \text{for an athermal solvent.} \tag{9.118}$$

The tube diameter in a θ-solvent is proportional to, but larger than, the distance between binary contacts

$$a(\phi) \approx a(1)\phi^{-2/3} \quad \text{for a } \theta\text{-solvent.} \tag{9.119}$$

The number of monomers in an entanglement strand has practically the same concentration dependence in all solvents $N_e \sim \phi^{-1.3}$. The number of monomers in an entanglement strand is usually determined from the plateau modulus G_e (the value of the stress relaxation modulus at the relaxation time of an entanglement strand τ_e):

$$G_e \approx \frac{kT}{v_0 N_e} \quad \text{for polymer melts.} \tag{9.120}$$

The plateau modulus in entangled polymer solutions has practically the same concentration dependence in all solvents, $G_e \sim \phi^{2.3}$.

The primary mode of motion of a linear chain along its confining tube is reptation, first proposed by de Gennes. Reptation is a snake-like diffusion of a chain, as a whole, along the contour of its tube, with a Rouse curvilinear diffusion coefficient. The relaxation time of the melt is the time it takes the chain to reptate out of its original tube, called the reptation time τ_{rep}. The reptation time and the viscosity of entangled polymers are predicted by the simple reptation model to be proportional to the cube of polymer molar mass

$$\tau_{rep} \sim \eta \sim M^3. \tag{9.121}$$

The three-dimensional diffusion coefficient is predicted by the simple reptation model to be reciprocally proportional to the square of polymer molar mass:

$$D \sim M^{-2}. \tag{9.122}$$

The probability distribution function of the tube length L for a chain with N monomers is approximately Gaussian, with mean-square fluctuation of the order of the mean-square end-to-end distance of the chain. The tube length fluctuates in time, leading to stronger molar mass dependences of relaxation time, viscosity, and diffusion coefficient resembling experimental observations over some range of molar masses:

$$\tau_{rep} \sim \eta \sim M^{3.4}, \tag{9.123}$$

$$D \sim M^{-2.4}. \tag{9.124}$$

Tube length fluctuations modify the rheological response of entangled polymers. Reptation dynamics adds a $t^{1/4}$ regime to the mean-square monomer displacement that was not present in the free Rouse model. This extra regime is a characteristic signature of Rouse motion of a chain confined to a tube.

Entangled star polymers relax by arm retractions with relaxation times and viscosities exponentially large in the number of entanglements per arm N_a/N_e [Eqs (9.58) and (9.61)]. This leads to exponentially small diffusion coefficients [Eq. (9.62)] for entangled star polymers.

Reptation and tube length fluctuations of surrounding chains release some of the entanglement constraints they impose on a given chain and lead to Rouse-like motion of its tube, called constraint release. Constraint release modes are important for stress relaxation, especially in polydisperse entangled solutions and melts.

Problems

Section 9.1

9.1 The molar mass of an entanglement strand in a PDMS melt at 25 °C is $M_e \cong 12\,000\,\mathrm{g\,mol^{-1}}$. Estimate the plateau modulus G_e if the density of the PDMS melt at 25 °C is $\rho = 0.97\,\mathrm{g\,cm^{-3}}$.

9.2 Consider a poly(methyl methacrylate) melt at 140 °C.

(i) Use the information in Table 2.1 to estimate the volume v_0 of a Kuhn monomer at 140 °C.

(ii) Calculate the number of Kuhn monomers per entanglement strand N_e and the molar mass M_e of an entanglement strand, assuming the number of entanglements strands $P_e = 20$ per confinement volume a^3 and compare your estimate with the experimental value $M_e \cong 10\,000\,\mathrm{g\,mol^{-1}}$.

(iii) What is the tube diameter a of PMMA at $140\,^\circ$C?

(iv) Calculate the plateau modulus G_e of poly(methyl methacrylate) melt at $140\,^\circ$C and compare it with the experimental value $G_e \cong 3.1 \times 10^5$ Pa.

Section 9.2

9.3 Consider a PDMS melt with molar mass $M = 6 \times 10^5$ g mol^{-1}. The relaxation time of a Kuhn monomer is $\tau_0 = 10^{-10}$ s. The molar mass of a Kuhn monomer is $M_0 \cong 381$ g mol^{-1} and the molar mass of an entanglement strand is $M_e \cong 12\,000$ g mol^{-1}.

(i) Estimate the reptation time τ_{rep} of chains. How much longer is it than the Rouse time τ_R of the chains?

(ii) Estimate the diffusion coefficient D of these chains if the Kuhn length is $b = 1.3$ nm.

(iii) Estimate the melt viscosity if the plateau modulus of a PDMS melt is $G_e = 2.0 \times 10^5$ Pa.

9.4 Consider a polystyrene melt with molar mass $M = 10^6$ g mol^{-1}.

(i) Estimate the width of the rubbery plateau (τ_{rep}/τ_e) if the molar mass of an entanglement strand $M_e \cong 17\,000$ g mol^{-1} assuming the cubic dependence of the reptation time on the molar mass [Eq. (9.121)].

(ii) Repeat the calculation of (i) using the 3.4 power law dependence of reptation time on the molar mass [Eq. (9.123)].

9.5 Consider an entangled polymer melt of N-mers with monomeric friction coefficient ζ, monomer size b, and tube diameter $a \approx bN_e^{1/2}$ under steady shear with shear rate $\dot\gamma$.

(i) Show that the relative three-dimensional velocity of two typical overlapping chains of size $R \approx b\sqrt{N}$ is $v \approx \dot\gamma R$.

In order for one chain to move a distance of the order of the tube diameter a, the other chain must move out of its way by the curvilinear distance of the order of the tube length $\langle L \rangle$.

(ii) Demonstrate that the relative velocity of the monomers of two overlapping chains is

$$v_c \approx \dot\gamma R \frac{N}{N_e}.$$

The rate of energy dissipation per unit volume $\eta\dot\gamma^2$ defines the viscosity η of the melt.

(iii) Calculate the viscosity of an entangled melt using the energy dissipation rate per monomer ζv_c^2.

9.6 Consider a segment s of the tube of an entangled N-mer at time $t = 0$. Assume that the chain moves along its tube by simple diffusion with curvilinear diffusion coefficient D_c.

(i) Show that the probability $\Psi(\xi, t; s)$ that the primitive path of the chain moves the distance ξ during time t, while its ends have not yet reached a segment s of the original tube is

$$\Psi(\xi, t; s) = \frac{2}{\langle L \rangle} \sum_{p=1}^{\infty} \sin\left(\frac{\pi s}{\langle L \rangle} p\right) \sin\left(\frac{\pi (s - \xi)}{\langle L \rangle} p\right) \exp\left(-\frac{p^2 t}{\tau_{rep}}\right), \quad (9.125)$$

where the reptation time is

$$\tau_{rep} = \frac{\langle L \rangle^2}{\pi^2 D_c}.$$

Hint: The probability $\Psi(\xi, t; s)$ is the solution of the diffusion equation

$$\frac{\partial \Psi}{\partial t} = D_c \frac{\partial^2 \Psi}{\partial \xi^2},$$

with initial condition $\Psi(s - \langle L \rangle, 0; s) = \delta(\xi)$ and two boundary conditions $\Psi(s, t; s) = 0 = \Psi(s - \langle L \rangle, t; s)$.

(ii) Show that the probability of the segment s of the original tube (at time $t = 0$) to still be the part of the tube at time t is

$$\psi(s, t) = \sum_{\text{odd } p} \frac{4}{\pi p} \sin\left(\frac{\pi s}{\langle L \rangle} p\right) \exp\left(-\frac{p^2 t}{\tau_{\text{rep}}}\right). \tag{9.126}$$

Hint: In order for segment s to still be part of the tube at time t is should not be reached by the ends of the tube. Therefore, the displacement ξ of the tube during time t should be between $s - \langle L \rangle$ and s. This means that

$$\psi(s, t) = \int_{s - \langle L \rangle}^{s} d\xi \Psi(\xi, t; s).$$

(iii) The normalized stress relaxation modulus for reptation is the fraction of original tube segments that have not been vacated between times 0 and t. Demonstrate that this normalized stress relaxation modulus is

$$\frac{G(t)}{G_e} = \frac{8}{\pi^2} \sum_{\text{odd } p} \frac{1}{p^2} \exp\left(-\frac{p^2 t}{\tau_{\text{rep}}}\right). \tag{9.127}$$

Hint: Note that the normalized stress relaxation modulus for reptation is the probability $\psi(s, t)$ that a segment s is still a part of the tube after time t averaged over all segments s:

$$\frac{G(t)}{G_e} = \frac{1}{\langle L \rangle} \int_0^{\langle L \rangle} ds \, \psi(s, t).$$

9.7 (i) Calculate the upper bound on the error made in replacing the reptation stress relaxation modulus [Eq. (9.20)] by a single exponential:

$$G(t) \cong G_e \exp\left(-\frac{t}{\tau_{\text{rep}}}\right).$$

Hint: The largest error arising from neglecting any term $p = 3, 5, 7, \ldots$ in Eq. (9.20) occurs at $t = 0$.

(ii) Calculate the upper bound on the error made in replacing the reptation stress relaxation modulus [Eq. (9.21)] by a sum of two exponentials:

$$G(t) \cong \frac{9}{10} G_e \left[\exp\left(-\frac{t}{\tau_{\text{rep}}}\right) + \frac{1}{9} \exp\left(-\frac{9t}{\tau_{\text{rep}}}\right) \right].$$

9.8   (i) Demonstrate that the storage modulus of the Doi–Edwards reptation model with stress relaxation modulus given by Eq. (9.20) is

$$G'(\omega) = \frac{8}{\pi^2} G_e \sum_{\text{odd } p} \frac{\omega^2 \tau_{\text{rep}}^2}{p^2 (p^4 + \omega^2 \tau_{\text{rep}}^2)}, \tag{9.128}$$

while the loss modulus is

$$G''(\omega) = \frac{8}{\pi^2} G_e \sum_{\text{odd } p} \frac{\omega \tau_{\text{rep}}}{p^4 + (\omega \tau_{\text{rep}})^2}. \tag{9.129}$$

(ii) Calculate the low frequency limiting behaviour of G' and G''.
(iii) Estimate the longest relaxation time as the reciprocal of the frequency at which the low-frequency power laws of G' and G'' cross.
(iv) What is the error in assuming this estimate of the longest relaxation time is τ_{rep}?

Section 9.3

9.9 Determine the number of other chains P_e in the entanglement volume a^3 as a function of concentration in semidilute solution for

(i) an athermal solvent;
(ii) a θ-solvent;
(iii) what happens at low concentrations in a θ-solvent?

9.10 Consider a semidilute polymer solution in a good solvent with excluded volume v. At different length scales an entanglement strand can be viewed as (see Figs 5.5 and 5.6) an ideal chain of g_T monomers in a thermal blob; a self-avoiding walk of g/g_T thermal blobs up to correlation length ξ, where g is the number of monomers in a correlation volume; and an ideal chain of N_e/g correlation blobs on the largest length scale, where N_e is the number of monomers in an entanglement strand. Assume that a certain fixed number of binary contacts is required in a confinement volume a^3.

(i) Estimate the number of binary contacts between two overlapping thermal blobs.
(ii) Assume that the number of binary contacts between two neighbouring correlation volumes is the same as between two overlapping thermal blobs. Estimate the number of monomers in an entanglement strand N_e from the assumption that the total number of contacts between monomers on different chains in the confinement volume a^3 is constant and equal to $(a(1)/b)^3$. Demonstrate that the number of monomers in an entanglement strand N_e in the concentration interval $\phi_e < \phi < \phi^{**}$ is

$$N_e(\phi) \approx N_e(1) \left(\frac{\phi^{**}}{\phi} \right)^{1/(3\nu-1)} (\phi^{**})^{-4/3}. \tag{9.130}$$

(iii) Estimate the entanglement concentration ϕ_e as a function of excluded volume v for the case with $\phi_e < \phi^{**}$. How does the width of unentangled semidilute regime ϕ_e/ϕ^* depend on the excluded volume v?
(iv) How does the number of monomers in an entanglement strand N_e depend on concentration at higher concentrations $\phi > \phi^{**}$?

(v) Determine the condition at which the entanglement concentration is in the concentrated regime $\phi_e > \phi^{**}$?

(vi) Estimate the width of the unentangled semidilute regime for the case with $\phi^{**} < \phi_e$.

9.11 (i) Consider long polystyrene chains in carbon disulphide at 25 °C (athermal solvent). Estimate the correlation length ξ and tube diameter a at volume fractions $\phi = 0.1$ and $\phi = 0.02$.

(ii) Consider long polystyrene chains in cyclohexane at 35 °C (θ-solvent). Estimate the correlation length ξ and tube diameter a at volume fractions $\phi = 0.1$ and $\phi = 0.02$.

(iii) What are the entanglement concentrations ϕ_e of polystyrene with molar mass $M = 500\,000$ in carbon disulphide at 25 °C (athermal solvent) and in cyclohexane at 35 °C (θ-solvent)?

9.12 Explain the length scales over which the reptation, Rouse, and Zimm models describe dynamics in semidilute entangled solutions of linear polymers.

9.13 Consider an entangled solution of N-mers in a good solvent with excluded volume v and Kuhn monomer length b, in a solvent with viscosity η_s at volume fraction $\phi_e < \phi < \phi^{**}$ and derive the following results for

(i) the relaxation time τ_ξ of the polymer strand of size equal to the correlation blob

$$\tau_\xi \approx \frac{\eta_s b^3}{kT} \left(\frac{b^3}{v}\right)^{3(2\nu-1)/(3\nu-1)} \phi^{-3\nu(3\nu-1)}; \qquad (9.131)$$

(ii) the reptation time of the chain

$$\tau_{rep} \approx \frac{\eta_s b^3}{kT} \frac{N^3}{N_e(1)} \left(\frac{v}{b^3}\right)^{2(15\nu-8)/[3(3\nu-1)]} \phi^{3(1-\nu)/(3\nu-1)}; \qquad (9.132)$$

(iii) the diffusion coefficient of the chain

$$D \approx \frac{kT}{\eta_s b} \frac{N_e(1)}{N^2} \left(\frac{b^3}{v}\right)^{(24\nu+13)/[3(3\nu-1)]} \phi^{-(2-\nu)/(3\nu-1)}; \qquad (9.133)$$

(iv) the specific viscosity of the solution

$$\eta_{sp} \approx \left(\frac{v}{b^3}\right)^{(42\nu-23)/[3(3\nu-1)]} \frac{N^3}{[N_e(1)]^2} \phi^{3/(3\nu-1)}. \qquad (9.134)$$

9.14 Repeat the calculations of Section 9.3 using the dependence of reptation time τ_{rep} on molar mass with power 3.4 instead of 3 and using the values of the scaling exponent $\nu \cong 0.588$ for athermal solvent and $\nu = 1/2$ for a θ-solvent.

(i) Derive the following results for the concentration dependence of the reptation time in athermal solvent and in θ-solvent:

$$\tau_{rep} \sim \phi^{2.1} \quad \text{for an athermal solvent;} \qquad (9.135)$$

$$\tau_{rep} \sim \phi^{2.9} \quad \text{for a } \theta\text{-solvent.} \qquad (9.136)$$

(ii) Derive the following results for the concentration dependence of the diffusion coefficient in athermal solvent and in θ-solvent:

$$D \sim \phi^{-2.4} \quad \text{for an athermal solvent;} \tag{9.137}$$
$$D \sim \phi^{-2.9} \quad \text{for a } \theta\text{-solvent.} \tag{9.138}$$

(iii) Derive the following results for the concentration dependence of viscosity in athermal solvent and in θ-solvent:

$$\eta \sim \phi^{4.5} \quad \text{for an athermal solvent;} \tag{9.139}$$
$$\eta \sim \phi^{5.2} \quad \text{for a } \theta\text{-solvent.} \tag{9.140}$$

9.15 Consider a solution of DNA with molar mass $M = 1.1 \times 10^8$ g mol^{-1} corresponding to $n = 1.64 \times 10^5$ base pairs. Assume that the ionic strength is high enough to ignore excluded volume interactions.

(i) What is the contour length $R_{\max}$ of this DNA if $l = 3.4$ Å per base pair?
(ii) Derive the following relation for the overlap concentration of a worm-like chain

$$c^* \approx M(R_{max}b)^{-3/2}/\mathcal{N}_{Av}$$

and determine the overlap concentration if the Kuhn length of DNA is $b = 100$ nm (approximately 300 base pairs).
(iii) Assuming that the entanglement concentration $c_e \approx 10c^*$ estimate the contour length of an entanglement strand at $c = 0.5$ mg mL^{-1}. What is the molar mass of an entanglement strand at this concentration?
(iv) Estimate the plateau modulus at the concentration 0.5 mg mL^{-1} at 30 °C.
(v) Estimate the relaxation time and the viscosity of the DNA solution at the concentration 0.5 mg mL^{-1} at 30 °C. Assume the solvent viscosity is $\eta_s = 10^{-3}$ Pa s.

9.16 Consider a solution of polystyrene with molar mass $M = 10^6$ g mol^{-1} in cyclohexane at 35 °C (θ-solvent with viscosity $\eta_s = 7.6 \times 10^{-4}$ Pa s). Estimate the relaxation time, plateau modulus, viscosity, and diffusion coefficient as functions of concentration in semidilute solution.

9.17 Demonstrate that the specific viscosity η_{sp} of entangled solutions in a good solvent can be expressed as a universal function (independent of molar mass and polymer species) of the ratio ϕ/ϕ^* and the number of Kuhn monomers in an entanglement strand in the melt.

9.18 Demonstrate that the ratio of specific viscosity η_{sp} and the 2/3 power of degree of polymerization N of entangled solutions in a θ-solvent can be expressed as a universal function of the ratio ϕ/ϕ^* and the number of Kuhn monomers in an entanglement strand in the melt.

9.19 How does the width of the plateau region of the stress relaxation modulus of entangled polymer solutions (the ratio of reptation time τ_{rep} to the Rouse time τ_e of a strand between entanglements) increase with concentration:

(i) for athermal solutions;
(ii) for θ-solutions.

9.20 Consider a linear chain of N Kuhn monomers of length b and monomer volume v_0 in a θ-solvent confined to an infinite cylindrical pore with diameter a and impenetrable walls. What average length $\langle L \rangle$ of the pore is occupied by the chain? Is this average length the same or different for the confining tube of diameter a in a semidilute entangled solution in a θ-solvent? Explain your answer.

9.21 Consider a linear chain of N monomers in a good solvent with excluded volume v.

(i) Find the condition on N for which the onset of entanglement concentration ϕ_e is in concentrated regime above ϕ^{**} if the number of monomers per entanglement strand in a melt is $N_e(1)$.

(ii) What is the concentration dependence of the diffusion coefficient for the case with $\phi_e > \phi^{**}$ in the three different concentration regimes: $\phi^* < \phi < \phi^{**}$; $\phi^{**} < \phi < \phi_e$; $\phi_e < \phi < 1$.

9.22 Consider a small non-adsorbing spherical particle of diameter d diffusing in an entangled polymer solution of linear chains with N Kuhn monomers of length b, with volume fraction ϕ in an athermal solvent with solvent viscosity η_s. Calculate the diffusion coefficient of the particle if its diameter is:

(i) smaller than the correlation length ξ of the solution $(d < \xi)$;
(ii) larger than ξ but smaller than the tube diameter a $(\xi < d < a)$;
(iii) larger than the diameter a of the confining tube $(d > a)$.

Section 9.4

9.23 The primitive path is defined as the shortest line with the same topology as the original chain. Provide a precise definition of what is meant by the 'same topology'.

9.24 The confining tube can be visualized by a long-exposure photograph of a wiggling chain (see Fig. 9.29). What is the shortest time for this photograph to clearly show the entire volume of the tube?

9.25 Consider a chain represented by an unrestricted random walk of N steps on a square lattice, as sketched in Fig. 9.37. Topological constraints are represented by obstacles placed in the middle of each cell. The chain is not allowed to cross any of these obstacles. The primitive path (thick line in Fig. 9.37) is then defined as the shortest walk on the lattice with the same end points and the same topology as the original walk with respect to obstacles.

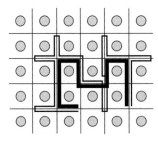

Fig. 9.37
Lattice model of a chain in an array of fixed topological obstacles. Thick line—primitive path.

(i) Show that the average number of segments in a primitive path $\langle K \rangle$ for a N-step random walk on a square lattice (Fig. 9.37) is

$$\langle K \rangle = \frac{N}{2}. \qquad (9.141)$$

(ii) Generalize this result to the N-step random walk on a cubic lattice (with coordination number $z = 6$) and show that the average number of segments in a primitive path $\langle K \rangle$ is

$$\langle K \rangle = \left(\frac{z-2}{z}\right) N = \frac{2}{3} N. \qquad (9.142)$$

Hint: Note that the primitive path is a non-reversible random walk (with no direct back-folding).

9.26 Imagine that one traditional student and another adventurous student are sent to randomly wander in two different cities (Fig. 9.38). The traditional student is sent to the city, called Squaros with a simple square grid of streets. The adventurous student is sent to a city, called Betheus on a high dimensional planet Cayleus in a recently discovered Universe. The streets in the city Betheus form a Bethe lattice with coordination number $z = 4$. The reason for this layout of streets is to avoid arguments about the shortest way of getting between any two locations in the city (there is only one path).

Fig. 9.38
Mapping of a primitive path onto a Cayley tree.

(i) Demonstrate that any sequence of loops and primitive path steps for a random walk in Squaros can be mapped onto a simple random walk in the streets of Betheus.

(ii) Show that loops formed by these two walks are identical, while the length of the primitive path of the walk in Squaros corresponds to the shortest path between the beginning and the end of the walk in Betheus.

9.27 Estimate the probability distribution for an N-step walk of the adventurous student on the streets of Betheus (each step corresponds to a block) with end-to-end distance of K blocks (see Fig. 9.38 and Problem 9.26).

(i) Show that for an arbitrary coordination number z this probability distribution function is

$$p(N, K) \cong \frac{1}{2} \left(\frac{1}{z}\right)^{(N-K)/2} \left(\frac{z-1}{z}\right)^{(N+K)/2} \frac{N!}{((N-K)/2)!((N+K)/2)!}.$$

(9.143)

(ii) Derive an approximate expression for the probability distribution function

$$p(N, K) = \sqrt{\frac{\gamma}{2\pi\langle K \rangle}} \exp\left[-\frac{\gamma}{2\langle K \rangle}(K - \langle K \rangle)^2\right],$$

(9.144)

where the average end-to-end distance (see Problem 9.25)

$$\langle K \rangle = \frac{z-2}{z} N$$

and the coefficient

$$\gamma = \frac{z(z-2)}{4(z-1)}.$$

(9.145)

Hint: Use Stirling's approximation [Eq. (2.75)] and see Section 2.5.

Note that Eq. (9.144) is a probability distribution for a primitive path of an N-step walk to consist of K steps and is an approximate expression for tube length fluctuations.

9.28 Consider an alternative mechanism for relaxation of entangled star polymers, studied by Klein. This relaxation process is analogous to reptation of linear chains with the branch point of a star moving up the tube of one of its f arms. All remaining arms follow the branch point along the primitive path of the chosen arm (Fig. 9.39). The tube length for each arm stays close to the equilibrium length.

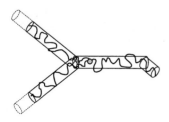

Fig. 9.39
Junction displacement mechanism of star dynamics.

(i) Show that the free energy cost of the branch point displacement by s/a primitive path steps is

$$F(s) \approx kT(f-2)\frac{s}{a}.$$

(9.146)

(ii) Compare this free energy cost with tube length fluctuations and estimate its importance for star polymers with different numbers of arms f.

9.29* Stress relaxation for polydisperse linear chains
Assume that the distribution of relaxation rates $P(\varepsilon)$ in Eq. (8.184) has a sharp cutoff at some characteristic lowest rate ε^* in the form of a stretched exponential with some exponent x. $A(\varepsilon)$ is a weaker pre-exponential function (such as a power law):

$$P(\varepsilon) = A(\varepsilon) \exp\left[-(\varepsilon^*/\varepsilon)^x\right].$$

(i) Show that the integral in Eq. (8.184) is dominated by the maximum of the exponential with

$$\varepsilon \approx \left[x \frac{(\varepsilon^*)^x}{t} \right]^{1/(x+1)} \approx \varepsilon^*(\varepsilon^* t)^{-1/(x+1)}$$

with the value of the exponent at this maximum

$$\varepsilon t \approx (\varepsilon^* t)^{x/(x+1)}.$$

Thus, the stress relaxation function can be approximated by a stretched exponential form

$$G(t) = B(t) \exp\left[-(\varepsilon_n t)^{x/(x+1)} \right], \qquad (9.147)$$

where $B(t)$ is a slowly varying function and $\varepsilon_n \approx \varepsilon^*$.

(ii) Consider the distribution of chains obtained in a linear condensation polymerization [Eq. (1.66)].

$$n_N \cong \frac{1}{N_n} \exp(-N/N_n).$$

In an unentangled melt, the number density of modes relaxing with rate ε corresponds to the number of chain sections containing K Kuhn segments such that

$$\frac{1}{\varepsilon} = \tau_0 K^2$$

Show that the number density of these sections of K monomers is

$$\left(\frac{N_n}{K} + 1 \right) \exp(-K/N_n)$$

and therefore the spectrum of relaxation rates for an unentangled melt can be represented by

$$P(\varepsilon)|\mathrm{d}\varepsilon| = \left(\sqrt{\varepsilon/\varepsilon_n} + 1 \right) \exp\left(-\sqrt{\varepsilon_n/\varepsilon} \right) \frac{|\mathrm{d}\varepsilon|}{2\tau_0^{1/2} \varepsilon^{3/2}},$$

where the relaxation rate of the N_n-mer is

$$\varepsilon_n = \frac{1}{\tau_0 N_n^2}.$$

(iii) Note that this distribution is of the stretched exponential form $P(\varepsilon) = A(\varepsilon) \exp(-(\varepsilon^*/\varepsilon)^x)$ with exponent $x = 1/2$. Show that the exponent in the stretched exponential form of $G(t)$ is $1/3$.

9.30* Consider a regular g-generation dendrimer entangled in an array of fixed obstacles. The functionality of each junction point is f and the number of

Kuhn monomers in each linear section N_a is sufficiently large that these sections overlap with many obstacles and entangle extensively. Describe the hierarchy of the relaxation processes of this entangled dendrimer. Estimate the relaxation time and the diffusion coefficient. Assume a harmonic arm retraction potential [Eq. (9.54)].

9.31*   Entangled rings

Consider the dynamics of an entangled ring polymer in an array of fixed obstacles [Fig. 9.40(a)]. The ring is not permanently trapped by the obstacles, but is able to diffuse. The ring does not have free ends and, therefore, 'classical' snake-like reptation is not expected for it. An ideal untrapped ring polymer in an array of fixed topological obstacles is an unentangled loop formed by double-folded strands of N_e monomers each, similar to an arm of a star at the moment of complete retraction.

(i) Show that the size of an ideal entangled untrapped ring polymer is

$$R_{\text{ring}} \approx b\sqrt[4]{NN_e}. \qquad (9.148)$$

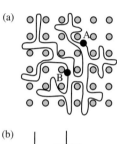

(a)

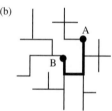

(b)

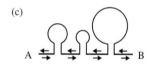

(c)

Fig. 9.40

(a) Ring polymer in an array of fixed topological obstacles. (b) Equivalent branched polymer with the 'trunk' of the branched tree between points A and B marked in bold. (c) Branches of the tree of the section AB, represented by circles of different size, act as reservoirs of kinks. Kinks diffuse between these reservoirs along the trunk, as indicated by arrows.

Even though the conformation of entangled ring polymers is similar to that of branched polymers, they do not have fixed branch points and therefore rings do not need to invoke the exponentially slow arm retraction process to change their conformations. The reason the arm retraction of branched polymers is an exponentially slow process is that they have to reduce their entropy to form exponentially unlikely unentangled loops. This high price was already paid when the ring polymer was squeezed into the array of fixed topological obstacles and formed conformations with double-folded strands. Motion of the ring polymer from one such conformation to another one does not carry any entropic penalty.

The dynamics of rings is qualitatively different from that of linear chains and branched polymers. Consider any section of a ring, such as the lower part between points A and B [Fig. 9.40(a)]: this section can be treated as a linear polymer consisting of a primitive path and several loops. The corresponding part AB of the branched polymer representation of the ring consists of a trunk [thick line in Fig. 9.40(b)], corresponding to the primitive path of the section, and side branches, corresponding to unentangled loops.

(ii) Demonstrate that the number of monomers n_l along the linear section (trunk) of a ring is proportional to the square root of the total number of monomers n in the section:

$$n_l \approx \sqrt{nN_e}.$$

Thus, most of the monomers are in the side branches, corresponding to the unentangled loops.

Consider the diffusion of kinks (small unentangled loops containing N_e monomers) along the portion AB of the molecule. Diffusion of kinks along large unentangled loops [branches of the equivalent tree in Fig. 9.41(b)] changes the conformation of these loops, but does not contribute significantly to the overall transport of the AB section. Mass transport, corresponding to this diffusion stays within larger unentangled loops, represented by circles in Fig. 9.40(c). These larger unentangled loops act as a reservoir of kinks that move along the primitive path (trunk) of the section AB [shown by arrows in Fig. 9.40(c)].

(iii) Prove that the mean-square displacement of the centre of mass of the section AB along the primitive path during the relaxation time τ_e of an entangled strand is

$$\Delta s_{\text{cm}}^2 \approx a^2 \left(\frac{N_e}{n}\right)^{3/2}$$

and the curvilinear diffusion coefficient of section AB along its primitive path is

$$D_c(n) \approx \frac{D_0}{N_e}\left(\frac{N_e}{n}\right)^{3/2},$$

where the monomeric diffusion coefficient is $D_0 \approx b^2/\tau_0$.

(iv) Show that the relaxation time of section AB is

$$\tau(n) \approx \tau_e\left(\frac{n}{N_e}\right)^{5/2}.$$

(v) Prove that the relaxation time of the whole ring with N monomers is

$$\tau_{ring} = \tau_e\left(\frac{N}{N_e}\right)^{5/2}. \tag{9.149}$$

(vi) Demonstrate that the diffusion coefficient of an ideal entangled untrapped ring is

$$D \approx D_0\frac{N_e}{N^2}. \tag{9.150}$$

9.32 Show that a stress relaxation modulus of an entangled but non-concatenated melt of rings on the basis of the single chain dynamic modes described in Problem 9.31 is

$$G(t) \approx G_e\left(\frac{t}{\tau_e}\right)^{-2/5} \exp(-t/\tau_{ring}), \tag{9.151}$$

where the relaxation time of a ring is given by Eq. (9.149).

9.33 What is the mean-square curvilinear displacement of monomers in an entangled melt along the confining tube on time scales between the relaxation time of an entanglement strand τ_e and the reptation time τ_{rep}? Sketch this time dependence of $\langle[s(t) - s(0)]\rangle^2$.

9.34 Estimate the time dependence of the mean-square displacement of the *centre of mass* of an entangled linear chain in a melt on time scales: (i) $t < \tau_e$; (ii) $\tau_e < t < \tau_R$; (iii) $\tau_R < t < \tau_{rep}$; (iv) $t > \tau_{rep}$.

9.35 Estimate the time dependence of the mean-square displacement of a monomer in an entangled polymer solution.

(i) On time scales $t < \tau_\xi$, where τ_ξ is the relaxation time of a chain section with size equal to the correlation length.

(ii) On time scales $\tau_\xi < t < \tau_e$, where τ_e is the relaxation time of a strand between entanglements.

(iii) On time scales $\tau_e < t < \tau_R$, where τ_R is the Rouse time of the polymer.

(iv) On time scales $\tau_R < t < \tau_{rep}$, where τ_{rep} is the reptation time of the polymer.

(v) On time scales $t > \tau_{rep}$.

9.36 Demonstrate that the loss modulus is predicted to have the following high-frequency behaviour for various models:

(i) $G''(\omega) \sim \omega^{-1}$ for the Maxwell model (if $\omega\tau \gg 1$);

(ii) $G''(\omega) \sim \omega^{-1/4}$ for the Doi fluctuation model (if $1/\tau_R \ll \omega \ll 1/\tau_e$);

(iii) $G''(\omega) \sim \omega^{-1/2}$ for the reptation model (if $\omega\tau_{rep} \gg 1$).

9.37* Calculate the storage and loss moduli corresponding to the Doi fluctuation model with stress relaxation modulus

$$G(t) = G_N^0 \int_0^{2\mu\sqrt{N_e/N}} d\xi \, \exp\left[-\frac{16\mu^2 N_e}{\xi^4 \tau_{rep} N} t\right]$$

$$+ G_N^0 \int_{2\mu\sqrt{N_e/N}}^1 d\xi \, \exp\left\{-\frac{t}{[\tau_{rep}(\xi - \mu\sqrt{N_e/N})^2]}\right\}. \tag{9.152}$$

9.38*

(i) Use the approximate expression for the time $\tau(s)$ for retraction of an arm of the entangled star polymer down to length s

$$\tau(s) \approx \tau_{arm} \, \exp\left[\frac{\gamma'}{2}\left(\frac{s^2}{N_a b^2} - \frac{N_a}{N_e}\right)\right] \tag{9.153}$$

to obtain a simple estimate of the stress relaxation modulus of a star polymer:

$$G(t) \approx G_N^0\left(1 - \sqrt{1 - \frac{N_e}{N_a}\frac{2}{\gamma'}\ln\frac{\tau_{arm}}{t}}\right) \quad \text{for } t < \tau_{arm} \tag{9.154}$$

(ii) A simple way to obtain the storage modulus of star polymers is to replace t in the stress relaxation modulus by $1/\omega$. Show that the storage modulus is

$$G'(\omega) \approx G_N^0\left(1 - \sqrt{1 - \frac{N_e}{N_a}\frac{2}{\gamma'}\ln \omega\tau_{arm}}\right).$$

9.39 Consider an entangled monodisperse melt of H-polymers with N_{bb} Kuhn monomers in the central backbone and N_a monomers in each of the four arms, with N_e monomers between entanglements.

(i) Estimate the terminal relaxation time of the H-polymer considering exclusively single-chain modes. Express this terminal time in terms of the single arm retraction time τ_{arm}.
(ii) What is the diffusion coefficient of the H-polymer? Ignore multi-chain contributions to dynamics.
(iii) What is the single-chain expression of the stress relaxation modulus of the H-polymer melt in terms of stress relaxation modulus $G_s(t)$ of individual arms and the reptation contribution of the central backbone?

9.40 Consider a molecule made out of two f-arm stars with N_a Kuhn segments per arm with junction points connected by a central linear strand of N_{bb} Kuhn monomers. This molecule is called a pom-pom polymer. If $f = 1$, this molecule is linear, while the H-polymer corresponds to $f = 2$. Estimate the f-dependence of relaxation time and diffusion coefficient of a melt of monodisperse pom-pom polymers for $f > 1$. Consider only single-chain modes and assume that the coordination number of an entanglement network is z.

9.41 Consider an asymmetric star with one short and two long arms in an array of fixed topological obstacles. The short arm contains N_S Kuhn monomers. The long arms contain N_L Kuhn monomers and the number of Kuhn monomers between entanglements is N_e.

(i) Estimate the relaxation time τ_S of the short arm.
(ii) How many entanglements of the long arms do not have time to disentangle during the relaxation time of the short arm τ_S.

(iii) Estimate the relaxation time of the asymmetric entangled star assuming that the junction point moves along the contour of the tubes of long arms with curvilinear diffusion constant $D_c \approx a^2/\tau_S$.

(iv) What is the diffusion coefficient of this asymmetric star in an array of fixed obstacles?

9.42 Ignoring many-chain effects, relate the stress relaxation modulus $G(t)$ of a comb polymer consisting of q branches with N_a monomers and backbone with N_{bb} monomers to the stress relaxation modulus of a star with N_a monomers per arm.

9.43 Curro–Pincus relaxation in networks

The stress relaxation of polymer networks on long time scales is believed to be due to arm retraction of dangling ends (Fig. 7.7). The polydispersity of dangling ends is determined during crosslinking. Assume the number fraction distribution of linear dangling ends [Eq. (1.52)]

$$n_N = p^N(1-p)$$

with the number-average number of monomers in a dangling end of the order of that of a network strand [Eq. (1.51)] $N_n = 1/(1-p)$. Assume that the contribution of dangling ends to the stress relaxation modulus is proportional to the unretracted part of these chains (similar to star polymers)

$$G(t) - G_\infty \sim \sum_{N=K(t)}^{\infty} \left(\frac{N}{N_e} - \frac{s(t)}{a} \right) n_N.$$

In the above equation, $a \approx bN_e^{1/2}$ is the tube diameter with N_e monomers between entanglements, $K(t)$ is the number of monomers in the dangling chain with retraction time equal to t

$$K(t) = \frac{2}{\gamma'} N_e \ln(t/\tau_1),$$

where we ignore weak (logarithmic) dependence of τ_1 on K [Eq. (9.58)]. All shorter dangling ends have already relaxed by time t. The sum in the equation above is only over longer dangling ends with $s(t)$ the length of the tube that has already been vacated by time t.

(i) Assume a simple logarithmic time dependence of the relaxed part of dangling ends $s(t) \approx a \ln(t/\tau_1)$ and show that the contribution of dangling ends to the stress relaxation modulus of the network decays as a power law in time.

(ii) What is wrong with the assumed time dependence of the relaxed part of dangling ends $s(t)$? What should be the corrected time dependence of the relaxed part of dangling ends? *Hint*: Is it consistent with a parabolic retraction potential [Eq. (9.56)]?

9.44 Entanglement during gelation

Consider vulcanization of a melt of N-mers. At some extent of reaction $\varepsilon = (p - p_c)/p_c$ above the gel point and beyond the Ginzburg point (in the mean-field regime) the network strands become entangled with each other.

(i) Show that the size of the network strand's linear backbone is

$$R_{bb} \approx bN^{1/2}\varepsilon^{-1/2}.$$

Hint: Remember that every network strand is branched, with structure similar to the characteristic polymer at each extent of reaction.

(ii) Elastically effective linear backbones of these network strands are random walks. Show that these linear backbones contain $N_{bb} \approx N/\varepsilon$ monomers.

(iii) In the mean-field region of the gelation transition, the gel fraction is $P_{gel} \approx \varepsilon$ and the number of monomers in a characteristic strand is $N^* \approx N/\varepsilon^2$ including both backbone and side branches. Show that the volume fraction of the backbone is $\phi_{bb} \approx \varepsilon^2$.

(iv) The confining tube diameter in the melt of N-mers before the vulcanization reaction was $a(1)$. Demonstrate that the tube diameter, due exclusively to entanglements between the backbones of the gel, is

$$a \approx a(1)\varepsilon^{-4/3} \qquad (9.155)$$

assuming all elastically ineffective side branches act as θ-solvent.

(v) Show that the number of monomers in the linear part of an entanglement strand is

$$N_e \approx \left(\frac{a(1)}{b}\right)^2 \varepsilon^{-8/3}. \qquad (9.156)$$

(vi) Demonstrate that the modulus of the network is

$$G \approx \frac{kT}{b[a(1)]^2}\varepsilon^{14/3}. \qquad (9.157)$$

Section 9.5

9.45 Tube dilation for a binary blend

Consider a melt consisting of a mixture of entangled long and short chains with degrees of polymerization $N_L \gg N_S > N_e$. Long chains are assumed to be entangled with each other and with short chains. The distance between entanglements of long chains a_L (if short chains are replaced by a θ-solvent) is larger than the average distance between all entanglements in the melt (tube length a_0). Ignore tube length fluctuations and consider only reptation of chains in their tubes with a single mode.

(i) Show the constraint release process for long chains due to motion of short chains takes place on time scale

$$\tau_S \approx \tau_{e0}\left(\frac{N_S}{N_{e0}}\right)^3$$

and the relaxation time of a long-chain strand between neighbouring entanglements with other long chains is

$$\tau_{eL} \approx \tau_S\left(\frac{N_{eL}}{N_{e0}}\right)^2 \approx \tau_S\left(\frac{a_L}{a_0}\right)^4.$$

(ii) Demonstrate that the average number of long-chain entanglements with other long chains is

$$\frac{N_L}{N_{eL}} \approx \frac{N_L}{N_{e0}}\left(\frac{a_0}{a_L}\right)^2$$

and the reptation time of a long chain in a dilated tube of diameter a_L with elementary friction due to constraint release is

$$\tau_{\text{dilated}} \approx \tau_{e0} \left(\frac{N_S}{N_{e0}} \right)^3 \left(\frac{N_L}{N_{e0}} \right)^3 \left(\frac{a}{a_L} \right)^2.$$

(iii) Prove that this reptation of long chains in dilated tubes is faster than their reptation τ_L in thin (original) tubes for relatively short surrounding chains

$$\frac{N_S}{N_{e0}} < \left(\frac{a_L}{a_0} \right)^{2/3}.$$

(iv) Recall the dependence of the number of monomers in an entanglement strand N_e on polymer concentration ϕ [Eq. (9.36)]:

$$N_e \sim \phi^{-\alpha},$$

where $\alpha = 4/3$ in θ-solvents. Polymer concentration ϕ is replaced in the tube dilation models by the fraction of unrelaxed chains. Show that the ratio of dilated and bare tube diameters is

$$\frac{a_L}{a_0} \approx \phi_L^{-\alpha/2}$$

and reptation in the dilated tubes is effective for small numbers of entanglements in the short chains:

$$\frac{N_S}{N_{e0}} < \phi_L^{-\alpha/3}.$$

(v) Demonstrate that for this inequality to be valid and for the volume fraction of long chains to be above the entanglement onset of long chains, it is necessary that the number of entanglements in the long chains is larger than the cube of the number of entanglements in the short chains:

$$\frac{N_L}{N_{e0}} > \left(\frac{N_S}{N_{e0}} \right)^3. \tag{9.158}$$

(vi) The dilated tube is effective as long as the lifetime of the corresponding constraint (the reptation time of the long chains) is longer than the Rouse time of constraint release of the corresponding entanglement strands $\tau_{eL} < \tau_L$. Prove that tube dilation can be observed for intermediate volume fraction ϕ_L of long chains in the melt:

$$\left(\frac{N_L}{N_{e0}} \right)^{-1/\alpha} < \phi_L < \left(\frac{N_S}{N_{e0}} \right)^{-3/\alpha}.$$

(vii) Estimate the composition range for the tube dilation approximation for a binary blend of long chains containing $N_L/N_e = 300$ entanglements and short chains containing $N_s/N_e = 5$ entanglements. Assume that the number of monomers in an entanglement strand N_e decreases with volume fraction ϕ as $N_e \sim \phi^{-4/3}$.

9.46 Tube dilation for a power law polydispersity

For a chain to be confined to a dilated tube, the mean-square displacement $\langle [r_\perp(t) - r_\perp(0)]^2 \rangle$ of its monomers in the direction perpendicular to the contour of the tube has to be restricted by the mean-square tube diameter $[a(t)]^2$.

Consider a melt with volume fraction of N-mers described by a power law distribution

$$\phi(N) \approx N^{-\beta} \quad \text{for } N > N_{e0}$$

with $\beta > 2$. Assume single-mode reptation of chains in their tubes.

(i) Show that polymers relaxing at time t have degree of polymerization

$$N(t) \approx N_e \left(\frac{t}{\tau_e}\right)^{1/3}.$$

Assume that all shorter chains with $N < N(t)$ have already relaxed, while all longer chains with $N > N(t)$ have not relaxed yet.

(ii) Show that the volume fraction of unrelaxed chains decreases with time as

$$\mu(t) \approx \left(\frac{t}{\tau_e}\right)^{(1-\beta)/3}.$$

(iii) Demonstrate that the mean-square diameter of dilated tubes grows with time as

$$\left(\frac{a(t)}{a_0}\right)^2 \approx \frac{N_e(t)}{N_e(0)} \approx \left(\frac{t}{\tau_e}\right)^{\alpha(\beta-1)/3}$$

assuming that all relaxed chains no longer constrain the unrelaxed chains.

(iv) Explain why the Rouse friction of the constraint release process at an entanglement with an N-mer with lifetime $\tau(N)$ is

$$\zeta(N) \approx \frac{kT}{a^2} \tau(N).$$

Prove that the average Rouse friction of the constraint release process at time t due to short N-mers with $N < N(t)$ is

$$\langle \zeta(t) \rangle \approx \begin{cases} kT/a^2 [N(t)]^{4-\beta} & \text{for } \beta < 4, \\ \zeta_0 & \text{for } \beta > 4. \end{cases}$$

Hint: Assume that the probability of entanglement with an N-mer is proportional to its volume fraction $\phi(N)$.

(v) Demonstrate that the time dependence of the average friction coefficient is

$$\langle \zeta(t) \rangle \sim \begin{cases} t^{(4-\beta)/3} & \text{for } \beta < 4, \\ t^0 & \text{for } \beta > 4. \end{cases}$$

(vi) Show that the subdiffusive mean-square monomer displacement due to constraint release increases with time as

$$\langle (r(t) - r(0))^2 \rangle \sim \begin{cases} t^{(\beta-1)/6} & \text{for } \beta < 4 \\ t^{1/2} & \text{for } \beta > 4 \end{cases}$$

and therefore the diluting tube is not effective in constraining the chain at all times and for all values of distribution exponent $\beta > 2$.

9.47 Stress relaxation modulus in tube dilation models

There is a conceptual difference between tube dilation and constraint release. The motion of the tube in *constraint release* does not affect single-chain modes inside the tube (reptation and tube length fluctuation). The tube diameter for these single chains in a tube processes is, on average, constant. In contrast, the tube diameter increases with time in the *tube dilation* process and dramatically affects the chain motion within the tube.

(i) Explain why the stress relaxation modulus in tube dilation models can be written as the product of the time-dependent modulus $G_e(t)$ times the single chain in a dilated tube relaxation function $\mu(t)$:

$$G(t) = G_e(t)\mu(t). \tag{9.159}$$

(ii) Demonstrate that the time-dependent modulus can be expressed in terms of the time-dependent number of monomers in an entanglement strand $N_e(t)$ of the dilated tube:

$$G_e(t) \approx \frac{kT}{b^3 N_e(t)}.$$

(iii) Use the dependence of the number of monomers in an entanglement strand $N_e(t)$ on the volume fraction of unrelaxed chains $\mu(t)$ to derive the stress relaxation modulus

$$G(t) \approx G_e(0)[\mu(t)]^{\alpha+1}$$

(iv) The double reptation model assumes that the stress relaxation modulus is

$$G(t) \approx G_e(0)[\mu(t)]^2.$$

What value of α corresponds to the double reptation model? In Section 9.3, we have presented theoretical arguments and experimental data supporting the value $\alpha = 4/3$ in θ-solvents (and melts with ideal chain statistics). What is the expression of the stress relaxation modulus of tube dilation models corresponding to $\alpha = 4/3$?

9.48 Constraint release vs. tube dilation

Consider an isolated long probe P-mer entangled in a melt of shorter N-mers. Tube dilation assumes that as soon as short chains relax, stress in the long P-mer drops to zero. In particular, a version of tube dilation called double reptation imposes an exact symmetry between single chains in a tube and multi-chain processes. As one chain reptates away, stress at a common entanglement (stress point) is relaxed *completely*. In constraint release models, this stress relaxes only partially due to connectivity of the P-mer.

(i) What is the relaxation time of a P-mer in the constraint release models if the reptation time of N-mers is $\tau_{rep}(N)$ and the number of monomers in an entanglement strand in the melt is N_e?

(ii) What is the relaxation time of an isolated long P-mer in the tube dilation model?

9.49 Consider a comb polymer consisting of a backbone with N_{bb} monomers and q branches with N_a monomers and arm retraction time τ_{arm}. Assume that backbones reptate along a dilated tube with entanglements only due to other backbones. Use θ-solvent scaling of the tube diameter with concentration of entangled polymers.

(i) Estimate the stress relaxation time of a comb.

(ii) What is the diffusion coefficient of a comb?

(iii) Relate the stress relaxation modulus $G(t)$ of a comb to the stress relaxation modulus $G_s(t)$ of a star with N_a monomers per arm.

9.50* Enormously long linear chains may have a many-chain effect that retards their reptation. The plateau modulus of an entanglement network is given by Eq. (9.18). In equilibrium there are small fluctuations of stress $\Delta\sigma(r)$

and strain $\Delta\varepsilon(r)$ in entanglement networks with elastic energy of order kT stored in these fluctuations on all length scales r.

(i) If the energy per unit volume due to strain $\Delta\epsilon(r)$ is $G_e[\Delta\epsilon(r)]^2$ show that a typical thermal fluctuation of stress is

$$\Delta\sigma(r) \approx \frac{kT}{\sqrt{v_0 N_e r^3}}.$$

(ii) A polymer segment with n monomers and size $r \approx bn^{1/2}$ entering this pre-stressed network will induce additional strain with energy

$$\Delta F(n) \approx v_0 n \Delta\sigma(r).$$

As long as this energy is less than the thermal energy kT, the segment will average over the whole volume r^3. Show that the largest unperturbed segment has degree of polymerization

$$n^* \approx \frac{b^6}{v_0^2} N_e^2.$$

(iii) Demonstrate that for smaller unperturbed segments the strain energy is

$$\Delta F(n) \approx kT \sqrt{\frac{v_0}{b^3}} \left(\frac{n}{N_e^2}\right)^{1/4} \quad \text{for } n < n^* \approx \frac{b^6}{v_0^2} N_e^2.$$

(iv) Longer segments $n > n^* \approx (b^6/v_0^2) N_e^2$ cause strain energy larger than kT and tend to localize in regions of lower stress with elastic energy proportional to the number of these localization blobs. Show that the strain energy of larger segments is

$$\Delta F(n) \approx kT \frac{n}{n^*} \approx kT \frac{v_0^2}{b^6} \frac{n}{N_e^2} \quad \text{for } n > n^* \approx \frac{b^6}{v_0^2} N_e^2.$$

(v) For a melt of very long chains with degree of polymerization N, the distance between chain ends is large ($r \approx bN^{1/3}$). The section of polymer that needs to penetrate the region devoid of chain ends of size r has a large number of monomers $n \approx N^{2/3}$. Using this result in the strain energy of part (iv) and assuming motion requires overcoming this potenial barrier, derive the following relation for the relaxation time of extremely long linear chains:

$$\tau \sim \exp\left(\text{const.} \frac{v_0^2}{b^6} \frac{N^{2/3}}{N_e^2}\right) \quad \text{for } N/N_e > \frac{b^9}{v_0^3} N_e^2. \tag{9.160}$$

Section 9.6

9.51 In a molecular dynamics simulation, the average value of any quantity A is estimated by averaging over a large number X of molecular dynamics time steps of length δt after a long initial equilibration of the system during Y molecular dynamics time steps:

$$\langle A \rangle = \frac{1}{X} \sum_{k=1}^{X} A(Y+k)\delta t.$$

Is the relative error of this estimate $1/\sqrt{X}$?

9.52 Which of the two computer simulation methods is more efficient in accurately simulating (i) static and (ii) dynamic properties of a single polymer chain in dilute solution: molecular dynamics with explicit solvent or Brownian dynamics without explicit solvent?

9.53 Determine the fraction of attempted moves of an entangled Evans–Edwards chain that actually move a monomer to a new lattice site for the:

(i) square lattice;
(ii) simple cubic lattice;
(iii) repeat both calculations for an unentangled Verdier–Stockmayer chain.

Hint: Write the final answers as functions of N.

9.54 Determine the assignment of reptons by starting at the opposite end of the chain in Fig. 9.34. Explain the differences between these assignments and the original assignments. Will they affect the diffusion coefficient?

9.55 Show that the minimum of the Lennard-Jones potential [Eq. (9.94)] is at $r = 2^{1/6}\sigma$. Find the value of the potential at the minimum.

9.56 (i) Consider particles of mass m interacting via a Lennard-Jones potential with energy parameter ε and distance parameter σ. Construct a time scale out of these parameters, called the Lennard-Jones time τ_{LJ}. What is the physical significance of this time scale?

(ii) Consider an example of a bead representing a Kuhn segment of polystyrene with $m = M_0 = 740\,\mathrm{g\,mol}^{-1}$, $\varepsilon = kT/2$, $\sigma = 18\,\text{Å}$, at temperature $T = 300\,\mathrm{K}$. Calculate the corresponding Lennard-Jones time τ_{LJ} and the molecular dynamics time step $\delta t = 0.01\tau_{\mathrm{LJ}}$, thereby taking it to be much smaller than the Lennard-Jones time.

(iii) Estimate the Einstein frequency of a Lennard-Jones crystal by approximating the Lennard-Jones potential near its minimum by a harmonic potential.

9.57 Consider a molecular dynamics simulation of a polymer system consisting of 20 chains, each with 100 Lennard-Jones monomers. Let us assume that the molecular dynamics step $\delta t = 0.01\tau_{\mathrm{LJ}}$ takes 10 ms of CPU time, where τ_{LJ} is the Lennard-Jones time (corresponding to the monomeric relaxation time):

(i) How long will it take to relax the system if we assume it is unentangled (obeying Rouse dynamics)?

(ii) How long would it take to obtain a 10% accuracy on an average quantity, such as the longest relaxation time.

9.58 Show that the Metropolis algorithm satisfies detailed balance.

9.59 Estimate the N-dependence of the success rate of a simple Monte Carlo simulation of a self-avoiding walk with $N = 100$ steps. Assume that random walks are generated on a cubic lattice. Each step of the walk is not allowed to step back (they can only go forward, up, down, turn left, or right with equal probability of 1/5). Walks that intersect themselves are discarded.

(i) Make a mean-field estimate of the N-dependence of the success rate.

(ii) Use the number of N-step self-avoiding walks

$$W_{\mathrm{sa}}(N) \approx \bar{z}^N N^{\gamma-1},$$

where $\gamma = 7/6$ and $\bar{z} = 4.68$ for a three-dimensional simple cubic lattice, to obtain a better estimate of the success rate. How does this success rate compare with the mean-field estimate?

9.60 Does the probability of successfully overlapping two independent three-dimensional self-avoiding walks increase or decrease with the number of steps N in these walks? Successful overlap of self-avoiding walks avoids direct overlap of any monomers. Answer this question for two different methods of overlapping walks:

(i) Dimerization—the end of one walk is adjacent (or coincides) with the end of the other walk.

(ii) Shared pervaded volume. The centres of mass of the two walks coincide.

Bibliography

Berry, G. C. and Fox, T. G. The viscosity of polymers and their concentrated solutions, *Adv. Polym. Sci.* **5**, 261 (1968).

Binder, K. *Monte Carlo Simulation in Statistical Physics*, 4th edition (Springer, New York, 2002).

Doi, M. *Introduction to Polymer Physics* (Clarendon Press, Oxford, 1996).

Doi, M. and Edwards, S. F. *The Theory of Polymer Dynamics* (Clarendon Press, Oxford, 1986).

Ferry, J. D. *Viscoelastic Properties of Polymers*, 3rd edition (Wiley, New York, 1980).

Graessley, W. W. The entanglement concept in polymer rheology, *Adv. Polym. Sci.* **16**, 1 (1974).

Graessley, W. W. Viscoelasticity and flow in polymer melts and concentrated solutions. In: *Physical Properties of Polymers* (American Chemical Society, Washington, 1984).

McLeish, T. C. B. and Milner, S. T. Entangled dynamics and melt flow of branched polymers, *Adv. Polym. Sci.* **143**, 195 (1999).

Pearson, D. S. Recent advances in the molecular aspects of polymer viscoelasticity, *Rubber Chem. Technol.* **60**, 439 (1987).

Notations

$\cong$	numerical approximation, e.g. $\pi \cong 3.14$
$\approx$	approximately equal, e.g. $R \approx bN^{0.588}$
$\sim$	proportional, e.g. $R \sim N^{1/2}$
$\langle \ldots \rangle$	ensemble average, p. 51
$!$	factorial $N! = 1 \times 2 \times 3 \times \ldots \times N$
A	entropic part of Flory interaction parameter χ, [dimensionless], p. 145
A_2	second virial coefficient, [m^3 kg^{-2} mol], p. 28
$A_{2,\,w}$	weight-average second virial coefficient, [m^3 kg^{-2} mol], p. 28
$A_{2,\,z}$	z-average second virial coefficient, [m^3 kg^{-2} mol], p. 33
A_{ij}	second virial coefficient between species i and j, [m^3 kg^{-2} mol], p. 28
a	Mark-Houwink exponent, [dimensionless], p. 34
a	tube diameter, [m], p. 265
a_n	degeneracy, [dimensionless], p. 206
a_T	time scale multiplicative shift factor, [dimensionless], p. 335
B	coefficient in enthalpic part of Flory interaction parameter χ, [K], p. 145
b	Kuhn length, [m], p. 54
b_T	modulus scale multiplicative shift factor, [dimensionless], p. 335
C	coefficient of molar mass dependence of glass transition, [kg K mol^{-1}], p. 340
C_1, C_2	Mooney-Rivlin coefficients, [kg m^{-1} s^{-2}], pp. 268–269
C_r, C_m	scaling factors, [dimensionless], p. 11
C_n, C_∞	Flory's characteristic ratio, [dimensionless], p. 53
c	polymer mass concentration, [kg m^{-3}], p. 13
$\mathbf{c}$	speed of light, [m s^{-1}], p. 30
c^*	overlap concentration, [kg m^{-3}], p. 13
c_N	mass concentration of N-mers, [kg m^{-3}], p. 16
c_n	monomer number density, [m^{-3}], p. 100
D	size of compression blob, [m], p. 108
D	diffusion coefficient, [m^2 s^{-1}], p. 309
$\mathcal{D}$	fractal dimension, [dimensionless], p. 10
D_c	curvilinear diffusion coefficient, [m^2 s^{-1}], p. 363
D_R	Rouse diffusion coefficient, [m^2 s^{-1}], p. 311
D_Z	Zimm diffusion coefficient, [m^2 s^{-1}], p. 313
d	space dimension, [dimensionless], p. 9
d	diameter of a cylindrical monomer, [m], p. 99
$\vec{E}$	electric field, [cm$^{-1/2}$ g$^{1/2}$ s^{-1}], p. 30

E	monomer-surface interaction energy, [kg m^2 s^{-2}], p. 112
E	Young's modulus, [kg m^{-1} s^{-2}], p. 296
E_a	activation energy for flow, [kg m^2 s^{-2}], p. 337
E_{cr}	critical adsorption energy per monomer, [kg m^2 s^{-2}], p. 112
E_i	incident electric field, [cm$^{-1/2}$ g$^{1/2}$ s^{-1}], p. 30
E_s	scattered electric field, [cm$^{-1/2}$ g$^{1/2}$ s^{-1}], p. 30
E_λ	Young's modulus due to wavelength λ, [kg m^{-1} s^{-2}], p. 334
ΔE	energy of vaporization, [kg m^2 s^{-2}], p. 143
ΔE	energy barrier between trans and gauche minima, [kg m^2 s^{-2}], p. 50
$\Delta E_A, \Delta E_B$	energy of vaporization for a molecule of species A or B, [kg m^2 s^{-2}], p. 144
e	elementary charge, [cm$^{3/2}$g$^{1/2}$s^{-1}]
F	Helmholtz free energy, [kg m^2 s^{-2}], p. 71
$\Delta \bar{F}_{mix}$	free energy of mixing per site, [kg m^2 s^{-2}], p. 140
ΔF_{mix}	free energy of mixing, [kg m^2 s^{-2}], p. 164
F_{conf}	confinement free energy, [kg m^2 s^{-2}], p. 108
F_{el}	elastic part of the free energy, [kg m^2 s^{-2}], p. 275
F_{ent}	entropic part of the free energy, [kg m^2 s^{-2}], p. 116
F_{int}	interaction part of the free energy, [kg m^2 s^{-2}], p. 100
$\vec{f}$	force, [kg m s^{-2}], p. 72
f	magnitude of force, [kg m s^{-2}], p. 72
f	functionality, [dimensionless], p. 206
f	free volume, [m^3], p. 337
$f_{j\alpha}$	α-component of the force acting on bead j, [kg m s^{-2}], p. 359
$f(r)$	Mayer f-function, [dimensionless], p. 99
$f_+(N/N^*)$	cutoff function above the gel point, [dimensionless], p. 227
$f_-(N/N^*)$	cutoff function below the gel point, [dimensionless], p. 227
f_E	energetic part of the force, [kg m s^{-2}], p. 254
f_g	free volume at the glass transition, [m^3], p. 338
f_S	entropic part of the force, [kg m s^{-2}], p. 255
G	shear modulus, [kg m^{-1} s^{-2}], p. 259, 282
$G(t)$	stress relaxation modulus, [kg m^{-1} s^{-2}], p. 284
$G'(\omega)$	storage modulus, [kg m^{-1} s^{-2}], p. 291
$G''(\omega)$	loss modulus, [kg m^{-1} s^{-2}], p. 291
$G^*(\omega)$	complex modulus, [kg m^{-1} s^{-2}], p. 292
G_0	Kuhn modulus, [kg m^{-1} s^{-2}], p. 364
G_e	plateau modulus, [kg m^{-1} s^{-2}], p. 266
G_{eq}	equilibrium shear modulus, [kg m^{-1} s^{-2}], p. 284
G_g	glassy modulus, [kg m^{-1} s^{-2}], p. 336
G_M	modulus in the Maxwell model, [kg m^{-1} s^{-2}], p. 283

G_x	crosslink contribution to the modulus, [kg m^{-1} s^{-2}], p. 263
g	number of monomers in a chain section, [dimensionless], p. 12
g	exponent of the end-to-end distribution function, [dimensionless], p. 122
$g(r)$	pair correlation function, [m^{-3}], p. 78
$g_{i \to j}$	probability of attempting a transition between states i and j, [dimensionless], p. 395
g_T	number of monomers in a thermal blob, [dimensionless], p. 113
H	diameter of a wire, [m], pp. 9, 10
H	height of the brush, [m], p. 187
$\bar{I}$	intensity scattered by molecules in a unit volume, [kg m^{-3} s^{-3}], p. 31
I_1, I_2, I_3	strain invariants, [dimensionless], p. 268
I_i	intensity of incident wave, [kg s^{-3}], p. 30
I_s	intensity of scattered wave, [kg s^{-3}], p. 30
$J(t)$	creep compliance, [kg^{-1} m s^2], p. 288
J_{eq}	steady state compliance, [kg^{-1} m s^2], p. 288
$J_R(t)$	recoverable compliance, [kg^{-1} m s^2], p. 290
K	optical constant, [m^2 kg^{-2} mol], p. 32
K	Mark-Houwink coefficient, [m^3 kg^{-1} (mol kg^{-1})^a], p. 34
K	number of chains in the scattering volume [dimensionless], p. 189
K	bulk modulus, [kg m^{-1} s^{-2}], p. 296
K	number of particles in a simulation box, [dimensionless], p. 392
k	Boltzmann constant, [kg m^2 s^{-2} K^{-1}], p. 27
k_p	spring constant of mode p, [kg s^{-2}], p. 359
k_H	Huggins coefficient, [dimensionless], p. 34
$\mathcal{L}$	Langevin function, [dimensionless], p. 76
L	contour length of the primitive path, [m], p. 361
L_a	contour length of a tube of an arm, [m], p. 377
L_x, L_y, L_z	dimensions of deformed network, [m], p. 256
L_{x0}, L_{y0}, L_{z0}	dimensions of undeformed network, [m], p. 256
l	length of a bond, [m], pp. 8, 12
l_0	scattering length of solvent, [m], p. 196
l_B	Bjerrum length, [m], p. 129
l_D	scattering length of deuterated monomers, [m], p. 196
l_H	scattering length of hydrogenated monomers, [m], p. 196
l_p	persistence length, [m], p. 57
M	molar mass, [kg mol^{-1}], p. 2
M_0	molar mass of a Kuhn monomer, [kg mol^{-1}], p. 54
M_e	molar mass of an entanglement strand, [kg mol^{-1}], p. 266

M_{mon}	molar mass of a chemical monomer, [kg mol^{-1}], p. 3
M_N	molar mass of N-mer, [kg mol^{-1}], p. 16
M_n	number-average molar mass, [kg mol^{-1}], p. 17
M_s	number-average molar mass of a network strand, [kg mol^{-1}], p. 259
M_w	weight-average molar mass, [kg mol^{-1}], p. 18
M_x	apparent molar mass of a network strand, [kg mol^{-1}], p. 263
M_z	z-average molar mass, [kg mol^{-1}], p. 18
M_{z+k}	$(z+k)$-average molar mass, [kg mol^{-1}], p. 18
m	mass of an object, [kg], pp. 9, 10
m	number of monomers in a chain section [dimensionless], p. 78
m_k	k-th moment of the number fraction distribution, [kg^k mol^{-k}], p. 17
N	degree of polymerization, [dimensionless], p. 2
N^*	characteristic degree of polymerization, [dimensionless], p. 210
N_a	number of Kuhn monomers in an arm of a star, [dimensionless], p. 377
N_A, N_B	number of lattice sites occupied by a molecule A and a molecule B, [dimensionless], p. 138
N_{bb}	number of monomers in a reptating backbone, [dimensionless], p. 381
N_e	number of monomers in an entanglement strand, [dimensionless], p. 266
N_n	number average degree of polymerization, [dimensionless], p. 17
$\mathcal{N}_{Av}$	Avogadro's number, [mol^{-1}], p. 3
N_{comb}	number of monomers in a combined chain, [dimensionless], p. 262
n	number of backbone bonds, [dimensionless], p. 3
n	refractive index of the medium, [dimensionless], p. 31
n	number of scatterers in the scattering volume, [dimensionless], p. 123
n	number of lattice sites in a mixture, [dimensionless], p. 138
n	number of strands in a network, [dimensionless], p. 257
n_0	refractive index of a solvent, [dimensionless], p. 31
$n(p, N)$	number of N-mers per monomer at extent of reaction p, [dimensionless], p. 21
n_A, n_B	number of molecules of species A and B in a mixture, [dimensionless], p. 139
n_N	number fraction of N-mers, [dimensionless], p. 16
$n_{tot}(p)$	number density of molecules, [dimensionless], p. 205
P	overlap parameter, [dimensionless], p. 14
$P(\vec{q})$	form factor, [dimensionless], p. 82

$P_{1d}(N, x)$	1-dimensional probability distribution function for the N-step walk, $[m^{-1}]$, pp. 69–70
$P_{3d}(N, x)$	3-dimensional probability distribution function for the N-step walk, $[m^{-3}]$, pp. 69–70
$P_{gel}(p)$	gel fraction, [dimensionless], p. 214
P_e	number of overlapping strands in an entanglement volume, [dimensionless], p. 362
P_i	probability of being in state i in a Monte Carlo simulation, [dimensionless], p. 395
$P_{sol}(p)$	sol fraction, [dimensionless], p. 214
p	extent of reaction, [dimensionless], pp. 20, 213
p	dipole moment, $[cm^{5/2}\ g^{1/2}\ s^{-1}]$, p. 30
p	bond probability, [dimensionless], p. 203
p_c	percolation threshold, [dimensionless], p. 203
$p_{i \to j}$	probability of accepting an attempted transition between states i and j, [dimensionless], p. 395
p_ξ	crossover mode index in semidilute solutions, [dimensionless], p. 328
Q	equilibrium swelling ratio, [dimensionless], p. 275
$Q(\vec{q})$	intermolecular contribution to scattering, [dimensionless], p. 195
q	charge, $[cm^{3/2}\ g^{1/2}\ s^{-1}]$, p. 74
q	magnitude of the scattering wavevector, $[m^{-1}]$, p. 81
$\vec{q}$	scattering wavevector, $[m^{-1}]$, p. 81
$\vec{q}_i$	incident wavevector, $[m^{-1}]$, p. 80
$\vec{q}_s$	scattered wavevector, $[m^{-1}]$, p. 80
R	chain size, [m], pp. 8, 11, 13
R	ball radius, [m], p. 9
$R_\parallel$	longitudinal size of a chain, [m], p. 108
R_0	root-mean-square end-to-end distance of an ideal chain, [m], p. 54
R_θ	Rayleigh ratio, $[m^{-1}]$, p. 31
R_g	radius of gyration, [m], p. 60
R_{gl}	size of a globule, [m], p. 114
R_h	hydrodynamic radius, [m], p. 311
$\vec{R}_i$	position vector of the i-th monomer, [m], p. 60
$\vec{R}_{ij}$	vector connecting monomers i and j, [m], p. 82
$R_{j\alpha}$	α-component of the position vector of monomer j, [m], p. 359
R_{max}	contour length, [m], p. 50
$\vec{R}_n$	end-to-end vector, [m], p. 51
$\vec{R}_{cm}$	center of mass position vector, [m], p. 60
R_{ref}	reference size of a network strand, [m], p. 275
$\mathcal{R}$	gas constant $\mathcal{R} = \mathcal{N}_{Av}k$, $[kg\ m^2\ s^{-2}\ K^{-1}\ mol^{-1}]$, p. 27
r	radius of a sphere enclosing an object, [m], pp. 9, 10
r	size of a chain section, [m], p. 12

r_{bond}	maximum bond extension in a FENE potential, [m], p. 394
r_{cut}	cut-off distance for the interaction potential, [m], p. 392
$\vec{r}_i$	bond vector, [m], p. 50
S	entropy, [kg m^2 s^{-2} K^{-1}], p. 70
$\Delta S_A, \Delta S_B$	entropy change on mixing of a molecule A and a molecule B, [kg m^2 s^{-2} K^{-1}], p. 138
$S_{DD}(\vec{q})$	labeled monomer pair contribution to scattering, [dimensionless], p. 196
$S_{HD}(\vec{q})$	labeled-unlabeled pair contribution to scattering, [dimensionless], p. 196
$S_{HH}(\vec{q})$	unlabeled monomer pair contribution to scattering, [dimensionless], p. 196
ΔS_{mix}	entropy of mixing, [kg m^2 s^{-2} K^{-1}], p. 139
$\Delta \bar{S}_{mix}$	entropy of mixing per site, [kg m^2 s^{-2} K^{-1}], p. 139
$S(\vec{q})$	scattering function, [dimensionless], p. 123
$S(\vec{q}, t)$	dynamic structure factor, [dimensionless], p. 348
s	curvilinear coordinate along the tube, [m], pp. 377, 382
s_p	number of main chain bonds in a persistence segment, [dimensionless], p. 56
T	absolute temperature, [K], p. 27
T_0	reference temperature, [K], p. 335
T_∞	Vogel temperature, [K], p. 338
T_b	temperature of a binodal, [K], p. 150
T_c	critical temperature, [K], p. 152
T_e	entanglement trapping factor, [dimensionless], p. 281
T_g	glass transition temperature, [K], p. 15
$T_{g\infty}$	glass transition temperature of high molar mass polymer, [K], p. 340
T_m	melting temperature, [K], p. 15
T_s	temperature of a spinodal, [K], p. 151
t	time, [s]
U	energy, [kg m^2 s^{-2}], p. 71
$U(r)$	effective interaction potential between a pair of monomers, [kg m^2 s^{-2}], p. 98
U_A	average interaction of an A monomer with one of its neighbors, [kg m^2 s^{-2}], p. 141
U_B	average interaction of a B monomer with one of its neighbors, [kg m^2 s^{-2}], p. 141
$\Delta \bar{U}_{mix}$	energy change on mixing per site, [kg m^2 s^{-2}], p. 142
U_λ	energy of a mode with wavelength λ, [kg m^2 s^{-2}], p. 332
$\vec{u}_i, \vec{u}_s$	unit vectors along incident and scattered directions, [dimensionless], p. 80
u_{AA}, u_{AB}, u_{BB}	pairwise interaction energies between adjacent lattice sites, [kg m^2 s^{-2}], p. 141
V	ball volume, [m^3], p. 9

V	pervaded volume, [m^3], p. 13
V	scattering volume, [m^3], p. 29
V_{dry}	network volume in a dry state, [m^3], p. 275
V_{eq}	volume of the equilibrium swollen state, [m^3], p. 275
v	excluded volume, [m^3], pp. 99, 156–157
v	velocity [m s^{-1}]
v_0	lattice site volume, [m^3], p. 137
v_A, v_B	molecular volume of species A and B, [m^3], p. 138
v_c	relative velocity of two overlapping chains, [m s^{-1}], p. 404
v_{mon}	occupied volume of a single chemical monomer, [m^3], p. 13
$W(N, x)$	number of N-step walks with displacement x, [dimensionless], p. 66
w	three-body interaction parameter, [m^6], pp. 100, 156
w_N	weight fraction of N-mers, [dimensionless], p. 16
$w(p, N)$	weight density of n-mers (per monomer), [dimensionless], p. 214
$\vec{X}_p$	normal mode p, [m], p. 359
$X_{p\alpha}$	α-component of normal mode p, [m], p. 359
x	fraction of labeled chains, [dimensionless], p. 189
Z	partition function, [dimensionless], p. 75
z	chain interaction parameter, [dimensionless], p. 103
z	coordination number, [dimensionless], p. 141
Γ	adsorbed amount per unit surface area, [m^{-2}], p. 188
Γ	decay rate, [s^{-1}], p. 349
$\Lambda_L(t)$	constraint release contribution to stress relaxation modulus of long chains, [dimensionless], p. 390
$\Lambda_S(t)$	constraint release contribution to stress relaxation modulus of short chains, [dimensionless], p. 390
Π	osmotic pressure, [kg m^{-1} s^{-2}], pp. 26, 155
$\Delta\varepsilon$	energy difference between trans and gauche minima, [kg m^2 s^{-2}], p. 50
Σ_k	k-th moment of the sum, [dimensionless], p. 208
Υ	numerical prefactor, [dimensionless], p. 375
Φ	Fox-Flory constant, [mol^{-1}], p. 316
$\Psi(\xi, t; s)$	probability for primitive path to move distance ξ in time t, while its end has not reached s, [m^{-1}], p. 404
Ω	number of states, [dimensionless], p. 70, 138
Ω_A	number of states of a molecule A in a pure A state, [dimensionless], p. 138
Ω_{AB}	number of states of a molecule in a homogeneous AB mixture, [dimensionless], p. 138
$\Omega(N, \vec{R})$	number of conformations of N-mer with end-to-end vector $\vec{R}$, [dimensionless], p. 70
α	polarizability, [cm^3], p. 30
α	angle between scattering wavevector and vector between monomers, [rad], p. 83

α_f	thermal expansion coefficient of the free volume, $[m^3\ K^{-1}]$, p. 338
γ	shear strain, [dimensionless], p. 282
γ, γ'	effective dimensionless spring constants, [dimensionless], pp. 375, 377
$\dot{\gamma}$	shear rate, $[s^{-1}]$, p. 283
γ_0	strain amplitude, [dimensionless], p. 290
γ_e, γ_v	elastic and viscous shear strains in the Maxwell model, [dimensionless], p. 283
δ	dimensionless adsorption energy per monomer, [dimensionless], p. 110
δ	exponent of the end-to-end distribution function, [dimensionless], p. 121
δ	phase angle, [rad], p. 291
$\delta, \delta_A, \delta_B$	solubility parameter, $[kg^{1/2}\ m^{-1/2}\ s^{-1}]$, p. 143
$\delta_{\alpha\beta}$	Kronecker delta, [dimensionless], p. 359
$\delta(t - t')$	Dirac delta function, [dimensionless], p. 359
δt	integration time step in a molecular dynamics simulation, [s], p. 393
$\delta\phi$	composition fluctuation, [dimensionless], p. 159
ϵ	dielectric constant, [dimensionless], p. 94
ϵ	Lennard-Jones interaction parameter, $[kg\ m^2\ s^{-2}]$, pp. 118 and 392
ε	relative extent of reaction, [dimensionless], p. 209
ε_G	Ginzburg relative extent of reaction, [dimensionless], p. 239
ζ	friction coefficient, $[kg\ s^{-1}]$, p. 309
ζ_0	friction coefficient at the reference temperature T_0, $[kg\ s^{-1}]$, p. 335
ζ_{br}	effective friction of a branch point, $[kg\ s^{-1}]$, p. 381
ζ_p	friction coefficient of normal mode p, $[kg\ s^{-1}]$, p. 359
ζ_R	Rouse friction coefficient of a chain, $[kg\ s^{-1}]$, p. 311
ζ_Z	Zimm friction coefficient of a chain, $[kg\ s^{-1}]$, p. 313
η	viscosity, $[kg\ m^{-1}\ s^{-1}]$, p. 33
η_M	viscosity in the Maxwell model, $[kg\ m^{-1}\ s^{-1}]$, p. 283
η_r	relative viscosity, [dimensionless], p. 314
η_s	solvent viscosity, $[kg\ m^{-1}\ s^{-1}]$, p. 33
η_{sp}	specific viscosity, [dimensionless], p. 315
$[\eta]$	intrinsic viscosity, $[m^3\ kg^{-1}]$, p. 34
θ	scattering angle, [rad], p. 29
θ	tertahedral angle, [rad], p. 49
θ	bending angle, [rad], p. 330
κ	exponent of the time dependence of stress relaxation modulus, [dimensionless], p. 351
κ	bond strength of the FENE potential, $[kg\ s^{-2}]$, p. 394
κ_λ	spring constant due to mode with wavelength λ, $[kg\ s^{-2}]$, p. 333
λ	wavelength of light, [m], p. 29

λ	wavelength of mode, [m], p. 331
$\lambda_x, \lambda_y, \lambda_z$	deformation factors, [dimensionless], p. 256
μ	number density of elastically effective crosslinks, [m^{-3}], p. 263
μ	coefficient in Doi fluctuation model, [dimensionless], p. 384
ν	frequency of light, [s^{-1}], p. 30
ν	scaling exponent, [dimensionless], p. 104
ν	number density of molecules, [m^{-3}], p. 156
ν	number density of elastically effective network strands, [m^{-3}], p. 259
ν	Poisson's ratio, [dimensionless], p. 296
ξ	blob size, [m], p. 72
ξ	correlation length, [m], p. 162
ξ_{ads}	adsorption blob size, [m], p. 110
ξ_h	hydrodynamic screening length, [m], p. 325
ξ_T	thermal blob size, [m], p. 113
ρ	polymer density, [kg m^{-3}], p. 13
σ	grafting density of a brush, [m^{-2}], p. 186
σ	shear stress, [kg m^{-1} s^{-2}], p. 282
σ	Lennard-Jones length, [m], pp. 118 and 392
σ_{ij}	stress tensor, [kg m^{-1} s^{-2}], p. 258
σ_{true}	true stress, [kg m^{-1} s^{-2}], p. 258
σ_{eng}	engineering stress, [kg m^{-1} s^{-2}], p. 259
σ_λ	stress due to mode with wavelength λ, [kg m^{-1} s^{-2}], p. 333
τ	turbidity, [m^{-1}], p. 45
τ^*	relaxation time of a characteristic cluster, [s], p. 342
τ_0	Kuhn monomer relaxation time, [s], p. 312
τ_{arm}	retraction time of an arm, [s], pp. 378, 381
τ_{chain}	relaxation time of a chain, [s], p. 327
τ_e	Rouse time of an entanglement strand, [s], p. 363
τ_g	relaxation time of the fastest stiff mode, [s], p. 334
τ_M	relaxation time in the Maxwell model, [s], p. 283
τ_p	relaxation time of the p-th mode, [s], p. 319
τ_R	Rouse time, [s], p. 311
τ_{rep}	reptation time, [s], p. 363
τ_{tube}	constraint release relaxation time of the confining tube, [s], p. 388
τ_Z	Zimm time, [s], p. 313
φ	torsion angle, [rad], p. 49
φ	phase difference, [rad], p. 80
ϕ	volume fraction, [dimensionless], p. 13
ϕ_0	volume fraction of a gel in a preparation state, [dimensionless], p. 274
$\overline{\phi}$	average volume fraction, [dimensionless], p. 159
ϕ_A, ϕ_B	volume fractions of components A and B of the mixture, [dimensionless], p. 137
ϕ_c	critical volume fraction, [dimensionless], p. 152

ϕ_e	entanglement volume fraction, [dimensionless], p. 369
ϕ_L	volume fraction of long chains, [dimensionless], p. 390
ϕ_S	volume fraction of short chains, [dimensionless], p. 390
ϕ^*	overlap volume fraction, [dimensionless], p. 13
ϕ_θ^*	overlap volume fraction for θ-solvents, [dimensionless], p. 172
ϕ^{**}	semidilute-concentrated crossover volume fraction, [dimensionless], p. 180
ϕ_{2body}	crossover volume fraction in mean-field theory, [dimensionless], p. 181
$\psi(s, t)$	probability for segment s to still be part of the tube at time t, [dimensionless], p. 405
χ	Flory interaction parameter, [dimensionless], p. 142
χ_b	Flory interaction parameter for a binodal, [dimensionless], p. 150
χ_c	critical interaction parameter, [dimensionless], p. 152
χ_s	Flory interaction parameter for a spinodal, [dimensionless], p. 151
ω	angular frequency, [rad s^{-1}], p. 291

Index

Bold page numbers refer to bold entries in the text, where terms are first introduced, Italic page numbers refer to homework problems.

Index

Conversion Factors

Length
$$1\,\text{m} = 10^2\,\text{cm} = 10^3\,\text{mm} = 10^6\,\mu\text{m} = 10^9\,\text{nm} = 10^{10}\,\text{Å}$$

Volume
$$1\,\text{Litre} = 10^{-3}\,\text{m}^3 = 10^3\,\text{cm}^3$$

Mass
$$1\,\text{kg} = 10^3\,\text{g}$$

Force
$$1\,\text{N} = 1\,\text{kg m s}^{-2} = 10^5\,\text{g cm s}^{-2} = 10^5\,\text{dyne}$$

Energy
$$1\,\text{J} = 1\,\text{kg m}^2\,\text{s}^{-2} = 1\,\text{N m} = 10^7\,\text{g cm}^2\,\text{s}^{-2} = 10^7\,\text{erg}$$
$$1\,\text{cal} = 4.18\,\text{J}$$
$$1\,\text{eV} = 1.602 \times 10^{-19}\,\text{J}$$
$$1\,\text{cm}^{-1}\ (\text{wavenumber unit of energy}) = 1.986 \times 10^{-23}\,\text{J}$$

Stress (also Modulus or Pressure)
$$1\,\text{Pa} = 1\,\text{N m}^{-2} = 1\,\text{J m}^{-3} = 1\,\text{kg m}^{-1}\,\text{s}^{-2} = 10\,\text{dyne cm}^{-2} = 10\,\text{g cm}^{-1}\,\text{s}^{-2}$$
$$1\,\text{bar} = 10^5\,\text{Pa}$$
$$1\,\text{atm} = 1.013 \times 10^5\,\text{Pa} = 1.013\,\text{bar}$$
$$1\,\text{torr} = 1\,\text{mm Hg} = 133.3\,\text{Pa}$$

Viscosity
$$1\,\text{Pa s} = 1\,\text{kg m}^{-1}\,\text{s}^{-1} = 10\,\text{g cm}^{-1}\,\text{s}^{-1} = 10\,\text{Poise}$$

Surace tension
$$1\,\text{N m}^{-1} = 1\,\text{J m}^{-2} = 10^3\,\text{dyne cm}^{-1}$$

Power
$$1\,\text{W} = 1\,\text{kg m}^2\,\text{s}^{-3} = 1\,\text{J s}^{-1} = 10^7\,\text{erg s}^{-1}$$
$$\text{Metric horse power} = 735.5\,\text{W}$$

Electric charge
$$1\,\text{C} = 2.998 \times 10^9\,\text{cm}^{3/2}\,\text{g}^{1/2}\,\text{s}^{-1}$$

Electric dipole moment
$$1\,\text{C m} = 2.998 \times 10^{11}\,\text{cm}^{5/2}\,\text{g}^{1/2}\,\text{s}^{-1} = 2.998 \times 10^{29}\,\text{Debye}$$

Fundamental Constants

Avogadro's number $\mathcal{N}_{Av} = 6.02 \times 10^{23}\,\text{mol}^{-1}$
Boltzmann constant $k = 1.38 \times 10^{-23}\,\text{J K}^{-1} = 1.38 \times 10^{7}\,\text{Pa \AA}^{3}\,\text{K}^{-1}$
Gas constant $\mathcal{R} = k\mathcal{N}_{Av} = 8.31\,\text{J mol}^{-1}\,\text{K}^{-1}$
Speed of light in vacuum $c = 2.998 \times 10^{8}\,\text{m s}^{-1}$
Elementary charge $e = 1.60 \times 10^{-19}\,\text{C}$
Gravitational constant $G = 6.67 \times 10^{-11}\,\text{m}^{3}\,\text{s}^{-2}\,\text{kg}^{-1}$
Atomic mass unit (1/12 of the mass of ^{12}C atom) $= 1.66 \times 10^{-27}\,\text{kg}$
Planck constant $h = 6.63 \times 10^{-34}\,\text{J s}$

Defined Constants

Zero of the Celsius scale (0° C) $\equiv 273.15\,\text{K}$
Standard gravitational acceleration $g \equiv 9.80665\,\text{m s}^{-2}$
$kT = 4.114 \times 10^{-21}\,\text{J}$ at 298 K
$\mathcal{R}T = 2.478\,\text{kJ mol}^{-1} = 0.592\,\text{kcal mol}^{-1}$ at 298 K

Greek Alphabet

Alpha	A	α	Iota	I	ι	Rho	P	ρ
Beta	B	β	Kappa	K	κ	Sigma	Σ	σ
Gamma	Γ	γ	Lambda	Λ	λ	Tau	T	τ
Delta	Δ	δ	Mu	M	μ	Upsilon	Υ	υ
Epsilon	E	ε, ϵ	Nu	N	ν	Phi	Φ	ϕ, φ
Zeta	Z	ζ	Xi	Ξ	ξ	Chi	X	χ
Eta	H	η	Omicron	O	o	Psi	Ψ	ψ
Theta	Θ	θ	Pi	Π	π	Omega	Ω	ω

SI Prefixes

10^{-1}	deci	d	10	deca	da
10^{-2}	centi	c	10^{2}	hecto	h
10^{-3}	milli	m	10^{3}	kilo	k
10^{-6}	micro	μ	10^{6}	mega	M
10^{-9}	nano	n	10^{9}	giga	G
10^{-12}	pico	p	10^{12}	tera	T
10^{-15}	fempto	f	10^{15}	peta	P
10^{-18}	atto	a	10^{18}	exa	E
10^{-21}	zepto	z	10^{21}	zetta	Z